Mathematics and Visualization

More information about this series at
http://www.springer.com/series/4562

Ingrid Hotz • Thomas Schultz
Editors

Visualization and Processing of Higher Order Descriptors for Multi-Valued Data

With 163 Figures, XXX in color

Springer

Editors
Ingrid Hotz
Linköping University
Norrköping
Sweden

Thomas Schultz
University of Bonn
Visualization and Medical Image Analysis
Bonn
Germany

ISSN 1612-3786 ISSN 2197-666X (electronic)
Mathematics and Visualization
ISBN 978-3-319-15089-5 ISBN 978-3-319-15090-1 (eBook)
DOI 10.1007/978-3-319-15090-1

Library of Congress Control Number: 2015944320

Mathematics Subject Classification (2010): 15-06, 68U10, 92-08

Springer Cham Heidelberg New York Dordrecht London

Cover illustration: Cover illustration from "A Survey of Illustrative Visualization Techniques for Diffusion-Weighted MRI Tractography" by Tobias Isenberg with kind permission.

Printed on acid-free paper

Springer International Publishing AG Switzerland is part of Springer Science+Business Media (www.springer.com)

Preface

Multi-valued data arises in many applications ranging from modern imaging techniques to computational simulations and from medicine to engineering. Its mathematical modeling frequently involves higher-order descriptors, such as tensors or basis functions that are indexed by multiple indices. Even though such descriptors have a long tradition of being used in the sciences to describe physical phenomena and they are widely studied in mathematics, it is only over the last few years that their significance for data and image analysis and a need for their visualization has been recognized. At the same time, most applications that involve higher-order descriptors require collaborations between diverse scientific communities such as applied mathematics, computer science, physics, engineering, neuroscience, and medicine.

This book is the fifth in a series that fosters a more active exchange between these communities by collecting recent research findings and survey chapters dealing with different aspects of this topic authored by researchers from different fields. Most chapters were contributed by the participants of a workshop on the Visualization and Processing of Higher Order Descriptors for Multi-Valued Data that was held in February 2014 in Dagstuhl, Germany, and we hope that it will convey some of the workshop's inspiring atmosphere of open intellectual exchange. We are pleased to also include a number of high-quality chapters that are relevant to the topic, but whose authors could not attend the workshop due to other obligations or to the very limited number of available places.

Comparing the range of chapters to previous books in the series, we can observe a particularly strong interest in the mathematical foundations of the field. Similarly, statistical analysis of higher-order descriptors and the use of machine learning techniques play a more prominent role than in previous years. A renewed interest in diffusion-weighted magnetic resonance imaging (dMRI) can be attributed to the wider availability of multi-shell and multi-parameter data, as well as the emergence of a completely new generation of dMRI techniques that use more complex gradient waveforms. At the same time, applications in engineering and the physical sciences continue to play an important role, and the workshop has witnessed promising collaborations in this domain.

The book is structured in five parts. The first is concerned with mathematical foundations, while the second brings together methods for the processing and interpolation of higher-order descriptors. The third part discusses questions of visualization. This is followed by a section on statistical analysis and a final part that presents solutions to specific application problems involving higher-order descriptors.

The **mathematical foundations** presented in Part I include the development of novel mathematical descriptions, the study and comparative analysis of existing ones as well as algorithms for their computation. A new generation of diffusion MRI techniques that make use of multiple gradient pulses or flexible gradient waveforms sparked a lot of interest at the workshop since they not only allow us to acquire information about tissue microstructure that is inaccessible to traditional diffusion MRI, but also produce data for which fully adequate models and visualization techniques are still to be developed. The first chapter unifies and compares two mathematical approaches to modeling this new type of data. The second chapter explores the use of Finsler geometry in the context of diffusion MRI by considering Brownian motion on Finsler manifolds. The third chapter relates the theory of orientation tensors and fabric tensors to parametric models of orientation distribution functions that are commonly used in dMRI. The fourth chapter studies the topology of linear symmetric tensor fields. Finally, the fifth chapter proposes two randomized algorithms for computing low-rank tensor approximations and applies them to an image-compression task.

The extension of data analysis methods from scalar fields to higher-order descriptors requires the generalization of the fundamental concepts of data **processing, filtering, and interpolation**. The first two chapters of Part II introduce morphological filters for higher-order descriptors. These are filters that are concerned with the detection and manipulation of shapes and structures in images with many applications. The topic of the first chapter is the extraction of long and thin structures, e.g. to find cracks. The second chapter proposes a partial ordering and a notion of maximum and minimum for color images using higher-order descriptors. The third chapter introduces a direction-controlled interpolation scheme to deal with tensor fields with conflicting orientations, e.g. to resolve fiber crossings in DTI fields. The last chapter summarizes the state of the art and challenges of tensor voting. The goal of tensor voting is to retrieve as much reliable information from various imaging data as possible, even where there is low resolution and in the presence of noise. The idea is to propagate local information encoded through tensors in a neighborhood following principles of proximity and similarity borrowed from Gestalt psychology.

The wealth of information present in higher-order descriptors poses significant challenges to **visualization**. The first three chapters in Part III address the visualization of data from different variants of diffusion MRI. The first presents direct volume rendering and glyph-based strategies for data from diffusion spectrum imaging, which estimates the diffusion propagator, a three-dimensional probability distribution, at each point of three-dimensional space. The second chapter focuses on diffusion kurtosis imaging, which approximates the diffusion propagator in its covariance and kurtosis. It describes a method that uses this information to segment and visualize tissue types. Reconstructing the trajectories of major nerve-fiber bundles is a common goal in many variants of diffusion MRI, and it is significant in graphically conveying their shapes, the larger-scale bundles they form and their relationships to other anatomical structures in a clear manner. The third chapter surveys illustrative techniques that tackle these challenges. Finally, the extraction of features is a common strategy to deal with complex data. The last chapter compares different ways of computing the heat kernel signature, a popular tool in computational geometry, for general symmetric tensor fields and proposes its use as a feature for visualization.

Part IV is devoted to a more formal **statistical analysis** of higher-order descriptors, generalizing tools from univariate statistics. The first chapter presents a comprehensive framework for analyzing diffusion MRI data while accounting for multiple fiber compartments throughout the whole pipeline of interpolation, filtering, spatial normalization, and statistical analysis. The second chapter summarizes the state of the art and current challenges in applying statistical hypothesis testing and predictive modeling through supervised machine learning to multivariate neuroimaging data, again with a special focus on diffusion MRI.

The last part (Part V) is a collection of three chapters dealing with specific **applications** of higher-order descriptors. The first application is the analysis of a turbulent combustion simulation. A tensor-based clustering method is used to define typical and atypical behavior in the field. The second application is a classical mechanical engineering problem. It describes a case study for the use of tensorline visualization to support the design process for mechanical parts. The last chapter is concerned with a clinical application of diffusion-weighted magnetic resonance imaging (dMRI). It proposes enhancing local features taking context information into account. The framework developed is used for more accurate neurosurgical planning.

We would like to thank the organizers of the Dagstuhl workshop, Bernhard Burgeth (Universität des Saarlandes, DE), Ingrid Hotz (Linköping University–Norrköping, Sweden), Anna Vilanova Bartroli (TU Delft, NL), and Carl-Fredrik Westin (Harvard Medical School—Boston, US), as well as the board and staff of Schloss Dagstuhl for creating a unique opportunity for interdisciplinary exchange. We are also grateful to all the authors and reviewers who contributed to this book and who ensured its scientific quality. Last but not least, we would like to thank the editors of the Springer book series Mathematics and Visualization, as well as Martin Peters and Ruth Allewelt (Springer, Heidelberg) for their support in publishing this collection as part of their series.

We hope that this book will further the scientific progress on higher-order descriptors by serving as a reference to those who work with applications that generate multi-valued data or could benefit from higher-order models, and by providing a source of inspiration to researchers who are working on novel methods in the areas of image processing and visualization.

Norrköping, Sweden Ingrid Hotz
Bonn, Germany Thomas Schultz
December 2014

Contents

Part I Mathematical Foundations

Diffusion-Weighted Magnetic Resonance Signal for General Gradient Waveforms: Multiple Correlation Function Framework, Path Integrals, and Parallels Between Them 3
Cem Yolcu and Evren Özarslan

Finslerian Diffusion and the Bloch–Torrey Equation 21
T.C.J. Dela Haije, A. Fuster, and L.M.J. Florack

Fiber Orientation Distribution Functions and Orientation Tensors for Different Material Symmetries 37
Maher Moakher and Peter J. Basser

Topology of 3D Linear Symmetric Tensor Fields 73
Yue Zhang, Jonathan Palacios, and Eugene Zhang

Random Projections for Low Multilinear Rank Tensors 93
Carmeliza Navasca and Deonnia N. Pompey

Part II Processing, Filtering and Interpolation

Path-Based Mathematical Morphology on Tensor Fields 109
Jasper J. van de Gronde, Mikola Lysenko, and Jos B.T.M. Roerdink

Processing Multispectral Images via Mathematical Morphology 129
Andreas Kleefeld and Bernhard Burgeth

Direction-Controlled DTI Interpolation 149
Luc Florack, Tom Dela Haije, and Andrea Fuster

Tensor Voting: Current State, Challenges and New Trends in the Context of Medical Image Analysis 163
Daniel Jörgens and Rodrigo Moreno

Part III Visualization

Visualization of Diffusion Propagator and Multiple Parameter Diffusion Signal 191
Olivier Vaillancourt, Maxime Chamberland, Jean-Christophe Houde, and Maxime Descoteaux

Visual Knowledge Discovery for Diffusion Kurtosis Datasets of the Human Brain 213
Sujal Bista, Jiachen Zhuo, Rao P. Gullapalli, and Amitabh Varshney

A Survey of Illustrative Visualization Techniques for Diffusion-Weighted MRI Tractography 235
Tobias Isenberg

Visualizing Symmetric Indefinite 2D Tensor Fields Using the Heat Kernel Signature 257
Valentin Zobel, Jan Reininghaus, and Ingrid Hotz

Part IV Statistical Analysis

A Framework for the Analysis of Diffusion Compartment Imaging (DCI) 271
Maxime Taquet, Benoit Scherrer, and Simon K. Warfield

Statistical and Machine Learning Methods for Neuroimaging: Examples, Challenges, and Extensions to Diffusion Imaging Data 299
Lauren J. O'Donnell and Thomas Schultz

Part V Applications

A Clustering Method for Identifying Regions of Interest in Turbulent Combustion Tensor Fields 323
Adrian Maries, Timothy Luciani, P.H. Pisciuneri, Mehdi B. Nik, S. Levent Yilmaz, Peyman Givi, and G. Elisabeta Marai

Tensor Lines in Engineering: Success, Failure, and Open Questions 339
Marc Schöneich, Andrea Kratz, Valentin Zobel, Gerik Scheuermann, Markus Stommel, and Ingrid Hotz

Contextual Diffusion Image Post-processing Aids Clinical Applications 353
Vesna Prčkovska, Magí Andorrà, Pablo Villoslada, Eloy Martinez-Heras, Remco Duits, David Fortin, Paulo Rodrigues, and Maxime Descoteaux

Index 379

Part I
Mathematical Foundations

Diffusion-Weighted Magnetic Resonance Signal for General Gradient Waveforms: Multiple Correlation Function Framework, Path Integrals, and Parallels Between Them

Cem Yolcu and Evren Özarslan

Abstract Effects of diffusion on the magnetic resonance (MR) signal carry a wealth of information regarding the microstructure of the medium. Characterizing such effects is immensely important for quantitative studies aiming to obtain microstructural parameters using diffusion MR acquisitions. Studies in recent years have demonstrated the potential of sophisticated gradient waveforms to provide novel information inaccessible by traditional measurements. There are mainly two approaches that can be used to incorporate the influence of restricted diffusion, particularly on experiments featuring general gradient waveforms. The multiple propagator framework essentially reduces the problem to a path integral, which can be evaluated analytically or approximated via a matrix representation. The multiple correlation function method tackles the Bloch–Torrey equation, and employs an alternative matrix formulation. In this work, we present the two techniques in a unified fashion and link the two approaches. We provide an explanation for why the multiple correlation function is computationally more efficient in the case of waveforms featuring piecewise constant gradients.

1 Introduction

Magnetic resonance (MR) measurements of the translational diffusion of spin bearing particles have been employed to characterize the microstructure of biological tissue, colloids, and other porous materials. To sensitize the MR signal to the diffusional motion of molecules, carefully devised profiles of magnetic

C. Yolcu
Department of Physics, Boğaziçi University, Bebek, 34342 İstanbul, Turkey

Present address: Dipartimento di Fisica e Astronomia, Università di Padova, 35131 Padova, Italy
e-mail: cem.yolcu@pd.infn.it

E. Özarslan (✉)
Department of Physics, Boğaziçi University, Bebek, 34342 İstanbul, Turkey
e-mail: evren.ozarslan@boun.edu.tr

I. Hotz, T. Schultz (eds.), *Visualization and Processing of Higher Order Descriptors for Multi-Valued Data*, Mathematics and Visualization,
DOI 10.1007/978-3-319-15090-1_1

field gradients are typically utilized. We shall refer to such profiles as gradient waveforms. As a particle diffuses during the course of the application of the waveform, it suffers a net phase shift, which is related to the local magnetic field it experiences throughout its movement. As such, the particle's motional history, along with the gradient waveform, determines the net phase acquired by the particle when an echo forms. The sum of the magnetic moments of all particles thus determines the signal intensity detected by the radiofrequency antenna.

The simplest gradient waveform involves a pair of identical gradient pulses applied effectively in opposite directions. This pulsed field gradient (PFG) waveform has been utilized since the 1960s with great success [45]. For example, characterization of diffusion anisotropy in fibrous tissues has made diffusion-weighted MR a powerful tool for mapping neural connections between different regions of the nervous system [9, 15, 21, 31, 36, 42]. In recent years, other important microstructural features of the tissue like cell size, size distribution, and complexity of the medium have been examined with MR techniques employing a pair of PFG pulses [1, 2, 35, 40, 41].

The power of MR diffusion measurements has prompted the research community to search for alternative encoding schemes that employ gradient waveforms that are more sophisticated than the traditional two pulse experiments described above. Most notably, oscillating gradient waveforms [46] have been successfully employed to probe the short time regime of the diffusion process [17], and multiple-PFG (e.g., double diffusion encoding, DDE) acquisitions [4, 16, 34] have been shown to provide sensitivity to diffusion anisotropy at different length scales [13, 26, 30, 32, 39].

All these endeavors mentioned above would benefit greatly from schemes that relate the MR signal intensity obtained from general gradient waveforms to microstructural parameters of the medium. Consider the case of the traditional PFG measurements. Although there are relatively easily obtained explicit expressions for measurements involving infinitesimally short pulses, those solutions are far from being accurate in real acquisitions that typically employ long pulses. Development of a theoretical framework is even more imperative for the case of more sophisticated pulse sequences.

To address this important issue, several different methods have been developed. In the first approach, commonly referred to as the "multiple propagator (MP)" method, a general gradient waveform is discretized and represented by a train of impulses [11] as shown in Fig. 1. Since all pulses are very short, diffusion is thought to take place in the absence of an impressed gradient field. Expressions have been derived for MR signal intensity by employing a sequence of propagators describing the movement between the application of the neighboring impulses. Based on this idea, a numerical scheme involving products of a series of matrices was developed [10, 14, 33, 47], and a few studies have validated the signal computations [3, 6]. This approach has been recognized as a path integral formulation of the diffusion process, and the matrix formulation was used as an analytical method by considering the signal in the limits of short intervals, and very large number of impulses [34]. For this reason, we use the phrases "multiple propagator" and "path integral" interchangeably.

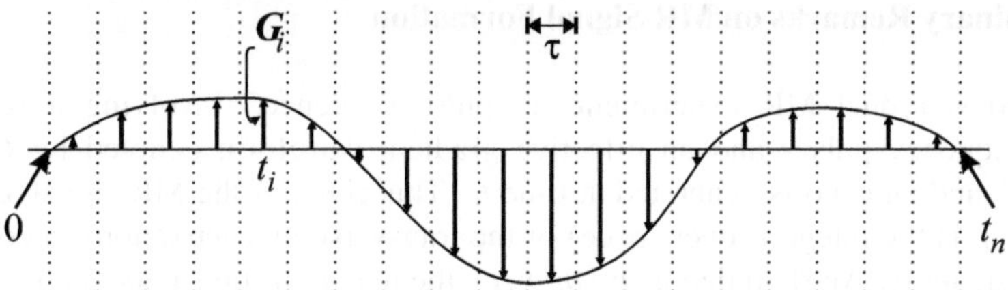

Fig. 1 A general gradient waveform $\boldsymbol{G}(t)$ is approximated by a series of impulses. The *continuous line* shows the effective gradient profile that accounts for any reversals due to the radiofrequency pulses

Another approach referred to as the "multiple correlation function (MCF)" method treats the general waveform as a piecewise constant function [7, 8, 43]. By directly tackling the Bloch–Torrey equation [48] using techniques that are employed routinely in quantum mechanics, explicit expressions for the signal intensity that involve products of exponentials of matrices have been derived. This technique was subsequently studied [5, 44] in the context of the Carr-Purcell-Meiboom-Gill experiment [12, 29]. In recent years, the technique has been revisited and restudied in depth [18]. Subsequently the method was adapted to geometries with layered structures [19, 20], capped cylinders [32], and triangular pores [28]. The technique was also generalized to waveforms with variations in the direction of the gradients [38] and structure-specific susceptibility gradients have been incorporated [27] into the formulation.

In this work, we present a general approach to the problem of estimating the signal intensity. In the next section, we first discuss the connection between probability concepts and the Bloch–Torrey equation characterizing the time evolution of the magnetization density. Next we show that both MP and MCF methods emerge naturally from a common approach, and are in a sense equivalent to each other. In Sect. 3 we discuss how the MP approach can be converted into a matrix product representation, and subsequently illustrate why the MCF technique is more efficient from a computational point of view for PFG measurements. Finally in Sect. 4, we treat the case of restricted diffusion at small diffusion sensitivity as an example for employing the MP approach as an analytical tool to calculate the path integral.

2 Theory

2.1 From Probability Concepts to Bloch–Torrey Equation

In this section, we provide a brief overview of the basics of diffusion MR signal, and invoke a theorem from probability theory to derive the Bloch–Torrey equation [48] that is fundamentally important for this work. In this section, we consider the one-dimensional problem for simplicity.

Preliminary Remarks on MR Signal Formation

In a conventional MR experiment, a "pulse sequence," involving a series of radiofrequency pulses and an effective gradient waveform denoted by $G(t)$, is applied, and an echo is generated at time t_e. This echo, or the MR signal, denoted by $E(t_e)$, is the integral (over space) of the transverse magnetization—a complex-valued quantity. We shall denote by $m(x,t)$, the magnetization at location x at time t. Thus,

$$E(t_e) = \int \mathrm{d}x \, m(x, t_e) \, . \tag{1}$$

Just prior to the start of the pulse sequence, all molecules can be assumed to be at their equilibrium states. At the time of the echo, however, there is a distribution of phases due to various phenomena. Here, we shall ignore all effects but diffusion. Let $p(\phi)$ represent the distribution of the phases in the sample (or voxel in image acquisitions). Another expression for the MR signal is then

$$E(t_e) = \int \mathrm{d}\phi \, p(\phi) \, e^{\mathrm{i}\phi} \, . \tag{2}$$

The expressions above describe the MR signal, which is a macroscopic quantity influenced by the random movements of many spin-bearing particles. We shall now consider only one such particle. The gradient waveform perturbs the main magnetic field by different amounts at different locations. Typically, linear gradients are employed, which induce a phase shift in the magnetic moment of the particle given by

$$\phi = -\gamma \int_0^{t_e} \mathrm{d}t \, x(t) \, G(t) \, , \tag{3}$$

where $x(t)$ denotes the trajectory of the particle during the course of the gradient waveform. The notion of x for a single trajectory is hence slightly different than that in the argument of magnetization Eq. (1), the latter being associated with the spatial variation of a *population* of spins.

A Relevant Theorem

Here, we invoke a theorem by Kac [22] in which the author considers a Wiener functional of the form

$$\alpha[V] = \int_0^{\upsilon} \mathrm{d}\upsilon' \, V(\xi(\upsilon')) \tag{4}$$

whose distribution function is $\sigma(\alpha;\nu)$. The author proves that the following relationship holds

$$\int_0^\infty \int_0^\infty e^{-u\alpha-s\nu} \mathrm{d}_\alpha\sigma(\alpha;\nu)\,\mathrm{d}\nu = \int_{-\infty}^{\infty} \psi(\xi,s)\,\mathrm{d}\xi \,, \tag{5}$$

where $\psi(\xi,s)$ is the fundamental solution to the differential equation

$$\frac{1}{2}\frac{\mathrm{d}^2\psi}{\mathrm{d}\xi^2} - (s + u\,V(\xi))\,\psi = 0 \,. \tag{6}$$

The last two equations can be expressed in an alternative form via the inverse Laplace transform:

$$\int_0^\infty e^{-u\alpha}\,\mathrm{d}_\alpha\sigma(\alpha;\nu) = \int_{-\infty}^{\infty} \mathrm{d}\xi\,\omega(\xi,\nu) \,, \tag{7}$$

where $\omega(\xi,\nu)$ is the inverse Laplace transform of $\psi(\xi,s)$, and obeys

$$\frac{\partial\omega}{\partial\nu} = \frac{1}{2}\frac{\partial^2\omega}{\partial\xi^2} - u\,V(\xi)\,\omega \,. \tag{8}$$

There are a few conditions for the above relationships to hold, which we don't include here for brevity.

It turns out that the above theorem is very closely related to the diffusion MR signal described in the previous subsection. We first note that the time at which the echo forms (t_e) can be taken as a characteristic time of the experiment, and further define a characteristic length $L = (2Dt_e)^{1/2}$, where D is the bulk diffusion coefficient. Next, we identify the variable $\xi = x/L$ as a dimensionless position, $\nu' = t/t_e$ as a dimensionless time variable, and $\nu = 1$ in Eq. (4). Further taking[1] $V = -\gamma\,t_e\,x(t)\,G(t)$, Eq. (4) turns into Eq. (3). Thus, $\mathrm{d}_\alpha\sigma(\alpha;\nu) = p(\phi)\,\mathrm{d}\phi$.

With the further identifications $u = -\mathrm{i}$ and $\omega = L\,m$, it is clear that the left and right hand sides of Eq. (7), correspond to the expressions for $E(t_e)$ in Eqs. (2) and (1), respectively. Finally, Eq. (8) becomes

$$\frac{\partial m}{\partial t} = D\frac{\partial^2 m}{\partial x^2} - \mathrm{i}\,\gamma\,x\,G(t)\,m \,. \tag{9}$$

This partial differential equation, referred to as the Bloch–Torrey equation, had been derived by Torrey using hydrodynamic arguments [48]. Multi-dimensional generalization of this equation will be employed in the subsequent section.

[1] Strictly speaking, the form of $V(\xi(\nu'))$ differs slightly from that in Eq. (4) due to the explicit time dependence. However, this difference doesn't appear to violate the applicability of Kac's theorem.

2.2 Parallels Between MCF and MP Methods

To illustrate the parallel between the path integral and the multiple correlator approaches, let us consider the echo attenuation at time t,

$$E(t) = \int \mathrm{d}\boldsymbol{r}\; m(\boldsymbol{r},t) = \int \mathrm{d}\boldsymbol{r}\; \langle \boldsymbol{r}|m(t)\rangle \,, \tag{10}$$

where $\langle \boldsymbol{r}|m(t)\rangle$ denotes the projection of the magnetization state $|m(t)\rangle$ onto a position eigenstate $|\boldsymbol{r}\rangle$ à la Dirac's bra-ket notation. We have also committed a slight abuse of notation by using $\mathrm{d}\boldsymbol{r}$ to denote the infinitesimal volume element. The transverse magnetization density at time t is related to that at the initial time (chosen here as $t = 0$) via $|m(t)\rangle = \mathrm{U}(t,0)|m(0)\rangle$, where $\mathrm{U}(t,0)$ could be called the Bloch–Torrey propagator. Consequently, we have

$$E(t) = \int \mathrm{d}\boldsymbol{r}\; \langle \boldsymbol{r}|\mathrm{U}(t,0)|m(0)\rangle \,. \tag{11}$$

Basically, the two approaches toward calculating this echo differ in the way they handle $\mathrm{U}(t,0)$, but before going into that, we must talk a little bit more about the propagator itself.

The propagator $\mathrm{U}(t,0)$ obeys the Bloch–Torrey equation[2]

$$\frac{\partial}{\partial t}\mathrm{U}(t,0) = -\left[D\mathbf{K}^2 + \mathrm{i}\gamma \boldsymbol{G}(t)\cdot\mathbf{R}\right]\mathrm{U}(t,0)\,, \tag{12}$$

with $\mathbf{K}$ and $\mathbf{R}$ being the wave vector and position operators,[3] respectively, and we also take this opportunity to define the operator

$$\mathrm{H}(t) = D\mathbf{K}^2 + \mathrm{i}\gamma \boldsymbol{G}(t)\cdot\mathbf{R}\,. \tag{13}$$

Due to the noncommutation of $\mathrm{H}(t)$ with $\mathrm{H}(t')$ for $t \neq t'$, the solution of Eq. (12) is not simply $\mathrm{U}(t,0) = \mathrm{e}^{-t\mathrm{H}}$. Rather, the time interval from 0 to t is sliced up into small intervals, and the propagation is carried out step-wise, i.e.,

$$\mathrm{U}(t,0) = \lim_{\substack{\tau\to 0\\ n\tau = t}} \mathrm{U}\big(n\tau,(n-1)\tau\big)\ldots\mathrm{U}(2\tau,\tau)\mathrm{U}(\tau,0)\,. \tag{14}$$

[2]With the boundary condition that $\mathrm{U}(0,0) = \mathbb{I}$, where $\mathbb{I}$ is the identity operator.

[3]The wave vector operator, when expressed in the position basis, is a derivative, i.e., $\langle \boldsymbol{r}|\mathbf{K} = -\mathrm{i}\nabla\langle \boldsymbol{r}|$. Its commutator with the position operator is $[\mathbf{K},\mathbf{R}] = -\mathrm{i}$.

Each of the "infinitesimal propagators" above *is* of the form $\mathrm{e}^{-\tau \mathrm{H}(t)}$, and we have[4]

$$\mathrm{U}(t,0) = \lim_{\substack{\tau\to 0\\ n\tau=t}} \mathrm{e}^{-\tau\mathrm{H}(n\tau-\tau)} \ldots \mathrm{e}^{-\tau\mathrm{H}(\tau)}\mathrm{e}^{-\tau\mathrm{H}(0)} \;. \tag{15}$$

It is important to keep in mind that the operator product is ordered from right to left in increasing time.

This "time-slicing" described above is common to both approaches, and more or less inevitable. Where the path integral (multiple propagator) and the multiple correlation function approaches (seem to) depart is how they proceed with extracting the number $E(t)$ through Eq. (11).

Multiple Correlation Function (MCF) Technique

The multiple correlator scheme adopts the eigenbasis $\{|\boldsymbol{k}\rangle\}$ of the wave vector operator $\mathbf{K}$ for computing $E(t)$. Using the completeness relation $\mathbb{I} = \sum_{\boldsymbol{k}} |\boldsymbol{k}\rangle\langle\boldsymbol{k}|$, Eq. (11) is rewritten as

$$E(t) = \sum_{\boldsymbol{k},\boldsymbol{k}'} \int \mathrm{d}\boldsymbol{r}\, \langle\boldsymbol{r}|\boldsymbol{k}\rangle\langle\boldsymbol{k}|\mathrm{U}(t,0)|\boldsymbol{k}'\rangle\langle\boldsymbol{k}'|m(0)\rangle \;. \tag{16}$$

Since $\langle\boldsymbol{k}'|m(0)\rangle \propto \delta_{\boldsymbol{k}',0}$ and $\int \mathrm{d}\boldsymbol{r}\, \langle\boldsymbol{r}|\boldsymbol{k}\rangle \propto \delta_{0,\boldsymbol{k}}$,[5] the echo simply becomes

$$E(t) = \langle 0|\mathrm{U}(t,0)|0\rangle \;. \tag{17}$$

In order to compute this "matrix element" of the operator $\mathrm{U}(t,0)$, one then employs the time-slicing, Eq. (15), inserting identity operators (in the $|\boldsymbol{k}\rangle$ basis) in between the infinitesimal propagators:

$$E(t) = \lim_{\substack{\tau\to 0\\ n\tau=t}} \sum_{\{\boldsymbol{k}\}} \langle 0|\mathrm{e}^{-\tau\mathrm{H}(n\tau-\tau)}|\boldsymbol{k}_{n-1}\rangle \ldots \langle\boldsymbol{k}_2|\mathrm{e}^{-\tau\mathrm{H}(\tau)}|\boldsymbol{k}_1\rangle\langle\boldsymbol{k}_1|\mathrm{e}^{-\tau\mathrm{H}(0)}|0\rangle \;. \tag{18}$$

---

[4] Also note that for a practical implementation where the limit $\tau \to 0$ is not *actually* taken, one might want to offset the argument of H by $\tau/2$ (like in Fig. 1) or some other amount, but we need not bother with that for our purposes.

[5] The eigenfunction corresponding to the $\boldsymbol{k} = 0$ eigenvalue is constant over the volume of interest: $\langle\boldsymbol{r}|0\rangle = V^{-1/2}$. Hence $\int \mathrm{d}\boldsymbol{r}\, \langle\boldsymbol{r}|\boldsymbol{k}\rangle = V^{1/2}\int \mathrm{d}\boldsymbol{r}\, \langle 0|\boldsymbol{r}\rangle\langle\boldsymbol{r}|\boldsymbol{k}\rangle = V^{1/2}\langle 0|\boldsymbol{k}\rangle = V^{1/2}\delta_{0,\boldsymbol{k}}$. On the other hand, the initial magnetization is in equilibrium, and therefore proportional to the $\boldsymbol{k} = 0$ eigenket, meaning $\langle\boldsymbol{k}'|m(0)\rangle = c\delta_{\boldsymbol{k}',0}$. For convenience, we assume a normalization for $m(\boldsymbol{r},t)$ such that $\int \mathrm{d}\boldsymbol{r}\, m(\boldsymbol{r},0) = 1$, whereby $c = V^{-1/2}$.

What needs to be computed, therefore, is the product of matrices of the (component-wise) form

$$\langle \boldsymbol{k}' | \mathrm{e}^{-\tau D\mathbf{K}^2 - \mathrm{i}\gamma\tau \boldsymbol{G}(t)\cdot\mathbf{R}} | \boldsymbol{k} \rangle \,, \tag{19}$$

or of exponentials of matrices of the (component-wise) form

$$-\tau\langle \boldsymbol{k}' | D\mathbf{K}^2 + \mathrm{i}\gamma \boldsymbol{G}(t)\cdot\mathbf{R} | \boldsymbol{k} \rangle = -\tau\omega_k \delta_{\boldsymbol{k}',\boldsymbol{k}} - \mathrm{i}\gamma\tau \boldsymbol{G}(t)\cdot\langle \boldsymbol{k}' | \mathbf{R} | \boldsymbol{k} \rangle \,, \tag{20}$$

where ω_k denotes the product of D with the kth eigenvalue of the operator $\mathbf{K}^2$.

We do not elaborate on the details of the (numerical) implementation of such matrix operations here.

Multiple Propagator Technique

The multiple propagator (path integral) approach, on the other hand, favors the eigenbasis $\{|\boldsymbol{r}\rangle\}$ of the position operator $\mathbf{R}$. Either in Eq. (11) or (17), one performs the time slicing, this time employing the completeness relation $\mathbb{I} = \int \mathrm{d}\boldsymbol{r}\, |\boldsymbol{r}\rangle\langle \boldsymbol{r}|$ in between the infinitesimal propagators. One obtains,

$$E(t) = \lim_{\substack{\tau\to 0\\ n\tau = t}} \int \mathrm{d}\boldsymbol{r}_n \ldots \mathrm{d}\boldsymbol{r}_0 \, \langle \boldsymbol{r}_n | \mathrm{e}^{-\tau \mathrm{H}(n\tau-\tau)} | \boldsymbol{r}_{n-1} \rangle \ldots$$
$$\ldots \langle \boldsymbol{r}_1 | \mathrm{e}^{-\tau\mathrm{H}(0)} | \boldsymbol{r}_0 \rangle \langle \boldsymbol{r}_0 | m(0) \rangle \,. \tag{21}$$

Now, $\mathrm{e}^{-\tau\mathrm{H}} = \mathrm{e}^{-\tau D\mathbf{K}^2 - \mathrm{i}\gamma\tau\boldsymbol{G}\cdot\mathbf{R}}$ is not equal to $\mathrm{e}^{-\tau D\mathbf{K}^2}\mathrm{e}^{-\mathrm{i}\tau\gamma\boldsymbol{G}\cdot\mathbf{R}}$ since $\left[\mathbf{K}^2, \mathbf{R}\right] = -2\mathrm{i}\mathbf{K} \neq 0$. However,

$$\mathrm{e}^{-\tau D\mathbf{K}^2 - \mathrm{i}\gamma\tau\boldsymbol{G}\cdot\mathbf{R}} = \mathrm{e}^{-\tau D\mathbf{K}^2}\mathrm{e}^{-\mathrm{i}\gamma\tau\boldsymbol{G}\cdot\mathbf{R}} + \mathcal{O}(\tau^2) \,, \tag{22}$$

which one may verify by simply considering the Taylor series expansions of the exponentials. Hence, to lowest nontrivial order in τ—which is nevertheless the limit of interest—any one of the matrix elements in Eq. (21) can be written as

$$\begin{aligned} \langle \boldsymbol{r}' | \mathrm{e}^{-\tau\mathrm{H}(t)} | \boldsymbol{r} \rangle &= \langle \boldsymbol{r}' | \mathrm{e}^{-\tau\left[D\mathbf{K}^2 + \mathrm{i}\gamma\boldsymbol{G}(t)\cdot\mathbf{R}\right]} | \boldsymbol{r} \rangle \\ &= \langle \boldsymbol{r}' | \mathrm{e}^{-\tau D\mathbf{K}^2}\mathrm{e}^{-\mathrm{i}\gamma\tau\boldsymbol{G}\cdot\mathbf{R}} | \boldsymbol{r} \rangle + \mathcal{O}(\tau^2) \\ &= \langle \boldsymbol{r}' | \mathrm{e}^{-\tau D\mathbf{K}^2} | \boldsymbol{r} \rangle \mathrm{e}^{-\mathrm{i}\tau\gamma\boldsymbol{G}\cdot\boldsymbol{r}} + \mathcal{O}(\tau^2) \,. \end{aligned} \tag{23}$$

Note that since $\mathrm{U}_0 = \mathrm{e}^{-tD\mathbf{K}^2}$ is the propagator for the diffusion equation[25], i.e., it satisfies $(\partial/\partial t)\mathrm{U}_0 = -D\mathbf{K}^2\mathrm{U}_0$, the ket $\mathrm{e}^{-tD\mathbf{K}^2}|\boldsymbol{r}\rangle$ is just the (probability) density profile at time t, when the initial density profile was proportional to $|\boldsymbol{r}\rangle$, i.e., localized at $\boldsymbol{r}$. Therefore, the matrix element

$$\langle \boldsymbol{r}' | \mathrm{e}^{-\tau D \mathbf{K}^2} | \boldsymbol{r} \rangle = P_\tau(\boldsymbol{r}' | \boldsymbol{r}) \,, \tag{24}$$

i.e., the probability of diffusing to $\boldsymbol{r}'$ a time τ after having been at $\boldsymbol{r}$. Additionally defining $\boldsymbol{q}(t) = \tau\gamma \boldsymbol{G}(t)$, and setting $\langle \boldsymbol{r} | m(0) \rangle = V^{-1}$ as was done earlier, we have

$$E(t) = V^{-1} \lim_{\substack{\tau \to 0 \\ n\tau = t}} \int \mathrm{d}\boldsymbol{r}_n \ldots \mathrm{d}\boldsymbol{r}_0 \, P_\tau(\boldsymbol{r}_n | \boldsymbol{r}_{n-1}) \ldots P_\tau(\boldsymbol{r}_1 | \boldsymbol{r}_0) \mathrm{e}^{-\mathrm{i} \sum_{j=0}^{n-1} \boldsymbol{q}_j \cdot \boldsymbol{r}_j} \,. \tag{25}$$

The summation in the exponent turns into an integral as $\tau \to 0$, and the sequence of n integrals essentially represents a path integral. Thus, the echo attenuation is given simply by

$$E(t) = V^{-1} \left\langle \mathrm{e}^{-\mathrm{i}\gamma \int_0^t \mathrm{d}t' \, \boldsymbol{G}(t') \cdot \boldsymbol{r}(t')} \right\rangle_{\text{paths}} \,. \tag{26}$$

3 Implementation Aspects

In the previous section, we showed that the multiple propagator and multiple correlator approaches are not fundamentally different. Nonetheless, we will refer to a strategy making explicit use of spatial coordinates as a path integral or multiple propagator approach, while reserving the multiple correlator terminology for when spatial coordinates are not used explicitly. Here, we briefly discuss implementation-related aspects of computing the echo, which was largely neglected in the previous section.

3.1 Matrix Product Formulation of the Multiple Propagator Technique

Let us denote the echo *before* the continuum limit ($\tau \to 0$) as

$$\tilde{E}(t) = V^{-1} \int \mathrm{d}\boldsymbol{r}_n \ldots \mathrm{d}\boldsymbol{r}_0 \, P_\tau(\boldsymbol{r}_n | \boldsymbol{r}_{n-1}) \ldots P_\tau(\boldsymbol{r}_1 | \boldsymbol{r}_0) \mathrm{e}^{-i \sum_{j=0}^{n-1} \boldsymbol{q}_j \cdot \boldsymbol{r}_j} \,. \tag{27}$$

Frequently, the stepwise path propagator (Eq. (24)) is neither available nor easy to derive. In this case, the usual strategy is to resort to an expansion in the eigenstates of $\mathrm{U}_0(t+\tau, t) = \mathrm{e}^{-\tau D \mathbf{K}^2}$ (recall that $\langle \boldsymbol{r} | \mathbf{K}^2 = -\nabla^2 \langle \boldsymbol{r} |$), since the eigenfunctions may be more easy to derive (or to look up). Denoting the eigenfunctions of ∇^2 with $u_k(\boldsymbol{r})$, one may write

$$P_\tau(\boldsymbol{r}'|\boldsymbol{r}) = \langle \boldsymbol{r}'|\mathrm{e}^{-\tau D\mathbf{K}^2}|\boldsymbol{r}\rangle = \sum_{\boldsymbol{k}} \mathrm{e}^{-\tau\omega_k} u_{\boldsymbol{k}}(\boldsymbol{r}') u_{\boldsymbol{k}}^*(\boldsymbol{r}) \,. \tag{28}$$

After employing this eigenfunction decomposition, matrices (indexed by the states $|\boldsymbol{k}\rangle$) of the form

$$\Lambda_{\boldsymbol{k},\boldsymbol{k}'}(\tau) \equiv \langle \boldsymbol{k}|\Lambda(\tau)|\boldsymbol{k}'\rangle = \mathrm{e}^{-\tau\omega_k}\delta_{\boldsymbol{k},\boldsymbol{k}'} \tag{29}$$

and

$$\Gamma_{\boldsymbol{k},\boldsymbol{k}'}(\boldsymbol{q}) \equiv \langle \boldsymbol{k}|\Gamma(\boldsymbol{q})|\boldsymbol{k}'\rangle = \int \mathrm{d}\boldsymbol{r}\; u_{\boldsymbol{k}}^*(\boldsymbol{r})\mathrm{e}^{-\mathrm{i}\boldsymbol{q}\cdot\boldsymbol{r}} u_{\boldsymbol{k}'}(\boldsymbol{r}) \tag{30}$$

crop up. Then, the echo becomes the "00" matrix element (in its wave vector space representation) of an operator, i.e.,

$$\tilde{E} = \langle 0|\Gamma(\boldsymbol{q}_n)\Lambda(\tau)\Gamma(\boldsymbol{q}_{n-1})\Lambda(\tau)\ldots\Lambda(\tau)\Gamma(\boldsymbol{q}_1)\Lambda(\tau)\Gamma(\boldsymbol{q}_0)|0\rangle \,. \tag{31}$$

The echo is then the continuous time limit of $\tilde{E}$.

In a previous article [34], one of the authors used this scheme to compute, analytically, the low gradient (small $\boldsymbol{q}$) limit of the echo in restricted diffusion experiments. More elaboration on this will follow after the next subsection.

3.2 The "Advantages" of the MCF Technique

Earlier, we discussed that the MCF technique does not rely on explicit projection of the propagators onto the spatial coordinate basis, but rather expresses them in the wave vector basis where the zero-gradient propagator is diagonal. Even though the required matrix elements can be computed analytically, the computation of the operator product is typically left to a computer, after which the continuum limit cannot be taken formally. However, when a gradient waveform $\boldsymbol{G}(t)$ consisting of subsequent plateaus of finite duration is considered, the continuum limit becomes insubstantial, and this approach seems to afford a more efficient calculation.

Recall that the reason behind the infinitesimal time-slicing of Eq. (15) was the fact that the commutator $[\mathrm{H}(t), \mathrm{H}(t')] \neq 0$ for $t \neq t'$. When the gradient sequence, and therefore (the "time translation generator") $\mathrm{H}(t)$ is a series of plateaus of finite duration, however, the commutator vanishes within these intervals, negating the need for slicing these intervals. That is, the product of propagators in the corresponding Eq. (15) need only include as many exponentials as there are plateaus, each being the result of summing up the *exponents* within each finite time interval.

Specifically, if we have a piecewise constant gradient profile, with subsequent plateaus labeled by $0 \leq j \leq p$, of width Δt_j, where $\boldsymbol{G}(t) = \boldsymbol{G}_j$ and $\mathrm{H}(t) = \mathrm{H}_j$, we have

$$E(t) = \langle 0| \prod_{j=0}^{p} \mathrm{e}^{-\Delta t_j \mathrm{H}_j} |0\rangle \,. \tag{32}$$

Thus, the above expression requires much fewer number of matrices to exponentiate and multiply in the case of a pulsed field gradient (PFG) experiment with a small number of pulses ($p \ll n$), as opposed to a general waveform situation.

The procedure then follows similarly to Eq. (18) onwards. The MCF technique is set up to take advantage of such a situation, while the multiple propagator (path integral) approach is not, because the isolation of the phase factor from the stepwise propagator in the latter relies on the width τ of the sampling interval being small—see Eq. (23).

4 Evaluation of the Path Integral: Effect of Restricted Diffusion at Small Attenuations

The effect of free (Gaussian) diffusion on the MR signal for general gradient waveforms has been quantified by solving the Bloch–Torrey equation using conventional means [23, 24]. The echo attenuation is given by

$$E(t) = \exp\left(-\gamma^2 D \int_0^t \left| \int_0^{t'} \boldsymbol{G}(t'')\, \mathrm{d}t'' \right|^2 \mathrm{d}t' \right) . \tag{33}$$

It is very desirable to obtain a similar expression for more complicated environments, e.g., when the water molecules are trapped inside isolated pores. To this end, the path integral derived above (Eq. (26)) can be studied, and in certain cases, evaluated analytically. In this section we shall summarize the derivation presented in [34].

We have shown earlier that the (discrete approximation of the) signal attenuation can be expressed as the "00" matrix element of an operator (product) as given in Eq. (31). We are primarily interested in evaluating the signal attenuation value up to terms of quadratic order in the $\boldsymbol{q}$'s. We henceforth denote the order in $\boldsymbol{q}$ of physical quantities with superscripts, and proceed by keeping good track of the order of $\boldsymbol{q}$ afforded by each operator $\Gamma(\boldsymbol{q})$—see Eq. (30).

The 0th order term of $\tilde{E}$ is easily found (all the Γ's are identity) as

$$\tilde{E}^{(0)} = \langle 0|\Lambda(n\tau)|0\rangle = 1 \,. \tag{34}$$

The last equality stems from the definition (29) of Λ. Sure enough, $E^{(0)} = \tilde{E}^{(0)}$ does not require the taking of the continuum limit.

The first order arises when in the product of Eq. (31) only one of the $n+1$ Γ's contribute its first order value $\Gamma^{(1)}(\boldsymbol{q}_j)$ while the rest are $\Gamma^{(0)}(\boldsymbol{q}_j) = \Gamma(0) = 1$. Joined with the recently exploited fact that the "top" eigenvalue of Λ is also $\mathrm{e}^0 = 1$, the first order correction to the echo consists of terms of the form

$$\langle 0|\Gamma^{(1)}(\boldsymbol{q}_j)|0\rangle = -\mathrm{i}\boldsymbol{q}_j V^{-1} \int \mathrm{d}\boldsymbol{r}\, \boldsymbol{r} =: -\mathrm{i}\boldsymbol{q}_j \cdot \boldsymbol{r}_{\mathrm{cm}} \,. \tag{35}$$

Here, $\boldsymbol{r}_{\mathrm{cm}}$ is the center of "mass" of the pore space. Hence,

$$\tilde{E}^{(1)}(n\tau) = -\mathrm{i}\boldsymbol{r}_{\mathrm{cm}} \cdot \sum_{j=0}^{n} \boldsymbol{q}_j \,, \tag{36}$$

and therefore (with $\boldsymbol{q} = \gamma\tau\boldsymbol{G}$, and $\tau\sum_j \to \int \mathrm{d}t$)

$$E^{(1)}(t) = -\mathrm{i}\gamma\boldsymbol{r}_{\mathrm{cm}} \cdot \int_0^t \mathrm{d}t'\, \boldsymbol{G}(t') \,. \tag{37}$$

The integral in the above expression vanishes at the echo time as required by the gradient echo condition. Thus, the MR signal typically does not include a first order term.

To calculate the second order term of the echo attenuation, the following quantities prove to be useful:

$$e_{20} = \langle 0|\Gamma^{(2)}(\boldsymbol{q}_j)|0\rangle = -\frac{1}{2V} \int \mathrm{d}\boldsymbol{r}\, (\boldsymbol{q}_j \cdot \boldsymbol{r})^2 \,, \tag{38}$$

and

$$\begin{aligned} e_{11} &= \langle 0|\Gamma^{(1)}(\boldsymbol{q}_i)\Lambda(t)\Gamma^{(1)}(\boldsymbol{q}_j)|0\rangle \\ &= V^{-1} \int \mathrm{d}\boldsymbol{r}_i\, \boldsymbol{q}_i \cdot \boldsymbol{r}_i \int \mathrm{d}\boldsymbol{r}_j\, \boldsymbol{q}_j \cdot \boldsymbol{r}_j\, P_t(\boldsymbol{r}_j|\boldsymbol{r}_i) \,. \end{aligned} \tag{39}$$

At this point, we consider three pore shapes of rectangular slabs, cylinders, and spheres commonly considered in the literature [14, 33]. We shall treat these three geometries simultaneously. To this end, we denote the number of restricted diffusion dimensions by d.

More explicitly, when $d = 1$, we refer to the slab geometry wherein diffusion takes place between two infinite plates separated by distance $2a$. All gradients of the sequence are applied in the orientation perpendicular to the plates. When this condition is not met, separation of variables could be employed, and the gradient vectors are decomposed into directions parallel and perpendicular to the plates.

The parallel components could then be used in Eq. (33) to calculate the attenuation due to free diffusion taking place in the parallel direction [37]. The product of the two attenuations yield the signal value.

The case $d = 2$ is for diffusion taking place within a cylindrical pore of radius a, and the derivation is valid for components of the gradients perpendicular to the lateral surface of the pore. When the gradients are applied along some arbitrary direction, the gradient vector can be decomposed similarly. The attenuation for $d = 2$ can be computed for the perpendicular component of the gradient. This attenuation can be multiplied with that along the cylinder direction; if Eq. (33) is employed for this direction, one obtains the aggregate signal for infinite cylinders [1]. If the solution for the slab geometry ($d = 1$) is used instead, the signal for a capped cylinder geometry [32] is obtained.

Finally, the case $d = 3$ represents diffusion taking place within a sphere of radius a.

For these shapes, the integrals in Eqs. (38) and (39) are given by

$$e_{20} = -\frac{q_j^2 a^2}{2(2+d)} , \tag{40}$$

and

$$e_{11} = 2a^2 \boldsymbol{q}_i \cdot \boldsymbol{q}_j \sum_{k=1}^{\infty} s_{dk} \, \mathrm{e}^{-\omega_{dk}(t_j - t_i)} . \tag{41}$$

where $\omega_{dk} = \alpha_{dk}^2 D/a^2$. Here, $\alpha_{1k} := (k - 1/2)\pi$, and α_{2k} and α_{3k} are the kth zero of the derivative of the first order Bessel and spherical Bessel functions, respectively. The quantity s_{dk}, given by

$$s_{dk} = \frac{1}{\alpha_{dk}^2 (\alpha_{dk}^2 - d + 1)} , \tag{42}$$

satisfies the relationship

$$\sum_{k=1}^{\infty} s_{dk} = \frac{1}{2(2+d)} . \tag{43}$$

The second order term of the expression for MR signal attenuation is then given by[6]

$$\tilde{E}^{(2)} = -a^2 \sum_{j=0}^{n} \sum_{i=0}^{n} \boldsymbol{q}_j \cdot \boldsymbol{q}_i \sum_{k=1}^{\infty} s_{dk} \, \mathrm{e}^{-\omega_{dk}|t_i - t_j|} . \tag{44}$$

[6] More details can be found in [34].

Taking the continuum limit of the above expression twice, one obtains

$$E^{(2)} = -2\gamma^2 a^2 \sum_{k=1}^{\infty} s_{dk} \int_0^t \mathrm{d}t' \, \mathrm{e}^{\omega_{dk} t'} \, \boldsymbol{G}(t') \cdot \boldsymbol{F}_{dk}(t') \,, \tag{45}$$

where

$$\boldsymbol{F}_{dk}(t') = \int_{t'}^{t} \mathrm{d}t'' \, \boldsymbol{G}(t'') \, \mathrm{e}^{-\omega_{dk} t''} \,. \tag{46}$$

For small attenuations (E values close to 1), the MR signal for diffusion within the considered pore shapes is thus given by

$$E(t) \approx \exp\left(-2\gamma^2 a^2 \sum_{k=1}^{\infty} s_{dk} \int_0^t \mathrm{d}t' \, \mathrm{e}^{\omega_{dk} t'} \, \boldsymbol{G}(t') \cdot \boldsymbol{F}_{dk}(t') \right) . \tag{47}$$

Note this expression's resemblance to that for free diffusion in Eq. (33).

The above derivation illustrates that the multiple propagator approach can be utilized as an analytical tool when the path integral it represents can be evaluated.

5 Conclusion

In this work, we revisited the main theoretical advances in relating the underlying diffusion process to the MR signal obtained via general gradient waveforms. Our main objective was to establish the connections of the probabilistic framework on which the multiple propagator (path integral) technique was developed to the multiple correlation function method that employs the Bloch–Torrey equation. A theorem from probability theory led to an alternative derivation of this equation. Formulating the two techniques using a common framework enabled us to draw parallels between them. Multiple propagator technique is more closely related to a path integral. As such, the technique can be considered an analytical method when the path integral can be evaluated; otherwise its matrix formulation provides a viable approximation. However, due to the commutation relations of the Bloch–Torrey propagator, the multiple correlation function method is computationally more efficient for many gradient waveforms employing pulsed field gradients.

Acknowledgements This research was supported by TÜBİTAK-EU Co-funded Brain Circulation Scheme (project number 114C015) and Boğaziçi University (project number 8521).

References

1. Assaf, Y., Freidlin, R.Z., Rohde, G.K., Basser, P.J.: New modeling and experimental framework to characterize hindered and restricted water diffusion in brain white matter. Magn. Reson. Med. **52**(5), 965–978 (2004)
2. Assaf, Y., Blumenfeld-Katzir, T., Yovel, Y., Basser, P.J.: AxCaliber: a method for measuring axon diameter distribution from diffusion MRI. Magn. Reson. Med. **59**(6), 1347–1354 (2008). doi:10.1002/mrm.21577. http://dx.doi.org/10.1002/mrm.21577
3. Avram, L., Özarslan, E., Assaf, Y., Bar-Shir, A., Cohen, Y., Basser, P.J.: Three-dimensional water diffusion in impermeable cylindrical tubes: theory versus experiments. NMR Biomed. **21**(8), 888–898 (2008). doi:10.1002/nbm.1277. http://dx.doi.org/10.1002/nbm.1277
4. Avram, A.V., Özarslan, E., Sarlls, J.E., Basser, P.J.: In vivo detection of microscopic anisotropy using quadruple pulsed-field gradient (qpfg) diffusion mri on a clinical scanner. NeuroImage **64**, 229–239 (2013). doi:10.1016/j.neuroimage.2012.08.048
5. Axelrod, S., Sen, P.N.: Nuclear magnetic resonance spin echoes for restricted diffusion in an inhomogeneous field: methods and asymptotic regimes. J. Chem. Phys. **114**, 6878–6895 (2001)
6. Bar-Shir, A., Avram, L., Özarslan, E., Basser, P.J., Cohen, Y.: The effect of the diffusion time and pulse gradient duration ratio on the diffraction pattern and the structural information estimated from q-space diffusion MR: experiments and simulations. J. Magn. Reson. **194**(2), 230–236 (2008). doi:10.1016/j.jmr.2008.07.009. http://dx.doi.org/10.1016/j.jmr.2008.07.009
7. Barzykin, A.V.: Exact solution of the Torrey-Bloch equation for a spin echo in restricted geometries. Phys. Rev. B **58**, 14171–14174 (1998)
8. Barzykin, A.V.: Theory of spin echo in restricted geometries under a step-wise gradient pulse sequence. J. Magn. Reson. **139**(2), 342–353 (1999). doi:10.1006/jmre.1999.1778. http://dx.doi.org/10.1006/jmre.1999.1778
9. Basser, P.J., Pajevic, S., Pierpaoli, C., Duda, J., Aldroubi, A.: In vivo fiber tractography using DT-MRI data. Magn. Reson. Med. **44**, 625–632 (2000)
10. Callaghan, P.T.: A simple matrix formalism for spin echo analysis of restricted diffusion under generalized gradient waveforms. J. Magn. Reson. **129**, 74–84 (1997)
11. Caprihan, A., Wang, L.Z., Fukushima, E.: A multiple-narrow-pulse approximation for restricted diffusion in a time-varying field gradient. J. Magn. Reson. A **118**, 94–102 (1996)
12. Carr, H.Y., Purcell, E.M.: Effects of diffusion on free precession in nuclear magnetic resonance experiments. Phys. Rev. **94**(3), 630–638 (1954)
13. Cheng, Y., Cory, D.G.: Multiple scattering by NMR. J. Am. Chem. Soc. **121**, 7935–7936 (1999)
14. Codd, S.L., Callaghan, P.T.: Spin echo analysis of restricted diffusion under generalized gradient waveforms: planar, cylindrical, and spherical pores with wall relaxivity. J. Magn. Reson. **137**, 358–372 (1999)
15. Conturo, T.E., Lori, N.F., Cull, T.S., Akbudak, E., Snyder, A.Z., Shimony, J.S., McKinstry, R.C., Burton, H., Raichle, M.E.: Tracking neuronal fiber pathways in the living human brain. Proc. Natl. Acad. Sci. **96**, 10422–10427 (1999)
16. Cory, D.G., Garroway, A.N., Miller, J.B.: Applications of spin transport as a probe of local geometry. Polym. Prepr. **31**, 149 (1990)
17. Gore, J.C., Xu, J., Colvin, D.C., Yankeelov, T.E., Parsons, E.C., Does, M.D.: Characterization of tissue structure at varying length scales using temporal diffusion spectroscopy. NMR Biomed. **23**(7), 745–56 (2010). doi:10.1002/nbm.1531
18. Grebenkov, D.S.: NMR survey of reflected Brownian motion. Rev. Mod. Phys. **79**, 1077–1137 (2007)
19. Grebenkov, D.S.: Analytical solution for restricted diffusion in circular and spherical layers under inhomogeneous magnetic fields. J. Chem. Phys. **128**(13), 134702 (2008). doi:10.1063/1.2841367. http://dx.doi.org/10.1063/1.2841367
20. Grebenkov, D.S.: Pulsed-gradient spin-echo monitoring of restricted diffusion in multilayered structures. J. Magn. Reson. **205**(2), 181–195 (2010). doi:10.1016/j.jmr.2010.04.017

21. Jian, B., Vemuri, B.C., Özarslan, E., Carney, P.R., Mareci, T.H.: A novel tensor distribution model for the diffusion-weighted MR signal. NeuroImage **37**(1), 164–176 (2007). doi:10.1016/j.neuroimage.2007.03.074. http://dx.doi.org/10.1016/j.neuroimage.2007.03.074
22. Kac, M.: On distributions of certain Wiener functionals. Trans. Am. Math. Soc. **65**, 1–13 (1949)
23. Karlicek, R.F., Lowe, I.J.: A modified pulsed gradient technique for measuring diffusion in the presence of large background gradients. J. Magn. Reson. **37**, 75–91 (1980)
24. Kenkre, V.M., Fukushima, E., Sheltraw, D.: Simple solutions of the Torrey-Bloch equations in the NMR study of molecular diffusion. J. Magn. Reson. **128**, 62–69 (1997)
25. Koay, C.G., Özarslan, E.: Conceptual foundations of diffusion in magnetic resonance. Concepts Magn. Reson. Part A **42A**, 116–129 (2013)
26. Komlosh, M.E., Özarslan, E., Lizak, M.J., Horkay, F., Schram, V., Shemesh, N., Cohen, Y., Basser, P.J.: Pore diameter mapping using double pulsed-field gradient MRI and its validation using a novel glass capillary array phantom. J. Magn. Reson. **208**(1), 128–135 (2011). doi:10.1016/j.jmr.2010.10.014. http://dx.doi.org/10.1016/j.jmr.2010.10.014
27. Laun, F.B.: Restricted diffusion in NMR in arbitrary inhomogeneous magnetic fields and an application to circular layers. J. Chem. Phys. **137**(4), 044704 (2012). doi:10.1063/1.4736849
28. Laun, F.B., Kuder, T.A., Wetscherek, A., Stieltjes, B., Semmler, W.: NMR-based diffusion pore imaging. Phys. Rev. E Stat. Nonlin. Soft Matter Phys. **86**(2 Pt 1), 021906 (2012)
29. Meiboom, S., Gill, D.: Modified spin-echo method for measuring nuclear relaxation times. Rev. Sci. Instrum. **29**, 688–691 (1958)
30. Mitra, P.P.: Multiple wave-vector extensions of the NMR pulsed-field-gradient spin-echo diffusion measurement. Phys. Rev. B **51**(21), 15074–15078 (1995)
31. Mori, S., Crain, B.J., Chacko, V.P., van Zijl, P.C.M.: Three-dimensional tracking of axonal projections in the brain by magnetic resonance imaging. Ann. Neurol. **45**, 265–269 (1999)
32. Özarslan, E.: Compartment shape anisotropy (CSA) revealed by double pulsed field gradient MR. J. Magn. Reson. **199**(1), 56–67 (2009). doi:10.1016/j.jmr.2009.04.002. http://dx.doi.org/10.1016/j.jmr.2009.04.002
33. Özarslan, E., Basser, P.J.: MR diffusion - "diffraction" phenomenon in multi-pulse-field-gradient experiments. J. Magn. Reson. **188**(2), 285–294 (2007). doi:10.1016/j.jmr.2007.08.002. http://dx.doi.org/10.1016/j.jmr.2007.08.002
34. Özarslan, E., Basser, P.J.: Microscopic anisotropy revealed by NMR double pulsed field gradient experiments with arbitrary timing parameters. J. Chem. Phys. **128**(15), 154511 (2008). doi:10.1063/1.2905765. http://dx.doi.org/10.1063/1.2905765
35. Özarslan, E., Basser, P.J., Shepherd, T.M., Thelwall, P.E., Vemuri, B.C., Blackband, S.J.: Observation of anomalous diffusion in excised tissue by characterizing the diffusion-time dependence of the MR signal. J. Magn. Reson. **183**(2), 315–323 (2006). doi:10.1016/j.jmr.2006.08.009. http://dx.doi.org/10.1016/j.jmr.2006.08.009
36. Özarslan, E., Shepherd, T.M., Vemuri, B.C., Blackband, S.J., Mareci, T.H.: Resolution of complex tissue microarchitecture using the diffusion orientation transform (DOT). NeuroImage **31**(3), 1086–1103 (2006). doi:10.1016/j.neuroimage.2006.01.024. http://dx.doi.org/10.1016/j.neuroimage.2006.01.024
37. Özarslan, E., Nevo, U., Basser, P.J.: Anisotropy induced by macroscopic boundaries: surface-normal mapping using diffusion-weighted imaging. Biophys. J. **94**(7), 2809–2818 (2008). doi:10.1529/biophysj.107.124081. http://dx.doi.org/10.1529/biophysj.107.124081
38. Özarslan, E., Shemesh, N., Basser, P.J.: A general framework to quantify the effect of restricted diffusion on the NMR signal with applications to double pulsed field gradient NMR experiments. J. Chem. Phys. **130**(10), 104702 (2009). doi:10.1063/1.3082078. http://dx.doi.org/10.1063/1.3082078
39. Özarslan, E., Komlosh, M., Lizak, M., Horkay, F., Basser, P.: Double pulsed field gradient (double-PFG) MR imaging (MRI) as a means to measure the size of plant cells. Magn. Reson. Chem. **49**, S79–S84 (2011). doi:10.1002/mrc.2797. http://dx.doi.org/10.1002/mrc.2797

40. Özarslan, E., Shemesh, N., Koay, C.G., Cohen, Y., Basser, P.J.: Nuclear magnetic resonance characterization of general compartment size distributions. New J. Phys. **13**, 15010 (2011). doi:10.1088/1367-2630/13/1/015010. http://dx.doi.org/10.1088/1367-2630/13/1/015010
41. Özarslan, E., Shepherd, T.M., Koay, C.G., Blackband, S.J., Basser, P.J.: Temporal scaling characteristics of diffusion as a new MRI contrast: findings in rat hippocampus. NeuroImage **60**(2), 1380–1393 (2012). doi:10.1016/j.neuroimage.2012.01.105. http://dx.doi.org/10.1016/j.neuroimage.2012.01.105
42. Özarslan, E., Koay, C.G., Shepherd, T.M., Komlosh, M.E., İrfanoğlu, M.O., Pierpaoli, C., Basser, P.J.: Mean apparent propagator (MAP) MRI: a novel diffusion imaging method for mapping tissue microstructure. NeuroImage **78**, 16–32 (2013). doi:10.1016/j.neuroimage.2013.04.016. http://dx.doi.org/10.1016/j.neuroimage.2013.04.016
43. Robertson, B.: Spin-echo decay of spins diffusing in a bounded region. Phys. Rev. **151**, 273–277 (1966)
44. Sen, P.N., André, A., Axelrod, S.: Spin echoes of nuclear magnetization diffusing in a constant magnetic field gradient and in a restricted geometry. J. Chem. Phys. **111**, 6548–6555 (1999)
45. Stejskal, E.O., Tanner, J.E.: Spin diffusion measurements: spin echoes in the presence of a time-dependent field gradient. J. Chem. Phys. **42**(1), 288–292 (1965)
46. Stepišnik, J.: Analysis of NMR self-diffusion measurements by a density matrix calculation. Physica B & C **104**, 350–364 (1981)
47. Sukstanskii, A.L., Yablonskiy, D.A.: Effects of restricted diffusion on MR signal formation. J. Magn. Reson. **157**(1), 92–105 (2002)
48. Torrey, H.C.: Bloch equations with diffusion terms. Phys. Rev. **104**(3), 563–565 (1956)

Finslerian Diffusion and the Bloch–Torrey Equation

T.C.J. Dela Haije, A. Fuster, and L.M.J. Florack

Abstract By analyzing stochastic processes on a Riemannian manifold, in particular Brownian motion, one can deduce the metric structure of the space. This fact is implicitly used in diffusion tensor imaging of the brain when cast into a Riemannian framework. When modeling the brain white matter as a Riemannian manifold one finds (under some provisions) that the metric tensor is proportional to the inverse of the diffusion tensor, and this opens up a range of geometric analysis techniques. Unfortunately a number of these methods have limited applicability, as the Riemannian framework is not rich enough to capture key aspects of the tissue structure, such as fiber crossings.

An extension of the Riemannian framework to the more general Finsler manifolds has been proposed in the literature as a possible alternative. The main contribution of this work is the conclusion that simply considering Brownian motion on the Finsler base manifold does not reproduce the signal model proposed in the Finslerian framework, nor lead to a model that allows the extraction of the Finslerian metric structure from the signal.

1 Introduction

By observing a Brownian motion in a Riemannian manifold one can extract the local geometric structure of the space. This fact was used implicitly in Diffusion Tensor Imaging (DTI) of the brain by O'Donnell et al. [32] to interpret the (inverse of the) diffusion tensor as the metric of a Riemannian manifold, and was first explicitly mentioned in this context by Lenglet et al. [24]. DTI is one of many diffusion weighted MRI techniques that allows one to measure the conglomerate stochastic behavior of water molecules, averaging the microscopic motion of the particles over *mm* scale voxels [31]. In their work, and the work of many authors afterwards, this manifold model of the brain white matter at the *mm* scale produced analyses of

T.C.J. Dela Haije (✉) • A. Fuster • L.M.J. Florack
Imaging Science and Technology Eindhoven (IST/e), Eindhoven University of Technology, 5612 AZ Eindhoven, The Netherlands
e-mail: T.C.J.Dela.Haije@tue.nl; A.Fuster@tue.nl; L.M.J.Florack@tue.nl

I. Hotz, T. Schultz (eds.), *Visualization and Processing of Higher Order Descriptors for Multi-Valued Data*, Mathematics and Visualization,
DOI 10.1007/978-3-319-15090-1_2

the data ranging from geometry-derived scalar maps [6, 14, 16, 32] and geodesic tractography [17, 20, 24], to segmentation algorithms [25]. Despite its many uses, the DTI model and the corresponding Riemannian framework suffer from the fact that they oversimplify the structure of the tissue to an extent that proves unrealistic in large parts of the brain [22]. Tissue structures such as crossing fibers cannot be modeled properly using DTI, and especially for applications such as tractography research is now focused mainly on more elaborate models [4, 21, 37, 45].

Recently Florack et al. [15] proposed a signal model that coincides with the generalized DTI model of Özarslan et al. [34, 35], with a geometric interpretation similar to the Riemannian interpretation of DTI. This new framework is based on Finsler geometry, a generalization of Riemannian geometry, and is hypothetically suitable for the modeling of diffusion MRI signals with an arbitrarily complex orientation dependence. A summary of Finsler geometry and the Finslerian diffusion MRI framework is given in Sect. 2. In this work we consider the question of whether the Finsler geometrical framework can be derived from Brownian motion in a Finsler manifold in the same way that the DTI model and its geometric interpretation are obtained when considering Brownian motion in a Riemannian manifold. To answer this question we replace the anisotropic diffusion generator in the Bloch–Torrey equation [41] that leads to the DTI model with the generator of Brownian motion in a Finsler manifold. This generator originates from Antonelli and Zastawniak's Finslerian diffusion theory [3, App. A], and is discussed in Sect. 3. In Sect. 4 we apply the Finsler diffusion generator to diffusion MRI, resulting in a signal model that differs from the one used in Florack et al. [15]. We show that the resulting model is evidently not suited for the modeling of complex biological tissues. We conclude in Sects. 5 and 6 with a discussion on the implications of these results for the Finsler diffusion MRI framework.

2 Preliminaries

Throughout the paper we employ the summation convention, implying summation over identical pairs of upper and lower indices. Occasionally we will write 'tensor' when factually referring to the coordinate representation of a tensor relative to a basis; the details should be clear from context. Unless otherwise stated dual elements will be designated as such by a circumflex diacritic, e.g. $\hat{\mathbf{y}}$ would be the dual element to $\mathbf{y}$. Boldface symbols refer to (co-)vectors with indexed components, e.g. the contravariant components of a vector $\mathbf{x} \in V$ would be $\{x^i\}_{i=1}^{\dim V}$.

2.1 A Brief Overview of Finsler Geometry

A Finsler manifold is a differentiable manifold M together with the central object in Finsler geometry, the eponymous Finsler function F. Let $\mathbf{x} \in M$ and denote $\mathbf{y} := \mathbf{y}|_{\mathbf{x}} \in T_{\mathbf{x}}M$, with $T_{\mathbf{x}}M$ the tangent space at $\mathbf{x}$, an element of the tangent bundle

$TM = \cup_{\mathbf{x}\in M} T_{\mathbf{x}}M$. Write T^*M for the dual tangent bundle to M. The Finsler function is then a mapping $F : TM \to \mathbb{R} : (\mathbf{x}, \mathbf{y}) \mapsto F(\mathbf{x}, \mathbf{y})$ that satisfies the following conditions[1]:

- **Positive-definiteness** For all $(\mathbf{x}, \mathbf{y}) \in TM$, $F(\mathbf{x}, \mathbf{y}) \geq 0$ and $F(\mathbf{x}, \mathbf{y}) = 0 \iff \mathbf{y} = \mathbf{0}$.
- **Homogeneity** F is positively homogeneous of degree one in $\mathbf{y}$, i.e., for all $(\mathbf{x}, \mathbf{y}) \in TM$ and $\lambda \in \mathbb{R}$, $F(\mathbf{x}, \lambda\mathbf{y}) = |\lambda| F(\mathbf{x}, \mathbf{y})$.
- **Strong convexity** The fundamental tensor $g_{ij}(\mathbf{x}, \mathbf{y}) = \frac{1}{2}\frac{\partial^2 F^2}{\partial y^i \partial y^j}(\mathbf{x}, \mathbf{y})$ is positive-definite on the slit tangent bundle $TM \setminus \mathbf{0}$.

From these conditions it follows that

$$F^2(\mathbf{x}, \mathbf{y}) = g_{ij}(\mathbf{x}, \mathbf{y}) y^i y^j \tag{1}$$

for $\mathbf{y} \neq \mathbf{0}$. Additionally, Euler's theorem on homogeneous functions shows that the fundamental tensor is homogeneous of degree zero in $\mathbf{y}$: $g_{ij}(\mathbf{x}, \lambda\mathbf{y}) = g_{ij}(\mathbf{x}, \mathbf{y})$ for all $\lambda \in \mathbb{R}$. The fundamental tensor can be interpreted as the Finslerian analogue of the metric tensor in Riemannian geometry, and is in the literature often referred to as the Finsler metric tensor.

The fundamental tensor can be used to define a correspondence between $\mathbf{y} \in TM$ and its dual $\hat{\mathbf{y}} \in T^*M$ analogous to Riemannian geometry:

$$\hat{y}_i = g_{ij}(\mathbf{x}, \mathbf{y}) y^j, \tag{2}$$

$$y^i = \hat{g}^{ij}(\mathbf{x}, \hat{\mathbf{y}}) \hat{y}_j. \tag{3}$$

Here the dual $\hat{g}^{ij}(\mathbf{x}, \hat{\mathbf{y}})$ of the fundamental tensor is defined for all $(\mathbf{x}, \hat{\mathbf{y}}) \in T^*M$ by

$$\hat{g}^{ik}(\mathbf{x}, \hat{\mathbf{y}}) g_{kj}(\mathbf{x}, \mathbf{y}) = \delta^i_j, \tag{4}$$

subject to Eqs. (2) and (3), and accordingly we have a dual Finsler function $\hat{F} =: H$ defined by

$$H^2(\mathbf{x}, \hat{\mathbf{y}}) = \hat{g}^{ij}(\mathbf{x}, \hat{\mathbf{y}}) \hat{y}_i \hat{y}_j. \tag{5}$$

Again analogous to Riemannian geometry, we can (informally) define a Finslerian geodesic as a C^∞ curve $\sigma(t) = \{\sigma^i(t)\}$ in M, $0 \leq t \leq T$, for which the first variation of its integral length

$$L(\sigma) := \int_0^T F\left(\sigma(t), \frac{d\sigma}{dt}(t)\right) dt$$

[1]We only consider reversible Finsler functions, meaning that for all $(\mathbf{x}, \mathbf{y}) \in TM$, we have $F(\mathbf{x}, -\mathbf{y}) = F(\mathbf{x}, \mathbf{y})$.

is equal to zero. Constant speed geodesics, i.e., those $\sigma(t)$ for which the integrand in the previous equation is 1 for all t, satisfy the geodesic equation

$$\frac{d^2\sigma^i}{dt^2} + \gamma^i_{jk}\frac{d\sigma^j}{dt}\frac{d\sigma^k}{dt} = 0. \tag{6}$$

In this equation the $\gamma^i_{jk} \equiv \gamma^i_{jk}\left(\sigma(t), \frac{d\sigma}{dt}(t)\right)$ are the formal Christoffel symbols of the second kind defined by

$$\gamma^i_{jk}(\mathbf{x},\mathbf{y}) = \frac{1}{2}g^{il}(\mathbf{x},\mathbf{y})\left(\frac{\partial g_{kl}}{\partial x^j}(\mathbf{x},\mathbf{y}) + \frac{\partial g_{jl}}{\partial x^k}(\mathbf{x},\mathbf{y}) - \frac{\partial g_{jk}}{\partial x^l}(\mathbf{x},\mathbf{y})\right), \tag{7}$$

evaluated at $\left(\sigma(t), \frac{d\sigma}{dt}(t)\right) \in TM \setminus \mathbf{0}$.

As an example we can consider Riemannian manifolds, which form a subset of Finsler manifolds. A Finsler manifold is Riemannian if the fundamental tensor coincides with a Riemannian metric tensor, which is the case if and only if the fundamental tensor has no $\mathbf{y}$ dependence. The Riemannian metric tensor $g(\mathbf{x}) = \{g_{ij}(\mathbf{x})\}$ then defines a Finsler function $F(\mathbf{x},\mathbf{y}) = \sqrt{g_{ij}(\mathbf{x})y^i y^j}$ based on Eq. (1). A second example is the family of locally Minkowskian manifolds. The structure of these manifolds is $\mathbf{x}$-independent but does depend on $\mathbf{y}$, i.e., at each point $\mathbf{x}$ there exists a chart with an induced coordinate basis on $T_\mathbf{x}M$ such that $F(\mathbf{x},\mathbf{y}|_\mathbf{x}) = F(\mathbf{0},\mathbf{y}|_\mathbf{0}) =: F(\mathbf{y})$.

Further information about Finsler geometry can be found in [8].

2.2 *The Riemannian Framework for DTI*

Before we look at the Finslerian framework for diffusion MRI, we revisit the Riemannian case. The Riemannian framework [24, 32] presents a geometric interpretation of the DTI model [10], in which the diffusion weighted signal $S(\mathbf{q})$ of a single voxel $\Omega \subset \mathbb{R}^3$ is expressed in terms of the diffusion tensor $D = \{D^{ij}\}^3_{i,j=1}$:

$$S(\mathbf{q}) = S(\mathbf{0}) \exp\left[-\left(\Delta - \frac{\delta}{3}\right) D^{ij} q_i q_j\right]. \tag{8}$$

Here Δ and δ are respectively the pulse separation time and the pulse width in the pulsed gradient spin echo sequence.[2] $\mathbf{q} = \{q_i\}^3_{i=1} = \gamma\delta\mathbf{G}$ is a wave (co-)vector[3] oriented parallel to the applied gradient $\mathbf{G}$, with γ the gyromagnetic ratio [19]. These various parameters and constants are typically given as a single value, the b-value

[2]Different gradient sequences can be used to find the coefficients $\{D^{ij}\}$, based on modified (but similar) versions of Eq. (8).

[3]In conformity with diffusion MRI literature we omit the diacritical mark above the covectors $\mathbf{q}$ and $\mathbf{G}$.

$b = \|\mathbf{q}\|^2 \left(\Delta - \frac{\delta}{3}\right)$ that we will refer to in Sect. 5. Equation (8) is derived from the Bloch–Torrey equation [41], which describes the dynamics within the voxel Ω of the net transverse magnetization vector $M_T(\mathbf{x}, t)$, $\mathbf{x} \in \Omega$, subject to a diffusion process generated by an infinitesimal diffusion generator [33] $\mathscr{L}$:

$$\frac{\partial M_T}{\partial t}(\mathbf{x}, t) = -i\gamma\, (\mathbf{G}(t) \cdot \mathbf{x})\, M_T(\mathbf{x}, t) - \frac{1}{T_2} M_T(\mathbf{x}, t) + \mathscr{L} M_T(\mathbf{x}, t). \tag{9}$$

Let $M_T(\mathbf{x}, t) = M(t) \exp\left[-i\gamma \int_0^t \mathbf{G}(\tau) \cdot \mathbf{x}\, d\tau - \frac{t}{T_2}\right]$, where $M(t)$ describes the magnetization amplitude without the effect of relaxation. Solving Eq. (9) for a pulsed gradient spin echo experiment with the assumption that $\mathscr{L} = D^{ij} \frac{\partial}{\partial x^i} \frac{\partial}{\partial x^j}$, the anisotropic diffusion generator, gives the DTI model [10]. Here the signal $S(\mathbf{q})$, measured with gradient settings defined in $\mathbf{q}$, is the amplitude of the magnetization $M(t)$ at the echo time TE.

In their 2002 paper, O'Donnell et al. [32] define a Riemannian manifold (M, g) with the components of the metric tensor g given by the inverse diffusion tensor, i.e., $g_{ij} = D_{ij}$, where $D^{-1} = D_{ij}$. As argued by Lenglet et al. [24] this definition in fact follows by assuming that the motion of the measured hydrogen nuclei can be described by a Brownian motion in a Riemannian manifold, generated by the Laplace–Beltrami operator

$$\mathscr{L} = \frac{1}{\sqrt{\det g(\mathbf{x})}} \frac{\partial}{\partial x^i} \left(\sqrt{\det g(\mathbf{x})}\, g^{ij}(\mathbf{x}) \frac{\partial}{\partial x^j} \right) =: \Delta_R, \tag{10}$$

where $g^{ik} g_{kj} = \delta^i_j$, along with the assumption that g is spatially homogeneous per voxel [17]. In other words, the assumption that the brain can be modeled as a Riemannian manifold leads to a signal model (the DTI model) that allows the extraction of the metric structure of this manifold [12].

2.3 The Finslerian Framework for Generalized DTI

One of the earlier extensions of the DTI model is Generalized DTI[4] (GDTI) [34], in which the signal model is given by

$$S(\mathbf{q}) = \begin{cases} S(\mathbf{0}) \exp\left[-\left(\Delta - \frac{\delta}{3}\right) \|\mathbf{q}\|^2 D(\mathbf{q}/\|\mathbf{q}\|)\right] & \text{if } \mathbf{q} \neq \mathbf{0} \\ S(\mathbf{0}) & \text{if } \mathbf{q} = \mathbf{0} \end{cases}, \tag{11}$$

where $\|.\|$ denotes the Euclidean norm and $D : S^2 \to \mathbb{R}^+$ is a general even function on the sphere that represents the apparent diffusion when measuring along a given

[4] There are two diffusion MRI models called generalized DTI, one by Özarslan et al. [34] and one by Liu et al. [27]. Whenever we refer to GDTI we mean the former.

direction $\mathbf{q} \sim \mathbf{G}$. GDTI has an improved angular resolution compared to DTI, while retaining the quadratically exponential scaling in the norm of $\mathbf{q}$.

As described in Florack et al. [15] Eq. (11) can be rewritten to

$$S(\mathbf{q}) = \begin{cases} S(\mathbf{0}) \exp\left[-\left(\Delta - \frac{\delta}{3}\right) D^{ij}(\mathbf{q}) q_i q_j\right] & \text{if } \mathbf{q} \neq \mathbf{0} \\ S(\mathbf{0}) & \text{if } \mathbf{q} = \mathbf{0} \end{cases},$$

where the domain of D is expanded to $\mathbb{R}^3 \setminus \mathbf{0}$, but with D constrained to be homogeneous of degree zero and thus effectively dependent on only the orientation of the argument $\mathbf{q}$. A comparison with Eq. (5) leads to the identification

$$D^{ij}(\mathbf{q}) q_i q_j = H^2(\mathbf{q}) \tag{12}$$

providing the relation to the fundamental tensor, where $\mathbf{q}$ is interpreted as a covector $\mathbf{q}$ in the cotangent bundle T^*M of a Finsler manifold M. A variant of Eq. (12) was already used by Melonakos et al. [29, 30] to perform Finsler tractography in human brain diffusion MRI data.

3 Laplace–Beltrami Operators in Finsler Geometry

3.1 The Generator of Finslerian Brownian Motion

The Riemannian framework for DTI is founded on the one-to-one connection between the diffusion tensor and the metric tensor of the presupposed Riemannian manifold. Recall that Δ_R, the generator for Brownian motion in Riemannian manifolds, could be used to derive the DTI model (Sect. 2.2). The question we now face is whether the assumed existence of a Finsler manifold likewise provides a direct link to the physical parameters describing the measured diffusion. In order to answer this, we need to substitute an expression for the generator of Finslerian Brownian motion for $\mathscr{L}$ in Eq. (9). The generator of Brownian motion in a Finsler manifold is due to Antonelli and Zastawniak [3, App. A], and extends the work by Pinsky [38] and Watanabe and Watanabe [44]. We will recapitulate the ideas in these works, moving from Euclidean space via Riemannian manifolds to the Finslerian case.

The Isotropic Transport Process in Euclidean Space

In the Watanabe paper [44] it is shown that the equations governing Brownian motion in n-dimensional Euclidean space can be derived from the intuitive notion of colliding particles moving in straight lines in-between collisions [13], extending proofs for one- and two-dimensional Brownian motion by others (see Watanabe [43] and references therein). This notion is formalized as a Markov process (X_t, Y_t), called the isotropic (scattering) transport process and defined as follows.

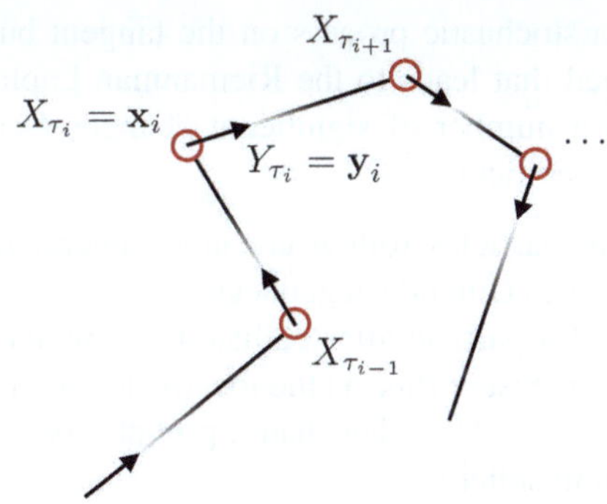

Fig. 1 An illustration of the isotropic transport process. A particle (represented by a *red circle*) moves along a straight line along $\mathbf{y}_i$ from the point $\mathbf{x}_i$, until at a time $t = \tau_{i+1}$ a (virtual) collision causes the particle to change direction. The collision times satisfy $P\{\tau_i - \tau_{i-1} > t\} = \exp(-t)$, where $\tau_0 = 0$. The transport process is defined by these paths X_t (*black lines*), and the direction of movement at each point Y_t, cf. Eq. (13)

Consider a particle initially at a position $\mathbf{x}_0 \in \mathbb{R}^n$ moving in a straight line with unit speed along the direction $\mathbf{y}_0 \in S^{n-1}$, and denote its path by $\xi(\mathbf{x}_0, \mathbf{y}_0)(t)$ with $0 \leq t \leq \tau_1$ and τ_1 the time at which the particle first collides. Next assume that the time between collisions $e_i = \tau_i - \tau_{i-1}$ (where we set $\tau_0 = 0$) satisfies $P\{e_i > t\} = \exp(-t)$ with $t \geq 0$ [26]. After the first collision the particle continues from the point of collision $\mathbf{x}_1 = \xi(\mathbf{x}_0, \mathbf{y}_0)(\tau_1)$ along a new direction $\mathbf{y}_1$ chosen according to the uniform probability measure on S^{n-1}. Continuing with this reasoning allows us to define the transport process (X_t, Y_t) as

$$\left.\begin{array}{l} X_t = \xi(\mathbf{x}_{i-1}, \mathbf{y}_{i-1})(t), \\ Y_t = \frac{dX_t}{dt}, \end{array}\right|_{\tau_{i-1} \leq t \leq \tau_i} \tag{13}$$

for all positive integers i. This process is illustrated in Fig. 1.

In order to show (weak) convergence of X_t to Brownian motion a parameter $\epsilon > 0$ is introduced such that the mean distance the particle traverses between collisions scales linearly with ϵ, while its speed scales with ϵ^{-1}. It can be shown that taking the limit $\epsilon \downarrow 0$ produces the Laplacian $\Delta = \frac{\partial^2}{\partial x^i \partial x^j}$ as the generator of this process, of which we will forego the technicalities discussed in [43, 44].

The Isotropic Transport Process in a Riemannian Manifold

The Laplace–Beltrami operator Δ_R is the unique generalization of the Laplacian Δ to Riemannian manifolds, and both Δ and Δ_R are defined as the divergence of the gradient (concepts that are well-defined in both Riemannian manifolds and Euclidean space, but not necessarily in Finsler manifolds). In his work, Pinsky [38] showed that in analogy to the isotropic transport process in Euclidean space of the previous section, a stochastic process on the tangent bundle of a Riemannian manifold can be defined that leads to the Riemannian Laplace–Beltrami operator. His adaption requires a number of significant changes (see Pinsky [38] and the references therein for specifics):

1. The trajectories that particles follow are now general Riemannian geodesics, rather than straight lines (Euclidean geodesics).
2. The new direction of a particle after collision is now based on the rotationally invariant probability measure $d\omega_{\mathbf{x}}$ on the locally defined Riemannian unit sphere $\{\mathbf{y} \in T_{\mathbf{x}}M \,|\, g_{ij}(\mathbf{x}) y^i y^j = 1\}$, rather than a probability measure on the globally defined Euclidean unit sphere.

Again a limit theorem on a parameter ϵ is invoked to show that this process leads to the result Eq. (10).

The Isotropic Transport Process in a Finsler Manifold

In the work by Antonelli and Zastawniak [3, App. A] the trajectories that particles follow are governed by the Finsler geodesic equation (Eq. (6)), while the direction after collision is distributed uniformly on the local Finslerian unit sphere, the so-called indicatrix $I_{\mathbf{x}}M$ at $\mathbf{x}$ of the n-dimensional Finsler manifold M, under the normalized Riemannian volume measure $d\mu_{\mathbf{x}}(\mathbf{y})$. The dependent variable $\mathbf{y}$ relates to the direction of motion in a sample path, analogous to Eqs. (6) and (13). Once more, details can be found in [3]. Similar limiting arguments as before lead to the generator Δ_F of Finslerian Brownian motion:

$$\Delta_F = n \int_{I_{\mathbf{x}}M} y^i y^j \left(\frac{\partial}{\partial x^i} \frac{\partial}{\partial x^j} - \gamma_{ij}^k(\mathbf{x}, \mathbf{y}) \frac{\partial}{\partial x^k} \right) d\mu_{\mathbf{x}}(\mathbf{y}). \tag{14}$$

The symbols $\gamma_{ij}^k(\mathbf{x}, \mathbf{y})$ are the formal Christoffel symbols introduced in Sect. 2.1, cf. Eq. (7).

As it should be, this expression reduces to the Riemannian Laplace–Beltrami operator Δ_R in the special case where the metric (and thus the volume element) is independent of the direction of particle motion $\mathbf{y}$. In that case M is equivalent to a Riemannian manifold, and the indicatrix bundle $IM = \cup_{\mathbf{x}\in M} I_{\mathbf{x}}M$ reduces to the unit tangent bundle $UTM = \cup_{\mathbf{x}\in M} UT_{\mathbf{x}}M$, with $UT_{\mathbf{x}}M = \{\mathbf{y} \in T_{\mathbf{x}}M \,|\, g_{ij}(\mathbf{x}) y^i y^j = 1\}$. The volume measure $d\mu_{\mathbf{x}}(\mathbf{y})$ becomes the rotationally invariant probability measure $d\omega_{\mathbf{x}}$ on the local Riemannian unit sphere. We find, using Eq. (14), that in this special case

$$\begin{aligned} \Delta_F &= n \textstyle\int_{UT_{\mathbf{x}}M} y^i y^j \, d\omega_{\mathbf{x}} \left(\frac{\partial}{\partial x^i} \frac{\partial}{\partial x^j} - \gamma_{ij}^k(\mathbf{x}) \frac{\partial}{\partial x^k} \right) \\ &= g^{ij}(\mathbf{x}) \left(\frac{\partial}{\partial x^i} \frac{\partial}{\partial x^j} - \Gamma_{ij}^k(\mathbf{x}) \frac{\partial}{\partial x^k} \right) \\ &\overset{[23]}{=} \Delta_R. \end{aligned}$$

In the second-to-last step we use the fact that the **y**-independent formal Christoffel symbols of the second kind $\gamma_{ij}^k(\mathbf{x})$ coincide with the connection coefficients $\Gamma_{ij}^k(\mathbf{x})$ of the Levi–Civita connection [23], as well as the identity

$$n \int_{UT_\mathbf{x}M} y^i y^j \, d\omega_\mathbf{x} = g^{ij}(\mathbf{x})$$

which follows from

$$n \int_{UT_\mathbf{x}M} y^i y^j \, d\omega_\mathbf{x} g_{ij}(\mathbf{x}) = n \int_{UT_\mathbf{x}M} g_{ij}(\mathbf{x}) y^i y^j \, d\omega_\mathbf{x} = n = \delta_i^i = g^{ij}(\mathbf{x}) g_{ij}(\mathbf{x}).$$

3.2 *Other Finslerian Laplace–Beltrami Operators*

While Δ_F as discussed above is representative of the intuition behind Brownian motion, it is by no means a unique definition of a Finslerian Laplace–Beltrami operator. Like with many counterparts of concepts in Riemannian geometry, different derivations of the Riemannian Laplace–Beltrami operator typically lead to completely different objects when translated to Finsler geometry. These operators have properties similar to the Laplace–Beltrami operator, and reduce to Δ_R when restricted to a Riemannian manifold. Examples of these can be found in Bao and Lackey [7], Barthelmé [9], and Centore [11]. Centore for instance derives a Laplace–Beltrami operator based on the defining property that a Laplacian, in a certain infinitesimal sense, locally measures the mean value of a function. The choice made for Δ_F made here allows us to retain the assumption that the behavior of the measured particles is governed by Brownian motion (albeit in a more complex space) as was done in the Riemannian framework.

4 Finslerian Adaptations of the Bloch–Torrey Equation

Application of the Finsler Laplace–Beltrami operator in the Bloch–Torrey equation proceeds according to Torrey [41]. We consider the previously given Bloch–Torrey equation, Eq. (9), and assume now that the generator $\mathscr{L}$ is given by Δ_F as presented in Eq. (14). As in DTI we assume that within a voxel, the generator may be considered spatially homogeneous, in which case $d\mu_\mathbf{x} = d\mu_\mathbf{0}$ and $\gamma_{ij}^k(\mathbf{x}, \mathbf{y}) = 0$. This leaves us with the generator for Brownian motion in a locally Minkowskian manifold:

$$\begin{aligned} \mathscr{L} &= n \textstyle\int_{I_0 M} y^i y^j \, d\mu_\mathbf{0}(\mathbf{y}) \frac{\partial}{\partial x^i} \frac{\partial}{\partial x^j} \\ &\overset{*}{=} A^{ij} \frac{\partial}{\partial x^i} \frac{\partial}{\partial x^j}, \end{aligned} \tag{15}$$

where in (∗) we define $A^{ij} := n \int_{I_0 M} y^i y^j \, d\mu_0(\mathbf{y})$. Next we substitute Eq. (15) in the Bloch–Torrey equation and again solve for a pulsed gradient spin echo sequence, giving the signal model

$$S(\mathbf{q}) = S(\mathbf{0}) \exp\left[-\left(\Delta - \frac{\delta}{3} \right) A^{ij} q_i q_j \right]. \tag{16}$$

The limitations of the resulting signal model are clear. The tissue descriptor A^{ij} has the same six degrees of freedom as the DTI tensor D, Eq. (8), and as a result will have the same problems with modeling complex tissue architecture. In the same vein, the integral relationship between A^{ij} and the fundamental tensor does not allow the resolution of the components g_{ij} describing the structure of the Finsler manifold unless the underlying manifold is actually Riemannian. Given Eq. (16) it is consequently not possible to extract the geometry of the Finsler manifold from diffusion MRI measurements, contrary to the Riemannian case.

One might wonder whether a different Finsler Laplace–Beltrami operator might produce more useful results, but this does not appear to be the case. The operators from the references mentioned above[7, 9, 11] all result in models with a limited number of degrees of freedom, as can be verified by solving the appropriate modifications of the Bloch–Torrey equation. Additionally these operators lack the physical interpretation that Δ_F has as described here, and generally have more complicated expressions.

5 Discussion

Regarding the hypothesized relation in Eq. (12) linking the GDTI model and the Finslerian framework, it is clear that the model does not follow from the assumption of a Brownian motion in a Finsler base manifold, nor does it seem likely that it is the result of a different diffusion process in this space. Of course the model follows when postulating a generator of the form $D^{ij}(\mathbf{q}) \frac{\partial}{\partial x^i} \frac{\partial}{\partial x^j}$ as was done in the original work by Özarslan et al. [34], but in this case relating the diffusion tensor D to structural properties of the tissue becomes more difficult.

What we have to let go of is the idea that the anisotropy that particles encounter is prescribed by the direction of their motion in the simple manner that we assumed here. Instead we can view the tissue as an abstract space whose anisotropy properties are locally parameterized by an externally imposed vector field $\sim \mathbf{q}$, which is actually more aligned with the original ideas in Özarslan et al. [34] and Florack et al. [15]. If stochastic processes on such a space could be linked to a process on a Finsler manifold, this would provide the assumptions that need to be validated for the geometric interpretation of GDTI to hold. This will be the subject of future investigations.

Earlier objections to the GDTI model raised by Liu et al. [28] consisted of a mathematical argument based on the Taylor expansion of the function $H^2(\mathbf{q}) = D^{ij}(\mathbf{q})q_i q_j$ (Eq. (5)) around the origin $\mathbf{q} = \mathbf{0}$, where the assumption of quadratic scaling in the norm of $\mathbf{q}$ was shown to constrict D to the diffusion tensor of the DTI model. The conclusion drawn by the authors was that the GDTI model of Özarslan et al. was self-inconsistent. However, the key step of writing the function as a series expansion requires analyticity in the origin; it is this assumption of analyticity, rather than the assumption of quadratic scaling, that imposes the a priori unwarranted restriction on the function space of D.

This discussion notwithstanding, the usability of GDTI is restricted to gradient strength regions in which the quadratic scaling assumption holds to a reasonable approximation. Assaf and Cohen [5] reported approximate validity up to $b = 5{,}000\,\mathrm{s/mm^2}$ for various combinations of δ and Δ. Additionally it is in principle possible to determine a Finsler function based on data for which the quadratic scaling assumption holds *locally*. In both cases though, the level sets of the reconstructed Finsler function are required to be convex in order to satisfy the strong convexity requirement mentioned in Sect. 2.1. This convexity can be checked with a simple convexity measure. A cursory investigation on a single data set from the Human Connectome Project [42] (subject ID 100307, see Fig. 2) suggests this might be satisfied to a large extent for gradient strengths up to at least[5] $b = 3{,}000\,\mathrm{s/mm^2}$, cf. Fig. 3. The experiment compares the apparent convexity of the level sets obtained from the data with 'typical' values of convex and non-convex level-sets, showing a fairly homogeneous and relatively high degree of convexity throughout the data set. The higher b-value shells of the data set show a higher degree of convexity. The specific acquisition protocol can be found in the overview paper by Sotiropoulos et al. [39], and further details are available in related references [2, 18, 40]. At this

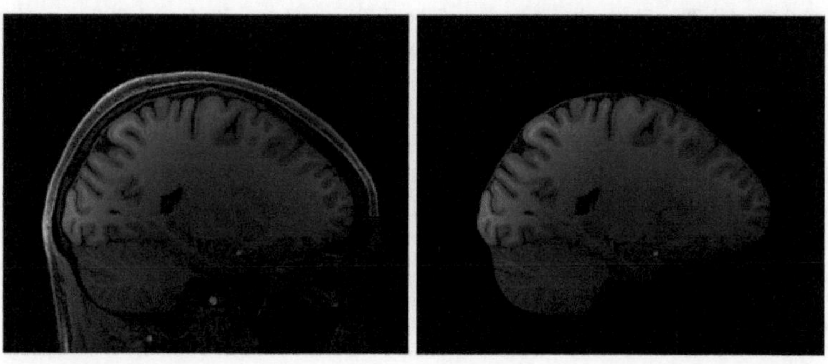

Fig. 2 A sagittal slice in a T_1 weighted image of a data set provided by the Human Connectome Project (subject ID 100307)

[5]Since we only report results of a single experiment we provide only the b-value. A more extensive analysis should consider the influence of the different parameters δ, Δ, and $\|\mathbf{G}\|$ separately.

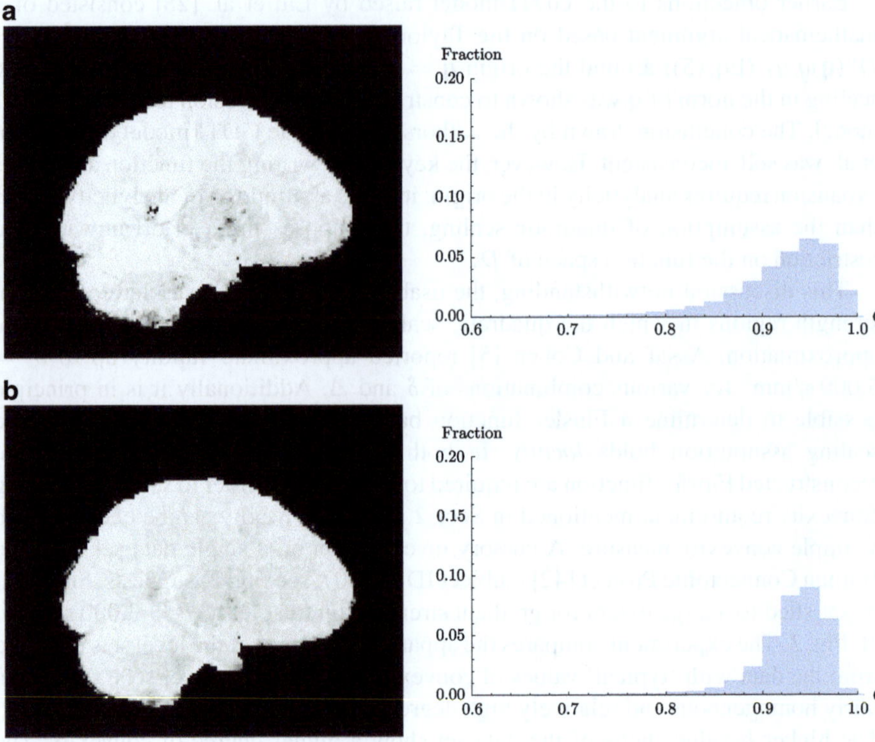

Fig. 3 A simple experiment showing a measure of convexity for a sagittal slice in a single data set. The convexity measure C is defined as the ratio between the volumes of the apparent unit level set of H^2, Eq. (5), and its convex hull. The level sets are defined per orientation according to Eq. (12). In the ventricles, where we expect the true underlying level sets to be convex, we typically find $C \approx 0.95$. (**a**) The convexity measure of the diffusion data with $b = 1{,}000\,\mathrm{s/mm^2}$ matching Fig. 2, and the corresponding histogram. (**b**) The convexity measure plot and histogram of diffusion data with $b = 3{,}000\,\mathrm{s/mm^2}$

stage the value in using the Finslerian framework in combination with GDTI has to be determined experimentally, see also [1, 36].

6 Conclusion

The commonly used Diffusion Tensor Imaging (DTI) model can be derived from the Bloch–Torrey equation by specifying the infinitesimal diffusion generator of Riemannian Brownian motion. This leads to the geometrical interpretation of DTI, where the (white) matter in the brain is viewed as a Riemannian manifold, with a metric tensor defined locally as the inverse of the diffusion tensor. A similar framework based on Finsler geometry has been suggested [15] as a geometric interpretation of the Generalized DTI (GDTI) model [34], but this framework lacks the relation to a stochastic process that makes the Riemannian framework so intuitive.

In this chapter we have discussed Brownian motion in Finsler manifolds, generated by a Finslerian Laplace–Beltrami operator due to Antonelli and Zastawniak [3, App. A], and we derived the corresponding diffusion MRI model. We have shown that the resulting signal model has the same general form as the DTI model, and will thus suffer the same limitations. Furthermore, it is not possible to extract the geometric structure of the proposed Finsler manifold from the signal based on this model. As a final remark, it should be noted that the model obtained through the assumption of Finslerian Brownian motion is entirely different from the GDTI model. The validity and value of the GDTI model and the corresponding geometric framework is the subject of future investigations.

Acknowledgements Tom Dela Haije gratefully acknowledges The Netherlands Organisation for Scientific Research (NWO) for financial support. The authors would like to thank Thomas Schultz and Remco Duits for their input regarding the quadratic scaling assumption. Data were provided by the Human Connectome Project, WU-Minn Consortium (Principal Investigators: David Van Essen and Kamil Ugurbil; 1U54MH091657) funded by the 16 NIH Institutes and Centers that support the NIH Blueprint for Neuroscience Research; and by the McDonnell Center for Systems Neuroscience at Washington University.

References

1. Aganj, I., Lenglet, C., Sapiro, G., Yacoub, E., Ugurbil, K., Harel, N.: Reconstruction of the orientation distribution function in single- and multiple-shell q-ball imaging within constant solid angle. Magn. Reson. Med. **64**(2), 554–566 (2010). doi:10.1002/mrm.22365
2. Andersson, J.L., Skare, S., Ashburner, J.: How to correct susceptibility distortions in spin-echo echo-planar images: application to diffusion tensor imaging. NeuroImage **20**(2), 870–888 (2003). doi:10.1016/S1053-8119(03)00336-7
3. Antonelli, P.L., Zastawniak, T.J.: Fundamentals of Finslerian Diffusion with Applications. Springer, Dordrecht (1999)
4. Assaf, Y., Basser, P.J.: Composite hindered and restricted model of diffusion (CHARMED) MR imaging of the human brain. NeuroImage **27**(1), 48–58 (2005). doi:10.1016/j.neuroimage.2005.03.042
5. Assaf, Y., Cohen, Y.: Non-mono-exponential attenuation of water and n-acetyl aspartate signals due to diffusion in brain tissue. J. Magn. Reson. **131**(1), 69–85 (1998)
6. Astola, L.J., Florack, L.M.J.: Finsler geometry on higher order tensor fields and applications to high angular resolution diffusion imaging. Int. J. Comput. Vis. **92**(3), 325–336 (2011)
7. Bao, D., Lackey, B.C.: A Hodge decomposition theorem for Finsler spaces. Comptes Rendus de l'Académie des Sciences. Série 1, Mathématique **323**(1), 51–56 (1996)
8. Bao, D., Chern, S.S., Shen, Z.: An Introduction to Riemann-Finsler Geometry. Springer, New York (2000)
9. Barthelmé, T.: A natural Finsler-laplace operator. Isr. J. Math. **196**(1), 375–412 (2013)
10. Basser, P.J., Mattiello, J., Lebihan, D.: Estimation of the effective self-diffusion tensor from the NMR spin echo. J. Magn. Reson. Ser. B **103**(3), 247–254 (1994). doi:10.1006/jmrb.1994.1037
11. Centore, P.: A mean-value Laplacian for Finsler spaces. In: Antonelli, P.L., Lackey, B.C. (eds.) The Theory of Finslerian Laplacians and Applications. Mathematics and Its Applications, vol. 459, pp. 151–186. Springer, Netherlands (1998)

12. de Lara, M.C.: Geometric and symmetry properties of a nondegenerate diffusion process. Ann. Probab. **23**(4), 1557–1604 (1995). doi:10.1214/aop/1176987794
13. Einstein, A.: Über die von der molekularkinetischen Theorie der Wärme geforderte Bewegung von in ruhenden Flüssigkeiten suspendierten Teilchen. Annalen der Physik **322**(8), 549–560 (1905). doi:10.1002/andp.19053220806
14. Fletcher, P.T., Tao, R., Jeong, W.K., Whitaker, R.T.: A volumetric approach to quantifying region-to-region white matter connectivity in diffusion tensor MRI. In: Information Processing in Medical Imaging, pp. 346–358. Springer, Berlin (2007)
15. Florack, L.M.J., Fuster, A.: Riemann-finsler geometry for diffusion weighted magnetic resonance imaging. In: Westin, C.F., Vilanova, A., Burgeth B. (eds.) Visualization and Processing of Tensors and Higher Order Descriptors for Multi-Valued Data. Mathematics and Visualization, pp. 189–208. vol. XV. Springer (2014, to appear)
16. Fuster, A., Astola, L., Florack, L.: A Riemannian scalar measure for diffusion tensor images. In: Computer Analysis of Images and Patterns, pp. 419–426. Springer, Berlin (2009)
17. Fuster, A., Tristan-Vega, A., Dela Haije, T.C.J., Westin, C.F., Florack, L.M.J.: A novel Riemannian metric for geodesic tractography in DTI. In: CDMRI, Nagoya, pp. 47–54 (2013)
18. Glasser, M.F., Sotiropoulos, S.N., Wilson, J.A., Coalson, T.S., Fischl, B., Andersson, J.L., Xu, J., Jbabdi, S., Webster, M., Polimeni, J.R., Van Essen, D.C., Jenkinson, M.: The minimal preprocessing pipelines for the human connectome project. NeuroImage **80**, 105–124 (2013). doi:10.1016/j.neuroimage.2013.04.127
19. Haacke, E.M.: Magnetic Resonance Imaging: Physical Principles and Sequence Design. Wiley, New York (1999)
20. Hao, X., Whitaker, R., Fletcher, P.: Adaptive Riemannian metrics for improved geodesic tracking of white matter. In: Information Processing in Medical Imaging, pp. 13–24. Springer, Heidelberg (2011)
21. Jensen, J.H., Helpern, J.A., Ramani, A., Lu, H., Kaczynski, K.: Diffusional kurtosis imaging: the quantification of non-gaussian water diffusion by means of magnetic resonance imaging. Magn. Reson. Med. **53**(6), 1432–1440 (2005). doi:10.1002/mrm.20508
22. Jeurissen, B., Leemans, A., Tournier, J.D., Jones, D.K., Sijbers, J.: Investigating the prevalence of complex fiber configurations in white matter tissue with diffusion magnetic resonance imaging: prevalence of multifiber voxels in WM. Hum. Brain Mapp. **34**(11), 2747–2766 (2013). doi:10.1002/hbm.22099
23. Jost, J.: Riemannian Geometry and Geometric Analysis. Springer, Berlin (2005)
24. Lenglet, C., Deriche, R., Faugeras, O.: Inferring white matter geometry from diffusion tensor MRI: application to connectivity mapping. In: Kanade, T., Kittler, J., Kleinberg, J.M., Mattern, F., Mitchell, J.C., Nierstrasz, O., Pandu Rangan, C., Steffen, B., Sudan, M., Terzopoulos D., Tygar, D., Vardi, M.Y., Weikum, G., Pajdla, T., Matas, J. (eds.) Computer Vision -ECCV 2004, vol. 3024, pp. 127–140. Springer, Berlin/Heidelberg (2004)
25. Lenglet, C., Rousson, M., Deriche, R., Faugeras, O., Lehericy, S., Ugurbil, K.: A Riemannian approach to diffusion tensor images segmentation. In: Information Processing in Medical Imaging, pp. 591–602. Springer, Heidelberg (2005)
26. Lewis, J.C.: Elementary statistical models for vector collision-sequence interference effects with Poisson-distributed collision times. Int. J. Spectrosc. **2010**, 1–5 (2010). doi:10.1155/2010/561697
27. Liu, C., Bammer, R., Acar, B., Moseley, M.E.: Characterizing non-gaussian diffusion by using generalized diffusion tensors. Magn. Reson. Med. **51**(5), 924–937 (2004). doi:10.1002/mrm.20071
28. Liu, C., Bammer, R., Moseley, M.E.: Limitations of apparent diffusion coefficient-based models in characterizing non-gaussian diffusion. Magn. Reson. Med. **54**(2), 419–428 (2005). doi:10.1002/mrm.20579
29. Melonakos, J., Mohan, V., Niethammer, M., Smith, K., Kubicki, M., Tannenbaum, A.: Finsler tractography for white matter connectivity analysis of the cingulum bundle. In: Medical Image Computing and Computer-Assisted Intervention - MICCAI 2007, pp. 36–43. Springer, Heidelberg (2007)

30. Melonakos, J., Pichon, E., Angenent, S., Tannenbaum, A.: Finsler active contours. IEEE Trans. Pattern Anal. Mach. Intell. **30**(3), 412–423 (2008). doi:10.1109/TPAMI.2007.70713
31. Novikov, D.S., Kiselev, V.G.: Effective medium theory of a diffusion-weighted signal. NMR Biomed. **23**(7), 682–697 (2010). doi:10.1002/nbm.1584
32. O'Donnell, L., Haker, S., Westin, C.F.: New approaches to estimation of white matter connectivity in diffusion tensor MRI: elliptic PDEs and geodesics in a tensor-warped space. In: Dohi, T., Kikinis, R. (eds.) Medical Image Computing and Computer-Assisted Intervention - MICCAI 2002, vol. 2488, pp. 459–466. Springer, Berlin/Heidelberg (2002)
33. Øksendal, B.K.: Stochastic Differential Equations: An Introduction with Applications. Springer, Berlin/New York (2003)
34. Özarslan, E., Mareci, T.H.: Generalized diffusion tensor imaging and analytical relationships between diffusion tensor imaging and high angular resolution diffusion imaging. Magn. Reson. Med. **50**(5), 955–965 (2003)
35. Özarslan, E., Vemuri, B.C., Mareci, T.H.: Fiber orientation mapping using generalized diffusion tensor imaging. In: IEEE International Symposium on Biomedical Imaging: Nano to Macro, pp. 1036–1039 (2004)
36. Özarslan, E., Shepherd, T.M., Vemuri, B.C., Blackband, S.J., Mareci, T.H.: Resolution of complex tissue microarchitecture using the diffusion orientation transform (DOT). NeuroImage **31**(3), 1086–1103 (2006). doi:10.1016/j.neuroimage.2006.01.024
37. Özarslan, E., Koay, C.G., Shepherd, T.M., Komlosh, M.E., İrfanoğlu, M.O., Pierpaoli, C., Basser, P.J.: Mean apparent propagator (MAP) MRI: a novel diffusion imaging method for mapping tissue microstructure. NeuroImage **78**, 16–32 (2013). doi:10.1016/j.neuroimage.2013.04.016
38. Pinsky, M.A.: Isotropic transport process on a Riemannian manifold. Trans. Am. Math. Soc. **218**, 353–360 (1976)
39. Sotiropoulos, S.N., Jbabdi, S., Xu, J., Andersson, J.L., Moeller, S., Auerbach, E.J., Glasser, M.F., Hernandez, M., Sapiro, G., Jenkinson, M., Feinberg, D.A., Yacoub, E., Lenglet, C., Van Essen, D.C., Ugurbil, K., Behrens, T.E.: Advances in diffusion MRI acquisition and processing in the Human Connectome Project. NeuroImage **80**, 125–143 (2013). doi:10.1016/j.neuroimage.2013.05.057
40. Sotiropoulos, S.N., Moeller, S., Jbabdi, S., Xu, J., Andersson, J.L., Auerbach, E.J., Yacoub, E., Feinberg, D., Setsompop, K., Wald, L.L., Behrens, T.E.J., Ugurbil, K., Lenglet, C.: Effects of image reconstruction on fiber orientation mapping from multichannel diffusion MRI: reducing the noise floor using sense: effects of dMRI image reconstruction on fiber orientations. Magn. Reson. Med. **70**(6), 1682–1689 (2013). doi:10.1002/mrm.24623
41. Torrey, H.C.: Bloch equations with diffusion terms. Phys. Rev. **104**(3), 563 (1956)
42. Van Essen, D.C., Smith, S.M., Barch, D.M., Behrens, T.E., Yacoub, E., Ugurbil, K.: The WU-Minn human connectome project: an overview. NeuroImage **80**, 62–79 (2013). doi:10.1016/j.neuroimage.2013.05.041
43. Watanabe, T.: Weak convergence of the isotropic scattering transport process with one speed in the plane to Brownian motion. Proc. Jpn. Acad. **44**(7), 677–680 (1968). doi:10.3792/pja/1195521091. MR: MR0236996 Zbl: 0177.45403
44. Watanabe, S., Watanabe, T.: Convergence of isotropic scattering transport process to Brownian motion. Nagoya Math. J. **40**, 161–171 (1970)
45. Wedeen, V.J., Wang, R.P., Schmahmann, J.D., Benner, T., Tseng, W.Y.I., Dai, G., Pandya, D.N., Hagmann, P., D'Arceuil, H., de Crespigny, A.J.: Diffusion spectrum magnetic resonance imaging (DSI) tractography of crossing fibers. NeuroImage **41**(4), 1267–1277 (2008). doi:10.1016/j.neuroimage.2008.03.036

Fiber Orientation Distribution Functions and Orientation Tensors for Different Material Symmetries

Maher Moakher and Peter J. Basser

Abstract In this paper we give closed-form expressions of the orientation tensors up to the order four associated with some axially-symmetric orientation distribution functions (ODF), including the well-known von Mises-Fisher, Watson, and de la Vallée Poussin ODFs. Each is characterized by a mean direction and a concentration parameter. Then, we use these elementary ODFs as building blocks to construct new ones with a specified material symmetry and derive the corresponding orientation tensors. For a general ODF we present a systematic way of calculating the corresponding orientation tensors from certain coefficients of the expansion of the ODF in spherical harmonics.

Mathematics Subject Classification (2010): 74A40, 74E10, 62H11, 92C10

1 Introduction

Fibrous composites are ubiquitous in nature and occur over a wide range of length scales. While there are familiar examples in engineering, materials sciences, and geophysics, in biology, they arise as nanoscale fibrous macromolecular systems, sub-microscopic fibrous bundles, and even macroscopic fibrous tissues and organs. There is an increasing appreciation and desire to describe, predict, and measure material and transport processes within these complex systems. Central to achieving this goal is developing a mathematical and statistical framework like the one

M. Moakher (✉)
Laboratory for Mathematical and Numerical Modeling in Engineering Science, National Engineering School at Tunis, University of Tunis El Manar, ENIT-LAMSIN, B.P. 37, 1002 Tunis-Belvédère, Tunisia
e-mail: maher.moakher@gmail.com

P.J. Basser
Section on Tissue Biophysics & Biomimetics, PPITS, NICHD, National Institutes of Health, 13 South Drive, Bldg. 13, Rm. 3W16, Bethesda, MD 20892-5772, USA
e-mail: pjbasser@helix.nih.gov

I. Hotz, T. Schultz (eds.), *Visualization and Processing of Higher Order Descriptors for Multi-Valued Data*, Mathematics and Visualization,
DOI 10.1007/978-3-319-15090-1_3

presented here, from which one can build constitutive laws and more precise, accurate, and predictive models of material behavior. In the medical imaging field, there has been much interest determining orientation distribution functions (ODF) in brain white matter. For this application, the ODFs we consider and propose here can be explicitly included in models of nerve fiber orientation within an imaging voxel, whose parameters could be measured or estimated from MRI data.

The microscopic description of fiber orientation in fibrous materials, which are made of a large collection of rod-like objects, is embodied in the ODF, which is a non-negative function ρ defined on the unit sphere S^2 of $\mathbb{R}^3$. It is normalized so that $\int_{S^2} \rho(\mathbf{n})\, d\sigma = 1$, where $d\sigma$ is the area element in S^2 and $\mathbf{n}$ is a generic vector on the unit sphere S^2 that can be parametrized by spherical coordinates as $\mathbf{n} = (\cos\theta \sin\phi,\ \sin\theta \sin\phi,\ \cos\phi)^T$ with $0 \le \theta \le 2\pi$ and $0 \le \phi \le \pi$. If ρ satisfies $\rho(\mathbf{n}) = \rho(-\mathbf{n})$, then ρ is said to be antipodally symmetric.

Since the work of Advani and Tucker [1], orientation tensors of even orders have been used to describe the orientation of fibers at the macroscopic scale. For a given positive integer k, the kth order orientation tensor is given by the expected value, with respect to the orientation distribution function ρ, over all orientations $\mathbf{n} \in S^2$:

$$\langle \mathbf{n}^{\otimes k} \rangle_\rho := \int_{S^2} \rho(\mathbf{n}) \mathbf{n}^{\otimes k}\, d\sigma, \tag{1}$$

where, $\mathbf{n}^{\otimes k}$ denotes the kth-power tensor product of $\mathbf{n}$ defined by

$$\mathbf{n}^{\otimes k} = \underbrace{\mathbf{n} \otimes \mathbf{n} \otimes \cdots \otimes \mathbf{n}}_{k \text{ times}}.$$

Orientation tensors enjoy certain properties that follow immediately from their definition (1). First, the kth order orientation tensor $\langle \mathbf{n}^{\otimes k} \rangle_\rho$ is totally symmetric, i.e., its components satisfy

$$(\langle \mathbf{n}^{\otimes k} \rangle_\rho)_{i_1,\cdots,i_k} = (\langle \mathbf{n}^{\otimes k} \rangle_\rho)_{i_{\sigma(1)},\cdots,i_{\sigma(k)}},$$

for all permutations $\sigma(\cdot)$ of the integers $1, \cdots, k$. Second, since $\mathbf{n}$ is a unit vector, all the components of $\langle \mathbf{n}^{\otimes k} \rangle_\rho$ are less than or equal to one in absolute value. Third, the contraction of the kth order orientation tensor with respect to any two indices is the $(k-2)$th order orientation tensor, i.e.,

$$(\langle \mathbf{n}^{\otimes k} \rangle_\rho)_{i_1,\cdots,i_{k-2},j,j} = (\langle \mathbf{n}^{\otimes k-2} \rangle_\rho)_{i_1,\cdots,i_{k-2}}.$$

Furthermore, the trace of even-order orientation tensors $\langle \mathbf{n}^{\otimes 2k} \rangle_\rho$ is equal to one

$$(\langle \mathbf{n}^{\otimes 2k} \rangle_\rho)_{i_1,\cdots,i_k,i_1,\cdots,i_k} = 1,$$

and the complete contraction of odd-order orientation tensors $\langle \mathbf{n}^{\otimes 2k+1} \rangle_\rho$ is equal to the orientation tensor of the first order, i.e.,

$$(\langle \mathbf{n}^{\otimes 2k+1} \rangle_\rho)_{i_1,\cdots,i_k,i_1,\cdots,i_k,j} = (\langle \mathbf{n} \rangle_\rho)_j .$$

Any ODF ρ can be (uniquely) decomposed into an antipodally symmetric part ρ^s and an antipodally skew-symmetric part ρ^a according to

$$\rho(\mathbf{n}) = \rho^s(\mathbf{n}) + \rho^a(\mathbf{n}), \tag{2}$$

where

$$\rho^s(\mathbf{n}) := \tfrac{1}{2}(\rho(\mathbf{n}) + \rho(-\mathbf{n})) \text{ and } \rho^a(\mathbf{n}) := \tfrac{1}{2}(\rho(\mathbf{n}) - \rho(-\mathbf{n})).$$

It should be noted that the antipodally symmetric part ρ^s is always an ODF, i.e., it is a non-negative function in S^2 and its integral over S^2 is equal to one. In general, the antipodally skew-symmetric part is not, however, an orientation distribution function.

By exploiting the fact that integration over S^2 is invariant under the change of variable $\mathbf{n} \to -\mathbf{n}$ it follows that for k even we have

$$\int_{S^2} \rho(\mathbf{n})\mathbf{n}^{\otimes k}\, d\sigma = \int_{S^2} \rho(-\mathbf{n})\mathbf{n}^{\otimes k}\, d\sigma = \int_{S^2} \tfrac{1}{2}(\rho(\mathbf{n}) + \rho(-\mathbf{n}))\mathbf{n}^{\otimes k}\, d\sigma,$$

and that for k odd we have

$$\int_{S^2} \rho(\mathbf{n})\mathbf{n}^{\otimes k}\, d\sigma = -\int_{S^2} \rho(-\mathbf{n})\mathbf{n}^{\otimes k}\, d\sigma = \int_{S^2} \tfrac{1}{2}(\rho(\mathbf{n}) - \rho(-\mathbf{n}))\mathbf{n}^{\otimes k}\, d\sigma.$$

Therefore, even-order orientation tensors depend only on the antipodally symmetric part of ρ and odd-order orientation tensors depend only on the antipodally skew symmetric part of ρ. As a consequence, when ρ is antipodally symmetric all odd-order orientation tensors vanish. The orientation tensor of order k, $\langle \mathbf{n}^{\otimes k} \rangle_\rho$, is also called a *fabric tensor of the first kind of rank k* [16].

For an ODF $\rho(\mathbf{n})$ defined on the unit sphere S^2, one can consider the kth order approximation

$$\tilde{\rho}(\mathbf{n}) := \mathbb{C}_0 + \mathbb{C}_1 \cdot \mathbf{n} + \mathrm{tr}(\mathbb{C}_2 \langle \mathbf{n}^{\otimes 2} \rangle_\rho) + \mathrm{tr}(\mathbb{C}_3 \langle \mathbf{n}^{\otimes 3} \rangle_\rho) + \cdots + \mathrm{tr}(\mathbb{C}_k \langle \mathbf{n}^{\otimes k} \rangle_\rho),$$

where the coefficients $\mathbb{C}_i$, $i = 0, \ldots, k$, each of which is a totally symmetric tensor of order i, are determined so that the least-squares functional

$$\mathscr{E}(\rho) := \int_{S^2} [\rho(\mathbf{n}) - \tilde{\rho}(\mathbf{n})]^2\, d\sigma,$$

is minimized. However, as explained in the seminal paper of Kanatani [16], these coefficients tensors are not uniquely defined. The reason is that the even-order tensors 1, $\mathbf{n}^{\otimes 2}$, $\mathbf{n}^{\otimes 4}$ ... are not linearly independent, and similarly, the odd-order tensors $\mathbf{n}$, $\mathbf{n}^{\otimes 3}$ ... are not linearly independent. In fact, since $\|\mathbf{n}\| = 1$, $\mathbf{n}^{\otimes l}$ can be obtained from $\mathbf{n}^{\otimes k}$, with $k \geq l + 2$, by repeated contractions.

We note that if $\rho(\mathbf{n})$ is antipodally symmetric, which is the condition that we will consider next, then all odd-order coefficient tensors vanish. Let V_l be the vector space of functions defined on S^2 spanned by $\mathbf{n}^{\otimes l}$. Then, because of what we stated earlier on the linear dependence of 1, $\mathbf{n}^{\otimes 2}$, $\mathbf{n}^{\otimes 4}$ ..., we have $V_0 \subset V_2 \subset V_4 \ldots$. Therefore, if one wants to approximate $\rho(\mathbf{n})$ up to order k (even), then it suffices to simply consider an approximation of the form

$$\rho(\mathbf{n}) \sim \frac{1}{4\pi} F_{i_1 \cdots i_k} n_{i_1} \cdots n_{i_k}, \tag{3}$$

where $F_{i_1 \cdots i_k}$ are the components of a kth order totally symmetric tensor called the *fabric tensor of the second kind of rank* k [16]. The fabric tensors of the second kind of rank 0, 2, and 4 are, respectively

$$\mathscr{F} = 1, \; \mathbf{F} = \frac{15}{2}\left[\langle \mathbf{n}^{\otimes 2}\rangle_\rho - \tfrac{1}{5}\mathbf{I}\right], \; \mathbb{F} = \frac{315}{8}\left[\langle \mathbf{n}^{\otimes 4}\rangle_\rho - \tfrac{1}{3}(\mathbf{I} \otimes \langle \mathbf{n}^{\otimes 2}\rangle_\rho)^s + \tfrac{1}{7}(\mathbf{I} \otimes \mathbf{I})^s\right].$$

Here and throughout the paper, $\mathbf{I}$ denotes the (second-order) identity tensor, and the superscript ^{s} on a tensor indicates taking the totally symmetric part of that tensor.

Because the approximation (3) has a compact form, the number of tensor components needed for the computation is minimal. However, to get a higher-order approximation one must recompute the tensor. By a Gram-Schmidt process, we can get the orthogonal decomposition of $V_k = V_k' \oplus V_{k-2}' \oplus \cdots \oplus V_0'$ where $V_0' = V_0$ and V_l' is the orthogonal complement of V_{l-2} in V_l with respect to the $L^2(S^2)$-inner product. Then, using this orthogonal decomposition, we obtain the more practical approximation of $\rho(\mathbf{n})$ [15, 16]:

$$\rho(\mathbf{n}) \sim \frac{1}{4\pi}\left[\mathscr{D} + \operatorname{tr}(\mathbf{D}\langle \mathbf{n}^{\otimes 2}\rangle_\rho) + \operatorname{tr}(\mathbb{D}\langle \mathbf{n}^{\otimes 4}\rangle_\rho) + \cdots\right]. \tag{4}$$

The coefficients $\mathscr{D}$, $\mathbf{D}$, and $\mathbb{D}$ are called the *fabric tensors of the third kind* of rank 0, 2, and 4, respectively. They are given by

$$\mathscr{D} = 1, \; \mathbf{D} = \frac{15}{2}\left[\langle \mathbf{n}^{\otimes 2}\rangle_\rho - \tfrac{1}{3}\mathbf{I}\right], \; \mathbb{D} = \frac{315}{8}\left[\langle \mathbf{n}^{\otimes 4}\rangle_\rho - \tfrac{3}{7}(\mathbf{I} \otimes \langle \mathbf{n}^{\otimes 2}\rangle_\rho)^s + \tfrac{1}{35}(\mathbf{I} \otimes \mathbf{I})^s\right].$$

In the field of diffusion MRI, high angular diffusion resolution imaging (HARDI) is a commonly used modality for non-invasively probing water diffusion in fibrous biological tissues such as muscle and brain white matter. HARDI encompasses several techniques such as Q-ball imaging [31], diffusion orientation transform MRI [24], and spherical deconvolution MRI [30]. In general, these techniques produce a function defined on the unit sphere. Various high-order tensor decompositions have been used for approximating such functions. The reader is referred to the recent review article [28] on the use of higher-order tensors in diffusion imaging. We here particularly mention the work of Özarslan and Mareci [23] who employed an approximation of the form (3), and the work of Florack and co-authors [9, 10] who used an approximation of the form (4).

We mention that orientation tensors of even order have been widely used in the macroscopic description of short-fiber composites [2, 14], fiber suspensions [7], damage mechanics [25, 33], etc. For classical solids and fluids, the even-order orientation tensors suffice for the macroscopic description of such media. However, there are natural and man-made materials that exhibit chiral behavior, i.e., they are not invariant under inversion. Such materials are called chiral, noncentrosymmetric, or hemitropic [17]. For instance, quartz, biological molecules such as the DNA double helix, and composites with helical or screw-shaped inclusions, polar chiral materials [6, 12, 22], and chiral metamaterials [34] all show different behaviors for opposite directions. Material properties such as piezoelectricity and pyroelectricity are represented by odd-order tensors [17]. It is therefore necessary to use odd-order orientation tensors as well for the macroscopic description of such properties for these types of media. We should also mention that odd-order orientation tensors are necessary for dealing with singularities in fiber arrangements and in fiber splaying and merging. Furthermore, in the work of Liu et al. [18], odd-order diffusion tensors are considered for the measurement of the phases of magnetic resonance signals.

In the remainder of the paper, we will give the expressions of the orientation tensors up to the fourth order for ODFs with some prescribed material symmetry classes. The notation that will be used is described below.

Let $\mathscr{E}^3$ denote the three-dimensional Euclidean space and let $\{\mathbf{e}_i\}_{i=1,2,3}$ be an orthonormal basis of it. Any vector $\mathbf{v}$ in $\mathscr{E}^3$ can be represented as $\mathbf{v} = v_i \mathbf{e}_i$. The inner product of two vectors $\mathbf{a}$ and $\mathbf{b}$ is denoted by $\mathbf{a} \cdot \mathbf{b}$.

A second-order tensor $\mathbf{T}$ of the three-dimensional space $\mathscr{E}^3$ is a linear map that assigns to each vector in $\mathscr{E}^3$ a vector in $\mathscr{E}^3$. We denote by $\mathbf{u} \otimes \mathbf{v}$ the second-order tensor that assigns to a vector $\mathbf{w}$ the vector $(\mathbf{v} \cdot \mathbf{w})\mathbf{u}$. A second-order tensor can thus be represented as $\mathbf{T} = T_{ij} \mathbf{e}_i \otimes \mathbf{e}_j$. A second-order tensor $\mathbf{T}$ is symmetric if $T_{ij} = T_{ji}$ for all $i, j = 1, 2, 3$. The tensor product $\mathbf{a} \otimes \mathbf{b}$ of two vectors is the tensor that assigns to each vector $\mathbf{u}$ the vector $(\mathbf{b} \cdot \mathbf{u})\mathbf{a}$. In components $(\mathbf{a} \otimes \mathbf{b})_{ij} = a_i b_j$.

A third-order tensor T can be seen as a linear map that assigns a vector to each second-order tensor. A third-order tensor admits the representation $\mathsf{T} = T_{ijk} \mathbf{e}_i \otimes \mathbf{e}_j \otimes \mathbf{e}_k$. This tensor is said to be symmetric in the last two indices if $T_{ijk} = T_{ikj}$ for all $i, j, k = 1, 2, 3$, and totally symmetric if in addition $T_{ijk} = T_{jik}$ for all $i, j, k = 1, 2, 3$.

A fourth-order tensor $\mathbb{T}$ can be seen as a linear map that assigns to each second-order tensor a second-order tensor. A fourth-order tensor admits the representation $\mathbb{T} = T_{ijkl} \mathbf{e}_i \otimes \mathbf{e}_j \otimes \mathbf{e}_k \otimes \mathbf{e}_l$. This tensor is said to possess the minor symmetries if $T_{ijkl} = T_{jikl} = T_{ijlk}$ for all $i, j, k, l = 1, 2, 3$, the major symmetry if $T_{ijkl} = T_{klij}$ for all $i, j, k, l = 1, 2, 3$, and total symmetry if in addition to the minor and major symmetries it satisfies $T_{ijkl} = T_{ikjl}$ for all $i, j, k, l = 1, 2, 3$. Any fourth-order tensor $\mathbb{T}$ possessing the minor and major symmetries can be decomposed, in a unique manner, into its totally symmetric part $\mathbb{T}^s$ and its asymmetric part $\mathbb{T}^a$ as: $\mathbb{T} = \mathbb{T}^s + \mathbb{T}^a$. The components of the totally symmetric and asymmetric parts are [3, 21]

$$T^s_{ijkl} = \tfrac{1}{3}(T_{ijkl} + T_{ikjl} + T_{ilkj}), \qquad T^a_{ijkl} = \tfrac{1}{3}(2T_{ijkl} - T_{ikjl} - T_{ilkj}). \tag{5}$$

Let $\mathbb{I}$ denote the fourth-order identity tensor whose components are given by $I_{ijkl} = \frac{1}{2}(\delta_{ik}\delta_{jl} + \delta_{il}\delta_{jk})$. Then the components of its totally symmetric part $\mathbb{I}^s$ are $I^s_{ijkl} = \frac{1}{3}(\delta_{ik}\delta_{jl} + \delta_{il}\delta_{jk} + \delta_{ij}\delta_{kl})$, and the components of its asymmetric part $\mathbb{I}^a$ are $I^a_{ijkl} = \frac{1}{6}(\delta_{ik}\delta_{jl} + \delta_{il}\delta_{jk} - 2\delta_{ij}\delta_{kl})$.

We note that vectors and second-order tensors are easily dealt with by using linear algebra operations. This is not the case for third- and fourth-order tensors. We therefore introduce a (non-physical) six-dimensional space $\hat{\mathscr{E}}$ so that the usual linear algebra operations can be used for the manipulation of third-order tensors that are symmetric with respect to the last two indices and fourth-order tensors possessing the minor symmetries.

Let $\hat{\mathbf{e}_1} = \mathbf{e}_1 \otimes \mathbf{e}_1$, $\hat{\mathbf{e}_2} = \mathbf{e}_2 \otimes \mathbf{e}_2$, $\hat{\mathbf{e}_3} = \mathbf{e}_3 \otimes \mathbf{e}_3$, $\hat{\mathbf{e}_4} = 2^{-1/2}(\mathbf{e}_2 \otimes \mathbf{e}_3 + \mathbf{e}_3 \otimes \mathbf{e}_2)$, $\hat{\mathbf{e}_5} = 2^{-1/2}(\mathbf{e}_1 \otimes \mathbf{e}_3 + \mathbf{e}_3 \otimes \mathbf{e}_1)$ and $\hat{\mathbf{e}_6} = 2^{-1/2}(\mathbf{e}_1 \otimes \mathbf{e}_2 + \mathbf{e}_2 \otimes \mathbf{e}_1)$. Then any third-order tensor $\mathsf{T} = T_{ijk}\mathbf{e}_i \otimes \mathbf{e}_j \otimes \mathbf{e}_k$ with $T_{ijk} = T_{ikj}$ can be represented by

$$\hat{\mathsf{T}} = \hat{T}_{i\alpha}\mathbf{e}_i \otimes \hat{\mathbf{e}}_\alpha,$$

where the Latin indices range from 1 to 3 and the Greek indices run from 1 to 6. Similarly, any fourth-order tensor $\mathbb{T} = T_{ijkl}\mathbf{e}_i \otimes \mathbf{e}_j \otimes \mathbf{e}_k \otimes \mathbf{e}_l$ with $T_{ijkl} = T_{jikl} = T_{ijlk}$ can be represented by

$$\hat{\mathbb{T}} = \hat{T}_{\alpha\beta}\hat{\mathbf{e}}_\alpha \otimes \hat{\mathbf{e}}_\beta.$$

In this way, the fourth-order identity tensor $\mathbb{I}$ is represented by, $\hat{\mathbf{I}}$, the second-order identity tensor in $\hat{\mathscr{E}}$ whose components are $\hat{I}_{\alpha\beta} = \delta_{\alpha\beta}$.

2 Isotropic, Transversely Isotropic, and Uniform ODFs

The isotropic ODF is given by

$$\rho^{\text{iso}}(\mathbf{n}) = \frac{1}{4\pi}. \tag{6}$$

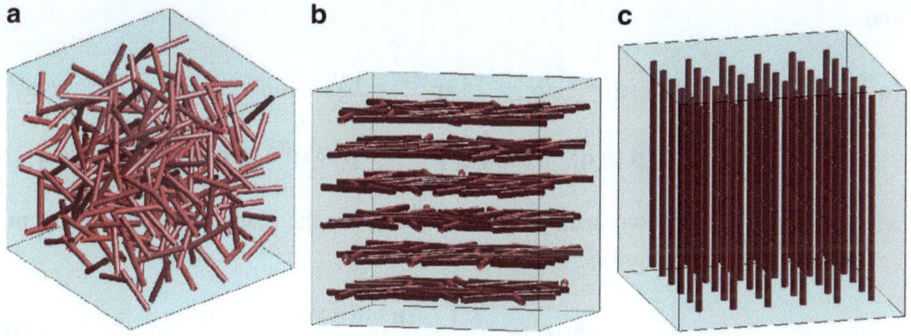

Fig. 1 (**a**) Spatial randomly oriented fibers, (**b**) planar randomly oriented fibers, (**c**) totally aligned fibers

The isotropic orientation averages of $\mathbf{n}^{\otimes 2}$ and $\mathbf{n}^{\otimes 4}$ are given by

$$\langle \mathbf{n}^{\otimes 2} \rangle^{\text{iso}} := \int_{S^2} \rho^{\text{iso}}(\mathbf{n}) \mathbf{n}^{\otimes 2} \, d\sigma = \tfrac{1}{3}\mathbf{I}, \tag{7}$$

and

$$\langle \mathbf{n}^{\otimes 4} \rangle^{\text{iso}} := \int_{S^2} \rho^{\text{iso}}(\mathbf{n}) \mathbf{n}^{\otimes 4} \, d\sigma = \tfrac{1}{5}\mathbb{I}^s, \tag{8}$$

where $\mathbb{I}^s$ is the totally symmetric part (defined by $(5)_1$) of the fourth-order identity tensor $\mathbb{I}$. The isotropic distribution (6) represents a uniform distribution of orientations on the unit sphere S^2. This distribution corresponds to fibers that are randomly oriented as depicted in Fig. 1a.

There are two other special distributions that need to be mentioned here. First, the distribution

$$\rho^{\text{iso},\mathbf{m}}(\mathbf{n}) = \frac{1}{2\pi}\delta(\mathbf{n} \cdot \mathbf{m}), \tag{9}$$

represents a uniform distribution of orientations in the plane perpendicular to $\mathbf{m}$, where $\mathbf{m}$ is a unit vector and $\delta(\cdot)$ denotes the Dirac delta function. This distribution corresponds to randomly oriented fibers in planes perpendicular to $\mathbf{m}$, see Fig. 1b. The second- and fourth-order orientation tensors associated with (9) are

$$\langle \mathbf{n}^{\otimes 2} \rangle^{\text{iso},\mathbf{m}} := \int_{S^2} \rho^{\text{iso},\mathbf{m}}(\mathbf{n}) \mathbf{n}^{\otimes 2} \, d\sigma = \tfrac{1}{2}(\mathbf{p}^{\otimes 2} + \mathbf{q}^{\otimes 2}),$$

and

$$\langle \mathbf{n}^{\otimes 4} \rangle^{\mathrm{iso},\mathbf{m}} := \int_{S^2} \rho^{\mathrm{iso},\mathbf{m}}(\mathbf{n}) \mathbf{n}^{\otimes 4} \, d\sigma = \tfrac{1}{8} \big[3\mathbf{p}^{\otimes 4} + 3\mathbf{q}^{\otimes 4} + \mathbf{p}^{\otimes 2} \otimes \mathbf{q}^{\otimes 2} + \mathbf{q}^{\otimes 2} \otimes \mathbf{p}^{\otimes 2}$$
$$+ (\mathbf{p} \otimes \mathbf{q})^{\otimes 2} + (\mathbf{q} \otimes \mathbf{p})^{\otimes 2} + \mathbf{p} \otimes \mathbf{q}^{\otimes 2} \otimes \mathbf{p} + \mathbf{q} \otimes \mathbf{p}^{\otimes 2} \otimes \mathbf{q} \big],$$

where **p** and **q** are any two orthogonal unit vectors in the plane perpendicular to **m**.

Second, the distribution

$$\rho^{\mathbf{m}}(\mathbf{n}) = \frac{\delta(1 - (\mathbf{n} \cdot \mathbf{m})^2)}{4\pi \sqrt{1 - (\mathbf{n} \cdot \mathbf{m})^2}}, \tag{10}$$

represents orientations totally aligned with **m**. This distribution corresponds to fibers oriented along **m**, see Fig. 1c. We note that this orientation distribution can simply be expressed as $\rho^{\mathbf{m}}(\mathbf{n}) = \boldsymbol{\delta}_{\mathbf{m}}(\mathbf{n})$, where $\boldsymbol{\delta}_{\mathbf{m}}(\cdot)$ is the spherical delta function defined such that for any functions f on S^2 we have

$$\int_{S^2} \boldsymbol{\delta}_{\mathbf{m}}(\mathbf{n}) f(\mathbf{n}) \, d\sigma = f(\mathbf{m}).$$

It is given by $\boldsymbol{\delta}_{\mathbf{m}}(\mathbf{n}) = \frac{1}{\sin\phi} \delta(\theta - \theta_0)\delta(\phi - \phi_0)$ where (θ, ϕ) and (θ_0, ϕ_0) are the spherical coordinates of **n** and **m**, respectively, see e.g., [29, p. 211]. The second- and fourth-order orientation tensors associated with (10) are

$$\langle \mathbf{n}^{\otimes 2} \rangle^{\mathbf{m}} := \int_{S^2} \rho^{\mathbf{m}}(\mathbf{n}) \mathbf{n}^{\otimes 2} \, d\sigma = \mathbf{m}^{\otimes 2}, \tag{11}$$

and

$$\langle \mathbf{n}^{\otimes 4} \rangle^{\mathbf{m}} := \int_{S^2} \rho^{\mathbf{m}}(\mathbf{n}) \mathbf{n}^{\otimes 4} \, d\sigma = \mathbf{m}^{\otimes 4}. \tag{12}$$

3 Axially-Symmetric ODFs

We say that an ODF $\varrho_{\mathbf{m}}(\cdot)$ is axially symmetric with respect to $\mathbf{m} \in S^2$ if $\varrho_{\mathbf{m}}(\mathbf{R}\mathbf{n}) = \varrho_{\mathbf{m}}(\mathbf{n})$ for all (proper) rotations **R** about the vector **m**. The ODF $\varrho_{\mathbf{m}}(\cdot)$ is also called *transversely hemitropic* with **m** as the direction of transverse hemitropy. In this case, there exists a positive function $\tilde{\varrho}(\cdot)$ defined on the interval $[0, \pi]$ such that

$$\varrho_{\mathbf{m}}(\mathbf{n}) = \tilde{\varrho}(\Phi), \tag{13}$$

where $\cos\Phi = \mathbf{m}\cdot\mathbf{n}$. As $\varrho_{\mathbf{m}}(\cdot)$ is normalized, the function $\tilde{\varrho}(\cdot)$ further satisfies

$$2\pi \int_0^\pi \tilde{\varrho}(\Phi)\sin\Phi\, d\Phi = 1.$$

If, in addition the ODF $\varrho_{\mathbf{m}}(\cdot)$ is antipodally symmetric, which is the case when $\tilde{\varrho}(\Phi) = \tilde{\varrho}(\pi - \Phi)$ for all $\Phi \in [0, \pi]$, then the ODF $\varrho_{\mathbf{m}}(\cdot)$ is called transversely isotropic with $\mathbf{m}$ as the direction of transverse isotropy.

The first-order orientation tensor, corresponding to the axially symmetric distribution $\varrho_{\mathbf{e}_3}$, is

$$\langle\mathbf{n}\rangle_{\varrho_{\mathbf{e}3}} := \int_{S^2} \varrho_{\mathbf{e}_3}(\mathbf{n})\mathbf{n}\, d\sigma = \begin{bmatrix} 0 \\ 0 \\ A \end{bmatrix},$$

and the corresponding second-order orientation tensor is

$$\langle\mathbf{n}^{\otimes 2}\rangle_{\varrho_{\mathbf{e}3}} := \int_{S^2} \varrho_{\mathbf{e}_3}(\mathbf{n})\mathbf{n}^{\otimes 2}\, d\sigma = \begin{bmatrix} B & 0 & 0 \\ 0 & B & 0 \\ 0 & 0 & 1-2B \end{bmatrix},$$

where

$$A = 2\pi \int_0^\pi \cos\Phi \sin\Phi\, \tilde{\varrho}(\Phi)\, d\Phi, \qquad B = \pi \int_0^\pi \sin^3\Phi\, \tilde{\varrho}(\Phi)\, d\Phi. \tag{14}$$

The corresponding third-order orientation tensor,

$$\langle\mathbf{n}^{\otimes 3}\rangle_{\varrho_{\mathbf{e}3}} := \int_{S^2} \varrho_{\mathbf{e}_3}(\mathbf{n})\mathbf{n}^{\otimes 3}\, d\sigma,$$

has the matrix representation

$$\begin{bmatrix} 0 & 0 & 0 & 0 & \sqrt{2}C & 0 \\ 0 & 0 & 0 & \sqrt{2}C & 0 & 0 \\ C & C & A-2C & 0 & 0 & 0 \end{bmatrix},$$

where

$$C = \pi \int_0^\pi \cos\Phi \sin^3\Phi\, \tilde{\varrho}(\Phi)\, d\Phi. \tag{15}$$

Similarly, the corresponding fourth-order orientation tensor,

$$\langle \mathbf{n}^{\otimes 4} \rangle_{\varrho_{\mathbf{e}_3}} := \int_{S^2} \rho_{\varrho_{\mathbf{e}_3}}(\mathbf{n}) \mathbf{n}^{\otimes 4} \, d\sigma,$$

has the 6D second-order tensor representation

$$\begin{bmatrix} 3D & D & F & 0 & 0 & 0 \\ D & 3D & F & 0 & 0 & 0 \\ F & F & E & 0 & 0 & 0 \\ 0 & 0 & 0 & 2F & 0 & 0 \\ 0 & 0 & 0 & 0 & 2F & 0 \\ 0 & 0 & 0 & 0 & 0 & 2D \end{bmatrix},$$

where

$$D = \tfrac{1}{4}\pi \int_0^{\pi} \sin^5 \Phi \, \tilde{\varrho}(\Phi) \, d\Phi, \tag{16a}$$

$$E = 2\pi \int_0^{\pi} \cos^4 \Phi \sin^5 \Phi \, \tilde{\varrho}(\Phi) \, d\Phi, \tag{16b}$$

$$F = \pi \int_0^{\pi} \cos^2 \Phi \sin^3 \Phi \, \tilde{\varrho}(\Phi) \, d\Phi. \tag{16c}$$

As the trace of the fourth-order orientation tensor is equal to one, one can easily verify that $8D + E + 4F = 2\pi \int_0^{\pi} \sin \Phi \, \tilde{\varrho}(\Phi) \, d\Phi = 1$.

In the following subsections we will give explicit expressions of the orientation tensors up to the order four for several uni-modal axially symmetric ODFs. All these orientation distributions are characterized with the modal vector (mean axis) $\mathbf{m}$ and a concentration parameter $\kappa \in [0, \kappa_\infty)$. When the concentration parameter κ is equal to zero, the orientation distribution reduces to the isotropic distribution (6). On the other hand, when the concentration parameter κ approaches κ_∞ the orientation distribution tends to the totally aligned orientation distribution (10).

3.1 The von Mises-Fisher ODF

The widely used uni-modal orientation distribution is the Fisher distribution (also known as the von Mises-Fisher distribution), with modal vector $\mathbf{m} \in S^2$ and concentration parameter $\kappa > 0$; it is given by (see e.g., [19, 35])

$$\rho^{\mathrm{MF}}_{\mathbf{m},\kappa}(\mathbf{n}) = \frac{\xi^{\mathrm{MF}}(\kappa)}{4\pi} \exp(\kappa \mathbf{m} \cdot \mathbf{n}), \tag{17}$$

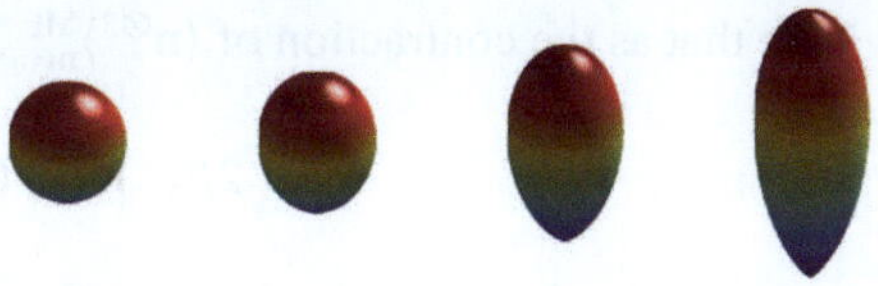

Fig. 2 Plots of the transversely hemitropic ODF (17) for $\kappa = 1, 2, 5,$ and 10

where

$$\xi^{\mathrm{MF}}(\kappa) = \frac{\kappa}{\sinh \kappa}.$$

In Fig. 2 we give plots of (17) for various values of the concentration parameter κ. It should be noted that this distribution is not invariant under the inversion of directions: $\mathbf{n} \mapsto -\mathbf{n}$, and hence it is not antipodally symmetric.

When $\mathbf{m} = \mathbf{e}_3$, the corresponding first-order orientation tensor is

$$\langle \mathbf{n} \rangle^{\mathrm{MF}}_{\mathbf{m}_3,\kappa} := \int_{S^2} \rho^{\mathrm{MF}}_{\mathbf{m}_3,\kappa}(\mathbf{n})\mathbf{n}\, d\sigma = \frac{1}{\kappa}\begin{bmatrix} 0 \\ 0 \\ \alpha(\kappa) \end{bmatrix},$$

and the corresponding second-order orientation tensor is

$$\langle \mathbf{n}^{\otimes 2} \rangle^{\mathrm{MF}}_{\mathbf{m}_3,\kappa} := \int_{S^2} \rho^{\mathrm{MF}}_{\mathbf{m}_3,\kappa}(\mathbf{n})\mathbf{n}^{\otimes 2}\, d\sigma = \frac{1}{\kappa^2}\begin{bmatrix} \alpha(\kappa) & 0 & 0 \\ 0 & \alpha(\kappa) & 0 \\ 0 & 0 & \kappa^2 - 2\alpha(\kappa) \end{bmatrix},$$

where

$$\alpha(\kappa) = \xi^{\mathrm{MF}}(\kappa)\cosh \kappa - 1.$$

The corresponding third-order orientation tensor

$$\langle \mathbf{n}^{\otimes 3} \rangle^{\mathrm{MF}}_{\mathbf{m}_3,\kappa} := \int_{S^2} \rho^{\mathrm{MF}}_{\mathbf{m}_3,\kappa}(\mathbf{n})\mathbf{n}^{\otimes 3}\, d\sigma,$$

has the matrix representation

$$\frac{1}{\kappa^3}\begin{bmatrix} 0 & 0 & 0 & 0 & \sqrt{2}\beta(\kappa) & 0 \\ 0 & 0 & 0 & \sqrt{2}\beta(\kappa) & 0 & 0 \\ \beta(\kappa) & \beta(\kappa) & \gamma(\kappa) & 0 & 0 & 0 \end{bmatrix},$$

where

$$\beta(\kappa) = (\kappa^2 + 3) - 3\xi^{\mathrm{MF}}(\kappa)\cosh \kappa,$$
$$\gamma(\kappa) = (\kappa^2 + 6)\xi^{\mathrm{MF}}(\kappa)\cosh \kappa - 3(\kappa^2 + 2).$$

Note that as the contraction of $\langle \mathbf{n}^{\otimes 3} \rangle^{\mathrm{MF}}_{\mathbf{m}_3,\kappa}$ is equal to $\langle \mathbf{n} \rangle^{\mathrm{MF}}_{\mathbf{m}_3,\kappa}$ we can verify that

$$2\beta(\kappa) + \gamma(\kappa) = \kappa^2 \alpha(\kappa).$$

Similarly, the corresponding fourth-order orientation tensor,

$$\langle \mathbf{n}^{\otimes 4} \rangle^{\mathrm{MF}}_{\mathbf{m}_3,\kappa} := \int_{S^2} \rho^{\mathrm{MF}}_{\mathbf{m}_3,\kappa}(\mathbf{n}) \mathbf{n}^{\otimes 4} \, d\sigma,$$

has the 6D second-order tensor representation

$$\frac{1}{\kappa^4} \begin{bmatrix} 3a(\kappa) & a(\kappa) & c(\kappa) & 0 & 0 & 0 \\ a(\kappa) & 3a(\kappa) & c(\kappa) & 0 & 0 & 0 \\ c(\kappa) & c(\kappa) & b(\kappa) & 0 & 0 & 0 \\ 0 & 0 & 0 & 2c(\kappa) & 0 & 0 \\ 0 & 0 & 0 & 0 & 2c(\kappa) & 0 \\ 0 & 0 & 0 & 0 & 0 & 2a(\kappa) \end{bmatrix},$$

where

$$\begin{aligned} a(\kappa) &= \kappa^2 + 3(1 - \xi^{\mathrm{MF}}(\kappa) \cosh \kappa), \\ b(\kappa) &= \kappa^4 + 12\kappa^2 + 24 - 4(\kappa^2 + 6)\xi^{\mathrm{MF}}(\kappa) \cosh \kappa, \\ c(\kappa) &= (\kappa^2 + 12)\xi^{\mathrm{MF}}(\kappa) \cosh \kappa - 5\kappa^2 - 12. \end{aligned}$$

Because the trace of even-order orientation tensors is equal to one, the three functions $a(\cdot)$, $b(\cdot)$, and $c(\cdot)$ are dependent, satisfying the relation

$$8a(\kappa) + b(\kappa) + 4c(\kappa) = \kappa^4.$$

As mentioned before, the von Mises-Fisher ODF (17) is not antipodally symmetric. Its antipodally symmetric part is given by

$$(\rho^{\mathrm{MF}}_{\mathbf{m},\kappa})^s(\mathbf{n}) = \frac{\xi^{\mathrm{MF}}(\kappa)}{4\pi} \cosh(\kappa \mathbf{m} \cdot \mathbf{n}), \tag{18}$$

and its antipodally skew-symmetric part is given by

$$(\rho^{\mathrm{MF}}_{\mathbf{m},\kappa})^a(\mathbf{n}) = \frac{\xi^{\mathrm{MF}}(\kappa)}{4\pi} \sinh(\kappa \mathbf{m} \cdot \mathbf{n}). \tag{19}$$

Plots of (18) for various values of the concentration parameter κ are shown in Fig. 3.

Fig. 3 Plots of the transversely isotropic ODF (18) for $\kappa = 1, 2, 5,$ and 10

All orientation tensors of even orders relative to the orientation distribution function (18) are equal to the orientation tensors of even orders relative to the orientation distribution function (17), whereas all orientation tensors of odd orders relative to the ODF (18) vanish.

3.2 *The Watson ODF*

The Watson distribution (also known as the Dimroth-Watson distribution) with modal vector $\mathbf{m} \in S^2$ and concentration parameter $\kappa > 0$ is given by [8]

$$\rho^{\mathrm{DW}}_{\mathbf{m},\kappa}(\mathbf{n}) = \frac{\xi^{\mathrm{DW}}(\kappa)}{4\pi} \exp(\kappa^2(\mathbf{m}\cdot\mathbf{n})^2), \tag{20}$$

where

$$\xi^{\mathrm{DW}}(\kappa) = \frac{2\kappa}{\sqrt{\pi}\,\mathrm{erfi}(\kappa)},$$

and $\mathrm{erfi}(\cdot)$ represents the imaginary error function defined by

$$\mathrm{erfi}(s) = -i\,\mathrm{erf}(is) = \frac{2s}{\sqrt{\pi}} \int_0^1 \exp(s^2 t^2)\,dt.$$

This distribution can be seen as a special case of the well-known Bingham (multi-modal) distribution for axial data

$$\rho_{\mathbf{K}}(\mathbf{n}) = \frac{1}{b(\mathbf{K})} \exp(\mathbf{n}\cdot\mathbf{K}\mathbf{n}),$$

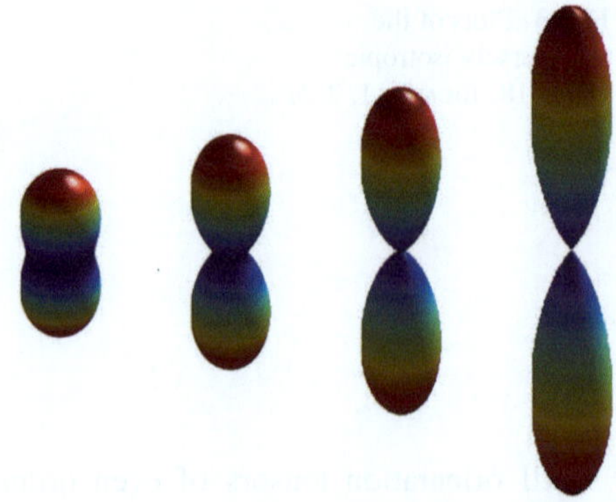

Fig. 4 Plots of the transversely isotropic ODF (20) for $\kappa = 1, \sqrt{2}, 2$, and 3

where $\mathbf{K}$ is a symmetric matrix and $b(\mathbf{K})$ is a normalization constant. The distribution $\rho^{\mathrm{DW}}_{\mathbf{m},\kappa}(\mathbf{n})$ is transversely isotropic. Like the von Mises-Fisher distribution, the Dimroth-Watson distribution is parametrized by the concentration parameter κ. Plots of (20) for various values of the concentration parameter κ are given in Fig. 4.

When $\mathbf{m} = \mathbf{e}_3$, the corresponding second-order orientation tensor $\langle \mathbf{n}^{\otimes 2} \rangle^{\mathrm{DW}}_{\mathbf{m}_3,\kappa} := \int_{S^2} \rho^{\mathrm{DW}}_{\mathbf{m}_3,\kappa}(\mathbf{n})\mathbf{n}^{\otimes 2}\, d\sigma$ is given by

$$\frac{1}{4\kappa^2}\begin{bmatrix} \tilde{\alpha}(\kappa) & 0 & 0 \\ 0 & \tilde{\alpha}(\kappa) & 0 \\ 0 & 0 & 2\left(2\kappa^2 - \tilde{\alpha}(\kappa)\right) \end{bmatrix},$$

where

$$\tilde{\alpha}(\kappa) = (2\kappa^2 + 1) - \xi^{\mathrm{DW}}(\kappa)e^{\kappa^2}.$$

The corresponding fourth-order orientation tensor,

$$\langle \mathbf{n}^{\otimes 4} \rangle^{\mathrm{DW}}_{\mathbf{m}_3,\kappa} := \int_{S^2} \rho^{\mathrm{DW}}_{\mathbf{m}_3,\kappa}(\mathbf{n})\mathbf{n}^{\otimes 4}\, d\sigma,$$

has the 6D second-order tensor representation

$$\frac{1}{32\kappa^4}\begin{bmatrix} 3\tilde{a}(\kappa) & \tilde{a}(\kappa) & \tilde{c}(\kappa) & 0 & 0 & 0 \\ \tilde{a}(\kappa) & 3\tilde{a}(\kappa) & \tilde{c}(\kappa) & 0 & 0 & 0 \\ \tilde{c}(\kappa) & \tilde{c}(\kappa) & \tilde{b}(\kappa) & 0 & 0 & 0 \\ 0 & 0 & 0 & 2\tilde{c}(\kappa) & 0 & 0 \\ 0 & 0 & 0 & 0 & 2\tilde{c}(\kappa) & 0 \\ 0 & 0 & 0 & 0 & 0 & 2\tilde{a}(\kappa) \end{bmatrix},$$

where

$$\tilde{a}(\kappa) = 4\kappa^2(\kappa^2+1) + 3 - (2\kappa^2+3)\xi^{\mathrm{DW}}(\kappa)e^{\kappa^2},$$
$$\tilde{b}(\kappa) = 8\left[3 + (2\kappa^2-3)\xi^{\mathrm{DW}}(\kappa)e^{\kappa^2}\right],$$
$$\tilde{c}(\kappa) = 4\left[3\xi^{\mathrm{DW}}(\kappa)e^{\kappa^2} - (2\kappa^2+3)\right].$$

We note that, since the trace of even-order orientation tensors is equal to one, the three functions $\tilde{a}(\cdot)$, $\tilde{b}(\cdot)$ and $\tilde{c}(\cdot)$ satisfy the relation

$$8\tilde{a}(\kappa) + \tilde{b}(\kappa) + 4\tilde{c}(\kappa) = 32\kappa^4.$$

3.3 The Singular-Kernel ODF

The uni-modal ODF with modal vector $\mathbf{m} \in S^2$ and concentration parameter $0 < K < 1$ given by

$$\rho^{\mathrm{SK}}_{\mathbf{m},K}(\mathbf{n}) = \frac{\xi^{\mathrm{SK}}(K)}{4\pi}\frac{1}{1-2K(\mathbf{m}\cdot\mathbf{n})+K^2}, \tag{21}$$

where

$$\xi^{\mathrm{SK}}(K) = \frac{K}{\tanh^{-1} K},$$

is called the singular-kernel distribution [32]. It should be noted that this distribution is not invariant under the inversion of directions: $\mathbf{n} \mapsto -\mathbf{n}$. In Fig. 5 we present plots of (21) for various values of the concentration parameter K.

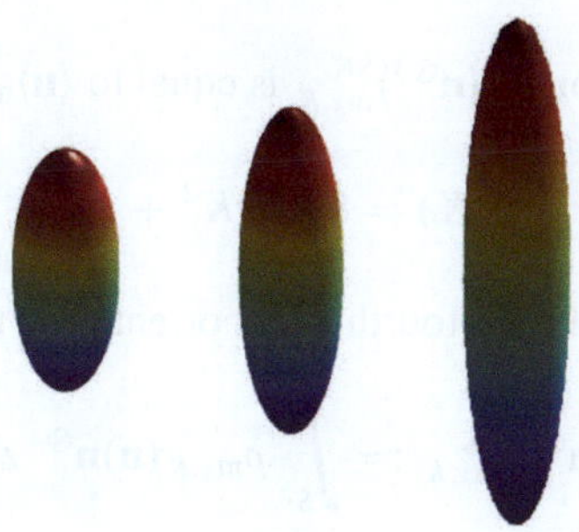

Fig. 5 Plots of the transversely hemitropic ODF (21) for $K = \frac{2}{3}, \frac{3}{4}$, and $\frac{5}{6}$

When $\mathbf{m} = \mathbf{e}_3$, the corresponding first-order orientation tensor is

$$\langle \mathbf{n} \rangle^{\mathrm{SK}}_{\mathbf{m}_3,K} := \int_{S^2} \rho^{\mathrm{SK}}_{\mathbf{m}_3,K}(\mathbf{n})\mathbf{n}\, d\sigma = \frac{1}{2K}\begin{bmatrix} 0 \\ 0 \\ K^2+1-\xi^{\mathrm{SK}}(K) \end{bmatrix},$$

and the corresponding second-order orientation tensor is

$$\langle \mathbf{n}^{\otimes 2} \rangle^{\mathrm{SK}}_{\mathbf{m}_3,K} := \int_{S^2} \rho^{\mathrm{SK}}_{\mathbf{m}_3,K}(\mathbf{n})\mathbf{n}^{\otimes 2}\, d\sigma = \frac{1}{8K^2}\begin{bmatrix} \hat{\alpha}(K) & 0 & 0 \\ 0 & \hat{\alpha}(K) & 0 \\ 0 & 0 & 2(4K^2-\hat{\alpha}(K)) \end{bmatrix},$$

where

$$\hat{\alpha}(K) = (K^2+1)\xi^{\mathrm{SK}}(K) - (K^2-1)^2.$$

The corresponding third-order orientation tensor,

$$\langle \mathbf{n}^{\otimes 3} \rangle^{\mathrm{SK}}_{\mathbf{m}_3,K} := \int_{S^2} \rho_{\mathbf{m}_3,K}(\mathbf{n})\mathbf{n}^{\otimes 3}\, d\sigma,$$

has the matrix representation

$$\frac{1}{16K^3}\begin{bmatrix} 0 & 0 & 0 & 0 & \sqrt{2}\hat{\beta}(K) & 0 \\ 0 & 0 & 0 & \sqrt{2}\hat{\beta}(K) & 0 & 0 \\ \hat{\beta}(K) & \hat{\beta}(K) & \hat{\gamma}(K) & 0 & 0 & 0 \end{bmatrix},$$

where

$$\hat{\beta}(K) = (K^4 - \tfrac{2}{3}K^2 + 1)\xi^{\mathrm{SK}}(K) - (K^6 - K^4 - K^2 + 1),$$
$$\hat{\gamma}(K) = 2\left[(K^6 + 3K^4 + 3K^2 + 1) - (K^4 + \tfrac{10}{3}K^2 + 1)\xi^{\mathrm{SK}}(K)\right].$$

Note that as the contraction of $\langle \mathbf{n}^{\otimes 3} \rangle^{\mathrm{SK}}_{\mathbf{m}_3,K}$ is equal to $\langle \mathbf{n} \rangle_{\mathbf{m}_3,K}$ we can verify that

$$2\hat{\beta}(K) + \hat{\gamma}(K) = 8K^2[(K^2+1) - \xi^{\mathrm{SK}}(K)].$$

Similarly, the corresponding fourth-order orientation tensor,

$$\langle \mathbf{n}^{\otimes 4} \rangle^{\mathrm{SK}}_{\mathbf{m}_3,K} := \int_{S^2} \rho_{\mathbf{m}_3,K}(\mathbf{n})\mathbf{n}^{\otimes 4}\, d\sigma,$$

has the 6D second-order tensor representation

$$\frac{1}{128K^4}\begin{bmatrix} 3\hat{a}(K) & \hat{a}(K) & \hat{c}(K) & 0 & 0 & 0 \\ \hat{a}(K) & 3\hat{a}(K) & \hat{c}(K) & 0 & 0 & 0 \\ \hat{c}(K) & \hat{c}(K) & \hat{b}(K) & 0 & 0 & 0 \\ 0 & 0 & 0 & 2\hat{c}(K) & 0 & 0 \\ 0 & 0 & 0 & 0 & 2\hat{c}(K) & 0 \\ 0 & 0 & 0 & 0 & 0 & 2\hat{a}(K) \end{bmatrix},$$

where

$$\hat{a}(K) = (K^8 - 4K^6 + 6K^4 - 4K^2 + 1) - (K^6 - \tfrac{11}{3}K^4 - \tfrac{11}{3}K^2 + 1)\xi^{\mathrm{SK}}(K),$$
$$\hat{b}(K) = 8\left[(K^8 + 4K^6 + 6K^4 + 4K^2 + 1) - (K^6 + \tfrac{13}{3}K^4 + \tfrac{13}{3}K^2 + 1)\xi^{\mathrm{SK}}(K)\right],$$
$$\hat{c}(K) = 4\left[(K^6 + \tfrac{1}{3}K^4 + \tfrac{1}{3}K^2 + 1)\xi^{\mathrm{SK}}(K) - (K^4 - 1)^2\right].$$

Because the trace of even-order orientation tensors is equal to one, the three functions $\hat{a}(\cdot)$, $\hat{b}(\cdot)$ and $\hat{c}(\cdot)$ are dependent. Indeed, they satisfy the relation

$$8\hat{a}(K) + \hat{b}(K) + 4\hat{c}(K) = 128K^4.$$

As mentioned before, the singular-kernel orientation distribution function (21) is not antipodally symmetric. Its antipodally symmetric part is given by

$$(\rho^{\mathrm{SK}}_{\mathbf{m},K})^s(\mathbf{n}) = \frac{\xi^{\mathrm{SK}}(K)}{4\pi}\frac{1 + K^2}{(1 + K^2)^2 - 4K^2(\mathbf{m}\cdot\mathbf{n})^2}, \tag{22}$$

and its antipodally skew-symmetric part is given by

$$(\rho^{\mathrm{SK}}_{\mathbf{m},K})^a(\mathbf{n}) = \frac{\xi^{\mathrm{SK}}(K)}{4\pi}\frac{2K\mathbf{m}\cdot\mathbf{n}}{(1 + K^2)^2 - 4K^2(\mathbf{m}\cdot\mathbf{n})^2}. \tag{23}$$

Plots of (22) for various values of the concentration parameter K are given in Fig. 6.

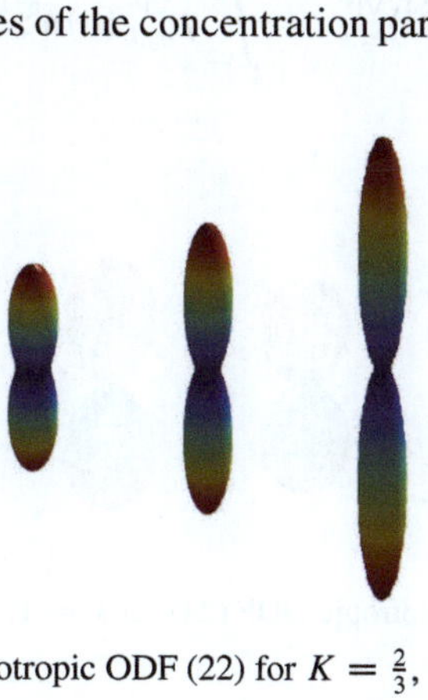

Fig. 6 Plots of the transversely isotropic ODF (22) for $K = \frac{2}{3}$, $\frac{3}{4}$, and $\frac{5}{6}$

All orientation tensors of even orders relative to the orientation distribution function (22) are equal to the orientation tensors of even orders relative to the orientation distribution function (21), whereas all orientation tensors of odd orders relative to the ODF (22) vanish.

For the comparison with the other distributions, we can compose the functions $\hat{\alpha}(K), \hat{\beta}(K), \hat{\gamma}(K), \hat{a}(K), \hat{b}(K)$, and $\hat{c}(K)$ with the function $K \mapsto \kappa = K/(1+K)$.

3.4 The de la Vallée Poussin ODF

The ODFs

$$\rho^{\mathrm{VP}}_{\mathbf{m},k}(\mathbf{n}) = \frac{2k+1}{4\pi}(\mathbf{n}\cdot\mathbf{m})^{2k}, \tag{24}$$

are a family (indexed by a positive integer k) of antipodally and axially symmetric ODFs. We note here that $\rho^{\mathrm{VP}}_{\mathbf{e}_3,k}(\mathbf{n}) = (\cos\phi)^{2k}$ has the same functional form (with ϕ replaced by $\frac{1}{2}\phi$) as the de la Vallée Poussin distribution in $SO(3)$, the group of rotations in $\mathbb{R}^3$ [26]. Accordingly, we call the family of ODFs (24) the de la Vallée Poussin ODF. The positive integer k acts as the concentration parameter of the Fisher and Watson distributions. Plots of (24) for various values of the concentration parameter k are given in Fig. 7.

For $\rho^{\mathrm{VP}}_{\mathbf{e}_3,k}$, the second-order orientation tensor is

$$\langle \mathbf{n}^{\otimes 2}\rangle^{\mathrm{VP}}_{\mathbf{e}_3,k} := \int_{S^2} \rho^{\mathrm{VP}}_{\mathbf{e}_3,k}(\mathbf{n})\mathbf{n}^{\otimes 2}\, d\sigma = \frac{1}{2k+3}\begin{bmatrix} 1 & 0 & 0 \\ 0 & 1 & 0 \\ 0 & 0 & 2k+1 \end{bmatrix}.$$

The corresponding fourth-order orientation tensor,

$$\langle \mathbf{n}^{\otimes 4}\rangle^{\mathrm{VP}}_{\mathbf{e}_3,k} := \int_{S^2} \rho^{\mathrm{VP}}_{\mathbf{e}_3,k}(\mathbf{n})\mathbf{n}^{\otimes 4}\, d\sigma,$$

Fig. 7 Plots of the transversely isotropic ODF (24) for $k = 1, 2, 3$ and 5

has the 6D second-order tensor representation

$$\frac{1}{(2k+3)(2k+5)}\begin{bmatrix} 3 & 1 & 2k+1 & 0 & 0 & 0 \\ 1 & 3 & 2k+1 & 0 & 0 & 0 \\ 2k+1 & 2k+1 & (2k+1)(2k+3) & 0 & 0 & 0 \\ 0 & 0 & 0 & 2(2k+1) & 0 & 0 \\ 0 & 0 & 0 & 0 & 2(2k+1) & 0 \\ 0 & 0 & 0 & 0 & 0 & 2 \end{bmatrix}.$$

Using a change of variables we can deduce that the family of ODFs

$$\rho^{\mathrm{VP}}_{\mathbf{e}_1,k}(\mathbf{n}) = \frac{2k+1}{4\pi}\cos^{2k}\theta\,\sin^{2k}\phi. \tag{25}$$

is transversely isotropic along the $\mathbf{e}_1$-axis. Similarly, the family of ODFs

$$\rho^{\mathrm{VP}}_{\mathbf{e}_2,k}(\mathbf{n}) = \frac{2k+1}{4\pi}\sin^{2k}\theta\,\sin^{2k}\phi. \tag{26}$$

is transversely isotropic along the $\mathbf{e}_2$-axis.

The de la Vallée Poussin ODF (24) can be generalized as

$$\rho^{\mathrm{VPm}}_{\mathbf{m},\kappa}(\mathbf{n}) = \frac{\kappa+1}{4\pi}|\mathbf{n}\cdot\mathbf{m}|^{\kappa}, \tag{27}$$

where the concentration parameter is now a positive real number κ. The corresponding orientation tensors have the same expressions as the ones for the de la Vallée Poussin ODF, we just need to replace $2k$ with κ.

For all the axially symmetric ODFs discussed in Sect. 3, when the concentration parameter κ is equal to zero, the second- and fourth-order orientation tensors are equal to the isotropic tensors (7) and (8), respectively. On the other hand, when κ goes to infinity, the second- and fourth-order orientation tensors are equal to the totally aligned orientation tensors (11) and (12) with $\mathbf{m} = \mathbf{e}_3$, respectively. For the comparison, in Fig. 8 we present plots of the 11 and 33 components of the second-order orientation tensors, and the 1111 and 3333 components of the fourth-order orientation tensor as functions of the concentration parameter κ and for the different ODFs.

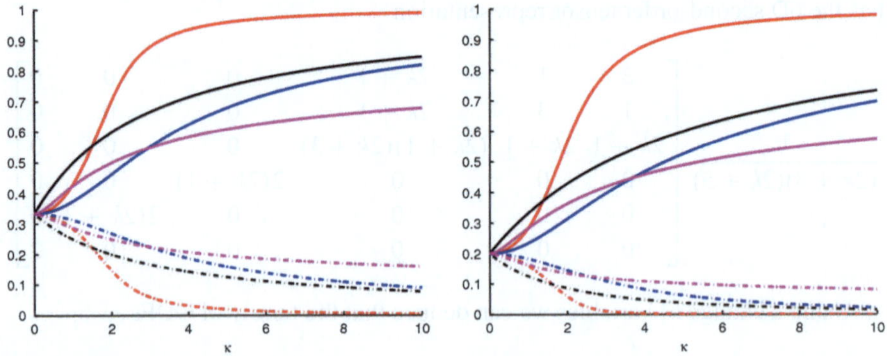

Fig. 8 Plots of the components of the second- and fourth-order orientation tensors as functions of κ for the different ODFs: von Mises-Fisher (*blue*), Watson (*red*), singular kernel (*magenta*), and de la Vallée Poussin (*black*). On the *left* the 11 (*dashed*) and 33 (*solid*) components of the second-order tensor, and on the *right* the 1111 (*dashed*) and 3333 (*solid*) components of the fourth-order tensor

4 Orthotropic ODFs

By adding two transversely isotropic ODFs with the same functional form but different modal vectors we obtain (after multiplication by $\frac{1}{2}$) an orthotropic orientation distribution. For example, by adding $\varrho_{\mathbf{m}^+}$ and $\varrho_{\mathbf{m}^-}$, where $\mathbf{m}^\pm = \cos\frac{1}{2}\psi\,\mathbf{e}_1 \pm \sin\frac{1}{2}\psi\,\mathbf{e}_2$, we obtain the orthotropic ODF

$$\varrho^{\text{orth}}(\mathbf{n}) = \tfrac{1}{2}\left(\varrho_{\mathbf{m}^+}(\mathbf{n}) + \varrho_{\mathbf{m}^-}(\mathbf{n})\right). \tag{28}$$

Plots of (28) based on the antipodally symmetric von Mises-Fisher, Watson, antipodally symmetric singular-kernel, and de la Vallée Poussin ODFs with various values of the concentration parameter are given in Fig. 9.

The first- and second-order orientation tensors are given by

$$\langle\mathbf{n}\rangle^{\text{orth}} = \tfrac{1}{2}\begin{bmatrix} 0 \\ 0 \\ A\cos\frac{1}{2}\psi \end{bmatrix}$$

and

$$\langle\mathbf{n}^{\otimes 2}\rangle^{\text{orth}} = \tfrac{1}{2}\begin{bmatrix} 1-B+(1-3B)\cos\psi & 0 & 0 \\ 0 & 1-B-(1-3B)\cos\psi & 0 \\ 0 & 0 & 2B \end{bmatrix},$$

where A and B are given in (14) for the chosen ODF $\varrho_{\mathbf{m}}$.

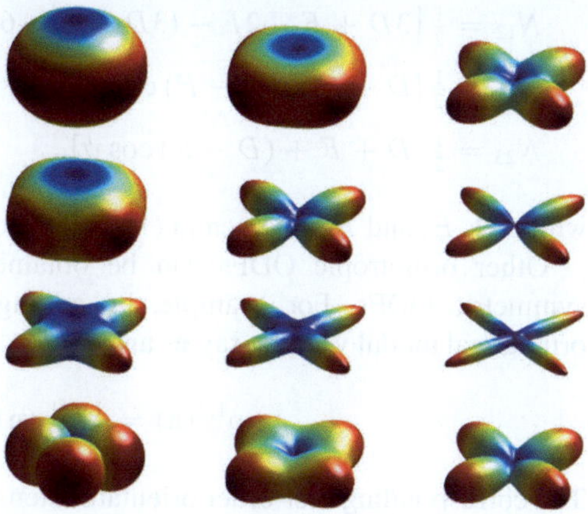

Fig. 9 Plots of orthotropic ODFs based on: antipodally symmetric von Mises-Fisher (*first row*), Watson (*second row*), antipodally symmetric singular kernel (*third row*), and de la Vallée Poussin (*fourth row*)

The third-order orientation tensor has the matrix representation

$$\langle \mathbf{n}^{\otimes 3} \rangle^{\text{orth}} = \frac{1}{4} \begin{bmatrix} n_{11} & n_{12} & n_{13} & 0 & 0 & 0 \\ 0 & 0 & 0 & 0 & 0 & \sqrt{2} n_{12} \\ 0 & 0 & 0 & 0 & \sqrt{2} n_{13} & 0 \end{bmatrix},$$

with

$$\begin{aligned} n_{11} &= A(3\cos\tfrac{1}{2}\psi + \cos\tfrac{3}{2}\psi) - C(3\cos\tfrac{1}{2}\psi - 5\cos\tfrac{3}{2}\psi), \\ n_{12} &= A(\cos\tfrac{1}{2}\psi - \cos\tfrac{3}{2}\psi) - C(\cos\tfrac{1}{2}\psi - 5\cos\tfrac{3}{2}\psi), \\ n_{13} &= C\cos\tfrac{1}{2}\psi, \end{aligned}$$

where C is given by (15) for the chosen ODF $\varrho_{\mathbf{m}}$.

The fourth-order orientation tensor has the 6D tensor representation

$$\langle \mathbf{n}^{\otimes 4} \rangle^{\text{orth}} = \begin{bmatrix} N_{11} & N_{12} & N_{13} & 0 & 0 & 0 \\ N_{12} & N_{22} & N_{23} & 0 & 0 & 0 \\ N_{13} & N_{23} & N_{33} & 0 & 0 & 0 \\ 0 & 0 & 0 & 2N_{23} & 0 & 0 \\ 0 & 0 & 0 & 0 & 2N_{13} & 0 \\ 0 & 0 & 0 & 0 & 0 & 2N_{12} \end{bmatrix}$$

with

$$\begin{aligned} N_{11} &= \tfrac{1}{4}(3D + E + 6F) - \tfrac{1}{2}(3D - E)\cos\psi + \tfrac{1}{4}(3D + E - 6F)\cos^2\psi, \\ N_{22} &= \tfrac{1}{4}(3D + E + 6F) + \tfrac{1}{2}(3D - E)\cos\psi + \tfrac{1}{4}(3D + E - 6F)\cos^2\psi, \\ N_{33} &= 3D, \end{aligned}$$

$$N_{12} = \tfrac{1}{2}\left[3D + E - 2F - (3D + E - 6F)\cos^2\psi\right],$$
$$N_{13} = \tfrac{1}{2}\left[D + F - (D - F)\cos\psi\right],$$
$$N_{23} = \tfrac{1}{2}\left[D + F + (D - F)\cos\psi\right],$$

where D, E, and F are given in (16) for the chosen ODF $\varrho_{\mathbf{m}}$.

Other orthotropic ODFs can be obtained by adding two different axially-symmetric ODFs. For example, by adding two axially-symmetric ODFs with orthogonal modal vectors, say $\mathbf{e}_1$ and $\mathbf{e}_2$,

$$\varrho^{1,2}(\mathbf{n}) = \tfrac{1}{2}(\varrho^1_{\mathbf{e}_1}(\mathbf{n}) + \varrho^2_{\mathbf{e}_2}(\mathbf{n})).$$

The corresponding first-order orientation tensor is

$$\langle\mathbf{n}\rangle_{\varrho^{1,2}} := \int_{S^2} \varrho^{1,2}(\mathbf{n})\mathbf{n}\, d\sigma = \tfrac{1}{2}\begin{bmatrix} A_1 \\ A_2 \\ 0 \end{bmatrix},$$

and the corresponding second-order orientation tensor is

$$\langle\mathbf{n}^{\otimes 2}\rangle_{\varrho^{1,2}} := \int_{S^2} \varrho^{1,2}(\mathbf{n})\mathbf{n}^{\otimes 2}\, d\sigma = \tfrac{1}{2}\begin{bmatrix} 1 - B_1 + B_2 & 0 & 0 \\ 0 & 1 + B_1 - B_2 & 0 \\ 0 & 0 & B_1 + B_2 \end{bmatrix},$$

where A_i and B_i are given in (14) for the chosen ODFs $\varrho^i_{\mathbf{e}_i}$, $i = 1, 2$.

The corresponding third-order orientation tensor

$$\langle\mathbf{n}^{\otimes 3}\rangle_{\varrho^{1,2}} := \int_{S^2} \varrho_{\mathbf{e}_3}(\mathbf{n})\mathbf{n}^{\otimes 3}\, d\sigma,$$

has the matrix representation

$$\tfrac{1}{2}\begin{bmatrix} A_1 - 2C_1 & C_1 & C_1 & 0 & 0 & \sqrt{2}C_2 \\ C_2 & A_2 - C_2 & C_2 & 0 & 0 & \sqrt{2}C_1 \\ 0 & 0 & 0 & \sqrt{2}C_2 & \sqrt{2}C_1 & 0 \end{bmatrix},$$

and the corresponding fourth-order orientation tensor,

$$\langle\mathbf{n}^{\otimes 4}\rangle_{\varrho^{1,2}} := \int_{S^2} \varrho^{1,2}(\mathbf{n})\mathbf{n}^{\otimes 4}\, d\sigma,$$

has the 6D second-order tensor representation

$$\frac{1}{2}\begin{bmatrix} E_1+3D_2 & F_1+F_2 & F_1+D_2 & 0 & 0 & 0 \\ F_1+F_2 & 3D_1+E_2 & D_1+F_2 & 0 & 0 & 0 \\ F_1+D_2 & D_1+F_2 & 3(D_1+D_2) & 0 & 0 & 0 \\ 0 & 0 & 0 & 2(D_1+F_2) & 0 & 0 \\ 0 & 0 & 0 & 0 & 2(F_1+D_2) & 0 \\ 0 & 0 & 0 & 0 & 0 & 2(F_1+F_2) \end{bmatrix},$$

where C_i D_i, E_i and F_i are given in (15) and (16) for the chosen ODFs $\varrho^i_{\mathbf{e}_i}$, $i = 1, 2$.

5 Cubic ODFs

By adding three transversely isotropic ODFs with the same functional form and mutually orthogonal modal vectors we obtain (after multiplication by $\frac{1}{3}$) a cubic orientation distribution. For example, by adding $\varrho_{\mathbf{e}_1}$, $\varrho_{\mathbf{e}_2}$ and $\varrho_{\mathbf{e}_3}$, we obtain the cubic orientation distribution

$$\varrho^{\mathrm{cub}}(\mathbf{n}) = \tfrac{1}{3}\left(\varrho_{\mathbf{e}_1}(\mathbf{n}) + \varrho_{\mathbf{e}_2}(\mathbf{n}) + \varrho_{\mathbf{e}_3}(\mathbf{n})\right). \tag{29}$$

Plots of (29) based on the antipodally symmetric von Mises-Fisher, Watson, antipodally symmetric singular kernel, and de la Vallée Poussin ODFs with various values of the concentration parameter are given in Fig. 10.

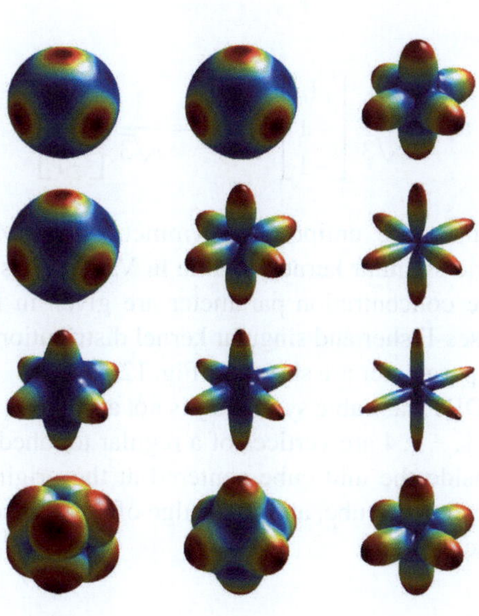

Fig. 10 Plots of cubic ODFs based on: antipodally symmetric von Mises-Fisher (*first row*), Watson (*second row*), antipodally symmetric singular kernel (*third row*), and de la Vallée Poussin (*fourth row*)

The first-order orientation tensor is $\frac{1}{3}A\left[1\ 1\ 1\right]^T$, where A is given in (14) for the chosen ODF $\varrho_{\mathbf{m}}$. The second-order orientation tensor is equal to $\frac{1}{3}\mathbf{I}$ and the third-order orientation tensor $\langle \mathbf{n}^{\otimes 3}\rangle^{\text{cub}}$ has the matrix representation

$$\frac{1}{3}\begin{bmatrix} A-2C & C & C & 0 & \sqrt{2}C & \sqrt{2}C \\ C & A-2C & C & \sqrt{2}C & 0 & \sqrt{2}C \\ C & C & A-2C & \sqrt{2}C & \sqrt{2}C & 0 \end{bmatrix},$$

where C is given in (15) for the chosen ODF $\varrho_{\mathbf{m}}$.

The fourth-order orientation tensor $\langle \mathbf{n}^{\otimes 4}\rangle^{\text{cub}}$ has the 6D tensor representation

$$\frac{1}{3}\begin{bmatrix} 6D+E & D+2F & D+2F & 0 & 0 & 0 \\ D+2F & 6D+E & D+2F & 0 & 0 & 0 \\ D+2F & D+2F & 6D+E & 0 & 0 & 0 \\ 0 & 0 & 0 & 2(D+2F) & 0 & 0 \\ 0 & 0 & 0 & 0 & 2(D+2F) & 0 \\ 0 & 0 & 0 & 0 & 0 & 2(D+2F) \end{bmatrix},$$

where D, E, and F are given in (16) for the chosen ODF $\varrho_{\mathbf{m}}$.

Another cubic ODF can be obtained by adding four axially symmetric ODFs (with the same functional form)

$$\varrho^{\text{thd}}(\mathbf{n}) = \tfrac{1}{4}\left(\varrho_{\mathbf{a}_1}(\mathbf{n}) + \varrho_{\mathbf{a}_2}(\mathbf{n}) + \varrho_{\mathbf{a}_3}(\mathbf{n}) + \varrho_{\mathbf{a}_4}(\mathbf{n})\right), \tag{30}$$

where

$$\mathbf{a}_1 = \frac{1}{\sqrt{3}}\begin{bmatrix}1\\1\\1\end{bmatrix}, \quad \mathbf{a}_2 = \frac{1}{\sqrt{3}}\begin{bmatrix}1\\-1\\-1\end{bmatrix}, \quad \mathbf{a}_3 = \frac{1}{\sqrt{3}}\begin{bmatrix}-1\\1\\-1\end{bmatrix}, \quad \mathbf{a}_4 = \frac{1}{\sqrt{3}}\begin{bmatrix}-1\\-1\\1\end{bmatrix}.$$

Plots of (30) based on the antipodally symmetric von Mises-Fisher, Watson, antipodally symmetric singular kernel, and de la Vallée Poussin distributions with various values of the concentration parameter are given in Fig. 11. Plots of (30) based on the von Mises-Fisher and singular kernel distributions with various values of the concentration parameter are shown in Fig. 12.

The fact that this ODF has cubic symmetry is not a surprise. Indeed, the tips of the unit vectors $\mathbf{a}_i$, $i = 1, \ldots, 4$ are vertices of a regular tetrahedron. This tetrahedron can be embedded inside the unit cube centered at the origin. Each vertex of the tetrahedron is a vertex of the cube, and each edge of the tetrahedron is a diagonal of one of the cube's faces.

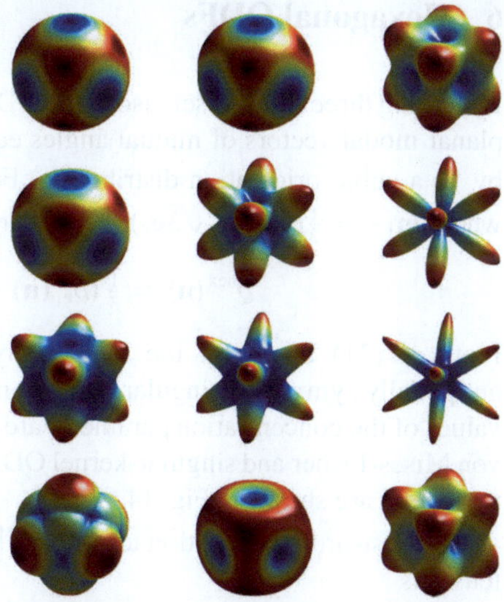

Fig. 11 Plots of tetrahedral ODFs based on: antipodally symmetric von Mises-Fisher (*first row*), Watson (*second row*), antipodally symmetric singular kernel (*third row*), and de la Vallée Poussin (*fourth row*)

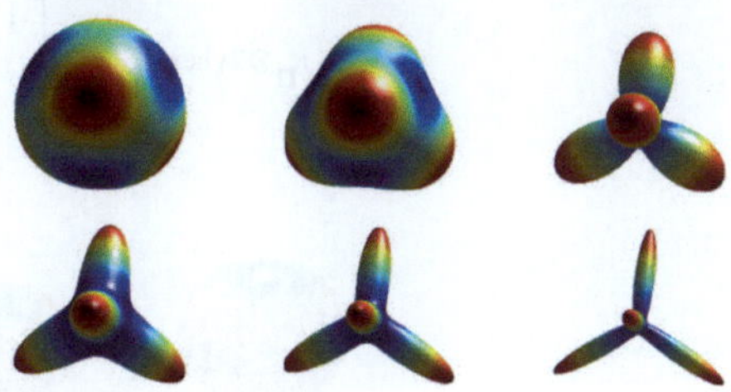

Fig. 12 Plots of tetrahedral ODFs based on: the von Mises-Fisher (*first row*) and the singular kernel (*second row*)

The first-order orientation tensor vanishes and the second-order orientation tensor is equal to $\frac{1}{3}\mathbf{I}$. The third-order orientation tensor has the matrix representation

$$\frac{\sqrt{3}(A-5C)}{9}\begin{bmatrix} 0 & 0 & 0 & \sqrt{2} & 0 & 0 \\ 0 & 0 & 0 & 0 & \sqrt{2} & 0 \\ 0 & 0 & 0 & 0 & 0 & \sqrt{2} \end{bmatrix},$$

while the fourth-order orientation tensor has the 6D tensor representation

$$\frac{1}{9}\begin{bmatrix} 12(D+F)+E & 6D+E & 6D+E & 0 & 0 & 0 \\ 6D+E & 12(D+F)+E & 6D+E & 0 & 0 & 0 \\ 6D+E & 6D+E & 12(D+F)+E & 0 & 0 & 0 \\ 0 & 0 & 0 & 2(6D+E) & 0 & 0 \\ 0 & 0 & 0 & 0 & 2(6D+E) & 0 \\ 0 & 0 & 0 & 0 & 0 & 2(6D+E) \end{bmatrix},$$

where D, E, and F are given in (16) for the chosen ODF $\varrho_{\mathbf{m}}$.

6 Hexagonal ODFs

By adding three transversely isotropic ODFs (having the same functional form) with planar modal vectors of mutual angles equal to $\frac{2}{3}\pi$ we obtain (after multiplication by $\frac{1}{3}$) a cubic orientation distribution. For example, by adding $\varrho_{\mathbf{e}_1}$, $\varrho_{\mathbf{m}_1}$ and $\varrho_{\mathbf{m}_2}$, where $\mathbf{m}_{1,2} = \frac{1}{2}(-\mathbf{e}_1 \pm \sqrt{3}\mathbf{e}_2)$, we obtain the hexagonal orientation distribution

$$\varrho^{\text{hex}}(\mathbf{n}) = \tfrac{1}{3}\left(\varrho_{\mathbf{e}_1}(\mathbf{n}) + \varrho_{\mathbf{m}_1}(\mathbf{n}) + \varrho_{\mathbf{m}_2}(\mathbf{n})\right). \tag{31}$$

Plots of (31) based on the antipodally symmetric von Mises-Fisher, Watson, antipodally symmetric singular kernel, and de la Vallée Poussin ODFs with various values of the concentration parameter are given in Fig. 13. Plots of (31) based on the von Mises-Fisher and singular-kernel ODFs with various values of the concentration parameter are shown in Fig. 14.

The first-order orientation tensor is $\frac{2}{3}\left[A\ 0\ 0\right]^T$ and the second-order orientation tensor is

$$\langle \mathbf{n}^{\otimes 2}\rangle^{\text{hex}} = \frac{1}{2}\begin{bmatrix} 1-B & 0 & 0 \\ 0 & 1-B & 0 \\ 0 & 0 & 2B \end{bmatrix},$$

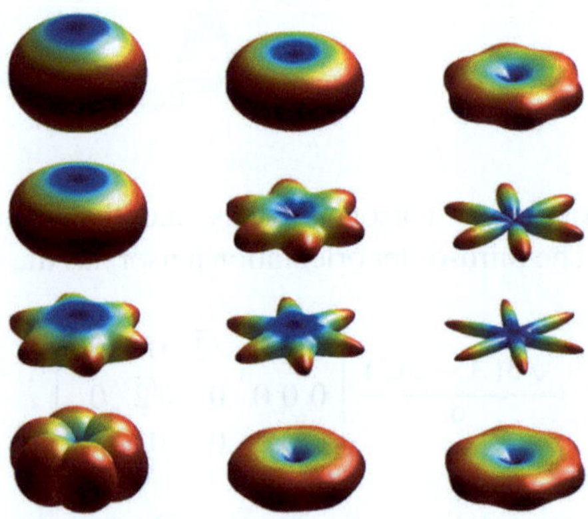

Fig. 13 Plots of hexagonal ODFs based on: antipodally symmetric von Mises-Fisher (*first row*), Watson (*second row*), antipodally symmetric singular kernel (*third row*), and de la Vallée Poussin (*fourth row*)

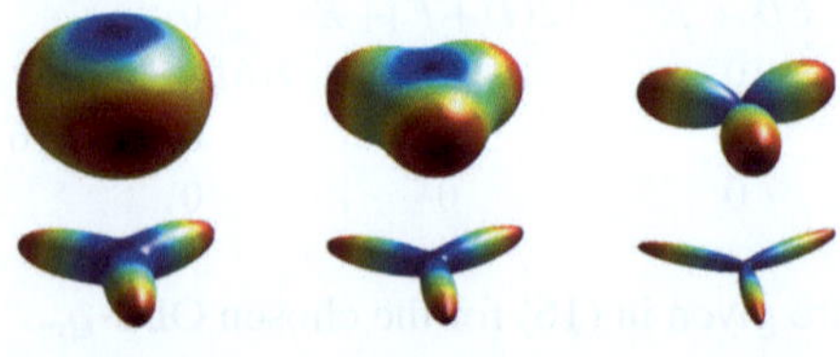

Fig. 14 Plots of hexagonal ODFs based on: the von Mises-Fisher (*first row*) and the singular kernel (*second row*)

where A and B are given in (14) for the chosen ODF $\varrho_{\mathbf{m}}$.

The third-order orientation tensor $\langle \mathbf{n}^{\otimes 3} \rangle^{\text{hex}}$ has the matrix representation

$$\frac{1}{12}\begin{bmatrix} 5A-C & 3A-7C & 8C & 0 & 0 & 0 \\ 0 & 0 & 0 & 0 & 0 & \sqrt{2}(3A-7C) \\ 0 & 0 & 0 & 0 & 8\sqrt{2}C & 0 \end{bmatrix},$$

and the fourth-order orientation tensor $\langle \mathbf{n}^{\otimes 4} \rangle^{\text{hex}}$ has the 6D matrix representation

$$\frac{1}{8}\begin{bmatrix} 3(3D+E+2F) & 3D+E+2F & 4(D+F) & 0 & 0 & 0 \\ 3D+E+2F & 3(3D+E+2F) & 4(D+F) & 0 & 0 & 0 \\ 4(D+F) & 4(D+F) & 24D & 0 & 0 & 0 \\ 0 & 0 & 0 & 8(D+F) & 0 & 0 \\ 0 & 0 & 0 & 0 & 8(D+F) & 0 \\ 0 & 0 & 0 & 0 & 0 & 2(3D+E+2F) \end{bmatrix},$$

where C, D, E, and F are given in (15) and (16) for the chosen ODF $\varrho_{\mathbf{m}}$.

7 Icosahedral ODFs

We have used (elementary) axially symmetric ODFs to construct new ODFs with different material symmetry classes. The eight symmetry classes of fourth-order tensors (isotropy, cubic, orthogonal, hexagonal, tetragonal, trigonal, monoclinic, and triclinic) can thus be obtained. There are more elaborate symmetries that can be detected only in higher order tensors. For example, the cubic symmetry is not detected at the second-order level. Furthermore, the following example exhibits a case where the second- and fourth-order orientation tensors are isotropic whereas the ODF is obviously not. Let $\mathbf{b}_i$, $i = 1, \ldots, 6$ be the unit vectors pointing to the opposite vertices of a regular icosahedron and given by

$$\mathbf{b}_{1,2} = \frac{1}{\sqrt{1+\tau^2}}\begin{bmatrix} 0 \\ 1 \\ \pm\tau \end{bmatrix}, \quad \mathbf{b}_{3,4} = \frac{1}{\sqrt{1+\tau^2}}\begin{bmatrix} \pm\tau \\ 0 \\ 1 \end{bmatrix}, \quad \mathbf{b}_{5,6} = \frac{1}{\sqrt{1+\tau^2}}\begin{bmatrix} 1 \\ \pm\tau \\ 0 \end{bmatrix},$$

where $\tau = (1+\sqrt{5})/2$. Let us consider the ODF obtained by adding the six axially symmetric ODFs with modal vectors $\mathbf{b}_i$:

$$\varrho^{\text{ico}}(\mathbf{n}) = \tfrac{1}{6}\left(\varrho_{\mathbf{b}_1}(\mathbf{n}) + \varrho_{\mathbf{b}_2}(\mathbf{n}) + \varrho_{\mathbf{b}_3}(\mathbf{n}) + \varrho_{\mathbf{b}_4}(\mathbf{n}) + \varrho_{\mathbf{b}_5}(\mathbf{n}) + \varrho_{\mathbf{b}_6}(\mathbf{n})\right). \quad (32)$$

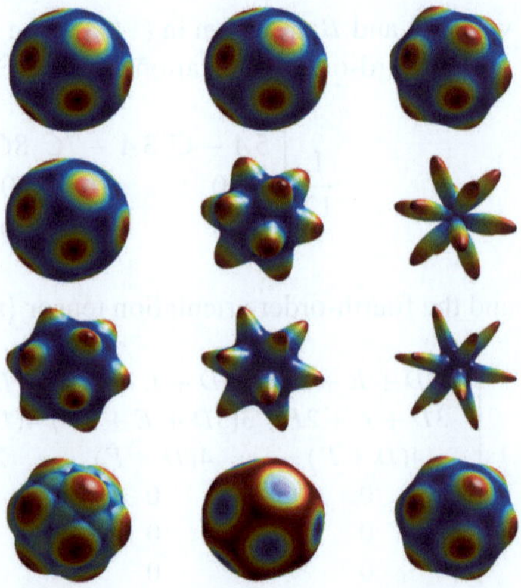

Fig. 15 Plots of icosahedral ODFs based on: antipodally symmetric von Mises-Fisher (*first row*), Watson (*second row*), antipodally symmetric singular kernel (*third row*), and de la Vallée Poussin (*fourth row*)

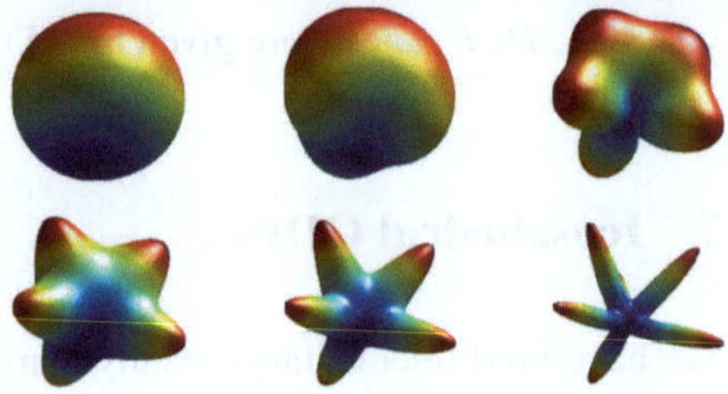

Fig. 16 Plots of icosahedral ODFs based on: the von Mises-Fisher (*first row*) and the singular kernel (*second row*)

Plots of (32) based on the antipodally symmetric von Mises-Fisher, Watson, antipodally symmetric singular-kernel, and de la Vallée Poussin ODFs with various values of the concentration parameter are given in Fig. 15. Plots of (32) based on the von Mises-Fisher and singular kernel ODFs with various values of the concentration parameter are shown in Fig. 16.

The first-order orientation tensor is equal to $A/(3\sqrt{1+\tau^2})\left[1\ 1\ 1\right]^T$, where A is given in (14) for the chosen ODF $\varrho_{\mathbf{m}}$. The second- and fourth-order orientation tensors are the isotropic tensors (7) and (8), respectively. The third-order orientation tensor has the representation

$$\frac{\sqrt{5}\tau}{30\sqrt{1+\tau^2}}\begin{bmatrix} u & v & w & 0 & \sqrt{2}v & \sqrt{2}w \\ w & u & v & \sqrt{2}w & 0 & \sqrt{2}v \\ v & w & u & \sqrt{2}v & \sqrt{2}w & 0 \end{bmatrix},$$

with

$$u = (3-\sqrt{5})A + 2\sqrt{5}C, \quad v = 2A - \sqrt{5}(\sqrt{5}+1)C, \quad w = \sqrt{5}(\sqrt{5}-1)C,$$

where C is given in (15) for the chosen ODF $\varrho_{\mathbf{m}}$.

8 General ODFs

As we have seen, in many cases it suffices to look at approximations (3) or (4) up to the order four. However, certain ODFs cannot be approximated by only keeping low-order terms and one has to consider higher-order approximations. Moreover, empirically, we may wish to measure or estimate a fiber ODF or its moments, so a convenient and general representation in terms of a complete and orthogonal set of basis functions is sought.

We claim that any ODF can be approximated by a sum of axially symmetric ODFs with different modal vectors and concentration parameters. For instance, one can use the uni-modal de la Vallée Poussin ODF as its related tensors are easily computed. In fact, the de la Vallée Poussin can be expressed in terms of the Cartesian coordinates as

$$\rho^{\mathrm{VP}}_{\mathbf{m},k}(\mathbf{n}) = R_{\mathbf{m},k}(x,y,z) := (m_1 x + m_2 y + m_3 z)^{2k} = (\mathbf{m}\cdot\mathbf{n})^{2k},$$

where, of course, the coordinates x, y, and z of $\mathbf{n}$ are constrained to satisfy the condition $x^2 + y^2 + z^2 = 1$.

Therefore, any square-integrable function on S^2 can be approximated by

$$f(\mathbf{n}) \sim \sum_{k=0}^{K}\sum_{l=1}^{L} f_{kl} R_{\mathbf{m}_l,k}(\mathbf{n}),$$

where $\mathbf{m}_l$ are unit vectors. For the determination of this (simple) approximation one can seek, for given K and L, unit vectors $\mathbf{m}_l$ and coefficients $f_{k,l}$ minimizers of the functional

$$\int_{S^2}\left[f(\mathbf{n}) - \sum_{k=1}^{K}\sum_{l=1}^{L} f_{k,l}(\mathbf{m}_l\cdot\mathbf{n})^{2k}\right]^2 d\sigma.$$

If we take only one value of the concentration parameter (power) in the de la Vallée Poussin kernel in the above approximation, i.e., we assume $f_{k,l} = 0$ for all $k < K$, then the above minimization problem reduces to finding unit vectors $\mathbf{m}_l$ and coefficients $\tilde{f}_l$ minimizers of the functional

$$\int_{S^2}\left[f(\mathbf{n}) - \sum_{l=1}^{L} \tilde{f}_l(\mathbf{m}_l\cdot\mathbf{n})^{2K}\right]^2 d\sigma.$$

This minimization problem is similar to the low-rank approximation approach of Schultz and Seidel [27] for estimating crossing fibers from an ODF generated by Q-ball imaging or spherical deconvolution. When the ODF is also estimated from diffusion weighed imaging signals, Gur et al. [13] presented a nonlinear method for the joint estimation of the ODF, extracting the fiber directions using low-rank approximations. We also mention that Ghosh et al. [11] and Megherbi et al. [20] used similar techniques for extracting fiber directions without any prior information about the number of fibers. These works employ the symmetric tensor decomposition algorithm proposed in [4] which is based on the decomposition of homogeneous polynomials as a sum of powers of linear forms.

Alternatively, if the coefficients of the expansion in spherical harmonics of the ODF are given (estimated from experimental data or computed from a given ODF), then one can compute the orientation tensors explicitly as shown below. Recall that any ODF can be expanded in real spherical harmonics

$$\rho(\mathbf{n}) = \sum_{l=0}^{\infty} \sum_{m=-l}^{l} c_{l,m} S_{l,m}(\theta, \phi).$$

Here $S_{l,m}(\cdot,\cdot)$ are the (normalized) real spherical harmonics of degree l and order m defined by

$$S_{l,m}(\theta,\phi) = \begin{cases} \frac{(-1)^m}{\sqrt{2}}(Y_{l,m}(\theta,\phi) + \bar{Y}_{l,m}(\theta,\phi)) & \text{for } m > 0, \\ Y_{l,0}(\theta,\phi) & \text{for } m = 0, \\ \frac{(-1)^m}{i\sqrt{2}}(Y_{l,-m}(\theta,\phi) - \bar{Y}_{l,-m}(\theta,\phi)) & \text{for } m < 0, \end{cases}$$

where the complex spherical harmonics $Y_{l,m}(\cdot,\cdot)$ (and its complex conjugate $\bar{Y}_{m,l}(\cdot,\cdot)$) are related to the associated Legendre polynomials $P_l^m(\cdot)$ by

$$Y_{l,m}(\theta,\phi) = (-1)^m \sqrt{\frac{2l+1}{4\pi}\frac{(m-l)!}{(m+l)!}} P_l^m(\cos\theta)\exp(im\phi).$$

For example, the non-vanishing spherical harmonic coefficients up to the order four for the von Mises-Fisher ODF (17) are

$$c_{0,0} = \frac{1}{2\sqrt{\pi}}, \qquad c_{1,0} = \frac{\sqrt{3}\,(\kappa\cosh\kappa - \sinh\kappa)}{2\sqrt{\pi}\kappa\sinh\kappa},$$

$$c_{2,0} = \frac{\sqrt{5}\left((3+\kappa^2)\sinh\kappa - 3\kappa\cosh\kappa\right)}{2\sqrt{\pi}\kappa^2\sinh\kappa},$$

$$c_{3,0} = \frac{\sqrt{7}\left(\kappa(\kappa^2+15)\cosh\kappa - 3(2\kappa^2+5)\sinh\kappa\right)}{2\sqrt{\pi}\kappa^3\sinh\kappa},$$

$$c_{4,0} = \frac{3\left((\kappa^4+45\kappa^2+105)\sinh\kappa - 5\kappa(2\kappa^2+21)\cosh\kappa\right)}{2\sqrt{\pi}\kappa^4\sinh\kappa}.$$

Similarly, the non-vanishing spherical harmonic coefficients up to order four for the Watson ODF (20) are

$$c_{0,0} = \frac{1}{2\sqrt{\pi}}, \quad c_{2,0} = \frac{\sqrt{5}\left(6\kappa e^{\kappa^2} - \sqrt{\pi}(2\kappa^2+3)\,\mathrm{erfi}\,\kappa\right)}{8\pi\kappa^2\,\mathrm{erfi}\,\kappa},$$

$$c_{4,0} = \frac{3\left(10\kappa(2\kappa^2-21)e^{\kappa^2} + 3\sqrt{\pi}(4\kappa^4+20\kappa^2+35)\,\mathrm{erfi}\,\kappa\right)}{64\pi\kappa^4\,\mathrm{erfi}\,\kappa}.$$

When the spherical harmonic coefficients $c_{l,m}$, $l = 0,\ldots,4$, $m = -l,\ldots l$, of an ODF ρ are given then we can compute the orientation tensors up to order four. The first-order orientation tensor is

$$\langle \mathbf{n} \rangle_\rho = 2\sqrt{\frac{\pi}{3}} \sum_{m=-1}^{1} c_{1,m}\mathbf{s}^{1,m}, \tag{33}$$

the second-order orientation tensor is

$$\langle \mathbf{n}^{\otimes 2} \rangle_\rho = 2\sqrt{\frac{\pi}{3}} \left(c_{0,0}\mathbf{S}^{0,0} + \sqrt{\frac{2}{5}} \sum_{m=-2}^{2} c_{2,m}\mathbf{S}^{2,m} \right), \tag{34}$$

the third-order orientation tensor is

$$\langle \mathbf{n}^{\otimes 3} \rangle_\rho = 2\sqrt{\frac{\pi}{5}} \left(\sum_{m=-1}^{1} c_{1,m}\mathsf{S}^{1,m} + \sqrt{\frac{2}{7}} \sum_{m=-3}^{3} c_{3,m}\mathsf{S}^{3,m} \right), \tag{35}$$

and the fourth-order orientation tensor is

$$\langle \mathbf{n}^{\otimes 4} \rangle_\rho = 2\sqrt{\frac{\pi}{3}} \left(c_{0,0}\mathbb{S}^{0,0} + \sqrt{\frac{2}{5}} \sum_{m=-2}^{2} c_{2,m}\mathbb{S}^{2,m} + \frac{2}{\sqrt{105}} \sum_{m=-4}^{4} c_{4,m}\mathbb{S}^{4,m} \right). \tag{36}$$

The vectors $\mathbf{s}^{l,m}$, second-order tensors $\mathbf{S}^{l,m}$, third-order tensors $\mathsf{S}^{l,m}$ and fourth-order tensors $\mathbb{S}^{l,m}$ are given in the Appendix.

9 Discussion and Concluding Remarks

We have presented explicit expressions of the orientation tensors up to order four for a hierarchy of ODFs with different material symmetries. We have given the coefficients of these orientation tensors in a natural coordinate system. However, we recognize that to use these ODF with experimental data, the mean direction vector may also need to be included in these distributions as a random variable and estimated. We have seen that by combining axially symmetric ODFs with different modal vectors and concentration parameters one can get an ODF with more complex material symmetry. Inversely, we claim that a given ODF can be well approximated by a convex combination of axially symmetric ones. However, finding such an approximation can be complicated due to non-uniqueness. Alternatively, we proposed using the approximation of this ODF by spherical harmonics up to a specified order. For each order, the orientation tensors are given by a linear combination of pre-computed tensors that form an orthonormal basis.

If from experimental data we can estimate the coefficients of the expansion in spherical harmonics of the ODF, then by using this framework we can compute the orientation tensors. From the orientation tensors we can infer the material symmetries (or direction of the fibers) by the method developed in [5].

Appendix

We give here the expressions for the normalized orientation-like tensors that appear in (33)–(36).

The vectors $\mathbf{s}^{1,m}$, $m = -1, \ldots, 1$, which are obtained from $\int_{S^2} S^{1,m}(\theta,\phi)\mathbf{n}\, d\sigma$ by normalization, are given by

$$\mathbf{s}^{1,-1} = \begin{bmatrix} -1 \\ 0 \\ 0 \end{bmatrix}, \quad \mathbf{s}^{1,0} = \begin{bmatrix} 0 \\ 0 \\ 1 \end{bmatrix}, \quad \mathbf{s}^{1,1} = \begin{bmatrix} 0 \\ -1 \\ 0 \end{bmatrix}.$$

The second-order tensors $\mathbf{S}^{l,m}$, $l = 0, 2$, $m = -l, \ldots, l$, which are obtained from $\int_{S^2} S^{l,m}(\theta,\phi)\mathbf{n}^{\otimes 2}\, d\sigma$ by normalization, are given by

$$\mathbf{S}^{0,0} = \frac{1}{\sqrt{3}}\begin{bmatrix} 1 & 0 & 0 \\ 0 & 1 & 0 \\ 0 & 0 & 1 \end{bmatrix}, \quad \mathbf{S}^{2,-2} = \frac{1}{\sqrt{2}}\begin{bmatrix} 1 & 0 & 0 \\ 0 & -1 & 0 \\ 0 & 0 & 0 \end{bmatrix}, \quad \mathbf{S}^{2,-1} = \frac{1}{\sqrt{2}}\begin{bmatrix} 0 & 0 & -1 \\ 0 & 0 & 0 \\ -1 & 0 & 0 \end{bmatrix},$$

$$\mathbf{S}^{2,0} = \frac{1}{\sqrt{6}}\begin{bmatrix} -1 & 0 & 0 \\ 0 & -1 & 0 \\ 0 & 0 & 2 \end{bmatrix}, \quad \mathbf{S}^{2,1} = \frac{1}{\sqrt{2}}\begin{bmatrix} 0 & 0 & 0 \\ 0 & 0 & -1 \\ 0 & -1 & 0 \end{bmatrix}, \quad \mathbf{S}^{2,2} = \frac{1}{\sqrt{2}}\begin{bmatrix} 0 & 1 & 0 \\ 1 & 0 & 0 \\ 0 & 0 & 0 \end{bmatrix}.$$

Note that $\mathbf{S}^{l,m}$ are traceless except for $\mathbf{S}^{0,0}$ which has unit trace. Furthermore, the set $\{\mathbf{S}^{l,m},\ l = 0, 2,\ m = -l, \ldots, l\}$ forms an orthonormal basis of the space of symmetric second-order tensors.

The third-order tensors $\mathsf{S}^{l,m}$, $l = 1, 3$, $m = -l, \ldots, l$, which are obtained from $\int_{S^2} S^{l,m}(\theta,\phi)\mathbf{n}^{\otimes 3}\, d\sigma$ by normalization, are given by

$$\mathsf{S}^{1,-1} = \frac{1}{\sqrt{15}}\begin{bmatrix} -3 & -1 & -1 & 0 & 0 & 0 \\ 0 & 0 & 0 & 0 & 0 & -\sqrt{2} \\ 0 & 0 & 0 & 0 & -\sqrt{2} & 0 \end{bmatrix}, \quad \mathsf{S}^{1,0} = \frac{1}{\sqrt{15}}\begin{bmatrix} 0 & 0 & 0 & 0 & \sqrt{2} & 0 \\ 0 & 0 & 0 & \sqrt{2} & 0 & 0 \\ 1 & 1 & 3 & 0 & 0 & 0 \end{bmatrix}, \quad \mathsf{S}^{1,1} = \frac{1}{\sqrt{15}}\begin{bmatrix} 0 & 0 & 0 & 0 & 0 & -\sqrt{2} \\ -1 & -3 & -1 & 0 & 0 & 0 \\ 0 & 0 & 0 & -\sqrt{2} & 0 & 0 \end{bmatrix},$$

$$\mathsf{S}^{3,-3} = \frac{1}{2}\begin{bmatrix} -1 & 1 & 0 & 0 & 0 & 0 \\ 0 & 0 & 0 & 0 & 0 & \sqrt{2} \\ 0 & 0 & 0 & 0 & 0 & 0 \end{bmatrix}, \quad \mathsf{S}^{3,-2} = \frac{1}{\sqrt{6}}\begin{bmatrix} 0 & 0 & 0 & 0 & \sqrt{2} & 0 \\ 0 & 0 & 0 & -\sqrt{2} & 0 & 0 \\ 1 & -1 & 0 & 0 & 0 & 0 \end{bmatrix}, \quad \mathsf{S}^{3,-1} = \frac{1}{2\sqrt{15}}\begin{bmatrix} 3 & 3 & -4 & 0 & 0 & 0 \\ 0 & 0 & 0 & 0 & 0 & \sqrt{2} \\ 0 & 0 & 0 & 0 & -4\sqrt{2} & 0 \end{bmatrix},$$

$$\mathsf{S}^{3,0} = \frac{1}{\sqrt{10}}\begin{bmatrix} 0 & 0 & 0 & 0 & -\sqrt{2} & 0 \\ 0 & 0 & 0 & -\sqrt{2} & 0 & 0 \\ -1 & -1 & 2 & 0 & 0 & 0 \end{bmatrix}, \quad \mathsf{S}^{3,1} = \frac{1}{2\sqrt{15}}\begin{bmatrix} 0 & 0 & 0 & 0 & 0 & \sqrt{2} \\ 1 & 1 & -4 & 0 & 0 & 0 \\ 0 & 0 & 0 & -4\sqrt{2} & 0 & 0 \end{bmatrix}, \quad \mathsf{S}^{3,2} = \frac{1}{\sqrt{6}}\begin{bmatrix} 0 & 0 & 0 & \sqrt{2} & 0 & 0 \\ 0 & 0 & 0 & 0 & \sqrt{2} & 0 \\ 0 & 0 & 0 & 0 & 0 & \sqrt{2} \end{bmatrix},$$

$$\mathsf{S}^{3,3} = \frac{1}{2}\begin{bmatrix} 0 & 0 & 0 & 0 & 0 & -\sqrt{2} \\ -1 & 1 & 0 & 0 & 0 & 0 \\ 0 & 0 & 0 & 0 & 0 & 0 \end{bmatrix}.$$

We remark that the set $\{\mathsf{S}^{l,m},\ l = 1,3,\ m = -l,\dots,l\}$ forms an orthonormal basis of the space of totally symmetric third-order tensors.

The fourth-order tensors $\mathbb{S}^{l,m}$, $l = 0,2,4$, $m = -l,\dots,l$, which are obtained from $\int_{S^2} S^{l,m}(\theta,\phi)\mathbf{n}^{\otimes 4}\, d\sigma$ by normalization, are given by

$$\mathbb{S}^{0,0} = \frac{1}{3\sqrt{5}}\begin{bmatrix} 3 & 1 & 1 & 0 & 0 & 0 \\ 1 & 3 & 1 & 0 & 0 & 0 \\ 1 & 1 & 3 & 0 & 0 & 0 \\ 0 & 0 & 0 & 2 & 0 & 0 \\ 0 & 0 & 0 & 0 & 2 & 0 \\ 0 & 0 & 0 & 0 & 0 & 2 \end{bmatrix}, \quad \mathbb{S}^{2,-2} = \frac{1}{2\sqrt{21}}\begin{bmatrix} 6 & 0 & 1 & 0 & 0 & 0 \\ 0 & -6 & -1 & 0 & 0 & 0 \\ 1 & -1 & 0 & 0 & 0 & 0 \\ 0 & 0 & 0 & -2 & 0 & 0 \\ 0 & 0 & 0 & 0 & 2 & 0 \\ 0 & 0 & 0 & 0 & 0 & 0 \end{bmatrix}, \quad \mathbb{S}^{2,-1} = \frac{1}{2\sqrt{21}}\begin{bmatrix} 0 & 0 & 0 & 0 & -3\sqrt{2} & 0 \\ 0 & 0 & 0 & 0 & -\sqrt{2} & 0 \\ 0 & 0 & 0 & 0 & -3\sqrt{2} & 0 \\ 0 & 0 & 0 & 0 & 0 & -2 \\ -3\sqrt{2} & -\sqrt{2} & -3\sqrt{2} & 0 & 0 & 0 \\ 0 & 0 & 0 & -2 & 0 & 0 \end{bmatrix},$$

$$\mathbb{S}^{2,0} = \frac{1}{2\sqrt{63}}\begin{bmatrix} -6 & -2 & 1 & 0 & 0 & 0 \\ -2 & -6 & 1 & 0 & 0 & 0 \\ 1 & 1 & 12 & 0 & 0 & 0 \\ 0 & 0 & 0 & 2 & 0 & 0 \\ 0 & 0 & 0 & 0 & 2 & 0 \\ 0 & 0 & 0 & 0 & 0 & -4 \end{bmatrix}, \quad \mathbb{S}^{2,1} = \frac{1}{2\sqrt{21}}\begin{bmatrix} 0 & 0 & 0 & -\sqrt{2} & 0 & 0 \\ 0 & 0 & 0 & -3\sqrt{2} & 0 & 0 \\ 0 & 0 & 0 & -3\sqrt{2} & 0 & 0 \\ -\sqrt{2} & -3\sqrt{2} & -3\sqrt{2} & 0 & 0 & 0 \\ 0 & 0 & 0 & 0 & 0 & -2 \\ 0 & 0 & 0 & 0 & -2 & 0 \end{bmatrix}, \quad \mathbb{S}^{2,2} = \frac{1}{2\sqrt{21}}\begin{bmatrix} 0 & 0 & 0 & 0 & 0 & 3\sqrt{2} \\ 0 & 0 & 0 & 0 & 0 & 3\sqrt{2} \\ 0 & 0 & 0 & 0 & 0 & \sqrt{2} \\ 0 & 0 & 0 & 0 & 2 & 0 \\ 0 & 0 & 0 & 2 & 0 & 0 \\ 3\sqrt{2} & 3\sqrt{2} & \sqrt{2} & 0 & 0 & 0 \end{bmatrix},$$

$$\mathbb{S}^{4,-4} = \frac{1}{2\sqrt{2}}\begin{bmatrix} 1 & -1 & 0 & 0 & 0 & 0 \\ -1 & 1 & 0 & 0 & 0 & 0 \\ 0 & 0 & 0 & 0 & 0 & 0 \\ 0 & 0 & 0 & 0 & 0 & 0 \\ 0 & 0 & 0 & 0 & 0 & 0 \\ 0 & 0 & 0 & 0 & 0 & -2 \end{bmatrix}, \quad \mathbb{S}^{4,-3} = \frac{1}{4}\begin{bmatrix} 0 & 0 & 0 & 0 & -\sqrt{2} & 0 \\ 0 & 0 & 0 & 0 & \sqrt{2} & 0 \\ 0 & 0 & 0 & 0 & 0 & 0 \\ 0 & 0 & 0 & 0 & 0 & 2 \\ -\sqrt{2} & \sqrt{2} & 0 & 0 & 0 & 0 \\ 0 & 0 & 0 & 2 & 0 & 0 \end{bmatrix}, \quad \mathbb{S}^{4,-2} = \frac{1}{\sqrt{14}}\begin{bmatrix} -1 & 0 & 1 & 0 & 0 & 0 \\ 0 & 1 & -1 & 0 & 0 & 0 \\ 1 & -1 & 0 & 0 & 0 & 0 \\ 0 & 0 & 0 & -2 & 0 & 0 \\ 0 & 0 & 0 & 0 & 2 & 0 \\ 0 & 0 & 0 & 0 & 0 & 0 \end{bmatrix},$$

$$\mathbb{S}^{4,-1} = \frac{1}{\sqrt{28}}\begin{bmatrix} 0 & 0 & 0 & 0 & 3\sqrt{2} & 0 \\ 0 & 0 & 0 & 0 & \sqrt{2} & 0 \\ 0 & 0 & 0 & 0 & -4\sqrt{2} & 0 \\ 0 & 0 & 0 & 0 & 0 & 2 \\ 3\sqrt{2} & \sqrt{2} & -4\sqrt{2} & 0 & 0 & 0 \\ 0 & 0 & 0 & 2 & 0 & 0 \end{bmatrix}, \quad \mathbb{S}^{4,0} = \frac{1}{2\sqrt{70}}\begin{bmatrix} 3 & 1 & -4 & 0 & 0 & 0 \\ 1 & 3 & -4 & 0 & 0 & 0 \\ -4 & -4 & 8 & 0 & 0 & 0 \\ 0 & 0 & 0 & -8 & 0 & 0 \\ 0 & 0 & 0 & 0 & -8 & 0 \\ 0 & 0 & 0 & 0 & 0 & 2 \end{bmatrix}, \quad \mathbb{S}^{4,1} = \frac{1}{2\sqrt{28}}\begin{bmatrix} 0 & 0 & 0 & \sqrt{2} & 0 & 0 \\ 0 & 0 & 0 & 3\sqrt{2} & 0 & 0 \\ 0 & 0 & 0 & -4\sqrt{2} & 0 & 0 \\ \sqrt{2} & 3\sqrt{2} & -4\sqrt{2} & 0 & 0 & 0 \\ 0 & 0 & 0 & 0 & 0 & 2 \\ 0 & 0 & 0 & 0 & 2 & 0 \end{bmatrix},$$

$$\mathbb{S}^{4,2} = \frac{1}{2\sqrt{14}}\begin{bmatrix} 0 & 0 & 0 & 0 & 0 & -\sqrt{2} \\ 0 & 0 & 0 & 0 & 0 & -\sqrt{2} \\ 0 & 0 & 0 & 0 & 0 & 2\sqrt{2} \\ 0 & 0 & 0 & 0 & 4 & 0 \\ 0 & 0 & 0 & 4 & 0 & 0 \\ -\sqrt{2} & -\sqrt{2} & 2\sqrt{2} & 0 & 0 & 0 \end{bmatrix}, \quad \mathbb{S}^{4,3} = \frac{1}{4}\begin{bmatrix} 0 & 0 & 0 & -\sqrt{2} & 0 & 0 \\ 0 & 0 & 0 & \sqrt{2} & 0 & 0 \\ 0 & 0 & 0 & 0 & 0 & 0 \\ -\sqrt{2} & \sqrt{2} & 0 & 0 & 0 & 0 \\ 0 & 0 & 0 & 0 & 0 & -2 \\ 0 & 0 & 0 & 0 & -2 & 0 \end{bmatrix}, \quad \mathbb{S}^{4,4} = \frac{1}{2\sqrt{2}}\begin{bmatrix} 0 & 0 & 0 & 0 & 0 & \sqrt{2} \\ 0 & 0 & 0 & 0 & 0 & -\sqrt{2} \\ 0 & 0 & 0 & 0 & 0 & 0 \\ 0 & 0 & 0 & 0 & 0 & 0 \\ 0 & 0 & 0 & 0 & 0 & 0 \\ \sqrt{2} & -\sqrt{2} & 0 & 0 & 0 & 0 \end{bmatrix}.$$

It should be noted that $\mathbb{S}^{l,m}$ are traceless except for $\mathbb{S}^{0,0}$ which has unit trace. The set $\{\mathbb{S}^{l,m},\ l = 0,2,4,\ m = -l,\dots,l\}$ forms an orthonormal basis of the space of totally symmetric fourth-order tensors.

References

1. Advani, S.G., Tucker III, C.L.: The use of tensors to describe and predict fiber orientation in short fiber composites. J. Rheol. **31**(8), 751–784 (1987)
2. Advani, S.G., Tucker III, C.L.: Closure approximations for three-dimensional structure tensors. J. Rheol. **34**(3), 367–386 (1990)
3. Backus, G.: A geometric picture of anisotropic elastic tensors. Rev. Geophys. Space Phys. **8**(3), 633–671 (1970)
4. Brachat, J., Comon, P., Mourrain, B., Tsigaridas, E.P.: Symmetric tensor decomposition. Linear Algebra Appl. **433**(11–12), 1851–1872 (2010)
5. Chebbi, Z.: Study of brain white matter fiber crossings using fourth-order diffusion tensors estimated from HARDI data. Master's thesis, National Engineering School at Tunis (2009)
6. Cieslinski, M.M., Steel, P.J., Lincoln, S.F., Easton, C.J.: Centrosymmetric and non-centrosymmetric packing of aligned molecular fibers in the solid state self assemblies of cyclodextrin-based rotaxanes. Supramol. Chem. **18**, 529–536 (2006)
7. Eberlea, A.P.R., Vélez-Garcíab, G.M., Bairda, D.G., Wapperomc P.: Fiber orientation kinetics of a concentrated short glass fiber suspension in startup of simple shear flow. J. Non-Newtonian Fluid Mech. **165**, 110–119 (2010)
8. Fisher, N.I., Lewis, T., Embleton, B.J.J.: Statistical Analysis of Spherical Data. Cambridge University Press, Cambridge (1987)
9. Florack, L., Balmashnova, E.: Two canonical representations for regularized high angular resolution diffusion imaging. In: Alexander, D., Gee, J., Whitaker, R. (eds.) MICCAI Workshop on Computational Diffusion MRI, New York, pp. 85–96 (2008)
10. Florack, L., Balmashnova, E., Astola, L., Brunenberg, E.: A new tensorial framework for single-shell high angular resolution diffusion imaging. J. Math. Imaging Vision **38**(3), 171–181 (2010)
11. Ghosh, A., Megherbi, T., Oulebsir-Boumghar, F., Deriche, R.: Fiber orientation distribution from non-negative sparse recovery. In: IEEE 10th International Symposium on Biomedical Imaging (ISBI), 2013, pp. 254–257 (2013)
12. Goldacker, T., Abetz, V., Stadler, R., Erukhimovich, I., Leibler, L.: Non-centrosymmetric superlattices in block copolymer blends. Nature **398**, 137–139 (1999)
13. Gur, Y., Jiao, F., Zhu, S.X., Johnson, C.R.: White matter structure assessment from reduced HARDI data using low-rank polynomial approximations. In: Panagiotaki, E., O'Donnell, L., Schultz, T., Zhang, G.H. (eds.) Proceedings of the Computational Diffusion MRI, pp. 186–197 (2012)
14. Jack, D.A., Smith, D.E.: Elastic properties of short-fiber polymer composites, derivation and demonstration of analytical forms for expectation and variance from orientation tensors. J. Compos. Mater. **42**(3), 277–308 (2008)
15. Jones, M.N.: Spherical Harmonics and Tensors for Classical Field Theory. Wiley, New York (1985)
16. Kanatani, K.-I.: Distribution of directional data and fabric tensors. Int. J. Eng. Sci. **22**(2), 149–164 (1984)
17. Lakes, R.: Elastic and viscoelastic behavior of chiral materials. Int. J. Mech. Sci. **43**(7), 1579–1589 (2001)
18. Liu, C., Bammer, R., Acar, B., Moseley, M.E.: Characterizing non-gaussian diffusion by using generalized diffusion tensors. Magn. Reson. Med. **51**, 924–937 (2004)
19. Mardia, K.V.: Statistics of directional data. J. R. Stat. Soc. Ser. B **37**(3), 349–393 (1975)
20. Megherbi, T., Kachouane, M., Oulebsir-Boumghar, F., Deriche, R.: Crossing fibers detection with an analytical high order tensor decomposition. Comput. Math. Methods Med. **2014**, 18 pp. (2014) [Article ID 476837]
21. Moakher, M.: Fourth-order Cartesian tensors: old and new facts, notions and applications. Q. J. Mech. Appl. Math. **61**(2), 181–203 (2008)

22. Ostroverkhov, V., Ostroverkhova, O., Petschek, R.G., Singer, K.D., Sukhomlinova, L., Twieg, R.J., Wang, S.-X., Chien, L.C.: Optimization of the molecular hyperpolarizability for second harmonic generation in chiral media. Chem. Phys. **257**, 263–274 (2000)
23. Özarslan, E., Mareci, T.H.: Generalized diffusion tensor imaging and analytical relationships between diffusion tensor imaging and high angular resolution diffusion imaging. Magn. Reson. Med. **50**(5), 955–965 (2003)
24. Özarslan, E., Shepherd, T.M., Vemuri, B.C., Blackband, S.J., Mareci, T.H.: Resolution of complex tissue microarchitecture using the diffusion orientation transform (DOT). NeuroImage **31**(3), 1086–1103 (2006)
25. Papenfuss, C., Ván, P.: Scalar, vectorial, and tensorial damage parameters from the mesoscopic background. Proc. Est. Acad. Sci. **57**(3), 132–141 (2008)
26. Schaeben, H.: A simple standard orientation density function: the hyperspherical de la Vallée Poussin kernel. Phys. Status Solidi (B) **200**(2), 367–376 (1997)
27. Schultz, T., Seidel, H.-P.: Estimating crossing fibers: a tensor decomposition approach. IEEE Trans. Vis. Comput. Graph. **14**(6), 1635–1642 (2008)
28. Schultz, T., Fuster, A., Ghosh, A., Deriche, R., Florack, L., Lim, L.-H.: Higher-order tensors in diffusion imaging. In: Westin, C.F., Vilanova, A., Burgeth, B. (eds.) Visualization and Processing of Tensors and Higher Order Descriptors for Multi-Valued Data. Mathematics and Visualization, pp. 129–161. Springer, Berlin/Heidelberg (2014)
29. Snieder, R.: A Guided Tour of Mathematical Methods for the Physical Sciences, 2nd edn. Cambridge University Press, Cambridge (2001)
30. Tournier, J.-D., Calamante, F., Gadian, D.G., Connelly, A.: Direct estimation of the fiber orientation density function from diffusion-weighted MRI data using spherical deconvolution. NeuroImage **23**(3), 1176–1185 (2004)
31. Tuch, D.S.: Q-ball imaging. Magn. Reson. Med. **52**(6), 1358–1372 (2004)
32. van der Boogaart, K.G., Hielscher, R., Prestin, J., Schaeben, H.: Kernel-based methods for inversion of the Radon transform on SO(3) and their applications to texture analysis. J. Comput. Appl. Math. **199**, 122–140 (2007)
33. Voyiadjis, G.Z., Kattan, P.I.: Evolution of fabric tensors in damage mechanics of solids with micro-cracks: Part I - Theory and fundamental concepts. Mech. Res. Commun. **34**(2), 145–154 (2007)
34. Wang, B., Zhou, J., Koschny, T., Kafesaki, M., Soukoulis, C.M.: Chiral metamaterials: simulations and experiments. J. Opt. A Pure Appl. Opt. **11**, 114003 (2009)
35. Watson, G.S.: Distributions on the circle and sphere. J. Appl. Probab. **19**, 265–280 (1982)

Topology of 3D Linear Symmetric Tensor Fields

Yue Zhang, Jonathan Palacios, and Eugene Zhang

Abstract There has been much research in 3D symmetric tensor fields, including recent work on tensor field topology. In this book chapter, we apply these research results to the most fundamental types of 3D tensor fields, i.e., linear tensor fields, and provide some novel insights on such fields. We also propose a number of hypotheses about linear tensor fields. We hope by studying linear tensor fields, we can gain more critical insights into the topology of general 3D tensor fields in the future.

1 Introduction

There has been much recent research in the analysis and visualization of second-order, symmetric tensor fields, especially on their topology; however, our understanding of 3D symmetric tensor fields are still far from mature, especially when compared to our knowledge of 2D symmetric tensor fields. In this chapter, we study the behaviors of 3D linear tensor fields, with a focus on topology.

As the first term in Taylor's expansion, the behaviors of a tensor field near the point of interest are usually dictated by its linearization at the point, which is a linear tensor field. Moreover, linear tensor fields are easier to study than more generic tensor fields due to their simpler forms. Finally, existing degenerate curve extraction methods often make the following assumptions: (1) that every degenerate curve intersecting a cell must also intersect the cell boundary, i.e., no degenerate loops exist completely inside a cell, and (2) there is at most one intersection per cell's face with all the degenerate curves. These assumptions, if invalid, can make

Y. Zhang (✉)
School of Electrical Engineering and Computer Science, 3117 Kelley Engineering Center, Oregon State University, Corvallis, OR 97331, USA
e-mail: zhangyue@eecs.oregonstate.edu; zhangyue@onid.oregonstate.edu

J. Palacios • E. Zhang
School of Electrical Engineering and Computer Science, 2111 Kelley Engineering Center, Oregon State University, Corvallis, OR 97331, USA
e-mail: palacijo@eecs.oregonstate.edu; zhange@eecs.oregonstate.edu

I. Hotz, T. Schultz (eds.), *Visualization and Processing of Higher Order Descriptors for Multi-Valued Data*, Mathematics and Visualization,
DOI 10.1007/978-3-319-15090-1_4

the extraction of degenerate curves incomplete, thus compromising their physical interpretations of applications. In this chapter we will discuss such tensor patterns in the context of linear tensor fields. We observe that a number of degenerate curves often intersect at a single degenerate point, and we discuss the minimum and maximum number of such curves.

Section 2 reviews past research in symmetric tensor field visualization. In Sect. 3 we review what we believe are the most relevant results on symmetric tensor fields. In Sect. 4 we provide a number of observations and hypotheses on 3D linear tensor fields before concluding in Sect. 5.

2 Previous Work

There has been much work on the topic of 2D and 3D tensor fields for medical imaging and scientific visualization. We refer the readers to the surveys by Zhang et al. [8] and Kratz et al. [5]. Here we only mention the research most relevant to this chapter.

Delmarcelle and Hesselink introduce the topology of 2D symmetric tensor fields in terms of *degenerate tensors* (with repeated eigenvalues) [1, 2]. This work is later extended to 3D symmetric tensor fields [4]. Zheng and Pang [11] point out that triple degeneracy, i.e., a tensor with three equal eigenvalues, cannot be extracted in a numerically stable fashion. They further show that double degeneracies, i.e., 3D tensors with only two equal eigenvalues, form lines in the domain. In this work and subsequent research [13], they provide a number of degenerate curve extraction methods based on the analysis of the discriminant function of the tensor field. Furthermore, Zheng et al. [12] point out that near degenerate curves the tensor field exhibits 2D degenerate patterns and define separating surfaces which are extensions of separatrices from 2D symmetric tensor field topology. Tricoche et al. [7] convert the problem of extracting degenerate curves in a 3D tensor field to that of finding the ridge and valley lines of an invariant of the tensor field, thus leading to a more robust extraction algorithm. Tricoche and Scheuermann [6] introduce a topological simplification operation which removes two degenerate points with opposite tensor indexes from the field. Zhang et al. [10] propose an algorithm to perform this *pair cancellation* operation by converting the tensor field to a vector field and reusing similar operations in vector field topological simplification [9].

3 Background on Tensors and Tensor Fields

In this section we review the most relevant background on tensors and tensor fields.

3.1 Tensors

A K-dimensional (symmetric) tensor $\mathbf{T}$ has K real-valued *eigenvalues*: $\lambda_1 \geq \lambda_2 \geq \ldots \geq \lambda_K$. The largest and smallest eigenvalues are referred to as the *major eigenvalue* and the *minor eigenvalue*, respectively. When $K = 3$, the middle eigenvalue is referred to as the *medium eigenvalue*. An eigenvector belonging to the major eigenvalue is referred to as a *major eigenvector*. Medium and minor eigenvectors can be defined similarly. Eigenvectors belonging to different eigenvalues are mutually perpendicular.

A tensor is *degenerate* when there are repeated eigenvalues. In this case, there exists at least one eigenvalue whose corresponding eigenvectors form a higher-dimensional space than a line. When $K = 2$ a degenerate tensor must be a multiple of the identity matrix. When $K = 3$, there are two types of degenerate tensors, corresponding to three repeated eigenvalues (*triple degenerate*) and two repeated eigenvalues (*double degenerate*), respectively. The latter can be further divided into *linear degenerate tensors*, i.e., $\lambda_1 > \lambda_2 = \lambda_3$, and *planar degenerate tensors*, i.e., $\lambda_1 = \lambda_2 > \lambda_3$. A tensor is degenerate if and only if it has a zero *discriminant* defined as $\prod_{1 \leq i < j \leq K} (\lambda_i - \lambda_j)^2$. A double degenerate tensor has one *repeated eigenvalue* and one *non-repeated eigenvalue*. The non-repeated eigenvalue for a linear degenerate tensor is its major eigenvalue, while for a planar degenerate tensor the non-repeated eigenvalue is the minor eigenvalue. The degenerate tensors are of essential importance in tensor field topology, and the study of degenerate tensors in 3D is made easier with the well-known *trace-deviator* decomposition, explained next.

The trace of a tensor $\mathbf{T} = (\mathbf{T}_{ij})$ is $trace(\mathbf{T}) = \sum_{i=1}^{K} \lambda_i$, which is invariant under the change of basis. $\mathbf{T}$ can be uniquely decomposed as $\mathbf{D} + \mathbf{A}$ where $\mathbf{D} = \frac{trace(\mathbf{T})}{K}\mathbb{I}$ ($\mathbb{I}$ is the K-dimensional identity matrix) and $\mathbf{A} = \mathbf{T} - \mathbf{D}$. The *deviator* $\mathbf{A}$ is a *traceless* tensor, i.e., $trace(\mathbf{A}) = 0$. Note that $\mathbf{T}$ and $\mathbf{A}$ have the same set of eigenvectors. Moreover, a tensor is degenerate if and only if its deviator is degenerate. Another nice property of the set of traceless tensors is that it is closed under matrix addition and scalar multiplication, making it a linear subspace of the set of tensors. These properties make it possible to study tensor field topology by focusing on traceless tensors, for which many tensor invariants have a simpler form. In the remainder of this chapter we will focus on only traceless tensors and therefore omit the mention of the word traceless.

There are a few other important tensor invariants, such as the determinant $|\mathbf{T}| = \prod_{i=1}^{K} \lambda_i$, magnitude $||\mathbf{T}|| = \sqrt{\sum_{1 \leq i,j \leq K} T_{ij}^2} = \sqrt{\sum_i^K \lambda_i^2}$, and $mode(\mathbf{T}) = \frac{3\sqrt{6}|\mathbf{T}|}{||\mathbf{T}||^3}$. The mode achieves its maximum value 1 when $\mathbf{T}$ is a linear degenerate tensor and its minimal value -1 when $\mathbf{T}$ is a planar degenerate tensor. Figure 1 shows the transition from planar degenerate tensors (left column) to linear degenerate tensors (right column) through non-degenerate tensors (middle columns). Given a 3D (traceless) symmetric tensor, the magnitude and the mode of the tensor uniquely

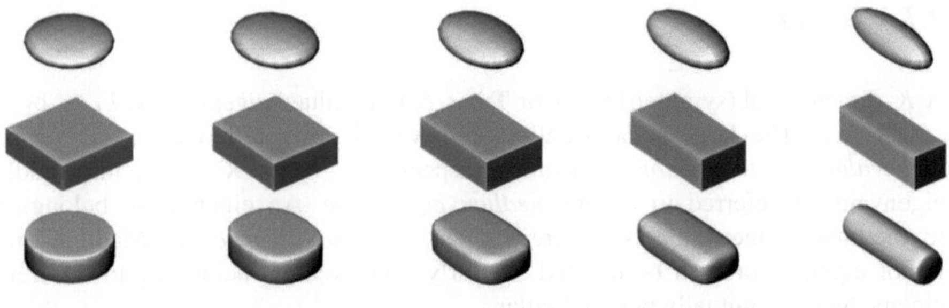

Fig. 1 Visualization of the impact of tensor mode using various types glyphs: ellipsoids (*top*), boxes (*middle*), and superquadrics (*bottom*). The *leftmost* and *rightmost columns* correspond to planar and linear degenerate tensors, respectively. Notice how planar degenerate tensors transition gradually towards linear degenerate tensors through non-degenerate tensors (*middle columns*)

determine the three eigenvalues, thus providing a more meaningful description than the three eigenvalues independently.

3.2 Tensor Fields

We now review *tensor fields*, which are continuous tensor-valued functions over some domain $\Omega \subset \mathbb{R}^K$. A tensor field can be thought of as K eigenvector fields, corresponding to the K eigenvalues. A *hyperstreamline* with respect to an eigenvector field $e_i(p)$ is a $3D$ curve that is tangent to e_i everywhere along its path. Two hyperstreamlines belonging to two different eigenvalues can only intersect at the right angle, since eigenvectors belonging to different eigenvalues must be mutually perpendicular.

Hyperstreamlines are usually curves; however, they can occasionally consist of only one point, where there are more than one choice of lines that correspond to the eigenvector field. This is precisely where the tensor field is degenerate. A point $p_0 \in \Omega$ is a *degenerate point* if $\mathbf{T}(p_0)$ is degenerate. The topology of a tensor field consists of its degenerate points.

In $2D$, the set of degenerate points of a tensor field are isolated points under numerically stable configurations, when the topology does not change given sufficiently small perturbation in the tensor field. An isolated degenerate point can be measured by its *tensor index* [10], defined in terms of the *winding number* of one of the eigenvector fields on a loop surrounding the degenerate point. The most fundamental types of degenerate points are *wedges* and *trisectors*, with a tensor index of $\frac{1}{2}$ and $-\frac{1}{2}$, respectively. Let $L\mathbf{T}_{p_0}(p)$ be the local linearization of $\mathbf{T}(p)$ at a degenerate point $p_0 = (x_0, y_0)$, i.e.,

$$L\mathbf{T}_{p_0}(p) = \begin{pmatrix} a_{11}(x - x_0) + b_{11}(y - y_0) & a_{12}(x - x_0) + b_{12}(y - y_0) \\ a_{12}(x - x_0) + b_{12}(y - y_0) & a_{22}(x - x_0) + b_{22}(y - y_0) \end{pmatrix} \quad (1)$$

The quantity $\delta = \left|\begin{pmatrix} \frac{a_{11}-a_{22}}{2} & a_{12} \\ \frac{b_{11}-b_{22}}{2} & b_{12} \end{pmatrix}\right|$, invariant under the change of basis, characterizes degenerate points as wedges or trisectors. Namely, a point p_0 is a wedge when $\delta > 0$ and a trisector when $\delta < 0$. When $\delta = 0$, p_0 is a higher-order degenerate point.

The total tensor index of a continuously tensor field over a two-dimensional manifold is equal to the *Euler characteristic* of the underlying manifold. Consequently, it is not possible to remove one degenerate point. Instead, a pair of degenerate points with opposing tensor indexes (a wedge and trisector pair) must be removed simultaneously [10].

The topology in 3D tensor fields is more challenging. While triple degeneracies can exist, they are numerically unstable, i.e., can disappear under arbitrarily small perturbations. Stable topological features in 3D tensor fields consists of double degenerate points that form curves. A curve consists of either purely linear degenerate points, or purely planar degenerate points. Furthermore, along a degenerate curve the tensor field exhibits 2D tensor degenerate patterns.

To be more precise, consider a degenerate curve γ and a point $p_0 \in \gamma$. The *repeated plane* at p_0 is the plane that is perpendicular to the non-repeated eigenvector at p_0. Recall that the non-repeated eigenvector is the eigenvector corresponding to the non-repeated eigenvalue of $\mathbf{T}(p_0)$, which is the major eigenvalue for linear degenerate points and the minor eigenvalue for the planar degenerate points.

Let (v_1, v_2) be an orthonormal basis for the repeated plane at p_0 and v_3 be a unit non-repeated eigenvector for $\mathbf{T}(p_0)$ such that (v_1, v_2, v_3) form a right-handed orthonormal basis. Under this basis the tensor field has the form:

$$\begin{pmatrix} \mathbf{M}_{11}(p) & \mathbf{M}_{12}(p) & \mathbf{M}_{13}(p) \\ \mathbf{M}_{12}(p) & \mathbf{M}_{22}(p) & \mathbf{M}_{23}(p) \\ \mathbf{M}_{13}(p) & \mathbf{M}_{23}(p) & \mathbf{M}_{33}(p) \end{pmatrix} \quad (2)$$

When $p \to p_0$, $\mathbf{M}_{13}(p)$ and $\mathbf{M}_{23}(p)$ approach 0. Moreover, if p_0 is a linear degenerate point, the difference between the eigenvectors of $\mathbf{M}(p) = \begin{pmatrix} \mathbf{M}_{11}(p) & \mathbf{M}_{12}(p) \\ \mathbf{M}_{12}(p) & \mathbf{M}_{22}(p) \end{pmatrix}$ and the projection of the eigenvectors corresponding to λ_2 and λ_3 approach 0 as p approaches p_0. A similar statement can be made for a planar degenerate point, except that we replace λ_2 and λ_3 with λ_1 and λ_2, respectively.

$\mathbf{M}(p)$ restricted to the repeated plane at p_0 has at least one degenerate point, p_0. The 2D tensor pattern near p_0 is typically either a wedge or a trisector. Recall that ∇f refers to the gradient vector of a scalar field f. The tangent of degenerate curve γ at p_0 is $\nabla(\mathbf{M}_{11}(p) - \mathbf{M}_{22}(p)) \times \nabla \mathbf{M}_{12}(p)$, which is the direction in which $\mathbf{M}(p)$ remains degenerate. When $\nabla(\mathbf{M}_{11}(p) - \mathbf{M}_{22}(p)) \times \nabla \mathbf{M}_{12}(p) = 0$, the tangent to the degenerate curve is parallel to the repeated plane at p_0. In this case, the 2D degenerate pattern of $\mathbf{M}$ inside the repeated plane is a higher-order degenerate point. It is referred to as a *transition point* as it is the boundary between wedge points on γ and trisector points on γ.

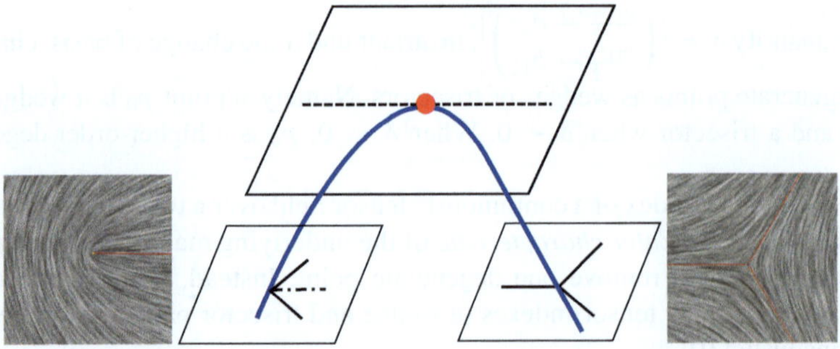

Fig. 2 Near a degenerate curve, 3D tensor fields exhibit 2D tensor patterns such as wedges (in the left plane, also the *inset* to the *left*) and trisectors (in the right plane, also the *inset* in the *right*). Such patterns persist along the curve until transition points (*top of the curve*), where the wedge and trisector cancel each other

Figure 2 illustrates this with an example. When taking a stack of cut planes moving upwards, we can observe two 2D degenerate patterns (wedge and trisector) in the projected tensor field moving closer and eventually canceling each other at the transition point (red).

4 3D Linear Tensor Fields

We wish to study the topological behaviors of 3D linear tensor fields, such as the number, location, and types of the degenerate curves contained in them. A 3D linear symmetric tensor field has the following form:

$$LT(x, y, z) = T_0 + xT_x + yT_y + zT_z \tag{3}$$

where

$$T_0 = \begin{pmatrix} T_{0,11} & T_{0,12} & T_{0,13} \\ T_{0,12} & T_{0,22} & T_{0,23} \\ T_{0,13} & T_{0,23} & -T_{0,11} - T_{0,22} \end{pmatrix}, T_x = \begin{pmatrix} T_{x,11} & T_{x,12} & T_{x,13} \\ T_{x,12} & T_{x,22} & T_{x,23} \\ T_{x,13} & T_{x,23} & -T_{x,11} - T_{x,22} \end{pmatrix}$$

$$T_y = \begin{pmatrix} T_{y,11} & T_{y,12} & T_{y,13} \\ T_{y,12} & T_{y,22} & T_{y,23} \\ T_{y,13} & T_{y,23} & -T_{y,11} - T_{y,22} \end{pmatrix}, T_z = \begin{pmatrix} T_{z,11} & T_{z,12} & T_{z,13} \\ T_{z,12} & T_{z,22} & T_{z,23} \\ T_{z,13} & T_{z,23} & -T_{z,11} - T_{z,22} \end{pmatrix}.$$

To understand their topological behaviors, we need to establish two facts about linear tensor fields: under the change of coordinate systems, and when projected onto a plane.

Lemma 1 *Given a linear symmetric tensor field $LT(x,y,z) = T_0 + xT_x + yT_y + zT_z$, its linearity is preserved under change of coordinate systems.*

Proof Let $C = (o, e_1, e_2, e_3)$ and $C' = (o', e_1', e_2', e_3')$ be two coordinate systems where o and o' are the respective origins and the e_i's and e_i''s are the basis vectors. Let (x, y, z) and (x', y', z') be the coordinates of a point p under C and C', respectively. The two sets of coordinates are related by

$$\begin{pmatrix} x - x_0 \\ y - y_0 \\ z - z_0 \end{pmatrix} = M \begin{pmatrix} x' \\ y' \\ z' \end{pmatrix} \tag{4}$$

where $\begin{pmatrix} x_0 \\ y_0 \\ z_0 \end{pmatrix}$ is the origin of C, and M is the unique linear transformation such that $M(e_i') = \sum_{j=1}^{3} M_{ij} e_j' = e_i$ for $1 \le i \le 3$.

Given a tensor field $T = T_0 + xT_X + yT_y + zT_z$ under C, we consider its formula under C'. For the point p, its tensor value using C is:

$$T_0 + xT_x + yT_y + zT_z \tag{5}$$

Under C' it becomes:

$$M^{-1}(T_0 + xT_x + yT_y + zT_z)M \tag{6}$$

Recall that

$$\begin{aligned} x &= M_{11}x' + M_{12}y' + M_{13}z' + x_0 \\ y &= M_{21}x' + M_{22}y' + M_{23}z' + y_0 \\ z &= M_{31}x' + M_{32}y' + M_{33}z' + z_0 \end{aligned}$$

Therefore, the tensor field has the following *linear* form under C':

$$LT(x', y', z') = T_0' + x'T_x' + y'T_y' + z'T_z' \tag{7}$$

where

$$T_0' = M^{-1}(T_0 + x_0T_x + y_0T_y + z_0T_z)M \tag{8}$$

$$T_x' = M^{-1}(M_{11}T_x + M_{21}T_y + M_{31}T_z)M \tag{9}$$

$$T_y' = M^{-1}(M_{12}T_x + M_{22}T_y + M_{32}T_z)M \tag{10}$$

$$T_z' = M^{-1}(M_{13}T_x + M_{23}T_y + M_{33}T_z)M \tag{11}$$

Note that T_0', T_x', T_y', and T_z' are symmetric, traceless tensors since the set of such tensors is closed under matrix addition. Furthermore, the linearity is preserved under the change of basis, i.e., the tensor field expressed under C' is still linear. □

Lemma 1 states that under a change of coordinate systems, a linear tensor field remains linear, a fact that is neither surprising nor difficult to prove; and Eqs. (8)–(11) establish how the coefficients of the linear tensor field vary as a result of change in the coordinate systems. These equations will be useful later when we understand the behaviors of the tensor field near degenerate curves.

We now consider the projection of a linear tensor field onto a plane.

Lemma 2 *Given a linear symmetric tensor field $LT(x, y, z) = T_0 + xT_x + yT_y + zT_z$, its projection onto a plane is a 2D symmetric, linear tensor field inside the plane.*

Proof Let N be a normal to the plane and $p_0 = (x_0, y_0, z_0)$ be a point on the plane. We construct a new coordinate system (p_0, X, Y, N) such that p_0 is the new origin and X and Y form a basis for the plane. Based on Lemma 1, the linear tensor field under the new basis has the form in Eqs. (8)–(11).

Given a point p in the plane. Under the new coordinate systems, p has the form

$$LT''(x', y') = T_0'' + x'T_x'' + y'T_y'' \tag{12}$$

where LT'' is the projected tensor field of LT on the plane, and T_0'', T_x'' and T_y'' are respectively the 2×2 subblock of T_0', T_x', and T_y' corresponding to the plane. It is clear that T'' remains a symmetric, linear tensor field. □

Again, the important message from Lemma 2 is how the projection tensor field relates to the original 3D tensor field. We now are ready to consider the topology of a 3D linear, symmetric tensor field. We begin with the simpler case when $T_0 = 0$. In this case, the origin is a triple degenerate point.

Theorem 1 *Given a linear symmetric tensor field $LT(x, y, z) = xT_x + yT_y + zT_z$, a point (x_0, y_0, z_0) is degenerate if and only if (kx_0, ky_0, kz_0) is also degenerate for any $k \neq 0$. Moreover, $k(x_0, y_0, z_0)$ is triple degenerate if and only if (x_0, y_0, z_0) is triple degenerate. If (x_0, y_0, z_0) is a degenerate point, then $k(x_0, y_0, z_0)$ is linear if $k > 0$ and planar if $k < 0$.*

Proof By the definition of linearity, $LT(kx_0, ky_0, kz_0) = kx_0T_x + ky_0T_y + kz_0T_z = k(x_0T_x + y_0T_y + z_0T_z) = kLT(x_0, y_0, z_0)$. The proof then follows directly from the fact that if the eigenvalues of $LT(x_0, y_0, z_0)$ are λ_1, λ_2, and λ_3, then $LT(kx_0, ky_0, kz_0)$ has $k\lambda_1$, $k\lambda_2$, and $k\lambda_3$ as the eigenvalues. □

An immediate result is:

Corollary 1 *Given a numerically stable linear tensor field $LT(x, y, z) = xT_x + yT_y + zT_z$, there are no degenerate loops.*

Proof Recall that if a point $p_0 = (x_0, y_0, z_0)$ is degenerate, then so are the points on the line passing through p_0 and the origin. So if there is a loop of degenerate points, then each point on the loop would lead to a line of degenerate points. Consequently, the loop would lead to a cylindrical surface of degenerate points, which is not numerically stable. □

The above corollary states that all degenerate curves in such a tensor field must end in infinity.

Besides the linear/planar classification, a degenerate point can be further classified into either a wedge, a trisector, or a transition point. This classification refers to the 2D degenerate pattern around the degenerate point inside the repeated plane at the point. A natural question is whether this classification even makes sense, i.e., when projecting the 3D tensor field onto the repeated plane, does the 2D degenerate pattern depend on the coordinate systems in the repeated plane or the choice of the plane normal? This is answered with the following theorem.

Theorem 2 *The wedge/trisector classification of a degenerate point p in a 3D linear, symmetric tensor field T is independent of the choice of the normal in the repeated plane and the coordinate system for the repeated plane.*

Proof Recall the definition of wedge/trisector classification of T under a right-handed orthonormal frame $O = (p_0, X, Y, N)$ where p_0 is a point in the space, N is a normal to the repeated plane at p and X and Y are parallel to the plane. Let the tensor field be expressed under O, which has the form $LT(x, y, z) = T_0 + xT_x + yT_y + zT_z$. The 2D projection onto the repeated plane with the basis (X, Y) is:

$$LT_p(x, y) = \begin{pmatrix} T_{0,11} + T_{x,11}x + T_{y,11}y & T_{0,12} + T_{x,12}x + T_{y,12}y \\ T_{0,21} + T_{x,21}x + T_{y,21}y & T_{0,22} + T_{x,22}x + T_{y,22}y \end{pmatrix} \tag{13}$$

The wedge/trisector classification is based on the quantity

$$\delta = (T_{x,22} - T_{x,11})T_{y,12} - (T_{y,22} - T_{y,11})T_{x,12} \tag{14}$$

The degenerate point p is a wedge, trisector, or transition point when δ is positive, negative, and zero, respectively. Note that δ depends only on the first 2×2 blocks of T_x and T_y. We now consider a second right-handed orthonormal frame $O' = (p'_0, X', Y', N')$. We wish to show that δ under O' is the same as that under O, i.e., δ is invariant under the change of basis. Note that O' can be obtained from O under the combination of three types of operations: translation, rotation, and reflection. Consequently, all we need is to show that δ is invariant under these three operations.

We first consider translation, i.e., $X' = X, Y' = Y$, and $N' = N$. Let (x_0, y_0, z_0) be the coordinates of p'_0 under O'. Let (x, y, z) and (x', y', z') be the coordinates of an arbitrary point p under O and O', respectively. Then

$$x = x' - x_0 \qquad y = y' - y_0 \qquad z = z' - z_0 \tag{15}$$

Consequently, the tensor field has the form $LT(x', y', z') = T_0' + x'T_x' + y'T_y' + z'T_z'$ where $T_x' = T_x$, $T_y' = T_y$, $T_z' = T_z$, and $T_0' = T_0 - x_0T_x - y_0T_y - z_0T_z$. Recall that δ only depends on the first 2×2 blocks of T_x' and T_y', which are identical to that of T_x and T_y, it does not change when translating the origin of the frame.

We now consider the case of reflection, i.e., $p_0 = p_0'$, $N' = -N$, $X' = X$, and $Y' = -Y$. In this case (x, y, z) and (x', y', z') are related by

$$x = x' \qquad y = -y' \qquad z = -z' \tag{16}$$

Furthermore, a matrix A under O and O' are related by $A' = MAM$ where $M = \begin{pmatrix} 1 & 0 & 0 \\ 0 & -1 & 0 \\ 0 & 0 & -1 \end{pmatrix}$. Consequently,

$$LT'(x', y', z') = M(LT(x, -y, -z))M = MT_0M + xMT_xM - yMT_yM - zMT_zM \tag{17}$$

It is straightforward to verify that δ computed for $LT'(x', y', z')$ is the same as that for $LT(x, y, z)$.

We now consider the case of rotation, i.e., $p_0 = p_0'$, $N' = N$, and X' and Y' are obtained by rotating X and Y counterclockwise by an angle of θ in the repeated plane. In this case (x, y, z) and (x', y', z') are related by

$$x = \cos\theta x' + \sin\theta y' \qquad y = -\sin\theta x' + \cos\theta y \tag{18}$$

This reduces to the case showing δ is invariant under 2D rotations for 2D linear symmetric tensor fields, which has been proven by Delmarcelle [1].

Consequently, δ is invariant under any combinations of translation, rotation, and reflection, and thus any change of basis. □

This theorem, while rather simple, is important in that it establishes the well-definedness of the wedge/trisector classification. Now a degenerate point can be classified as a linear wedge, a linear trisector, a planar wedge, or a planar trisector. Next we state two facts on linear tensor fields.

Theorem 3 *Given a linear tensor field* $LT(x, y, z) = T_0 + xT_x + yT_y + zT_z$, *a point* $p_0 = (x_0, y_0, z_0)$ *is a wedge if and only if* $p_0' = (kx_0, ky_0, kz_0)$ *(for any* $k \neq 0$*) is also a wedge. Similarly, a point* $p_0 = (x_0, y_0, z_0)$ *is a trisector if and only if* $p_0' = (kx_0, ky_0, kz_0)$ *is also a trisector (for any* $k \neq 0$*). Moreover, the orientations of the degenerate patterns remain constant regardless of* k.

Proof Recall that the wedge/trisector classification and the orientation of the 2D degenerate pattern are solely dependent on the matrix $\begin{pmatrix} T_{x,22} - T_{x,11} & T_{y,22} - T_{y,11} \\ T_{x,12} & T_{y,12} \end{pmatrix}$. We construct a coordinate system using the point p_0 and the normal N to the plane where the wedge or trisector is on, and we denote this system by (p_0, X, Y, N).

Without loss of generality, we align N with Z. In this system, our tensor field is $LT(x, y, z) = T_0 + xT_x + yT_y + zT_z$. We then translate the plane to where the point p_0' is, and the tensor field becomes $LT'(x, y, z) = T_0' + xT_x' + yT_y' + zT_z'$. As the repeated plane at p_0' is parallel to that at p_0, $T_x' = T_x$, $T_y' = T_y$ and

$$\begin{pmatrix} T'_{x,22} - T'_{x,11} & T'_{y,22} - T'_{y,11} \\ T'_{x,12} & T'_{y,12} \end{pmatrix} = \begin{pmatrix} T_{x,22} - T_{x,11} & T_{y,22} - T_{y,11} \\ T_{x,12} & T_{y,12} \end{pmatrix}.$$

As a result, p_0 and p_0' have the same wedge/trisector type. Moreover, the local orientation of the degenerate patterns at p_0 and p_0' are the same. □

The above results indicate that if a point $p = (x, y, z)$ is a degenerate point, then any point on the ray emanating from the origin and containing p is also a degenerate point. Moreover, the linear/planar classification and wedge/trisector classification along the ray does not change. In contrast, the ray in the opposite direction from the origin has the same wedge/trisector classification but opposite linear/planar classification. Figure 3 (middle) indicates this with an example tensor field $LT(x, y, z) = \begin{pmatrix} 0 & x & y \\ x & 0 & z \\ y & z & 0 \end{pmatrix}$. The origin is a triple degenerate point, colored in white. Linear wedge points are colored green, while planar wedge points in yellow. Trisector points are colored blue if linear and red if planar. This field has eight degenerate rays:

1. $\{k(1, 1, 1)|k > 0\}$ (linear wedge)
2. $\{k(1, -1, 1)|k > 0\}$ (planar wedge)
3. $\{k(-1, -1, 1)|k > 0\}$ (linear trisector)
4. $\{k(-1, 1, 1)|k > 0\}$ (planar trisector)
5. $\{k(1, 1, -1)|k > 0\}$ (planar trisector)
6. $\{k(1, -1, -1)|k > 0\}$ (linear trisector)
7. $\{k(-1, -1, -1)|k > 0\}$ (planar wedge)
8. $\{k(-1, 1, -1)|k > 0\}$ (linear wedge)

The top and bottom rows of Fig. 3 show the projected tensor pattern inside the non-repeated planes at some degenerate points. Notice that along a degenerate ray pair, the wedge/trisector classification does not change, nor does the orientation of the wedges and trisectors.

A fundamental question is how many such degenerate ray pairs exist given a linear tensor field where $T_0 = 0$. We will consider the generic case in which degenerate points form curves instead of isolated points, surfaces, or volumes. For example, when T_x, T_y, and T_z span a *one*-dimensional space in the set of 3D symmetric, traceless tensors, $LT(x, y, z)$ is degenerate either everywhere, i.e., T_x, T_y and T_z are degenerate, or only at the origin, i.e., T_x, T_y and T_z are non-degenerate.

One way to consider the problem is to ask how many degenerate points exist on the set of unit tensors, i.e., with unit magnitude. Notice that each such point

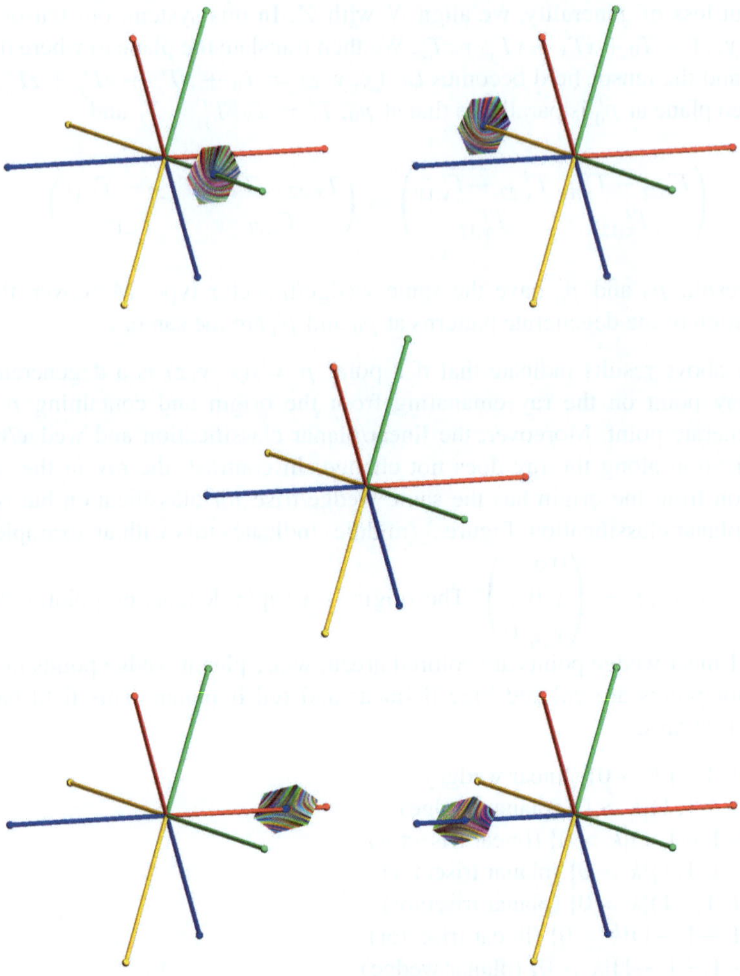

Fig. 3 The *middle-row image* shows a tensor field $LT(x, y, z) = xT_x + yT_y + zT_z$, which has a triple degenerate point at the origin, from which degenerate rays emanate. *Yellow points* are planar wedges, *red points* are planar trisectors, *green points* are linear wedges, and *blue points* are linear trisectors. Notice that along each degenerate ray, the linear/planar classification does not change, while opposite rays have opposing linear/planar classification (*green/yellow* and *blue/red*). Along each ray pair (two opposing rays), wedge/trisector classification does not change. Moreover, the orientation of the wedge/trisector patterns do not change along each ray pair, such as the wedge pattern along the *green/yellow* ray pair (*top row*) and the trisector pattern along the *red/blue* ray pair (*bottom row*)

corresponds to a degenerate ray in 3D. In generic cases, the set of such points (degenerate points with unit magnitude) should be isolated points instead of curves or surfaces, for the latter would correspond to degenerate surfaces and volumes, respectively.

Let D be the dimension of tensors spanned by T_x, T_y, and T_z. Notice that the higher the dimension D, the more the number of degenerate rays. The most generic case is when T_x, T_y, and T_z are linearly independent and form a three-dimensional subspace in the set of 3×3 symmetric, traceless tensors.

Theorem 4 *Given a linear tensor field* $LT(x, y, z) = xT_x + yT_y + zT_z$*, there are at most* 18 *degenerate rays, forming nine pairs of opposing curves.*

Proof Degenerate points of a tensor field correspond to mode ± 1, i.e.,

$$3\sqrt{6}\frac{|T|}{||T||^3} = \pm 1 \tag{19}$$

On the unit tensor surface, i.e., the levelset of $||T|| = 1$, degenerate points are the global minima and maxima. Consequently, $\bigtriangledown |T| \times \bigtriangledown ||T||^2 = 0$.

Since $T(x, y, z)$ is linear, $|T|$ and $||T||^2$ are cubic and quadratic in terms of x, y, and z, respectively. Thus, $\bigtriangledown |T|$ and $\bigtriangledown ||T||^2$ are quadratic and linear in terms of x, y, and z, respectively, and $\bigtriangledown |T| \times \bigtriangledown ||T||^2$ is cubic in terms of the coordinates. Notice that while $\bigtriangledown |T| \times \bigtriangledown ||T||^2$ is a 3-vector, leading to three equations, only two of which are linearly independent. Denote these equations as

$$F(x, y, z) = 0, \quad G(x, y, z) = 0 \tag{20}$$

Furthermore,

$$||T||^2 = 1 \tag{21}$$

Since F and G are both cubic and $||T||^2$ is quadratic, Bézout's Theorem states that there are at most $3 \times 3 \times 2 = 18$ solutions [3]. □

The above theorem establishes that there are up to nine degenerate lines in a linear tensor field $LT(x, y, z) = xT_x + yT_y + zT_z$.

We now consider the most generic linear tensor fields, i.e., $LT(x, y, z) = T_0 + xT_x + yT_y + zT_z$. While this is clearly a more complex situation, notice that the asymptotic behavior of the tensor field towards infinity is the same as $LT(x, y, z) = xT_x + yT_y + zT_z$, i.e., by setting $T_0 = 0$. Consequently, there exists $r \in \mathbb{R}$ such that given a sphere centered at the origin with the radius r, the tensor field $LT(x, y, z) = T_0 + xT_x + yT_y + zT_z$ has up to eighteen degenerate curves outside the sphere, with a one-to-one correspondence to the degenerate rays for the field $LT(x, y, z) = xT_x + yT_y + zT_z$. Let $2N$ be the number of such curves. Notice that N curves are of the linear type and the other N curves are of the planar types.

Assume that these curves do not converge onto a finite point or limit cycle. Then two of such curves must be connected, leading to N pairs of degenerate curves. Note that an L and P type curve pair cannot be connected. Consequently, N needs to be even. We thus conclude that the maximum number of curves emanating from the infinity must be even and at most eight pairs, instead of nine pairs as stated in the

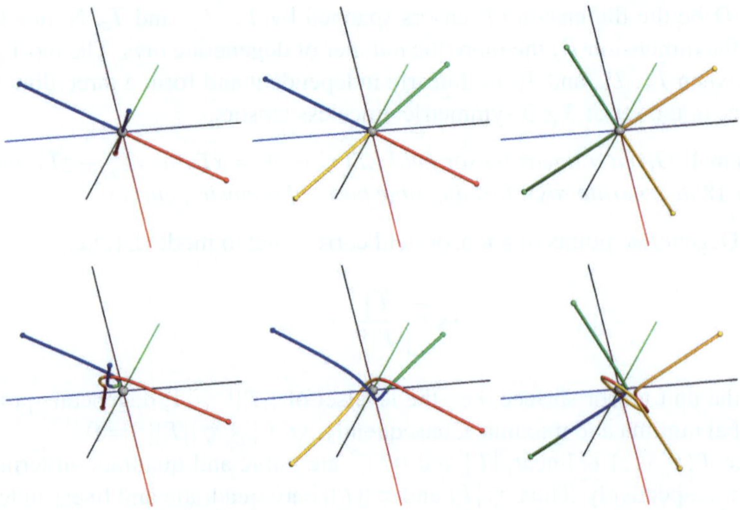

Fig. 4 The asymptotic behavior of the a linear tensor field $LT(x, y, z) = T_0 + xT_x + yT_y + zT_z$ (*top*) is the same as the case when $T_0 = 0$ (*bottom*). There appears to be no degenerate loops in a linear tensor field when $T_0 \neq 0$, and the number of degenerate curves in the tensor field does not seem to depend on T_0; however, these degenerate curves are paired up based on their linear/planar classification

theoretical upper bound. Figure 4 compares three pairs of tensor fields (top: $T_0 = 0$; bottom: $T_0 \neq 0$). Notice the same asymptotic behaviors of these fields towards infinity and how in the non-zero T_0 case, these degenerate curves are pairwise connected.

However, it is not clear to us whether eight is the real upper bound, and whether for the case $T_0 \neq 0$, additional degenerate loops would appear as a result of adding T_0. We discuss these issues next.

4.1 Observations and Hypotheses

We have implemented a system in which the user can design a 3D linear, traceless, symmetric tensor field $LT(x, y, z) = T_0 + xT_x + yT_y + zT_z$ by specifying T_0, T_x, T_y, and T_z either numerically or graphically. The degenerate curves in the tensor field are then extracted and classified based on its linear/planar and wedge/trisector properties. For degenerate curve extraction, we utilize the method of Zheng et al. [11]. Figure 5 shows the user interface.

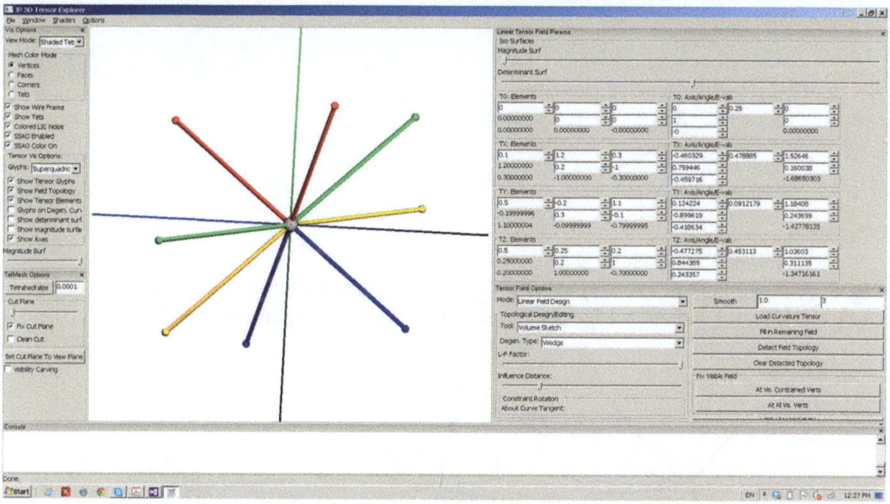

Fig. 5 The interface of our linear tensor field design system. The user can design a 3D linear traceless tensor field by specifying either the coefficients of the tensor field directly or by specifying the desired eigenvalue and eigenvectors. The degenerate curves are extracted and displayed

In the default configuration,

$$T_0 = \begin{pmatrix} 0 & 0 & 0 \\ 0 & 0 & 0 \\ 0 & 0 & 0 \end{pmatrix}, T_x = \begin{pmatrix} 0 & 1 & 0 \\ 1 & 0 & 0 \\ 0 & 0 & 0 \end{pmatrix}, T_y = \begin{pmatrix} 0 & 0 & 1 \\ 0 & 0 & 0 \\ 1 & 0 & 0 \end{pmatrix}, T_z = \begin{pmatrix} 0 & 0 & 0 \\ 0 & 0 & 1 \\ 0 & 1 & 0 \end{pmatrix}$$

To specify a matrix (e.g., T_0), the user can specify either its entries or its eigenvalues and eigenvectors. Our system automatically ensures that the resulting matrix is traceless. While entries of a matrix are specified directly using textboxes provided in the system, the eigenvalues can be specified either directly or by deriving from specified tensor magnitude and mode. The eigenvectors are specified using the graphical interface in which the user can rotate a 3D orthonormal frame, whose axes give the major, medium, and minor eigenvectors.

Through experiments using our system, we make the following observations.

Conjecture 1 Given a generic linear tensor field $LT(x, y, z) = xT_x + yT_y + zT_z$, i.e., T_x, T_y, and T_z are linearly independent, there are either four degenerate rays (two pairs) or eight degenerate rays (four pairs). Any number of these rays can be wedges.

Figures 6 and 7 illustrate the observation with example tensor fields. There are either two or four pairs of degenerate rays intersecting at the origin (a triple degenerate point). It is possible to have any subset of these curves being wedges and trisectors, respectively. Seemingly, this observation is contradicting to the Poincaré-Hopf theorem which states that on a sphere the number of wedges is four more than

Fig. 6 A tensor field $LT(x, y, z) = xT_x + yT_y + zT_z$ may have two trisector degenerate curves (*left*), one trisector and one wedge curve (*middle*), and two wedge curves (*right*)

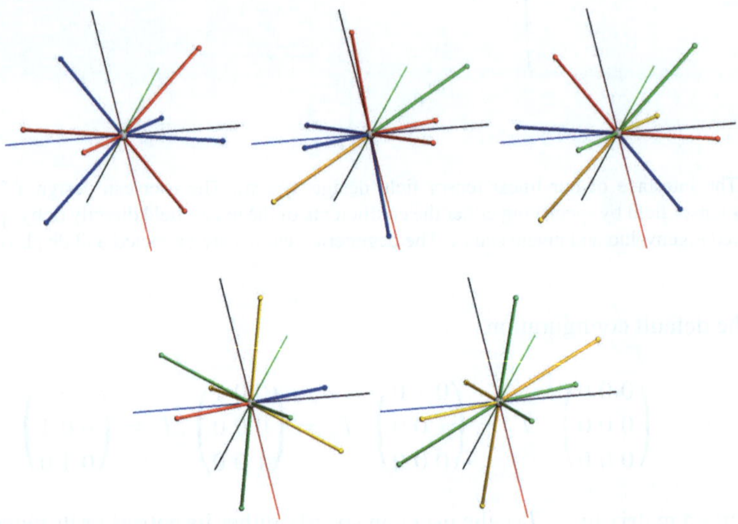

Fig. 7 A tensor field $LT(x, y, z) = xT_x + yT_y + zT_z$ may also have four trisector degenerate curves (*upper-left*), three trisector and one wedge curve (*upper-middle*), two trisector and two wedge curves (*upper-right*), one trisector and three wedge curves (*lower-left*), and four wedge curves (*lower-right*)

that of the trisectors. In fact, the degenerate rays of a linear tensor field is not related to the degenerate points of that field projected on a plane. This plane may not be the repeated plane. A degenerate point in 3D does not need to be a degenerate point in 2D. As a consequence a degenerate point in 3D does not need to be a degenerate point in 2D, and a non-degenerate 3D tensor can become a degenerate 2D tensor when projected on a plane. To illustrate, the following examples are provided:

$$T = \begin{pmatrix} 2 & 0 & 0 \\ 0 & -1 & 0 \\ 0 & 0 & -1 \end{pmatrix} \quad T = \begin{pmatrix} 1 & 0 & -2 \\ 0 & 1 & -2 \\ -2 & -2 & -2 \end{pmatrix}.$$

Conjecture 2 Given a generic linear tensor field $LT(x, y, z) = xT_x + yT_y + zT_z$, i.e., T_x, T_y, and T_z are linearly independent, the following are the only possible bifurcations: (1) a wedge ray pair and a trisector ray pair can merge into a ray of transition points before disappearing when there were four degenerate ray pairs; (2) a wedge ray pair and a trisector pair can merge into a ray of transition points before splitting again into a wedge and trisector ray pair when there were two degenerate ray pairs; (3) a wedge degenerate ray pair becomes a ray of transition points before turning into a trisector ray pair.

Figures 8, 9, and 10 show examples of these bifurcations. While the first type of bifurcations seems normal, we were surprised to observe the last two bifurcations. The second bifurcation seems related to observations that in a generic case, there are minimal of two degenerate ray pairs (instead of zero). When there were only two degenerate ray pairs to start with, they cannot cancel each other as it would lead to zero degenerate ray pairs.

Conjecture 3 Given a numerically stable linear tensor field $LT(x, y, z) = T_0 + xT_x + yT_y + zT_z$, there are no degenerate loops. Consequently, the tensor field has the same number of degenerate rays as when $T_0 = 0$, i.e., the number of degenerate curves in a linear tensor field is independent of its constant term T_0.

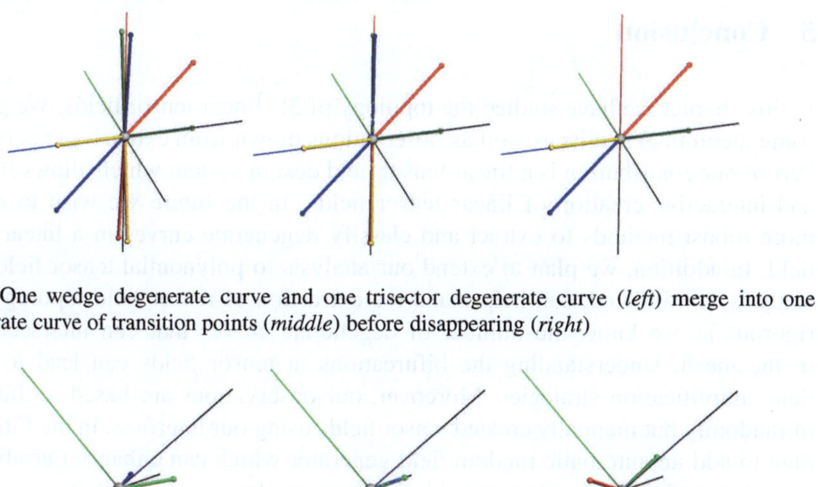

Fig. 8 One wedge degenerate curve and one trisector degenerate curve (*left*) merge into one degenerate curve of transition points (*middle*) before disappearing (*right*)

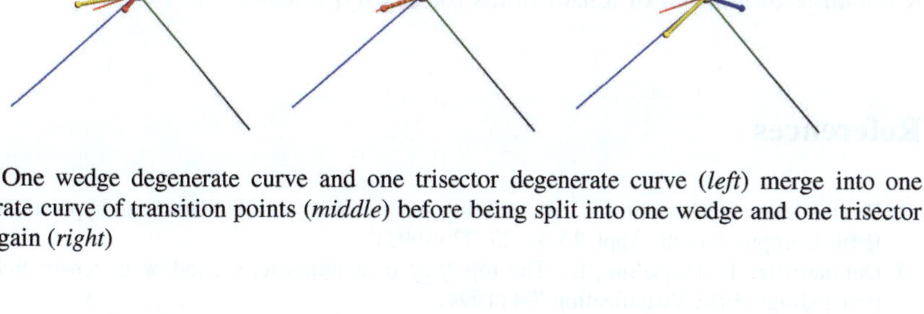

Fig. 9 One wedge degenerate curve and one trisector degenerate curve (*left*) merge into one degenerate curve of transition points (*middle*) before being split into one wedge and one trisector curve again (*right*)

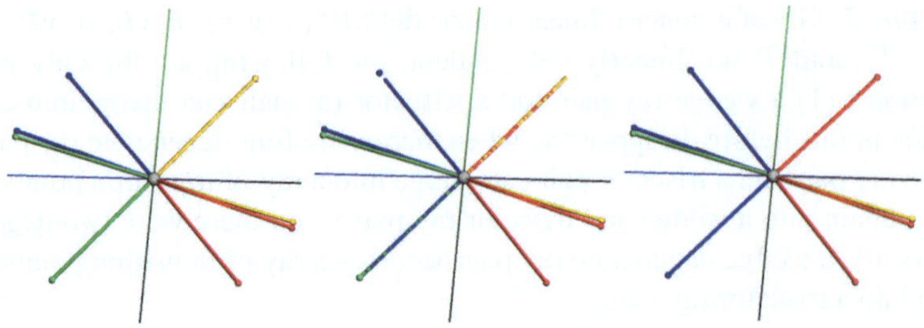

Fig. 10 A wedge degenerate curve (*left*) becomes a degenerate curve of transition points (*middle*) before becoming a trisector degenerate curve (*right*)

We had anticipated the appearance of degenerate loops when $T_0 \neq 0$; however, we have never observed them with extensive experiments. This observation, if true, would allow us to draw conclusions about linear tensor fields with a non-zero T_0 by studying the simpler case where $T_0 = 0$, such as the number of wedge and trisector curves in the tensor field as well as possible bifurcations. Figure 4 (bottom) show three tensor fields with a non-zero T_0. No degenerate loops exist for these fields.

5 Conclusion

In this chapter we have studied the topology of 3D linear tensor fields. We provide some theoretical results as well as observations drawn from extensive experiments. Part of our contribution is a linear tensor field design system which allows intuitive and interactive creation of linear tensor fields. In the future we wish to explore more robust methods to extract and classify degenerate curves in a linear tensor field. In addition, we plan to extend our analysis to polynomial tensor fields. Our analysis, if validated, can help make the extraction of tensor field topology more rigorous as we know the number of degenerate curves that can intersect a cell in the mesh. Understanding the bifurcations in tensor fields can lead to tensor field simplification strategies. Moreover, our observations are based on hundreds of randomly but manually created tensor fields using our interface. In the future, we plan to add an automatic random field generator which can enhance our ability to test a large of number of tensor fields for any hypotheses.

References

1. Delmarcelle, T., Hesselink, L.: Visualizing second-order tensor fields with hyperstream lines. IEEE Comput. Graph. Appl. **13**(4), 25–33 (1993)
2. Delmarcelle, T., Hesselink, L.: The topology of symmetric, second-order tensor fields. In: Proceedings IEEE Visualization '94 (1994)

3. Fulton, W.: Algebraic Curves: An Introduction to Algebraic Geometry. Advanced Book Classics. Addison-Wesley, Redwood City, CA (1989)
4. Hesselink, L., Levy, Y., Lavin, Y.: The topology of symmetric, second-order 3D tensor fields. IEEE Trans. Vis. Comput. Graph. **3**(1), 1–11 (1997)
5. Kratz, A., Auer, C., Stommel, M., Hotz, I.: Visualization and analysis of second-order tensors: moving beyond the symmetric positive-definite case. Comput. Graph. Forum **32**(1), 49–74 (2013). http://dblp.uni-trier.de/db/journals/cgf/cgf32.html#KratzASH13
6. Tricoche, X., Scheuermann, G.: Topology simplification of symmetric, second-order 2d tensor fields. In: Geometric Modeling Methods in Scientific Visualization. Springer, Berlin (2003)
7. Tricoche, X., Kindlmann, G., Westin, C.F.: Invariant crease lines for topological and structural analysis of tensor fields. IEEE Trans. Vis. Comput. Graph. **14**(6), 1627–1634 (2008). http://doi.ieeecomputersociety.org/10.1109/TVCG.2008.148
8. Zhang, S., Kindlmann, G., Laidlaw, D.H.: Diffusion tensor MRI visualization. In: Visualization Handbook. Academic, Amsterdam (2004). http://www.cs.brown.edu/research/vis/docs/pdf/Zhang-2004-DTM.pdf
9. Zhang, E., Mischaikow, K., Turk, G.: Vector field design on surfaces. ACM Trans. Graph. **25**(4), 1294–1326 (2006)
10. Zhang, E., Hays, J., Turk, G.: Interactive tensor field design and visualization on surfaces. IEEE Trans. Vis. Comput. Graph. **13**(1), 94–107 (2007)
11. Zheng, X., Pang, A.: Topological lines in 3d tensor fields. In: Proceedings IEEE Visualization 2004, VIS '04, pp. 313–320. IEEE Computer Society, Washington, DC (2004). doi:10.1109/VISUAL.2004.105. http://dx.doi.org/10.1109/VISUAL.2004.105
12. Zheng, X., Parlett, B., Pang, A.: Topological structures of 3D tensor fields. In: Proceedings IEEE Visualization 2005, pp. 551–558 (2005)
13. Zheng, X., Parlett, B.N., Pang, A.: Topological lines in 3d tensor fields and discriminant hessian factorization. IEEE Trans. Vis. Comput. Graph. **11**(4), 395–407 (2005)

Random Projections for Low Multilinear Rank Tensors

Carmeliza Navasca and Deonnia N. Pompey

Abstract We propose two randomized tensor algorithms for reducing multilinear tensor rank. The basis of these randomized algorithms is from the work of Halko et al. (SIAM Rev 53(2):217–288, 2011). Here we provide some random versions of the higher order SVD and the higher order orthogonal iteration. Moreover, we provide a sharp probabilistic error bound for the matrix low rank approximation. In consequence, we provide an error bound for the tensor case. Moreover, we give several numerical examples which includes an implementation on a MRI dataset to test the efficacy of these randomized algorithms.

1 Introduction

The problem of approximating a given matrix $M \in \mathbb{R}^{n\times m}$ with a low rank matrix $\hat{M} \in \mathbb{R}^{n\times k}$ is the optimization,

$$\min_{\text{rank}(\hat{M})\leq k} \|M - \hat{M}\|_F^2$$

where $\|\cdot\|$ is the matrix Frobenius norm. The optimal solution due to Eckart-Young [8] is

$$M^* = \text{argmin}_{\text{rank}(\hat{M})\leq k} \|M - \hat{M}\|_F^2 = U_k \Sigma_k V_k^T$$

where $U_k \Sigma_k V_k^T = \sum_{i=1}^{k} \sigma_i u_i \otimes v_i$ is the first k leading terms in the SVD. Similarly, we ask a similar question for tensors: for a given nth order tensor $\mathscr{T}$, can one find the best rank k approximant $\hat{\mathscr{T}}$ via

$$\min \|\mathscr{T} - \hat{\mathscr{T}}\|_F^2 \tag{1}$$

C. Navasca (✉) • D.N. Pompey
Department of Mathematics, University of Alabama at Birmingham, 1300 University Blvd, Birmingham, AL 35209-1170, USA
e-mail: cnavasca@uab.edu; cnavasca@gmail.com; nicholep@uab.edu

I. Hotz, T. Schultz (eds.), *Visualization and Processing of Higher Order Descriptors for Multi-Valued Data*, Mathematics and Visualization,
DOI 10.1007/978-3-319-15090-1_5

where $\|\cdot\|_F^2 = \mathrm{tr}\langle\cdot,\cdot\rangle$ is a Frobenius norm. Here the best rank k approximant is defined as $\hat{\mathcal{T}} = \sum_{i=1}^{k} \sigma_i u_i^{(1)} \otimes u_i^{(2)} \otimes \cdots \otimes u_i^{(k)}$. The tensor case is more complicated and challenging. The best rank k approximant may exist, but it may not have orthogonal factors like in the SVD of a matrix [16]. Or the best k rank-one terms may simply not exist [6]. Also, a minimizer in (1) is not guaranteed to exist unlike in the matrix case where the best rank k solution always exists. Moreover, this is well-known that this problem is NP-hard [12].

These tensor rank reducing questions are important since they have practical applications across many disciplines, namely in signal processing [3], PDEs [2, 15], geophysics [14], environmental sciences [18], brain connectome [24] and etc. Other interesting applications are in bioinformatics [21] and biomedical imaging [13]; see more applications in [17] and the references therein.

In this work, we propose some algorithms for approximating a low multilinear rank tensor $\hat{\mathcal{T}}$ from a given tensor $\mathcal{T} \in \mathcal{R}^{n_1 \times n_2 \times n_3}$; i.e

$$\min \|\mathcal{T} - \hat{\mathcal{T}}\|_F^2$$

where $\hat{\mathcal{T}} = \sum_{l=1}^{r_1} \sum_{m=1}^{r_2} \sum_{k=1}^{r_3} (\mathcal{S})_{lmn} (U_1)_{il}, (U_2)_{jm}, (U_3)_{kn}$ (Tucker format) with orthogonal matrices $U_1 \in \mathbb{R}^{n_1 \times r_1}$, $U_2 \in \mathbb{R}^{n_2 \times r_2}$ and $U_3 \in \mathbb{R}^{n_3 \times r_3}$. Here the core tensor is $\mathcal{S} \in \mathbb{R}^{r_1 \times r_2 \times r_3}$. It follows that the tensor $\hat{\mathcal{T}}$ has rank-(r_1, r_2, r_3). In these algorithms, we use random projections on matrices which are based on the work of Halko et al. [11]. We also prove a sharper probabilistic error bound found in [11] for the matrix case. Then, this error bound was applied to error bounds in the tensor case. The tensor extension of these randomized projections was first proposed by Mahoney et al. [19].

In addition, we demonstrate the efficacy of the proposed randomized algorithms with MRI data compression. The numerical experiments are compared to tensor based methods for compression (e.g. quasi-Newton methods on Grassmannian manifolds [23]). We restrict our comparison to tensor based methods for three-dimensional datasets even though there are well-known methods for compression based on wavelet analysis that has been successful for two-dimensional data. In the work of Wu et al. [27], they have shown that the higher-order SVD within the hierarchical tensor framework has some advantages over wavelet analysis for compression. The advantages are the following: (a) it can achieve far higher quality than wavelet transform at large compression ratios (b) the tensor framework facilitates progressive or partial data transmission and visualization; i.e. the receiver can quickly view the low resolution versions first and decide whether it is worthwhile to wait for higher resolution details.

2 Preliminaries

We denote the scalars in $\mathbb{R}$ with lower-case letters $(\alpha, \beta, \ldots)$ and the vectors with lower-case letters $(a, b, \ldots)$. The matrices are written as upper-case letters $(A, B, \ldots)$ and the symbol for tensors are calligraphic letters $(\mathcal{A}, \mathcal{B}, \ldots)$. The subscripts represent the following scalars: $(\mathscr{A})_{ijk} = a_{ijk}$, $(A)_{ij} = a_{ij}$, $(a)_i = a_i$. The superscripts indicate the length of the vector or the size of the matrices. For example, b^K is a vector with length K and $B^{N\times K}$ is a $N \times K$ matrix. In addition, the lower-case superscripts on a matrix indicate the mode in which it has been matricized. For example, R_n is the mode-n matricization of the tensor $\mathscr{R} \in \mathbb{R}^{I\times J\times K}$ for $n = 1, 2, 3$.

Definition 1 The Kronecker product of matrices A and B is defined as

$$A \otimes B = \begin{bmatrix} a_{11}B & a_{12}B & \dots \\ a_{21}B & a_{22}B & \dots \\ \vdots & \vdots & \ddots \end{bmatrix}.$$

Definition 2 (Mode-*n* Vector) Given a tensor $\mathscr{T} \in \mathbb{R}^{I\times J\times K}$, there are three types of mode vectors, namely, mode-1, mode-2, and mode-3. There are $J \cdot K$ mode-1 vectors that are of length I which are obtained by fixing the indices (j, k) while varying i. Similarly, the mode-2 vector (mode-3 vector) is of length J (K) obtained from the tensor by varying j (k) with fixed (k, i) (i, j).

Definition 3 (Mode-*n* Rank) The mode-n $rank$ of a tensor $\mathscr{T}$ is the dimension of the subspace spanned by the mode-n vectors.

The order of a tensor refers to the cardinality of the index set. A matrix is a second-order tensor and a vector is a first-order tensor.

Definition 4 (Rank-(L,M,N)) A third-order tensor $\mathscr{T} \in \mathbb{R}^{I\times J\times K}$ is rank-(L, M, N) if the mode-1 rank is L, the mode-2 rank is M and the mode-3 rank is N.

In the case when a third-order tensor has rank-$(1, 1, 1)$, it is simply called a *rank-1* tensor.

Definition 5 (Tucker Mode Product) Given a tensor $\mathscr{T} \in \mathbb{R}^{I\times J\times K}$ and the matrices $A \in \mathbb{R}^{\hat{I}\times I}$, $B \in \mathbb{R}^{\hat{J}\times J}$ and $C \in \mathbb{R}^{\hat{K}\times K}$, then the Tucker mode-$n$ products are as follows:

$$(\mathscr{T} \bullet_1 A)_{\hat{i},j,k} = \sum_{i=1}^{I} (\mathscr{T})_{ijk} a_{\hat{i}i}, \;\forall \hat{i}, j, k \quad \text{(mode-1 product)}$$

$$(\mathscr{T} \bullet_2 B)_{\hat{j},i,k} = \sum_{j=1}^{J} (\mathscr{T})_{ijk} b_{\hat{j}j}, \;\forall \hat{j}, i, k \quad \text{(mode-2 product)}$$

$$(\mathscr{T} \bullet_3 C)_{\hat{k},i,j} = \sum_{k=1}^{K} (\mathscr{T})_{ijk} c_{\hat{k}k}, \;\forall \hat{k}, i, j \quad \text{(mode-3 product)}$$

In general, we have the following definition.

Definition 6 Given a kth order tensor $\mathcal{T} \in \mathbb{R}^{n_1 \times n_2 \cdots \times n_k}$ and matrices $U_i \in \mathbb{R}^{\hat{n}_l \times n_l}$, then the Tucker mode-$n_l$ is

$$(\mathcal{T} \bullet_l U_l)_{i_1 i_2 \cdots \hat{i}_l \cdots i_k} = \sum_{i_l} (\mathcal{T})_{i_1 i_2 \cdots i_l \cdots i_k} (U)_{\hat{i}_l i_l}$$

Definition 7 (Matricization) Matricization is the process of reordering the elements of an Nth order tensor into a matrix. The mode-n matricization of a tensor $\mathcal{T} \in \mathbb{R}^{n_1 \times n_2 \times \cdots \times n_k}$ is denoted by T_l and arranges the mode-l fibers to be the columns of the resulting matrix. The mode-l fiber, $t_{n_1 \cdots n_{l-1} : n_{l+1} \cdots n_k}$, is a vector obtained by fixing every index with the exception of the lth index.

If we use a map to express such matricization process for any Nth order tensor $\mathcal{T} \in \mathbb{R}^{n_1 \times n_2 \times \cdots \times n_k}$, that is, the tensor element $(n_1, n_2, \ldots, n_k)$ maps to matrix element (n_l, j), then there is a formula to calculate j:

$$j = 1 + \sum_{\substack{l=1 \\ l \neq k}}^{k} (n_l - 1) J_l \quad \text{with} \quad J_l = \prod_{\substack{m=1 \\ m \neq l}}^{l-1} n_m.$$

Then, given a third-order tensor $\mathcal{X} \in \mathbb{R}^{I \times J \times K}$, the mode-1, mode-2 and mode-3 matricizations of $\mathcal{X}$, respectively, are:

$$\begin{aligned} X_1 &= [x_{:11}, \ldots, x_{:J1}, x_{:12} \ldots, x_{:J2}, \ldots, x_{:1K}, \ldots, x_{:JK}], \\ X_2 &= [x_{1:1}, \ldots, x_{I:1}, x_{1:2} \ldots, x_{I:2}, \ldots, x_{1:K}, \ldots, x_{I:K}], \\ X_3 &= [x_{11:}, \ldots, x_{I1:}, x_{12:} \ldots, x_{I2:}, \ldots, x_{1J:}, \ldots, x_{IJ:}]. \end{aligned} \tag{2}$$

3 Theoretical Error Bounds

In this section, we introduce a randomization technique for calculating low rank matrices. It is well known from Eckart-Young Theorem [8] that the low rank k matrix approximation is attained from calculating the leading first k rank-one terms in the SVD; i.e.

$$\operatorname{argmin}_{\operatorname{rank}(\hat{A}) \leq k} \|A - \hat{A}\|_F = U_k \Sigma_k V_k^T = \hat{A}$$

where $\operatorname{rank}(\hat{A}) \leq k$.

Recent results [11] show that the randomized versions of classical numerical linear algebra techniques give fast, efficient and accurate algorithms. Here we build on the theoretical error bounds found in [11] which will be essential for the error bounds in the tensor case.

Table 1 Fixed low matrix rank [11]

Input: A, rank k, oversampling parameter p
Output: $Q \in \mathbb{R}^{n \times k+p}$
Draw a random $n \times k + p$ test matrix Ω
Form the matrix product $Y = A\,\Omega$
Compute a QR: $[Q, R]$=qr($A\Omega$)

Table 2 Randomized SVD [11]

Input: A, orthogonal matrix Q
Output: orthogonal matrices U, V and diagonal matrix Σ; i.e $A = U\Sigma V^T$
Form $B = Q^T A$
Compute a small SVD: $[\hat{U}, \Sigma, V^T]$=svd(B)
Set $U = Q\hat{U}$

The goal is to create an algorithm for SVD with randomness. For simplicity, we fix a specific low rank k in mind. Given a matrix $A \in \mathbb{R}^{n \times m}$, a desired rank k, and an oversampling parameter p, we want to construct $Q \in \mathbb{R}^{n \times k+p}$ with orthonormal columns such that

$$\|A - QQ^T A\| \approx \min_{\text{rank}(\hat{A}) \leq k} \|A - \hat{A}\|.$$

Essentially, we are constructing a random orthogonal projection such that the residual $\|A - QQ^T A\| \leq \epsilon$. A Gaussian test matrix Ω is used such that the columns of $A\Omega$ are orthonormalized; i.e. $A\Omega = QR$. See Table 1. Thus, Q captures the first k columns as the left singular vectors of A. Then, the matrix $B = Q^T A$ of size $k + p \times m$ is formed and factored into its SVD; i.e. $B = \hat{U}\Sigma V^T$. It follows that $A = QB = Q\hat{U}\Sigma V^T = U\Sigma V^T$ where $U = Q\hat{U}$. SVD is performed on a smaller matrix B of size $k + p \times m$ as opposed to A of size $n \times m$. See Table 2 for the randomized SVD.

Now we discuss a deterministic error bound for calculating the range space of **A**. Let A be an $m \times n$ that has an SVD of $A = U\ \Sigma\ V^T$; i.e.

$$A = U \begin{bmatrix} \Sigma_1 & \\ & \Sigma_2 \end{bmatrix} \begin{bmatrix} V_1^T \\ V_2^T \end{bmatrix} \tag{3}$$

where Σ_1 is an $k \times k$ matrix and Σ_2 is an $n - k \times n - k$ matrix. Let Ω be an $n \times l$ test matrix where $l \geq k$ in the coordinate system determined by the right polar decomposition of A via

$$\Omega_1 = V_1^T \Omega \quad \text{and} \quad \Omega_2 = V_2^T \Omega. \tag{4}$$

A deterministic bound is the following:

Theorem 1 ([11]) *Let A be an $m \times n$ matrix with SVD $A = U\Sigma V^T$ (3). Let Ω be a test matrix and construct the matrix $Y = A\Omega$ with Ω_1 and Ω_2 are defined in (4). Assume Ω_1 has full row rank, then*

$$\|A - QQ^T A\|_F^2 \leq \|\Sigma_2\|_F^2 + \|\Sigma_2 \Omega_2 \Omega_1^\dagger\|_F^2.$$

Let V_s be an $n \times s$ matrix whose entries are i.i.d. $N(0,1)$ random variables and let $M_s = \frac{1}{s} V_s V_s^T$. The random matrix M_s is called the Wishart matrix.

Theorem 2 ([9]) *Let $M_s = \frac{1}{s} V_s V_s^T$ where V_s is an $n \times s$ matrix with i.i.d. entries with $\sigma_{ij} \sim N(0,1)$. For $0 < \frac{n}{s} < \infty$, the largest singular values of M_s converges a.s. to $\left(1 + \sqrt{\frac{n}{s}}\right)$ as $s \to \infty$.*

Theorem 3 ([25]) *Let $M_s = \frac{1}{s} V_s V_s^T$ where V_s is an $n \times s$ matrix with i.i.d. entries with $\sigma_{ij} \sim N(0,1)$. For $0 < \frac{n}{s} < 1$, the smallest singular values of M_s converges a.s. to $\left(1 - \sqrt{\frac{n}{s}}\right)$ as $s \to \infty$.*

Here we improve the theoretical bound in Theorem 1.

Theorem 4 *Let A be an $m \times n$ matrix with SVD $A = U\Sigma V^T$ (3). Let Ω be an $n \times l$ matrix with i.i.d. entries with $a_{ij} \sim N(0,1)$ and construct the matrix $Y = A\ \Omega$ with Ω_1 and Ω_2 are defined in (4). Assume Ω_1 has full row rank, then*

$$\|A - QQ^T A\|_F^2 \leq \left(1 + \frac{1 + \sqrt{\frac{n-k}{l}}}{1 - \sqrt{\frac{k}{l}}}\right) \|\Sigma_2\|_F^2$$

as $l \to \infty$ where $0 < \frac{n-k}{l} < \infty$ and $0 < \frac{k}{l} < 1$.

Proof Theorem 1 states this bound:

$$\|A - QQ^T A\|_F^2 \leq \|\Sigma_2\|_F^2 + \|\Sigma_2 \Omega_2 \Omega_1^\dagger\|_F^2.$$

We now calculate a bound for $\|\Sigma_2 \Omega_2 \Omega_1^\dagger\|_F^2$. The following is true:

$$\|\Sigma_2 \Omega_2 \Omega_1^\dagger\|_F^2 = \operatorname{tr}((\Omega_1^\dagger)^T \Omega_2^T \Sigma_2^2 \Omega_2 \Omega_1^\dagger) = \operatorname{tr}(\Omega_2^T \Sigma_2^2 \Omega_2 \Omega_1^\dagger (\Omega_1^\dagger)^T).$$

Let $P_1 = \Omega_1^\dagger (\Omega_1^\dagger)^T$ and $P_2 = \Omega_2 \Omega_2^T$ with orthonormal bases $\{\phi_j\}$ and $\{\psi_k\}$ [20], respectively. Then

$$\begin{aligned} \operatorname{tr}(\Omega_2^T \Sigma_2^2 \Omega_2 P_1) = \operatorname{tr}\langle I, \Omega_2^T \Sigma_2^2 \Omega_2 P_1 \rangle &= \sum_i \langle \phi_i, \Omega_2^T \Sigma_2^2 \Omega_2 P \phi_i \rangle \\ &\leq \sum_i \|P\|_2 \langle \phi_i, \Omega_2^T \Sigma_2^2 \Omega_2 \phi_i \rangle \leq \|P_1\|_2 \operatorname{tr}\langle I, \Omega_2^T \Sigma_2^2 \Omega_2 \rangle \end{aligned}$$

$$\leq \|P_1\|_2 \operatorname{tr}\langle I, \Sigma_2^2 \Omega_2 \Omega_2^T \rangle \leq \|P_1\|_2 \sum_i \langle \psi_k, \Sigma_2^2 P_2 \psi_i \rangle$$

$$\leq \|P_1\|_2 \|P_2\|_2 \sum_i \langle \psi_k, \Sigma_2^2 P_2 \psi_i \rangle \leq \|P_1\|_2 \|P_2\|_2 \|\Sigma_2\|_F^2$$

Using Theorem 2, the largest singular value of P_2 converges almost surely to $\sqrt{l}\left(1+\sqrt{\frac{n-k}{l}}\right)$ as $l \to \infty$ for $0 < \frac{n-k}{l} < \infty$. Similarly, using Theorem 3, the largest singular value of P_1 is the reciprocal of the smallest singular value of $\Omega_1 \Omega_1^T$; i.e. the largest singular value of P_1 converges almost surely to $\frac{1}{\sqrt{l}\left(1-\sqrt{\frac{k}{l}}\right)}$ as $l \to \infty$ for $0 < \frac{k}{l} < 1$. Thus,

$$\begin{aligned}\|A - QQ^T A\|_F^2 &\leq \|\Sigma_2\|_F^2 + \|\Sigma_2 \Omega_2 \Omega_1^\dagger\|_F^2 \\ &\leq (1 + \|P_1\|_2 \|P_2\|_2) \|\Sigma_2\|_F^2 \\ &\leq \left(1 + \frac{1+\sqrt{\frac{n-k}{l}}}{1-\sqrt{\frac{k}{l}}}\right) \|\Sigma_2\|_F^2\end{aligned}$$

as $l \to \infty$ where $0 < \frac{n-k}{l} < \infty$ and $0 < \frac{k}{l} < 1$. □

This is a probabilistic bound which sharpens the result of [11] and [1]. The deterministic bound of [11] is an improvement on the result of [1]. In the next section, we will apply Theorem 4 for the error bounds in the tensor case.

4 Low Multilinear Rank Tensor Approximation

In this section, we will describe two low multilinear rank tensor approximations as well apply the probabilistic theoretical bounds. An extension of SVD to tensor is called the multilinear SVD [4] (or Higher-order SVD). Recall the Tucker mode-n_l (Definition 6). Given a kth order tensor $\mathcal{T} \in \mathbb{R}^{n_1 \times n_2 \cdots \times n_k}$ and matrices $U_i \in \mathbb{R}^{\hat{n}_l \times n_l}$ for $i = 1, \cdots$, then the Tucker mode-n_l is

$$(\mathcal{T} \bullet_l U_l)_{i_1 i_2 \cdots \hat{i}_l \cdots i_k} = \sum_{i_l} \mathcal{T}_{i_1 i_2 \cdots i_l \cdots i_k} U_{\hat{i}_l i_l}$$

Definition 8 (Multilinear SVD) A third order tensor $\mathcal{T} \in \mathbb{R}^{n_1 \times n_2 \cdots \times n_k}$ can be factored into a product of a core third order tensor and three orthogonal matrices; i.e.

$$\mathcal{T} = \mathcal{S} \bullet_1 U_1 \bullet_2 U_2 \cdots \bullet_l U_l \cdots \bullet_k U_k$$

where $\mathscr{S} \in \mathbb{R}^{n_1 \times n_2 \cdots \times n_k}$ is the core tensor and $U_i U_i \in \mathbb{R}^{n_l \times n_l}$ are orthogonal matrices. The core tensor satisfies:

- an all-orthogonality constraint for each mode l: $\langle \mathscr{S}_{i_1 i_2 \cdots i_l = \alpha \cdots i_k}, \mathscr{S}_{i_1 i_2 \cdots i_l = \beta \cdots i_k} \rangle = (\sigma^l)^2 \delta_{\alpha,\beta}$ with $\alpha, \beta = 1, \cdots, n_l$
- ordering

$$\|S_{i_l=1}\|_F \geq \|S_{i_l=2}\|_F \geq \|S_{i_l=3}\|_F \geq \cdots \geq \|S_{i_l=n_l}\|_F$$

where $S_{i_l=\alpha} = \mathscr{S}_{i_1 i_2 \cdots i_l = \alpha \cdots i_k}$ and $\|S_{i_l=\alpha}\|_F = \sigma^l_\alpha$ for $\alpha = 1, \cdots, n_l$.

A kth order tensor has k set of singular values $\{\sigma^l_\alpha\}$ and singular vectors $(U_l)_\alpha$ for $\alpha = 1, \cdots, n_l$.

Definition 9 (Low Multilinear SVD) A given third order tensor $\mathscr{T} \in \mathbb{R}^{n_1 \times n_2 \cdots \times n_k}$ can be factored into a product of a core third order tensor and three orthogonal matrices; i.e.

$$\mathscr{T} = \mathscr{S} \bullet_1 U_1 \bullet_2 U_2 \cdots \bullet_l U_l \cdots \bullet_k U_k$$

where $\mathscr{S} \in \mathbb{R}^{r_1 \times r_2 \cdots \times r_k}$ is the core tensor and $U_i \in \mathbb{R}^{n_l \times r_l}$ are orthogonal matrices. The core tensor satisfies

- an all-orthogonality constraint for each mode l: $\langle \mathscr{S}_{i_1 i_2 \cdots i_l = \alpha \cdots i_k}, \mathscr{S}_{i_1 i_2 \cdots i_l = \beta \cdots i_k} \rangle = (\sigma^l)^2 \delta_{\alpha,\beta}$ with $\alpha, \beta = 1, \cdots, r_l$
- ordering

$$\|S_{i_l=1}\|_F \geq \|S_{i_l=2}\|_F \geq \|S_{i_l=3}\|_F \geq \cdots \geq \|S_{i_l=n_l}\|_F$$

where $S_{i_l=\alpha} = \mathscr{S}_{i_1 i_2 \cdots i_l = \alpha \cdots i_k}$ and $\|S_{i_l=\alpha}\|_F = \sigma^l_\alpha$ for $\alpha = 1, \cdots, r_l$.

A kth order tensor has k set of singular values $\{\sigma^l_\alpha\}$ and singular vectors $(U_l)_\alpha$ for $\alpha = 1, \cdots, r_l$.

There are methods for computing (low) multilinear SVD [4, 5, 26]. One approach is to flatten (Definition 2) the tensor and apply matrix SVD. A third order tensor $\mathscr{T}$ is matricized into $T_1 = U_1 S_1 (U_2 \otimes U_3)^T$, $T_2 = U_2 S_2 (U_3 \otimes U_1)^T$ and $T_3 = U_3 S_3 (U_1 \otimes U_2)^T$. To obtain orthogonal matrices U_i, each matrix T_i is decomposed into its SVD. The low rank core tensor $\mathscr{S} \in \mathbb{R}^{r_1 \times r_2 \times r_3}$ is built via $\mathscr{S} = \mathscr{T} \bullet_1 U_1^T \bullet_2 U_2^T \bullet_3 U_3^T$. Low rank orthogonal matrices are constructed by taking the first r_i columns of $U_i \in \mathbb{R}^{n_i \times r_i}$. Moreover, the randomized SVD in Table 1 which is based on random projections can applied to each T_i. See Table 3 and Fig. 1. In the next section, we describe an improvement of this method.

Table 3 Randomized multilinear SVD (or rand HOSVD)

Input: $\mathcal{T} \in \mathbb{R}^{n_1 \times n_2 \times n_3}$, rank $-$ (r_1, r_2, r_3), oversampling parameter p_1, p_2, p_3
Output: orthogonal matrices U_l and diagonal tensor $\mathcal{S}$ in $\mathcal{T} = \mathcal{S} \bullet_1 U_1 \bullet_2 U_2 \bullet_3 U_3$
Reshape $\mathcal{T}$ into matrices T_1, T_2, T_3
$U_1 \leftarrow randsvd(T_1, r_1, p_1)$
$U_2 \leftarrow randsvd(T_2, r_2, p_2)$
$U_3 \leftarrow randsvd(T_3, r_3, p_3)$
Form $\mathcal{S} = \mathcal{T} \bullet_1 U_1^T \bullet_2 U_2^T \bullet_3 U_3^T$

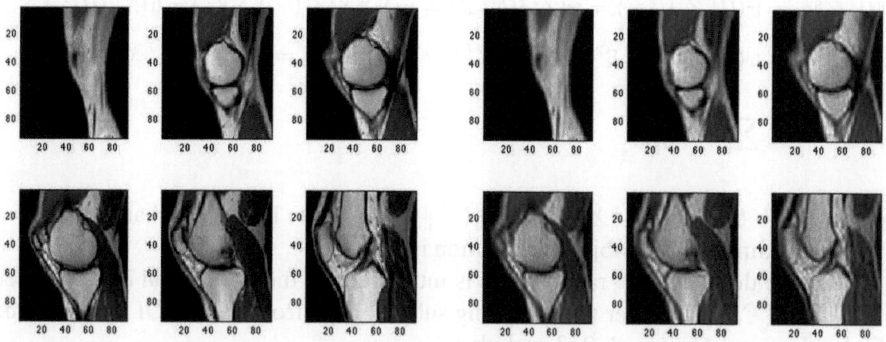

Fig. 1 Randomized HOSVD Example. (*left*) Six frames of the original data of size 92 × 92 × 26. (*right*) Reconstructed data using Randomized HOSVD with low multilinear rank-(35, 35, 26)

4.1 A Randomized Multilinear Orthogonal Iteration (MOI)

The HOOI method is based on minimizing the Frobenius norm squared of the residual:

$$\min_{U_1,U_2,U_3} \|\mathcal{T} - \mathcal{S} \bullet_1 U_1 \bullet_2 U_2 \bullet_3 U_3\|_F^2 = \min_{U_1,U_2,U_3} \|\mathcal{T}\|_F^2 - 2\langle \mathcal{S}, \mathcal{S}\rangle + \|\mathcal{S}\|_F^2$$
$$= \max_{U_1,U_2,U_3} \|\mathcal{S}\|_F^2$$

where $\mathcal{S} = \mathcal{T} \bullet_1 U_1^T \bullet_2 U_2^T \bullet_3 U_3^T$. In general, the orthogonal matrices are obtained through

$$\max_{U_1,\cdots,U_k} \|\mathcal{T} \bullet_1 U_1^T \bullet_2 U_2^T \cdots \bullet_k U_k^T\|_F^2$$

The implementation of this maximization problem is an alternating maximization of the matricized subproblems:

$$\max_{U_l} \|U_l^T Z\|_F^2 \tag{5}$$

where $Z = S_l(U_1^T \otimes U_2^T \cdots \otimes U_{l-1}^T \otimes U_{l+1}^T \cdots U_k)$. Orthogonal U_l is attained by setting the leading r_l singular vectors of Z [10]. Here we show an alternative proof in the spirit of Regalia [22] on how the maximum is attained in (5); i.e.

$$\max_{U_l} \|U_l^T Z\|_F^2$$

where $Z = S_l(U_1^T \otimes U_2^T \cdots \otimes U_{l-1}^T \otimes U_{l+1}^T \cdots U_k)$.

Assume Z has an SVD; i.e. $Z = X \Sigma Y^T$. We calculate

$$\begin{aligned}
\|U_l^T Z\|_F^2 &= \mathrm{tr}\langle U_l^T Z, U_l^T Z\rangle = \mathrm{tr}(Z^T U U^T Z) = \mathrm{tr}(Y \Sigma X^T U U^T X \Sigma Y^T) = \mathrm{tr}(X^T U U^T X \Sigma^2) \\
&= \mathrm{tr}\langle X^T U U^T X, \Sigma^2\rangle = \mathrm{tr} \sum_{i,j} (X^T U U^T X)_{ij} (\Sigma^2)_{ji} = \sum_i (X^T U U^T X)_{ii} (\sigma(\Sigma^2))_i \\
&\leq \sum_i (\sigma(\Sigma^2))_i
\end{aligned}$$

since $\|X^T U U^T X\|_2 \leq \|X^T\|_2 \|U\|_2 \|U^T\|_2 \|X\|_2 \leq 1$. If follows that when $U = X$, the maximum of the objective function is attained.

Here we describe how randomness is introduced to the orthogonal iteration. Let $\mathcal{T} \in \mathbb{R}^{n_1 \times n_2 \times n_3}$. Consider the following subproblems from the HOOI formulation: find orthogonal Q_i, $i = 1, 2, 3$ such that

$$\begin{aligned}
\|T_1 - Q_1 Q_1^T T_1\|_F^2 &= \min_{\mathrm{rank}(U_1) \leq r_1} \|T_1 - U_1 S_1 (U_2 \otimes U_3)^T\|_F^2 \\
\|T_2 - Q_2 Q_2^T T_2\|_F^2 &= \min_{\mathrm{rank}(U_2) \leq r_2} \|T_2 - U_2 S_2 (U_3 \otimes U_1)^T\|_F^2 \\
\|T_3 - Q_3 Q_3^T T_3\|_F^2 &= \min_{\mathrm{rank}(U_3) \leq r_3} \|T_3 - U_3 S_3 (U_1 \otimes U_2)^T\|_F^2
\end{aligned}$$

Thus, we can iteratively calculate $Q_i^{(k)}$ and $U_i^{(k)}$ until a stopping criteria is satisfied. Moreover, we calculate the theoretical bounds for each mode. We call this method randomized Multilinear Orthogonal Iteration (randomized MOI).

Theorem 5 *Let $T_l \in \mathbb{R}^{n_l \times n_1 \cdots n_{l-1} n_{l+1} \cdots n_k}$ be a matricization of a kth-order tensor with SVD $T_l = U_l \Sigma (U_1 \otimes \cdots U_{l-1} \otimes U_{l+1} \cdots \otimes U_k)^T$ (3). Let Ω_l be an $N_l \times L_l$ matrix with i.i.d. entries with $a_{ij} \sim N(0,1)$ and construct the matrix $Y_l = T_l\, \Omega$ with Ω_1 and Ω_2 are defined in (4) where $N = n_1 \cdots n_{l-1} n_{l+1} \cdots n_k$. Assume Ω_1 has full row rank, then*

$$\|T_l - Q_l Q_l^T T_l\|_F^2 \leq \left(1 + \frac{1 + \sqrt{\frac{N_l - r_l}{L_l}}}{1 - \sqrt{\frac{r_l}{L_l}}}\right) \|\Sigma_2\|_F^2$$

as $L_l \to \infty$ where $0 < \frac{N_l - r_l}{L_l} < \infty$ and $0 < \frac{r_l}{L_l} < 1$.

Remark 1 It follows that we can bound (1) but the theorem above; i.e.

$$\|\mathscr{T} - \hat{\mathscr{T}}\|_F^2 \le \left(1 + \frac{1 + \sqrt{\frac{N_l - r_l}{L_l}}}{1 - \sqrt{\frac{r_l}{L_l}}}\right) \|\Sigma_2\|_F^2$$

for any mode l since $\|\mathscr{T} - \hat{\mathscr{T}}\|_F^2 = \|T_l - Q_l Q_l^T T_l\|_F^2$ by rearrangements of elements.

5 Numerical Examples

In the first two numerical experiments, the knee MRI dataset was obtained from OsiriX [7]. The first numerical experiment is the implementation of the randomized HOSVD (rand HOSVD) described in Table 3. The dataset of size $92 \times 92 \times 26$ is compressed with a core tensor of size $35 \times 35 \times 26$. Here for each orthogonal matrix U_i, we took the first r_i columns of U_i where $r_1 = 35, r_2 = 35, r_3 = 26$. The calculated errors are $\|S_1\|_F = 1333.3$, $\|S_2\|_F = 58.2$ and $\|S_3\|_F = 5.1$. See Fig. 1. Here $l_1 = l_2 = 85$ and $p_1 = p_2 = 50$.

The second experiment is the implementation of the randomized MOI in Table 4. The dataset of size $92\times92\times26$ is compressed with a core tensor of size $35\times35\times26$. Here for each orthogonal matrix U_i, we took the first r_i columns of U_i where $r_1 = 35, r_2 = 35, r_3 = 26$. The calculated errors are $\|S_1\|_F = 5.1$, $\|S_2\|_F = 5.1$ and $\|S_3\|_F = 5.1$ after two iterations. See Fig. 2. Here $l_1 = l_2 = 85$ and $p_1 = p_2 = 50$.

The third experiment is a comparison study of multilinear rank reduction using HOOI, randomized HOOI and Quasi-Newton [23]. In Fig. 3, we find a low multilinear rank of (3,3,3) from a tensor $\mathscr{A}$ of size $5 \times 5 \times 5$ while having a

Table 4 Randomized multilinear orthogonal iteration (randomized MOI)

Input: $\mathscr{T} \in \mathbb{R}^{n_1 \times n_2 \times n_3}$, rank – (r_1, r_2, r_3), oversampling parameter p_1, p_2, p_3
Output: orthogonal matrices U_l and diagonal tensor $\mathscr{S}$ in $\mathscr{T} = \mathscr{S} \bullet_1 U_1 \bullet_2 U_2 \bullet_3 U_3$
Reshape $\mathscr{T}$ into matrices T_1, T_2, T_3
for $k = 1, \cdots, MAXit$
$Z_1 \leftarrow T_1(U_2^{k-1} \otimes U_3^{k-1})$
$U_1^k \leftarrow randsvd(Z_1, r_1, p_1)$
$Z_2 \leftarrow T_2(U_3^{k-1} \otimes U_1^k)$
$U_2^k \leftarrow randsvd(Z_2, r_2, p_2)$
$Z_3 \leftarrow T_3(U_1^k \otimes U_3^k)$
$U_3^k \leftarrow randsvd(Z_3, r_3, p_3)$
Form $\mathscr{S} = \mathscr{T} \bullet_1 U_1^T \bullet_2 U_2^T \bullet_3 U_3^T$
end

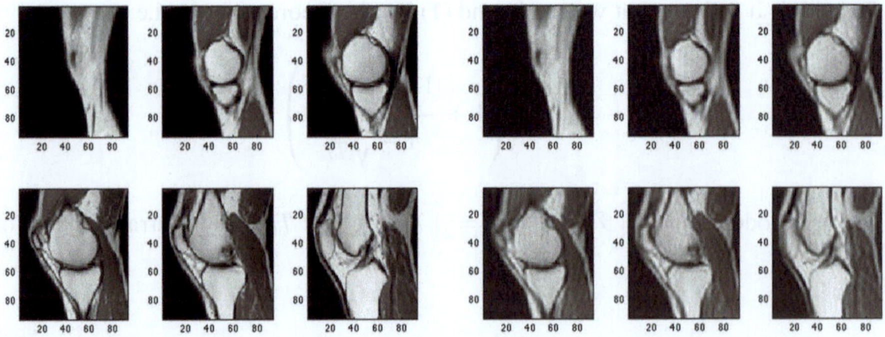

Fig. 2 Randomized HOOI Example. (*left*) Six frame of the original data of size 92 × 92 × 26. (*right*) Reconstructed data using Randomized HOOI with low multilinear rank-(35, 35, 26)

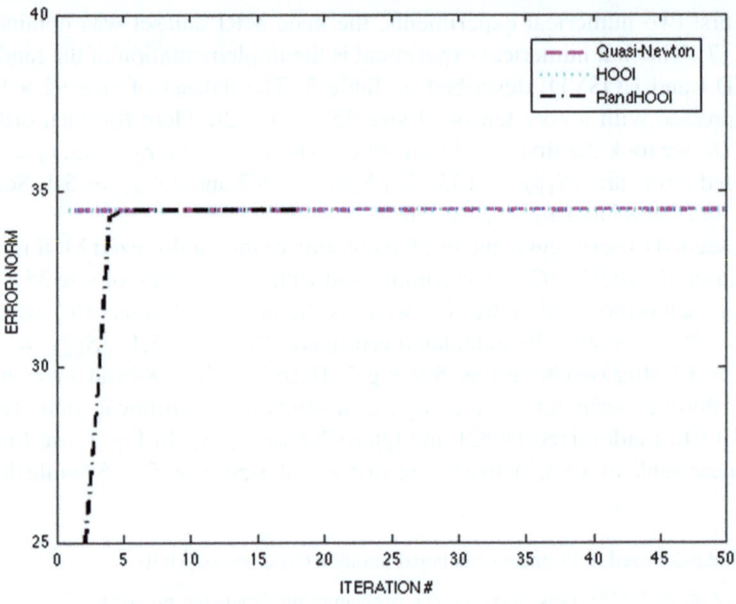

Fig. 3 Comparison of HOOI, randomized MOI and quasi-Newton methods

stopping criteria of a maximum number of iterations of 100 and the error norm, $\|U_3 A_3 (U_1 \otimes U_2)\|_F^2$, to be within 10^{-6}. HOOI maxed out at 100 iterations and Quasi-Newton required 99 iterations while Randomized MOI needed 18 iterations. For the randomized MOI, we take $L_3 = 5$ (with an oversampling parameter of $p = 2$) and the desired rank of $r_3 = 3$.

Acknowledgements C.N. would like to thank Shannon Starr for pointing out some important references. We would also like to thank the reviewers for their valuable suggestions and comments.

References

1. Boutsidis, C., Mahoney, M.W., Drineas, P.: An improved approximation algorithm for the column subset selection problem. In: Proceedings of the 20th Annual ACM-SIAM Symposium on Discrete Algorithms (SODA), pp. 968–977 (2009)
2. Brazell, M., Li, N., Navasca, C., Tamon, C.: Solving multilinear systems via tensor inversion. SIAM J. Matrix Anal. Appl. **34**(2), 542–570 (2013)
3. Comon, P., Jutten, C.: Handbook of Blind Source Separation: Independent Component Analysis and Applications. Academic, New York (2010)
4. De Lathauwer, L., De Moor, B., Vandewalle, J.: A multilinear singular value decomposition. SIAM J. Matrix Anal. Appl. **21**, 1253–1278 (2000)
5. De Lathauwer, L., De Moor, B., Vandewalle, J.: On the best rank-1 and rank-$(R_1, R_2, \ldots, R_N)$ approximation of higher-order tensors. SIAM J. Matrix Anal. Appl. **21**, 1324–1342 (2000)
6. De Silva, V., Lim L.-H.: Tensor rank and the ill-posedness of the best low-rank approximation problem. SIAM J. Matrix Anal. Appl. **30**(3), 1084–1127 (2008)
7. DICOM sample image sets repository. http://www.osirix-viewer.com, http://pubimage.hcuge.ch:8080/ (2013–2014)
8. Eckart, C., Young, G.: The approximation of one matrix by another of lower rank. Psychometrika **1**(3), 211–218 (1936)
9. Geman, S.: A limit theorem for the norm of random matrices. Ann. Probab. **8**, 252–261 (1980)
10. Golub, G., Van Loan, C.: Matrix Computations. Johns Hopkins University Press, Baltimore (1996)
11. Halko, N., Martinsson, P.G., Tropp, J.A.: Finding structure with randomness: probabilistic algorithms for constructing approximate matrix decompositions. SIAM Rev. **53**(2), 217–288 (2011)
12. Hillar, C.J., Lim, L.-H.: Most tensor problems are NP-hard. J. ACM **60**(6), 39 pp. (2013). Article 45
13. Jiao, F., Gur, Y., Johnson, C.R., Joshi, S.: Detection of crossing white matter fibers with high-order tensors and rank-k decompositions. In: Proceedings of the International Conference on Information Processing in Medical Imaging (IPMI 2011). Lecture Notes in Computer Science, vol. 6801, pp. 538–549 (2011). doi: 10.1007/978-3-642-22092-044
14. Kreimer, N., Sacchi, M.D.: A tensor higher-order singular value decomposition for prestack seismic data noise reduction and interpolation. Geophysics **77**(3), V113–V122 (2012)
15. Kressner, D., Tobler, C.: Low-rank tensor Krylov subspace methods for parametrized linear systems. approximation of higher-order tensors. SIAM J. Matrix Anal. Appl. **32**(4), 1288–1316 (2011)
16. Kolda, T.G.: Orthogonal tensor decompositions. Approximation of higher-order tensors. SIAM J. Matrix Anal. Appl. **23**, 243–255 (2001)
17. Kolda, T., Bader, B.: Tensor decompositions and applications. SIAM Rev. **51**(3), 455–500 (2009)
18. Li, N., Hopke, P., Pramod, K., Cliff, S., Zhao, Y., Navasca, C.: Source apportionment of time- and size-resolved ambient particulate matter. Chemom. Intell. Lab. Syst. **129**, 15–20 (2013)
19. Mahoney, M.W., Maggioni, M., Drineas, P.: Tensor-CUR decompositions for tensor-based data. SIAM J. Matrix Anal. Appl. **30**(2), 957–987 (2008)
20. Metha, M.L.: Random Matrices. Elsevier, Amsterdam (2004)
21. Ponnapalli, S.P., Saunders, M.A., Van Loan, C.F., Alter, O.: A higher-order generalized singular value decomposition for comparison of global mRNA expression from multiple organisms. Public Libr. Sci. **6**(12) (2011). Article e28072. doi:10.1371/journal.pone.0028072
22. Regalia, P.A.: Monotonically convergent algorithms for symmetric tensor approximation. Linear Algebra Appl. **438**(2), 875–890 (2013)
23. Savas, B., Lim, L.-H.: Quasi-Newton methods on Grassmannians and multilinear approximations of tensors. SIAM J. Sci. Comput. **32**(6), 3352–3393 (2010)

24. Schultz, T., Seidel, H.-P.: Estimating crossing fibers: a tensor decomposition approach. IEEE Trans. Vis. Comput. Graph. **14**(6), 1635–1642 (2008)
25. Silverstein, J.: The smallest eigenvalue of a large dimensional Wishart matrix. Ann. Probab. **13**(4), 1364–1368 (1985)
26. Tucker, L.R.: Some mathematical notes on three-mode factor analysis. Psychometrika **31**, 279–311 (1966)
27. Wu, Q., Xia, T., Chen, C., Lin, H.-Y.S., Wang, H., Yu, Y.: Hierarchical tensor approximation of multidimensional visual data. IEEE Trans. Vis. Comput. Graph. **14**(1), 186–199 (2008)

Part II
Processing, Filtering and Interpolation

Path-Based Mathematical Morphology on Tensor Fields

Jasper J. van de Gronde, Mikola Lysenko, and Jos B.T.M. Roerdink

Abstract Traditional path-based morphology allows finding long, approximately straight, paths in images. Although originally applied only to scalar images, we show how this can be a very good fit for tensor fields. We do this by constructing directed graphs representing such data, and then modifying the traditional path opening algorithm to work on these graphs. Cycles are dealt with by finding strongly connected components in the graph. Some examples of potential applications are given, including path openings that are not limited to a specific set of orientations.

1 Introduction

An image can be considered as a function whose argument is a position. However, with increasing computing power and increasing sensor capabilities, it is becoming more and more common to acquire data that varies as a function of both position *and* orientation. For example: orientation scores (and similar schemes) [9, 15–17, 29, 32], flow fields, diffusion MRI data [28, 48] and seismic anisotropy data [40]. When filtering such data this additional structure can be taken into account to extract more relevant data.

Path openings [23, 24] are a class of morphological filters that are able to extract long and thin structures. They essentially allow preserving pixels that are part of a long path, while suppressing pixels that are only part of short paths (this allows these filters to be interpreted as hyperconnected filters [53]). Applied to traditional binary or greyscale images this can be useful for finding cracks, roads, fibres, and other thin elongated structures [10, 34, 37, 45, 49].

J.J. van de Gronde • J.B.T.M. Roerdink (✉)
Johann Bernoulli Institute for Mathematics and Computer Science, University of Groningen, P.O. Box 407, 9700 AK Groningen, The Netherlands
e-mail: j.j.van.de.gronde@rug.nl; j.b.t.m.roerdink@rug.nl

M. Lysenko
Department of Computer Sciences, University of Wisconsin - Madison, 1210 W. Dayton St., Madison, WI 53706-1613, USA
e-mail: mikolalysenko@gmail.com

I. Hotz, T. Schultz (eds.), *Visualization and Processing of Higher Order Descriptors for Multi-Valued Data*, Mathematics and Visualization,
DOI 10.1007/978-3-319-15090-1_6

In this work we motivate why the concept of a path is natural to consider in the context of (symmetric) tensor fields. We then explain how path openings traditionally work on directed *acyclic* graphs, and how they can be generalized to allow for cycles. This leads to an efficient algorithm, which we use to demonstrate that our method can indeed be used to produce sensible results on (tensorial) orientation scores and diffusion MRI data.

1.1 Related Work

Certain fibre tractography methods already use a graph-based approach [6, 25, 44, 47]. However, these are typically undirected graphs, and are primarily used for finding tracts (using *shortest* paths), and not for filtering the data. The use of directed graphs in this work allows us to prevent a path from doubling-up on itself unnecessarily, while still allowing very efficient filtering of the *entire* data set. Still, the graph building methods employed by these tractography methods could be a source of inspiration.

Some of the work on mathematical morphology on tensor fields by Duits et al. [18] also considered dilations along what can be considered as "paths" following the local orientation. But rather than using path openings, more traditional dilations and erosions were used (it is not immediately clear whether these operations could be used to implement path openings). And instead of using a graph-based approach, a PDE-based approach was used. Franken and Duits [20] applied a similar approach to orientation scores (although without focussing on morphological operators).

Bismuth et al. [4] developed a method to at least partially deal with curved features in a close relative of path openings, by iteratively linking up approximately linear segments. However, they only support a limited number of changes of direction, and still rely on being able to define suitable directed acyclic graphs. In contrast, here we simply forgo the requirement that the graph is acyclic. In addition, our method could be used to link up approximately linear segments with different orientations too, without the need to explicitly build chains of a certain length. It is not immediately clear how some of their other refinements could translate to our work though.

The technique recently developed by Morard et al. [38] could in theory also be adapted to compute path openings on tensor fields, as it simply finds a number of (potential) paths and then filters along those using a 1D algorithm (instead of implicitly filtering along *all* paths). However, we would still need a suitable method for finding paths. Also, it is not yet clear how well it approximates the "true" path opening.

2 Definitions and Notation

2.1 *Mathematical Morphology and Openings*

Mathematical morphology is a framework for non-linear image processing based on (algebraic) lattices [22]. A lattice is a partially ordered set such that every two elements have a uniquely defined least upper bound and a uniquely defined greatest lower bound. These are called the join/supremum ('$\vee$') and meet/infimum ('$\wedge$'), respectively. Typically, one works with lattices of images, using a partial order that compares images in a pixel-wise fashion: an image is less than or equal to another image if every pixel value in it is less than or equal to the corresponding pixel value in the other image. We will work mostly with the lattice of graphs, in which one graph is less than or equal to another graph if it is a subgraph of the other graph.

Central in the theory of mathematical morphology is a categorization of operators based on the properties they have. For example, an operator is called an erosion if it distributes over taking the meet ($\varepsilon(a \wedge b) = \varepsilon(a) \wedge \varepsilon(b)$). Another important class of operator is the *opening*. An (algebraic) opening is an operator from a lattice to another (possibly the same) lattice that is [22, 41]:

increasing	a smaller input implies a smaller output,
anti-extensive	the output is less than or equal to the input, and
idempotent	the output is a fixed point of the operator.

So more input means more output, you cannot get more out than you put in, and applying an opening twice does the same as applying it once. If an opening depends on some parameter to control its "strength", it is often possible to produce an intermediate data structure called an *opening transform* that contains information on which pixels to keep for what parameter values.

2.2 *Tensors*

We consider a tensor field to be an image whose pixel values are tensors based on the (local) tangent space. For example, even though a vector can be considered a tensor, we do not consider a colour image to describe a tensor field, as a colour vector does not describe a direction in the image domain.

We will only concern ourselves with real, symmetric tensors based on (tangent spaces of) Euclidean spaces. Most crucially, the tensors are based on some vector space $\mathcal{V}$ with an inner product '$\cdot$'. We consider symmetric tensors to be built from the symmetrized tensor product '$\odot$' [7, 31], so a rank-r, degree-n (or order-n) tensor $\mathbf{A}$ is any tensor that can be written as a weighted sum of r (but not less than r) tensors of the form [11]

$$\mathbf{a}^{\odot n} = \underbrace{\mathbf{a} \odot \cdots \odot \mathbf{a}}_{n \text{ times}}, \text{ with } \mathbf{a} \in \mathcal{V}.$$

The terms in such a weighted sum form a so-called "rank-one decomposition" of a tensor, analogous to the eigendecomposition of a symmetric matrix. Due to its linearity, the inner product on tensors is fully determined by

$$\mathbf{a}^{\odot n} \cdot \mathbf{b}^{\odot n} = (\mathbf{a} \cdot \mathbf{b})^n.$$

The *identity tensor* $\mathbf{I}_n$ is the unique degree-n symmetric tensor that satisfies $\mathbf{I}_n \cdot \mathbf{a}^{\odot n} = \|\mathbf{a}\|^n$, with n even. See our previous work [21, 50] for details.

2.3 Graphs

A directed graph G can be identified with a pair (V, E) containing a set of vertices and a set of edges, with all of the edges being pairs of vertices. A graph is a subgraph of another graph if both its vertex set and its edge set are subsets of the vertex set and the edge set, respectively, of the other. A cycle is a sequence of edges of the form $(v_1, v_2), (v_2, v_3), \ldots, (v_n, v_1)$ (that is, if an edge ends at vertex v then the next edge starts at vertex v, and the first and last vertex are also the same). If a directed graph does not contain any cycles it is called a directed acyclic graph, or DAG. The vertices in a directed acyclic graph can always be ordered in such a way that if there is an edge from a vertex to another vertex, then this other vertex comes after the first in the ordering. Such an ordering is called a topological ordering of the graph.

3 From Tensor Fields to Paths in Graphs

Previously [21, 50, 51], we looked at generalizing mathematical morphology to vector-valued images and tensor fields. For this, it was found that it is important to construct rotation-invariant operators. This is in line with some older research [2, 8, 42], as well as the more recent work by Angulo [1] for example. We accomplished this by "lifting" to rotation-invariant representations. For tensor fields, this boiled down to taking a function that maps a position to a tensor (describing a tensor field), and turning it into a function that maps a tangent vector[1] to a scalar (for example: if f maps positions to degree-n tensors, then one could construct a function f' on tangent vectors defined by $f'(x, \mathbf{v}) = f(x) \cdot \mathbf{v}^{\odot n}$, where the pair $(x, \mathbf{v})$ represents a tangent vector). This makes it much easier to apply traditional morphological concepts, as these are typically already suited to scalar images.

[1]A tangent vector can be considered an element in the tangent bundle, and is a *combination* of a position and a vector describing an orientation/direction. Physically, a tangent vector can be considered to describe the position and velocity of a particle, for example.

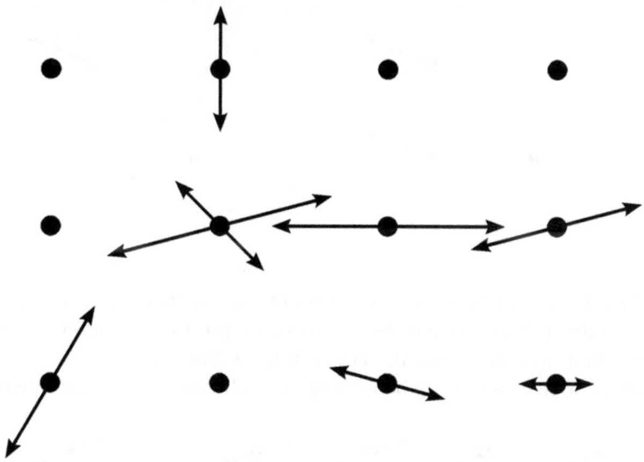

Fig. 1 Illustration of what (spherically deconvolved) diffusion MRI data looks like. There is a regular grid of points, and for each of those points there are zero or more *antipodal pairs* of directions with weights (the *arrow length* corresponds to the weight). Any two antipodal directions have the same weight. The idea is that vectors that roughly line up could be indicative of a fibre bundle running through those vectors

By adapting the theorem that showed the usefulness of lifting to a rotation-invariant representation [51, Thm. 1], we can also consider operations like tensor decomposition to be admissible lifting operators. That is, we can consider the tensor decomposition as a rotation invariant map from the tensor space to a rotation invariant lattice. Although we will not prove this here, this *is* part of our (intuitive) justification for using tensor decompositions for morphological filters on tensor fields.

Once we view a tensor field as a function on tangent vectors—or as a sparse set of (weighted) tangent vectors (see Fig. 1)—it becomes natural to look for paths in the data. This is like looking for streamlines [35] in a flow field, or performing tractography on diffusion tensor fields [3, 14, 36, 39]. More fundamentally, orientations really only make sense if they line up somehow, naturally giving rise to the concept of a path.

In some cases, like gradients of functions, it might be more natural to consider something like a "hyperplane/sheet" opening (extracting surfaces or other structures that are more than one dimensional), but this is somewhat outside the scope of the current work. We do show an example (using orientation scores) in 2D though, where a hyperplane is also a line, so that it still makes sense to use a path opening.

In our examples we build graphs in which the vertices correspond to tangent vectors. There is an edge from a tangent vector to another tangent vector with the same orientation at a neighbouring position if and only if the neighbouring position is within 0.65 times the cell spacing of the *ray* spanned by the first tangent vector. More formally, if we denote the set of all *possible* edges on the current grid by $\mathcal{N}$ (so regardless of which tangent vectors are in any particular data set), then (with unit cell spacing, x_1, x_2 in some Euclidean space, $\mathbf{v}$ a unit vector in the tangent space, and ignoring the weights associated with the tangent vectors):

$$((x_1,\mathbf{v}),(x_2,\mathbf{v})) \in \mathcal{N} \iff (x_2 - x_1)\cdot\mathbf{v} \geq 0$$
$$\text{and } \|(x_2 - x_1) - ((x_2 - x_1)\cdot\mathbf{v})\mathbf{v}\| \leq 0.65.$$

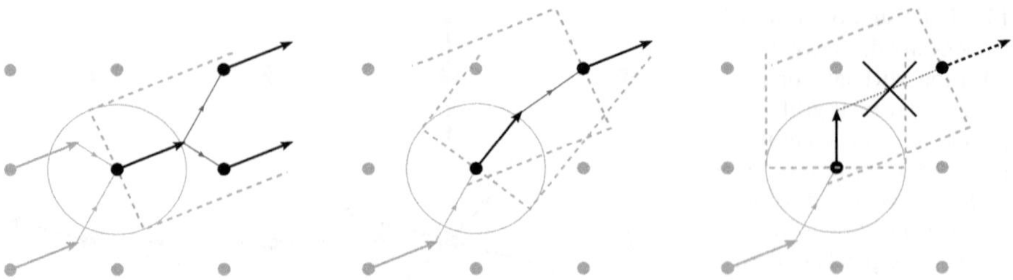

Fig. 2 A tangent vector is connected to another tangent vector with the same direction at a neighbouring position if the neighbouring position is within a set distance from the ray spanned by the first tangent vector. For connecting to different orientations, positions are chosen only if both orientations "agree" on being able to reach/come from each other's position

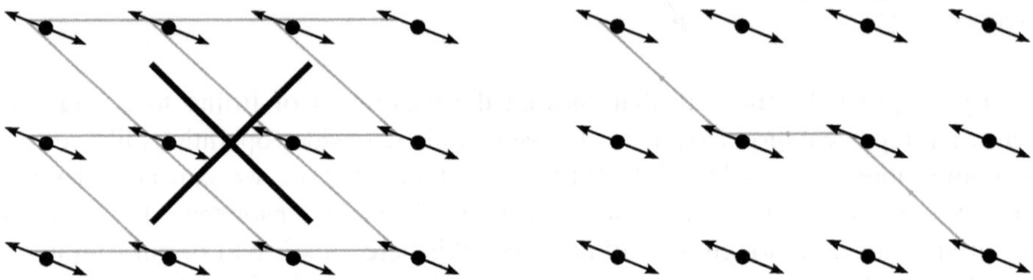

Fig. 3 If we connect vectors to other vectors in a bidirectional way (*left*), we quickly link everything together and lose the idea behind finding paths (the *grey line* could be a single path!). That is why we only connect vectors to other vectors that lie roughly in the direction the vector is pointing to (*right figure* shows an example path). Note that the *grey line* segments connect the midpoints of the *arrows* representing tangent vectors

We also connect tangent vectors to tangent vectors with a different orientation if they "agree" on being able to reach each other's position (see Fig. 2):

$$((x_1, \mathbf{v}_1), (x_2, \mathbf{v}_2)) \in \mathcal{N} \iff ((x_1, \mathbf{v}_1), (x_2, \mathbf{v}_1)) \in \mathcal{N}$$
$$\text{and } ((x_1, \mathbf{v}_2), (x_2, \mathbf{v}_2)) \in \mathcal{N}.$$

If $G = (V, E)$ is the graph we build for a particular data set, with V containing all the tangent vectors, then $E = (V \times V) \cap \mathcal{N}$. See Fig. 3 for why it is important that the above relation is not symmetric.

The above connectivity scheme strikes a balance that seems to work reasonably well in practice. However, it would definitely be interesting to examine more closely what kind of connectivity makes sense for this kind of data. In particular, though it might be tempting to try to fit "trajectories" through tangent vectors to determine the connectivity between neighbouring positions and orientations, in practice the discrete grid (of positions) works against this. Also, a good balance should be struck between being able to track sharp turns and preventing spurious connections.

Note that since in our applications we always have antipodal pairs of tangent vectors, it is useful to have the concept of an "opposite". For each graph $G = (V, E)$ we have a map of opposites $o : V \to V$ that gives the antipodal partner of each vertex. We have: $(v_1, v_2) \in E \iff (o(v_2), o(v_1)) \in E$, for all $v_1, v_2 \in V$.

4 Path Openings on Graphs with Cycles

Traditionally [23, 24], binary path openings are computed by constructing several directed acyclic graphs (DAGs), one for each direction, like in Fig. 4. These graphs are then used to determine the lengths of the longest paths running through each point using a dynamic programming scheme. Said paths are constrained to form a (connected) path in one of the graphs. So one can have an approximately horizontal path, an approximately vertical path, etc. (depending on the number of directions used). Some enhancements have been developed [34], but the basic idea remains the same. This means that path openings are not well suited to extract curved features.

In most work on path openings, weighted graphs are considered, with the weights corresponding to—binary—pixel values. In our setting it is more convenient to simply consider the graph induced by those pixels that are "on". We define a path a in a graph $G = (V, E)$ to be a sequence of vertices such that any pair of consecutive vertices is an edge in E. The set of vertices in a is denoted by $\sigma(a)$, its length by $|a|$, and the set of all paths in G by $\Pi(G)$. A path opening on a directed *acyclic* graph G, with path length threshold L, can then be defined as:

$$\alpha_L(G) = \bigcup \{\sigma(a) \mid a \in \Pi(G) \text{ and } |a| \geq L\}.$$

Introducing cycles in the graph poses a problem: there is no such thing as a *longest* path in the presence of a cycle. One solution is to constrain paths to never visit the same *vertex* twice. However, finding the length of the longest path through each vertex would then be NP-hard (even approximations are hard [5, 27]). Constraining paths to never traverse the same *edge* twice might be simpler in some cases, but it is not immediately clear in what cases, nor how we could (easily) find such paths. Also, if the idea is to find the *longest* (or, rather, largest) path, why should we avoid cycles in the path?

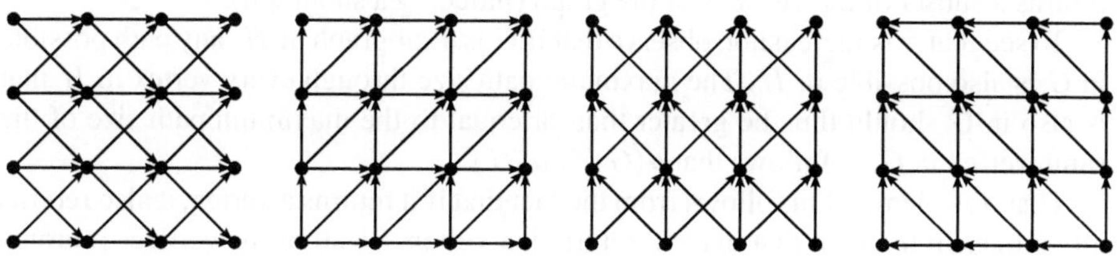

Fig. 4 A set of DAGs that has been used for (traditional) 2D path openings

The key is to focus on path *size* ($|\sigma(a)|$) instead of path *length* ($|a|$). For a DAG, there is no difference, as a vertex can only occur once in a path in a DAG anyway, but in the presence of cycles it can make a big difference. In particular, traversing a cycle multiple times does not make the path larger (in the sense of the size of the set of vertices covered by the path). In the interest of brevity, we will, from now on, say "path size" instead of "the size of the set of vertices covered by a path". So rather than finding longest paths through vertices, we find, for each vertex, the maximum *size* of all paths through that vertex. We thus (re)define the path opening as follows:

$$\alpha_S(G) = \bigcup \{\sigma(a) \mid a \in \Pi(G) \text{ and } |\sigma(a)| \geq S\}. \tag{1}$$

The subscript S will be dropped whenever it is immaterial to the matter at hand. Also, when convenient we equate $\alpha_S(G)$ with the subgraph of $G = (V, E)$ it induces: $(\alpha_S(G), \{(a_1, a_2) \in E \mid a_1, a_2 \in \alpha_S(G)\})$.

Note that the above allows us to apply path openings to oriented data and find curved paths, as we demonstrate in Sect. 5. Also, we give an efficient algorithm to compute α_S in Sect. 4.2. But first we show that α_S is indeed worthy of the title path *opening*.

4.1 α_S is an Algebraic Opening

The above develops a generalization of the traditional path opening, but it is not immediately clear that the result is in fact still an opening in the algebraic sense of the word. That is, it should be increasing, anti-extensive and idempotent. Here we first show that the operator does in fact still have these properties in an abstract graph-based setting, and then go on to discuss how this applies to some more concrete settings.

Theorem 1 *A path opening α on directed graphs is an opening: increasing, anti-extensive and idempotent. In particular, for all directed graphs G and H, if G is a subgraph of H, then $\alpha(G)$ is a subset of $\alpha(H)$ (increasing). Also, $\alpha(G)$ induces a subgraph of G (anti-extensive), and $\alpha(\alpha(G)) = \alpha(G)$ (idempotent), where we equate $\alpha(G)$ with the subgraph of G that it induces.*

Proof We start by observing that α is almost trivially anti-extensive: by definition it returns a subset of the vertices of the graph (inducing a subgraph).

To see that α is increasing, observe that if G is a subgraph of H, any path possible in G is also possible in H. The maximum path size through every vertex in H that is also in G should thus be greater than or equal to the maximum path size of the same vertex in G. It follows that $\alpha(G) \subseteq \alpha(H)$.

That α is idempotent follows from the fact that if it returns a vertex, it also returns all vertices in the path that made it return that vertex. So all of the vertices returned by the path opening will have large enough paths in the induced subgraph to be preserved by a second application of the path opening: $\alpha(\alpha(G)) \supseteq \alpha(G)$. And since we already concluded that α is anti-extensive, implying that $\alpha(\alpha(G)) \subseteq \alpha(G)$, we have that α must be idempotent.

In the above, filtering is done by removing vertices from the graph rather than "turning them off" (as in traditional path openings). However, if we have a graph with boolean weights, then we can simply consider the subgraph that contains only those vertices that are turned on. Since paths do not propagate through vertices that are turned off, this has no effect on the result. Using this trick the above can be interpreted as a(nother) proof that traditional path openings are indeed openings.

When applying the above to orientation scores (see Sect. 5.1), where we use tensor decompositions to compute sparse sets of tangent vectors, it is not clear yet to what extent the whole procedure (decomposition, filtering, reconstruction) can be considered an opening. The main issues are picking a suitable order, and the possible ambiguity of tensor decompositions. Since the path opening itself is anti-extensive, we suspect the Loewner order [8, 50] might be sufficient to prove anti-extensivity, but it is unlikely to be of use for proving increasingness. As for tensors with ambiguous decompositions (analogous to having eigenvalues with multiplicity greater than one), a more detailed study of how often and where they occur would be needed. As it is, we consider this firmly outside the scope of the current work, but look forward to future developments.

4.2 Implementation

The first step in realizing an efficient algorithm to find the maximum path sizes through all vertices, is to observe that all vertices within a cycle share the same longest path size, and that this size is greater than or equal to the cycle size. Also, since for counting the path size it does not matter *where* we enter the loop, we can contract any loop of n vertices into a single vertex that we count as n vertices without affecting the results (see Fig. 5). If we do this with all cycles, then we clearly end up

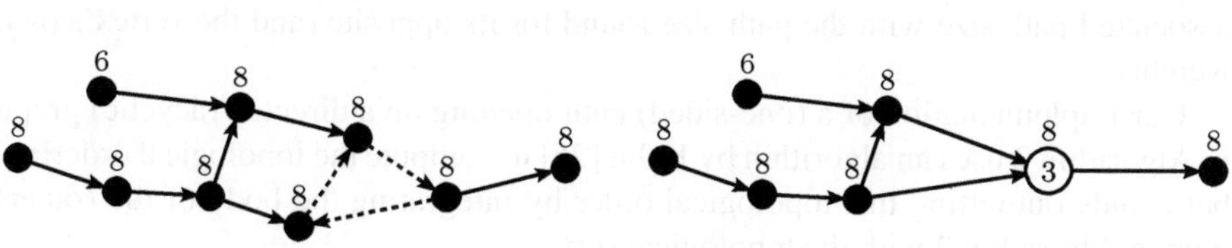

Fig. 5 *Left*: original directed graph containing a cycle (*dashed*), along with the maximum path "lengths" through each vertex. *Right*: same graph after contracting the cycle into a single vertex with weight 3 (rather than 1)

Algorithm 1: Path opening on graphs, allowing for cycles

Input : A directed vertex-weighted graph $G = (V, E, w)$, with $w : V \to \mathbb{R}$, a map of opposites $o : V \to V$, and a threshold S.
Output: A set $A \subseteq V$.

$H = (V', E', w') \leftarrow$ graph of strongly connected components in G
construct $o' : V' \to V'$ such that $v \in v'$ and $u \in u'$ and $o(v) = u \implies o'(v') = u'$
find $\lambda_+ : V' \to \mathbb{R}$ using Algorithm 2
$A \leftarrow \emptyset$
for $v' \in V'$ **do**
  $\lambda \leftarrow \lambda_+(v') + w'(v') + \lambda_+(o'(v'))$
  if $\lambda \geq S$ **then**
    $A \leftarrow A \cup v'$

with a directed acyclic graph, allowing the application of a traditional path opening algorithm.

Can we easily find the graph resulting from contracting all cycles? (Known as the condensation of a graph.) Yes: by finding strongly connected components. A strongly connected component in a directed graph is a (maximal) set of vertices such that there is a path in the graph from every vertex to every other vertex in the set. The set of all vertices in a directed graph can be partitioned into strongly connected components. As all vertices in a cycle clearly belong to the same strongly connected component, contracting the cycle does not change the partition. So after we have contracted all cycles, each vertex of the contracted graph corresponds to a strongly connected component in the original graph. We use Tarjan's algorithm [46] to find all the strongly connected components in a graph in linear time.

Instead of just counting the number of vertices in a path, Algorithm 1 uses the sum of all vertex (and edge) weights associated with the set of vertices (and edges) covered by the path. The weight of a strongly connected component is the sum of the weights of the vertices (and edges) in the component. Also, we only find the sizes of paths that *end* in each vertex (up to, but not including, the weight of the vertex itself). Since in our context the vertices (tangent vectors) come in antipodal pairs, the maximum size of a path through a vertex is then determined by combining its associated path size with the path size found for its opposite (and the vertex's own weight).

Our implementation of a (one-sided) path opening on a directed (acyclic) graph in Algorithm 2 uses an algorithm by Kahn [26] to compute the topological ordering, but avoids outputting this topological order by integrating the body of the (outer) loop in Algorithm 2 with the topological sort.

Algorithm 2: One-sided path opening transform

Input : A directed acyclic vertex-weighted graph $G = (V, E, w)$, with $w : V \to \mathbb{R}$.
Output: A map $\lambda : V \to \mathbb{R}$ giving the maximum lengths of paths ending in each vertex (excluding the weight of the vertex itself).

initialize $\lambda : V \to \mathbb{R}$ to zero
$O \leftarrow$ topological sort of V
for o *in* O **do**
  $l \leftarrow \lambda(o) + w(o)$
  for $s \in V \mid (o, s) \in E$ **do**
    $\lambda(s) \leftarrow \max\{\lambda(s), l\}$

5 Examples

We now apply the above to two types of oriented data: tensorial orientation scores and diffusion MRI data. These proof-of-concepts illustrate how our methods can be applied to oriented data. We use tensor decomposition and spherical deconvolution to obtain sparse sets of tangent vectors and associated weights that together describe the data. We then build graphs on these tangent vectors using simple rules for deciding what vectors can be connected. Finally, we apply the above to filter the data, and show the results.

5.1 *Orientation Scores*

Our first example[2] takes a 2D image and computes an orientation score inspired on work by Duits and Franken [15–17, 19, 20], except that we use a fourth order tensor to represent the orientation score rather than a discrete number of directions based on a discrete approximation of a tensor-valued filter defined in the Fourier domain by $F(\xi) = \frac{\xi^{\odot n}}{\|\xi\|^n}$. We then use the (symmetric) higher order power method [30] in combination with multiplicative update rules for non-negative least squares [43] to find a decomposition of the tensors into positively weighted rank-one tensors and the identity tensor (in as far as this is possible). Since it proved slightly simpler to extract positive components from a tensor than negative components, we ignore negative components; our example has a dark background and light foreground to ensure that the main features are represented by positive components. We also ignore the weight of the identity tensor, as we are only interested in oriented features. The compound effect of these measures is shown in Fig. 6.

Having found directions and weights, we rotate the directions by 90° so they point along the edge/line, and turn them into antipodal pairs. For building a graph

[2]Code available at http://bit.ly/1zpfIXf.

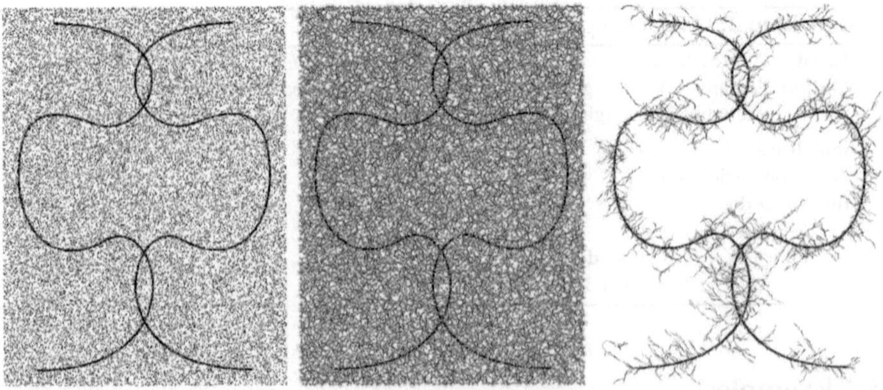

Fig. 6 From *left* to *right*: the original image; the reconstruction from the vectors and values extracted from the orientation tensors (ignoring any negatively weighted components, as well as the identity component); the reconstruction after path opening with a threshold of 400 pixels. Since the image dimensions are 387 × 517, this threshold is way too large for a traditional path opening to have picked up any part of the curves. Note that for the path opening the weights are ignored (and kept unchanged). Grey values have been inverted, so the filter was run on an image with a dark background and light curve

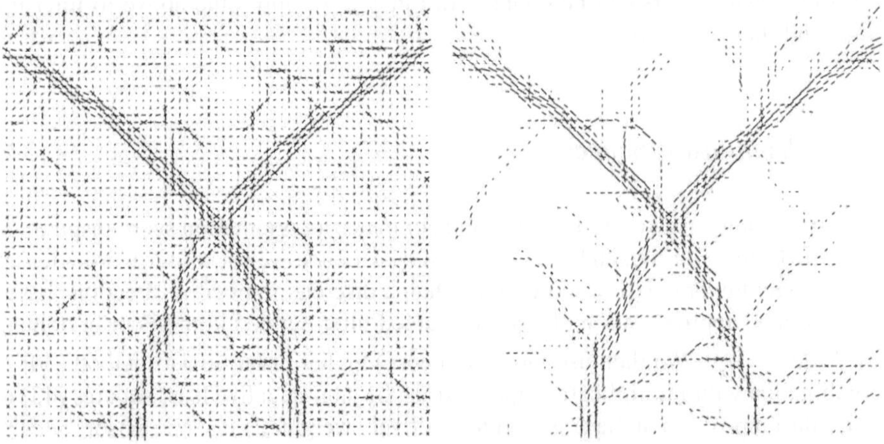

Fig. 7 *Left*: the original orientations found (near the top-most crossing in Fig. 6). *Right*: the output of the path opening. The orientations are shown as small line segments extending symmetrically from the center of each pixel (the length corresponding to the weight)

on the orientation score data, we use the rules already discussed in Sect. 3. For this example we use one additional constraint though: we only connect tangent vectors to other tangent vectors if they differ in angle by less than 30°. We compute a path opening on the graph, and then reconstruct the data from the output graph. Figure 7 shows what the intermediate sets of tangent vectors look like.

Figure 6 shows that we can extract a long curved path from a noisy background, even though we use a threshold that is in fact higher than the width of the image, and definitely higher than any of the approximately straight paths that could have been picked up by a traditional path opening. Also, our method is not limited to fixed set of orientations, it works with whatever orientations are found by the tensor decompositions. This is in stark contrast to traditional path openings, which have to process the entire image once per orientation, and require relatively complicated schemes to be able to distinguish between more than eight orientations (in 2D) [52].

5.2 *Diffusion MRI*

Our spherically deconvolved diffusion MRI dataset has a (short) list of directions and associated weights for each point in a regular (although slightly anisotropic) 3D grid. Because of the physics of diffusion, these directions again come in antipodal pairs, with the same weight for both directions. To build a graph on the diffusion MRI data, we use the same rules as explained in Sect. 3.

ExploreDTI [33] was used for motion and eddy currents correction, while StarTrack [12] was used for spherical deconvolution. In principle tensor decomposition could be considered a kind of spherical deconvolution, but for diffusion MRI it is often appropriate to use a different kernel (fibre response function) to deconvolve with. Also, we typically do not start with a tensor, but rather with measurements in many discrete directions. The resulting path opening is demonstrated in Fig. 8. Figure 9 shows the largest strongly connected component.

In this example we remove tangent vectors whose weight is below a threshold. However, at crossings, the values tend to be smaller than in regions with a single

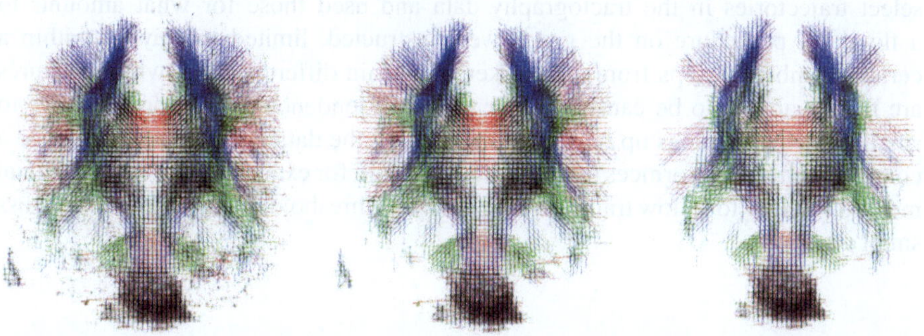

Fig. 8 The result of performing a path opening on spherically deconvolved (and thresholded) diffusion MRI data. From *left* to *right*: the original graph; a path opening with length threshold five; and a path opening with length threshold 100. The most striking differences are visible along the periphery. Path length was measured in voxels; the complete volume contains 128 × 128 × 69 voxels. Interactive version available at http://bit.ly/1zpfIXf

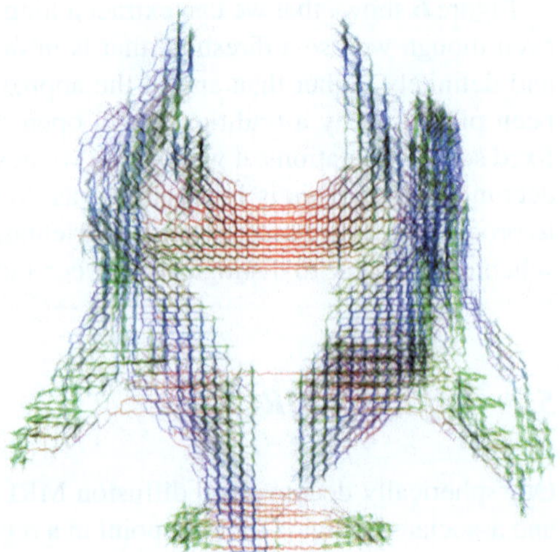

Fig. 9 Largest strongly connected component present in the (thresholded) diffusion MRI data (containing 16,540 elements). Apart from this strongly connected component there was one other of a size of several thousands (corresponding to the noisy blob at the bottom of the data visible in Fig. 8), and two with a size of only four elements. All 668,966 remaining strongly connected components had size one (so were not part of a cycle)

fibre orientation.[3] Our attempt at compensating for this assumes that all directions suffer the same attenuation, and that this factor is equal to the maximum weight for a voxel divided by the sum of all the weights in a voxel (compensating for the fact that vectors come in pairs).

A more traditional way of finding paths/fibres in diffusion MRI is using tractograpy. In a common variant of tractography the diffusion MRI data is used to extract a finite set of trajectories/fibres representing the layout of the actual nerve fibre bundles. Here, tractography was performed using StarTrack [13]. Figure 10 shows some examples of how the constructed graph compares to more traditional tractography. For comparison purposes we took the markers used to select trajectories in the tractography data and used those for what amounts to a flood fill procedure on the graph we constructed, limited to staying within a certain number of steps from the marker. The main differences (in what structures are found) appear to be caused by our method's tendency to be more sensitive to small branches (it picks up on what, when taking the data at face value, could be a connection from the fornices to the corpus callosum for example), and the traditional method's ability to follow trajectories at a low(er) threshold (at least in order to cross small gaps).

[3]This makes some intuitive sense, as every voxel contains the same amount of space, but at a crossing it is shared by fibres of multiple orientations.

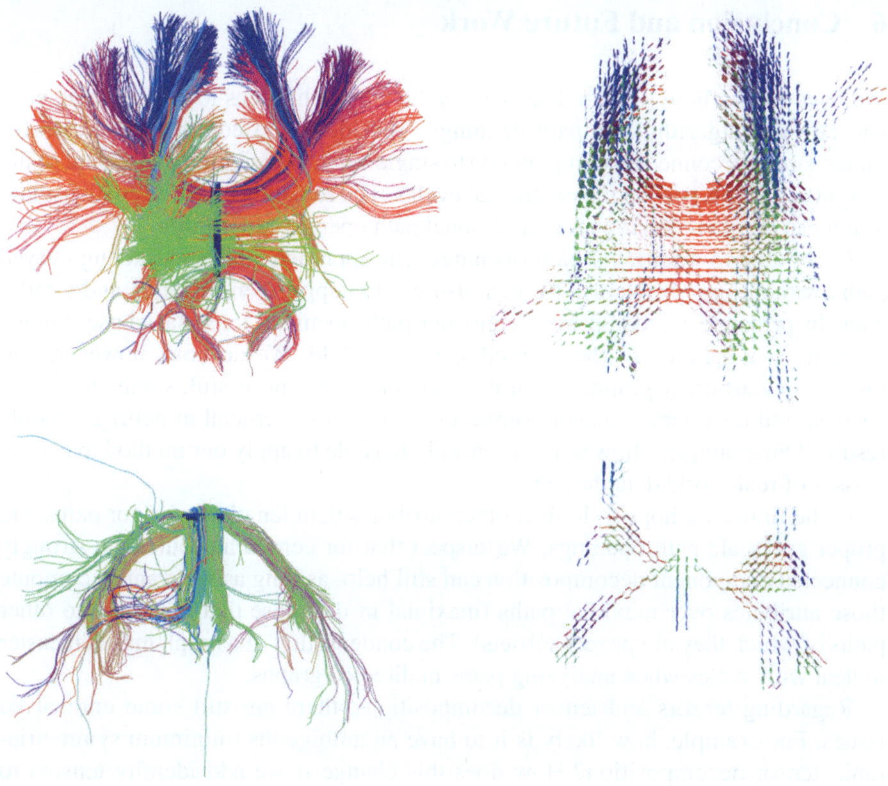

Fig. 10 Tractography of the corpus callosum (*top*) and the fornices (*bottom*), and propagation of the same markers (seedpoints) on the graph we constructed (on the *right*, showing all tangent vectors within 20 steps of the markers). Note that the views are all at roughly the same scale and look onto the anterior of the brain, but that the graph-based images are much sparser due to only plotting tangent vectors at voxel positions (as well lines between connected tangent vectors)

Although our method is perhaps not really suitable as a drop-in replacement for more traditional tractography methods, it does have some interesting characteristics:

- Symmetry of the connections. By construction, if our method finds a path from position and orientation A to position and orientation B, it will also find a reversed path. Tractography methods based on numerically integrating PDEs will tend to miss small branches of larger tracts.
- Output data has the same format as input data (it is just sparser). In fact, as the result of an algebraic opening, the output is a subset of the input.

Also, in principle, we could imagine first thinning the data (within the same kind of framework), and *then* filtering it. If we record the thickness of the bundles, this could then be taken into account during filtering. Essentially this would mean working with bundles rather than paths.

6 Conclusion and Future Work

We have shown how we can deal with cycles in graphs so as to still have an easy and efficient algorithm for path openings. The idea is to contract all cycles (or rather, strongly connected components) to single vertices weighted by the sum of the weights of the vertices in the contracted cycles. The remaining *acyclic* condensation graph can be filtered using more traditional path opening algorithms.

In contrast to traditional path openings, our approach allows extracting curved paths without any problems. We can also easily support tensor fields and similar data. In principle we could even construct path openings for data whose "image domain" is a sphere or other curved space/manifold. We do note, however, that picking any arbitrary graph structure is unlikely to be successful, some amount of sparsity and a tendency to avoid (unnecessary) cycles is crucial in getting sensible results. Our examples show that it is entirely feasible to apply our method to certain classes of real-world data though.

In the future we hope to look at other attributes than length or size for paths and proper greyscale path openings. We suspect that for certain attributes the strongly connected component decomposition can still help, as long as we want to compute those attributes over maximal paths (maximal in the sense that there are no other paths of which they are proper subsets). The condensation of a graph makes it easier to deal with cycles when analyzing paths in directed graphs.

Regarding tensors and tensor decompositions, there are still some unresolved issues. For example, how likely is it to have an ambiguous (minimum symmetric-rank) tensor decomposition? How does this change if we add identity tensors to the possible "basis" tensors? What if we put certain constraints on the tensors? Ultimately, we hope to be able to consider the path openings we have created here as algebraic openings on tensor fields (and not just graphs built on top of those tensor fields), but so far the possibility or impossibility of this is still an open question. In addition, it would be interesting to explore tensor decompositions that can also deal with negative components, for use with orientation scores.

We are also highly interested in looking at *bundles* (paths with a width) rather than paths (without width). In 3D it might also be worthwhile to attempt finding sheets rather than paths. Finally, we would like to develop versions of our path openings that are more robust to noise [10, 45], and experiment with "greyscale" path openings on (decomposed) tensor fields.

Acknowledgements The first and third author were funded by the Netherlands Organisation for Scientific Research (NWO), project no. 612.001.001. Also, we thank Remco Renken and Jelmer Kok from the NeuroImaging Center Groningen for supplying us with the diffusion MRI data (including regions of interest and tractography results), as well as for some inspiring discussions.

References

1. Angulo, J.: Supremum/infimum and nonlinear averaging of positive definite symmetric matrices. In: Nielsen, F., Bhatia, R. (eds.) Matrix Information Geometry, pp. 3–33. Springer, Berlin/Heidelberg (2013). doi:10.1007/978-3-642-30232-9_1
2. Astola, J., Haavisto, P., Neuvo, Y.: Vector median filters. Proc. IEEE **78**(4), 678–689 (1990). doi:10.1109/5.54807
3. Bastiani, M., Shah, N.J., Goebel, R., Roebroeck, A.: Human cortical connectome reconstruction from diffusion weighted MRI: the effect of tractography algorithm. NeuroImage **62**(3), 1732–1749 (2012). doi:10.1016/j.neuroimage.2012.06.002
4. Bismuth, V., Vaillant, R., Talbot, H., Najman, L.: Curvilinear structure enhancement with the polygonal path image - application to guide-wire segmentation in X-ray fluoroscopy. In: Ayache, N., Delingette, H., Golland, P., Mori, K. (eds.) Medical Image Computing and Computer-Assisted Intervention. Lecture Notes in Computer Science, vol. 7511, pp. 9–16. Springer, Berlin/Heidelberg (2012). doi:10.1007/978-3-642-33418-4_2
5. Björklund, A., Husfeldt, T., Khanna, S.: Approximating longest directed paths and cycles. In: Díaz, J., Karhumäki, J., Lepistö, A., Sannella, D. (eds.) Automata, Languages and Programming. Lecture Notes in Computer Science, vol. 3142, pp. 222–233. Springer, Berlin/Heidelberg (2004). doi:10.1007/978-3-540-27836-8_21
6. Booth, B.G., Hamarneh, G.: Multi-region competitive tractography via graph-based random walks. In: Proceedings IEEE Workshop on Mathematical Methods in Biomedical Image Analysis, pp. 73–78 (2012). doi:10.1109/mmbia.2012.6164747
7. Bourbaki, N.: Algebra I. Elements of Mathematics. Springer, Berlin (1989)
8. Burgeth, B., Bruhn, A., Didas, S., Weickert, J., Welk, M.: Morphology for matrix data: ordering versus PDE-based approach. Image Vis. Comput. **25**(4), 496–511 (2007). doi:10.1016/j.imavis.2006.06.002
9. Citti, G., Sarti, A.: A cortical based model of perceptual completion in the Roto-translation space. J. Math. Imaging Vision **24**(3), 307–326 (2006). doi:10.1007/s10851-005-3630-2
10. Cokelaer, F., Talbot, H., Chanussot, J.: Efficient robust d-dimensional path operators. IEEE J. Sel. Top. Signal Process. **6**(7), 830–839 (2012). doi:10.1109/jstsp.2012.2213578
11. Comon, P., Golub, G., Lim, L.H., Mourrain, B.: Symmetric tensors and symmetric tensor rank. SIAM J. Matrix Anal. Appl. **30**(3), 1254–1279 (2008). doi:10.1137/060661569
12. Dell'Acqua, F., Scifo, P., Rizzo, G., Catani, M., Simmons, A., Scotti, G., Fazio, F.: A modified damped Richardson–Lucy algorithm to reduce isotropic background effects in spherical deconvolution. NeuroImage **49**(2), 1446–1458 (2010). doi:10.1016/j.neuroimage.2009.09.033
13. Dell'Acqua, F., Simmons, A., Williams, S.C.R., Catani, M.: Can spherical deconvolution provide more information than fiber orientations? Hindrance modulated orientational anisotropy, a true-tract specific index to characterize white matter diffusion. Hum. Brain Mapp. **34**(10), 2464–2483 (2013). doi:10.1002/hbm.22080
14. Descoteaux, M., Deriche, R., Knosche, T.R., Anwander, A.: Deterministic and probabilistic tractography based on complex fibre orientation distributions. IEEE Trans. Med. Imaging **28**(2), 269–286 (2009). doi:10.1109/tmi.2008.2004424
15. Duits, R.: Perceptual organization in image analysis: a mathematical approach based on scale, orientation and curvature. PhD thesis, Eindhoven University of Technology (2005)
16. Duits, R., Franken, E.: Left-invariant parabolic evolutions on SE(2) and contour enhancement via invertible orientation scores. Part I: Linear left-invariant diffusion equations on SE(2). Q. Appl. Math. **68**(2), 255–292 (2010). doi:10.1090/s0033-569x-10-01172-0
17. Duits, R., Franken, E.: Left-invariant parabolic evolutions on SE(2) and contour enhancement via invertible orientation scores. Part II: Nonlinear left-invariant diffusions on invertible orientation scores. Q. Appl. Math. **68**(2), 293–331 (2010). doi:10.1090/s0033-569x-10-01173-3
18. Duits, R., Dela Haije, T.C.J., Creusen, E., Ghosh, A.: Morphological and linear scale spaces for fiber enhancement in DW-MRI. J. Math. Imaging Vision **46**(3), 326–368 (2013). doi:10.1007/s10851-012-0387-2

19. Franken, E.M.: Enhancement of crossing elongated structures in images. PhD thesis, Eindhoven University of Technology (2008)
20. Franken, E., Duits, R.: Crossing-preserving coherence-enhancing diffusion on invertible orientation scores. Int. J. Comput. Vis. **85**(3), 253–278 (2009). doi:10.1007/s11263-009-0213-5
21. van de Gronde, J.J., Roerdink, J.B.T.M.: Frames for tensor field morphology. In: Nielsen, F., Barbaresco, F. (eds.) Geometric Science of Information. Lecture Notes in Computer Science, vol. 8085, pp. 527–534. Springer, Berlin/Heidelberg (2013). doi:10.1007/978-3-642-40020-9_58
22. Heijmans, H.J.A.M.: Morphological Image Operators. Academic, Boston (1994)
23. Heijmans, H., Buckley, M., Talbot, H.: Path-based morphological openings. In: IEEE International Conference on Image Processing, vol. 5, pp. 3085–3088 (2004). doi:10.1109/icip.2004.1421765
24. Heijmans, H., Buckley, M., Talbot, H.: Path openings and closings. J. Math. Imaging Vision **22**(2), 107–119 (2005). doi:10.1007/s10851-005-4885-3
25. Iturria-Medina, Y., Canales-Rodríguez, E.J., Melie-García, L., Valdés-Hernández, P.A., Martínez-Montes, E., Alemán-Gómez, Y., Sánchez-Bornot, J.M.: Characterizing brain anatomical connections using diffusion weighted MRI and graph theory. NeuroImage **36**(3), 645–660 (2007). doi:10.1016/j.neuroimage.2007.02.012
26. Kahn, A.B.: Topological sorting of large networks. Commun. ACM **5**(11), 558–562 (1962). doi:10.1145/368996.369025
27. Karger, D., Motwani, R., Ramkumar, G.D.S.: On approximating the longest path in a graph. Algorithmica **18**(1), 82–98 (1997). doi:10.1007/bf02523689
28. Kingsley, P.B.: Introduction to diffusion tensor imaging mathematics: Part I. Tensors, rotations, and eigenvectors. Concepts Magn. Reson. **28A**(2), 101–122 (2006). doi:10.1002/cmr.a.20048
29. Knutsson, H., Westin, C.F., Andersson, M.: Structure tensor estimation: introducing monomial quadrature filter sets. In: Laidlaw, D.H., Vilanova, A. (eds.) New Developments in the Visualization and Processing of Tensor Fields. Mathematics and Visualization, pp. 3–28. Springer, Berlin/Heidelberg (2012). doi:10.1007/978-3-642-27343-8_1
30. Kofidis, E., Regalia, P.A.: On the best rank-1 approximation of higher-order supersymmetric tensors. SIAM J. Matrix Anal. Appl. **23**(3), 863–884 (2002). doi:10.1137/s0895479801387413
31. Kostrikin, A.I., Manin, I.I.: Linear Algebra and Geometry. Algebra, Logic and Applications, vol. 1. Gordon and Breach, Amsterdam (1997)
32. Köthe, U.: Edge and junction detection with an improved structure tensor. In: Michaelis, B., Krell, G. (eds.) Pattern Recognition. Lecture Notes in Computer Science, vol. 2781, pp. 25–32. Springer, Berlin/Heidelberg (2003). doi:10.1007/978-3-540-45243-0_4
33. Leemans, A., Jeurissen, B., Sijbers, J., Jones, D.K.: ExploreDTI: a graphical toolbox for processing, analyzing, and visualizing diffusion MR data. In: ISMRM 17th Scientific Meeting & Exhibition, p. 3537 (2009)
34. Luengo Hendriks, C.L.: Constrained and dimensionality-independent path openings. IEEE Trans. Image Process. **19**(6), 1587–1595 (2010). doi:10.1109/tip.2010.2044959
35. McLoughlin, T., Laramee, R.S., Peikert, R., Post, F.H., Chen, M.: Over two decades of integration-based, geometric flow visualization. Comput. Graph. Forum **29**(6), 1807–1829 (2010). doi:10.1111/j.1467-8659.2010.01650.x
36. Melhem, E.R., Mori, S., Mukundan, G., Kraut, M.A., Pomper, M.G., van Zijl, P.C.M.: Diffusion tensor MR imaging of the brain and white matter tractography. Am. J. Roentgenol. **178**(1), 3–16 (2002). doi:10.2214/ajr.178.1.1780003
37. Morard, V., Dokladal, P., Decencière, E.: One-dimensional openings, granulometries and component trees in O(1) per pixel. IEEE J. Sel. Top. Signal Process. **6**(7), 840–848 (2012). doi:10.1109/jstsp.2012.2201694
38. Morard, V., Dokládal, P., Decencière, E.: Parsimonious path openings and closings. IEEE Trans. Image Process. **23**(4), 1543–1555 (2014). doi:10.1109/tip.2014.2303647
39. Mori, S., van Zijl, P.C.M.: Fiber tracking: principles and strategies – a technical review. NMR Biomed. **15**(7–8), 468–480 (2002). doi:10.1002/nbm.781
40. Nolet, G.: A Breviary of Seismic Tomography. Cambridge University Press, New York (2008)

41. Serra, J. (ed.): Theoretical Advances, Image Analysis and Mathematical Morphology, vol. 2. Academic, London (1988)
42. Serra, J.: Anamorphoses and function lattices. In: Dougherty, E.R., Gader, P.D., Serra, J.C. (eds.) Image Algebra and Morphological Image Processing IV, SPIE Proceedings, vol. 2030, pp. 2–11 (1993). doi:10.1117/12.146650
43. Sha, F., Lin, Y., Saul, L.K., Lee, D.D.: Multiplicative updates for nonnegative quadratic programming. Neural Comput. **19**(8), 2004–2031 (2007). doi:10.1162/neco.2007.19.8.2004
44. Sotiropoulos, S.N., Bai, L., Morgan, P.S., Constantinescu, C.S., Tench, C.R.: Brain tractography using Q-ball imaging and graph theory: improved connectivities through fibre crossings via a model-based approach. NeuroImage **49**(3), 2444–2456 (2010). doi:10.1016/j.neuroimage.2009.10.001
45. Talbot, H., Appleton, B.: Efficient complete and incomplete path openings and closings. Image Vis. Comput. **25**(4), 416–425 (2007). doi:10.1016/j.imavis.2006.07.021
46. Tarjan, R.: Depth-first search and linear graph algorithms. SIAM J. Comput. **1**(2), 146–160 (1972). doi:10.1137/0201010
47. Terajima, K., Nakada, T.: EZ-tracing: a new ready-to-use algorithm for magnetic resonance tractography. J. Neurosci. Methods **116**(2), 147–155 (2002). doi:10.1016/s0165-0270(02)00039-0
48. Tournier, J.D., Mori, S., Leemans, A.: Diffusion tensor imaging and beyond. Magn. Reson. Med. **65**(6), 1532–1556 (2011). doi:10.1002/mrm.22924
49. Valero, S., Chanussot, J., Benediktsson, J.A., Talbot, H., Waske, B.: Advanced directional mathematical morphology for the detection of the road network in very high resolution remote sensing images. Pattern Recognit. Lett. **31**(10), 1120–1127 (2010). doi:10.1016/j.patrec.2009.12.018
50. van de Gronde, J.J., Roerdink, J.B.T.M.: Frames, the Loewner order and eigendecomposition for morphological operators on tensor fields. Pattern Recognit. Lett. (2014). doi:10.1016/j.patrec.2014.03.013
51. van de Gronde, J.J., Roerdink, J.B.T.M.: Group-invariant colour morphology based on frames. IEEE Trans. Image Process. **23**(3), 1276–1288 (2014). doi:10.1109/tip.2014.2300816
52. Vincent, L.: Minimal path algorithms for the robust detection of linear features in gray images. In: Heijmans, H.J.A.M., Roerdink, J.B.T.M. (eds.) Mathematical Morphology and Its Applications to Image and Signal Processing, ISMM '98, pp. 331–338. Kluwer Academic Publishers, Norwell, MA (1998)
53. Wilkinson, M.H.F.: Hyperconnectivity, attribute-space connectivity and path openings: theoretical relationships. In: Wilkinson, M.H.F., Roerdink, J.B.T.M. (eds.) Mathematical Morphology and Its Application to Signal and Image Processing. Lecture Notes in Computer Science, vol. 5720, Chap. 5, pp. 47–58. Springer, Berlin/Heidelberg (2009). doi:10.1007/978-3-642-03613-2_5

Processing Multispectral Images via Mathematical Morphology

Andreas Kleefeld and Bernhard Burgeth

Abstract In this chapter, we illustrate how to process multispectral and hyper-spectral images via mathematical morphology. First, according to the number of channels the data are embeded into a sufficiently high dimensional space. This transformation utilizes a special geometric structure, namely double hypersimplices, for further processing the data. For example, RGB-color images are transformed into points within a specific double hypersimplex. It is explained in detail how to calculate the supremum and infimum of samples of those transformed data to allow for the meaningful definition of morphological operations such as dilation and erosion and in a second step top hats, gradients, and morphological Laplacian. Finally, numerical results are presented to explore the advantages and shortcomings of the new proposed approach.

1 Introduction

Mathematical morphology is concerned with the detection, extraction and manipulation of shapes, contours, and structures in image data. Scientists working in this field may look back at almost half a century of successful developments and applications of powerful methods for image processing in, roughly speaking, medical and engineering sciences. Since the path-breaking work of Matheron and Serra [12, 13, 17, 18], a vast amount of literature (e.g. [3, 10, 11, 16, 20–22, 25]) provides testimony of this story of success. The operations of dilation and erosion

A. Kleefeld (✉)
Faculty of Mathematics, Natural Sciences and Computer Science, Brandenburg University of Technology Cottbus, 03046 Cottbus, Germany
e-mail: kleefeld@tu-cottbus.de

B. Burgeth
Faculty of Mathematics and Computer Science, Saarland University, 66041 Saarbrücken, Germany
e-mail: burgeth@math.uni-sb.de

I. Hotz, T. Schultz (eds.), *Visualization and Processing of Higher Order Descriptors for Multi-Valued Data*, Mathematics and Visualization,
DOI 10.1007/978-3-319-15090-1_7

are the cornerstones of mathematical morphology typically applied to gray-scale images endowed with a (complete) lattice structure. However, the recent decade has seen ever rising efforts to develop powerful morphological tools for data types that do not allow for such a lattice structure. There the main focus certainly lies on color images [1, 4, 9, 19, 24], that is, vectorial data, see the excellent survey [2] and the literature cited therein, but, to a much lower extend, on matrix valued images as well (e.g. [8]). The most promising and recent approach to color morphology, presented by van de Gronde and Roerdink [24] relies on the very versatile concept called frames. The main idea of Burgeth and Kleefeld [5, 6] consists of establishing a one-to-one correspondence of a RGB-image with a matrix-valued image of 2×2 real symmetric matrices for which morphological techniques already have been developed, see [7]. This correspondence is inspired by the striking similarity of the Ostwald color bi-cone [14] in the well known HSL-model and the Loewner order cone for 2×2 symmetric matrices.

In essence, a color image is transformed into a matrix valued image, undergoes morphological processing and the result is transformed back into a color image. However, this approach has two disadvantages: One is that it is geared towards images with three channels which is the degree of freedom in a 2×2 symmetric matrix. The other one is that due to its non-polyhedral structure the Loewner bi-cone (e.g. the transformed Ostwald bi-cone) is not stable under taking the Loewner-supremum (or -infimum). Precisely, taking the supremum can lead to an element outside the bi-cone. In this chapter an approach is presented that overcomes both obstacles; it utilizes an embedding into polyhedral double (hyper-)simplexes instead of Loewner cones making them both stable under a sup-, inf-operation (now the operations satisfy the closure condition) and adaptable to an arbitrary number of channels in a multivariate image. The simplicial structure is important since, as it turns out, mappings into other polyhedra entails non-uniqueness of the supremum/infimum of a finite set of multivariate data. For example, draw the two squares with edge length 0.1 with lower left vertex $(0.1, 0.1)$ and $(0.2, 0.1)$, respectively. The smallest enclosing square enclosing the two given squares has edge length 0.2 and the lower left coordinate is given by $(0.1, y)$ with $0 \leq y \leq 0.1$.

Note also that the Loewner order for symmetric matrices has another flaw; it is not associative. In general, it holds $\sup(\sup(A, B), C) \neq \sup(A, \sup(B, C))$, likewise for the inf-operation. This makes the calculation of a supremum/infimum of three or more matrices cumbersome. This disadvantage alone makes it worthwhile to look for an alternative approach such as the double hypersimplex approach.

The chapter is organized as follows. First, it is explained in Sect. 2 how to map the data of a multispectral/hyperspectral image into a suitable space. The space under consideration will be a double hypersimplex. Additionally, the inverse of the map introduced in Sect. 2.1 is constructed in Sect. 2.2. Then, it is illustrated in Sect. 2.3 how to find the supremum and infimum of a set of points located in the double hypersimplex using a geometric approach which is needed to define mathematical morphological operation such as dilation and erosion. In Sect. 2.4, we explain how to subtract two points located in the double hypersimplex to make mathematical morphological operations such as top hats, gradients, and morphological Laplacian meaningful. In Sect. 3, we give a glimpse at scalar-valued morphology operations and explain how to use those operations for multispectral/hyperspectral images. Numerical results are presented for a variety of multispectral and hyperspectral images in Sect. 4. A short summary including possible future research concludes this article in Sect. 5.

2 Mathematical Morphology for Multispectral/Hyperspectral Images

A typical multispectral image consists of data in the form depth×width×height, $n \times w \times h$ for short, where $3 \leq n \leq 10$. For $n > 10$, we call it a hyperspectral image, which can contain as many as 200 (or more) depth information. Note that in the sequel we assume $n \geq 3$. For example, an RGB-image can be considered as a multispectral image of size $3 \times w \times h$. We call a slice in the n-direction of a multispectral/hyperspectral image a *pixel*. Hence, a pixel is a vector of size n containing normalized data; i.e., data in the range $[0, 1]$. If this is not the case, we can always normalize them assuming that the range is finite. Precisely, a multispectral/hyperspectral image has $w \times h$ pixels of the form $[0, 1]^n$. To apply mathematical morphology, we need to define an appropriate supremum and infimum of a given set of pixels for the operations such as dilation and erosion. As a first step, we convert a pixel datum $[0, 1]^n$ (a hypercube, $\square^n$ for short) into a double hypersimplex, $\diamondsuit^n$ for short. As a second step, we define an appropriate supremum and infimum of a set of points located in the double hypersimplex.

2.1 Map from the Hypercube to the Double Hypersimplex

In this subsection, we first illustrate how to convert a point $\mathbf{x} = (x_1, \ldots, x_n) \in \square^n$ into a point $\mathbf{y} = (y_1, \ldots, y_n) \in \Delta^n$, where the hypersimplex is defined as

$$\Delta^n := \left\{ \mathbf{y} \in \mathbb{R}^n \ : \ y_i \geq 0 \,, \quad \sum_{i=1}^{n} y_i \leq 1 \right\}$$

for $n \geq 2$. Define the map $\psi : \square^n \to \mathbb{R}^n$ by

$$\psi(\mathbf{x}) = \left(x_1 \frac{m}{s}, \ldots, x_{n-1} \frac{m}{s}, x_n \frac{m}{s} \right) =: \mathbf{y} \,, \tag{1}$$

where $m = \max\{x_1, \ldots, x_{n-1}, x_n\}$ and $s = x_1 + \ldots + x_{n-1} + x_n$. If $\mathbf{x} = \mathbf{0}$, then we define $\mathbf{y} = \mathbf{0}$. We have $\psi(\square^n) = \Delta^n$. To see this, let $\mathbf{x} \neq \mathbf{0}$ with $x_i \in [0, 1]$, $\forall\, i = 1, \ldots, n$, then clearly $y_i = x_i \frac{m}{s} \geq 0$, since $m \geq 0$ and $s \geq 0$ and we have

$\sum_{i=1}^{n} y_i = \sum_{i=1}^{n} x_i \frac{m}{s} = \frac{m}{s} \sum_{i=1}^{n} x_i = \frac{m}{s} \cdot s = m \leq 1$, since the maximal m is one.

Example 1 Application of the map (1) to the point $(1/2, 1/10, 3/5, 4/5, 1/20) \in \square^5$ yields $(8/41, 16/410, 48/205, 64/201, 4/205) \in \Delta^5$, since $s = 41/20$ and $m = 4/5$.

Next, we define a map from the hypercube to the double hypersimplex $\Diamond^n$ which is given by the union of Δ^n and Δ^n mirrored at the $x_1 - \ldots - x_{n-1}$–plane. The desired map is constructed by transforming the last coordinate by using $2x_n - 1$ and then applying the mapping (1) with $m = \max\{x_1, \ldots, |x_n|\}$ and $s = x_1 + \ldots + |x_n|$. It is noteworthy that we could pick any coordinate position and transform the entry as before, but we decided to change the last coordinate to keep things simple.

Precisely, we have the map $\theta : \square^n \to \mathbb{R}^n$ given by

$$\theta(\mathbf{x}) = \left(x_1 \frac{m}{s}, \ldots, x_{n-1} \frac{m}{s}, (2x_n - 1) \frac{m}{s}\right) =: \mathbf{y}, \tag{2}$$

where $m = \max\{x_1, \ldots, x_{n-1}, |2x_n - 1|\}$ and $s = x_1 + \ldots + x_{n-1} + |2x_n - 1|$. If $\mathbf{x} = (0, \ldots, 0, 1/2)$, then we define $\mathbf{y} = \mathbf{0}$. We clearly have $\psi(\square^n) = \Diamond^n$ by construction.

Example 2 Application of the map (2) to the point $(1/2, 1/10, 3/5, 4/5, 1/20) \in \square^5$ yields $(8/41, 16/410, 48/205, 64/201, -72/205) \in \Diamond^5$, since $s = 41/20$ and $m = 4/5$.

With the two maps (1) and (2) we are able to convert any point located in the hypercube to a point located in the hypersimplex and the double hypersimplex, respectively. The following two examples illustrate the result of the map ψ and θ given by (1) and (2) applied to 1,000 points located in the unit square and unit cube, respectively.

Example 3 In Figs. 1a and 2a, we show 1,000 randomly chosen points in the unit square. The result of the mappings ψ and θ given by (1) and (2) for $n = 2$ applied to those 1,000 points is shown in Figs. 1b and 2b, respectively. As we can see, the 1,000 randomly chosen points from the unit square are mapped into the two-dimensional simplex and into the two-dimensional double simplex, respectively.

Example 4 In Figs. 3a and 4a, we show 1,000 randomly chosen points in the unit cube. The result of the mapping ψ and θ given by (1) and (2) for $n = 3$ applied to those 1,000 points is shown in Figs. 3b and 4b, respectively. As we can see, the 1,000 randomly chosen points from the unit cube are mapped into the three-dimensional simplex and three-dimensional double simplex, respectively.

Now, we are able to convert an *rgb*-datum into a point located in $\Diamond^3$.

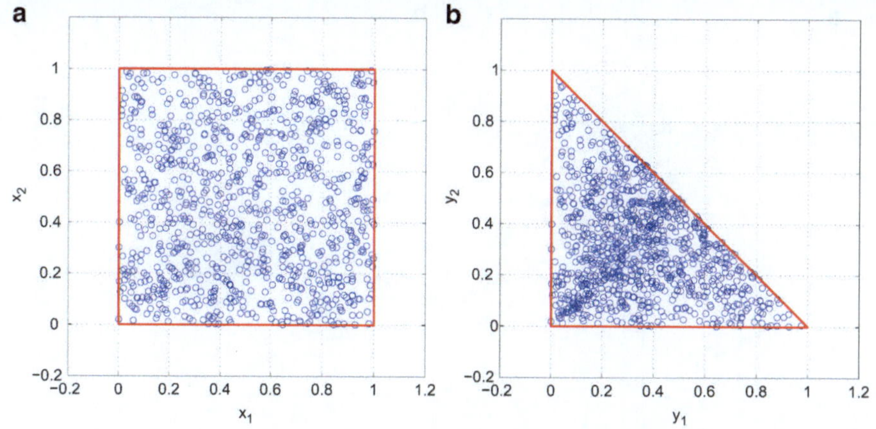

Fig. 1 Thousand randomly chosen points in $\square^2$ and the result of the mapping ψ applied to those 1,000 points. (**a**) 1,000 randomly chosen points in $\square^2$. (**b**) 1,000 points in $\triangle^2$

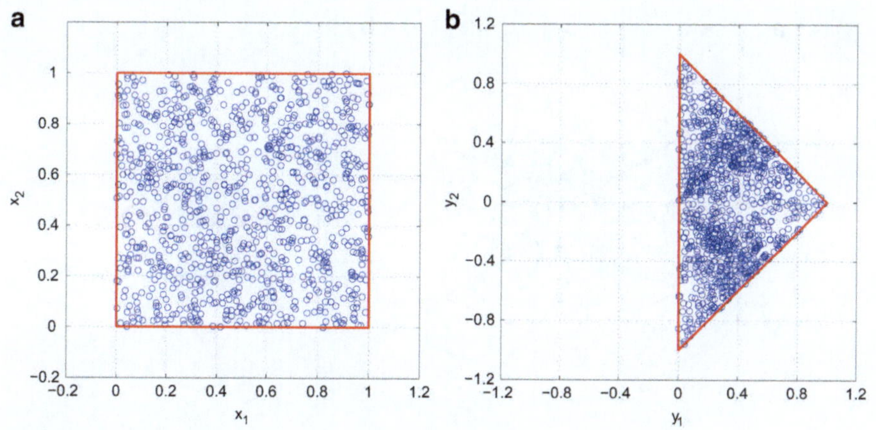

Fig. 2 Thousand randomly chosen points in $\square^2$ and the result of the mapping θ applied to those 1,000 points. (**a**) 1,000 randomly chosen points in $\square^2$. (**b**) 1,000 points in $\diamondsuit^2$

2.2 Map from the Hypersimplex to the Hypercube

In this subsection, we define the inverse of the map ψ given by (1). The map $\psi^{-1} : \triangle^n \to \mathbb{R}^n$ is given by

$$\psi^{-1}(\mathbf{y}) = \left(y_1 \frac{s}{m}, \ldots, y_{n-1} \frac{s}{m}, y_n \frac{s}{m}\right) =: \mathbf{x}\,, \tag{3}$$

where $m = \max\{y_1, \ldots, y_{n-1}, y_n\}$ and $s = y_1 + \ldots + y_{n-1} + y_n$. If $\mathbf{y} = \mathbf{0}$, then we define $\mathbf{x} = \mathbf{0}$. We have $\psi^{-1}(\triangle^n) = \square^n$. To see this, let $\mathbf{x} \in \square^n$, then

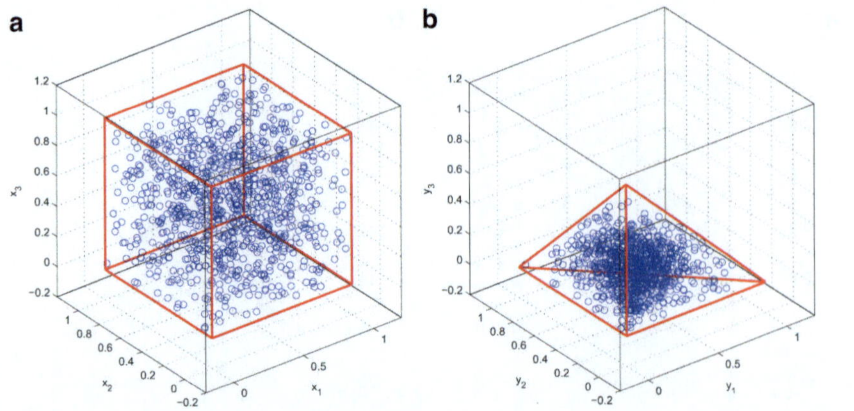

Fig. 3 Thousand randomly chosen points in $\square^3$ and the result of the mapping ψ applied to those 1,000 points. (**a**) 1,000 randomly chosen points in $\square^3$. (**b**) 1,000 points in Δ^3

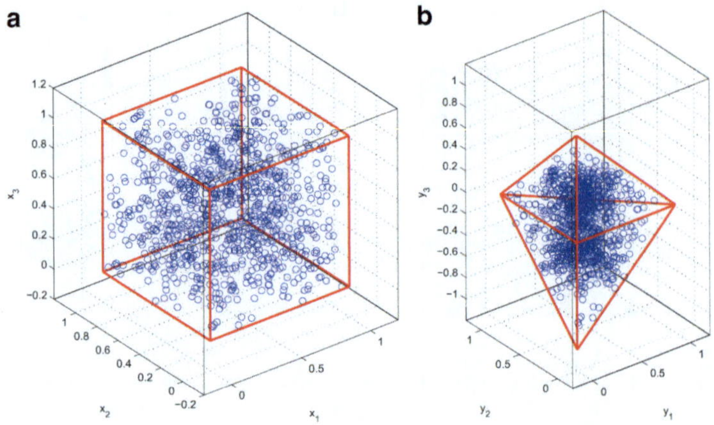

Fig. 4 Thousand randomly chosen points in $\square^3$ and the result of the mapping θ applied to those 1,000 points. (**a**) 1,000 randomly chosen points in $\square^3$. (**b**) 1,000 points in $\diamondsuit^3$

$\mathbf{y} = \psi(\mathbf{x}) \in \Delta^n$; i.e., $y_i = x_i \frac{m}{s} = x_i \frac{\max\{x_1,\ldots,x_{n-1},x_n\}}{x_1+\ldots+x_{n-1}+x_n}$ by definition, which can be rewritten as

$$y_i = x_i \frac{\max\{\frac{m}{s}x_1, \ldots, \frac{m}{s}x_{n-1}, \frac{m}{s}x_n\}}{\frac{m}{s}x_1 + \ldots + \frac{m}{s}x_{n-1} + \frac{m}{s}x_n} = x_i \frac{\max\{y_1, \ldots, y_{n-1}, y_n\}}{y_1 + \ldots + y_{n-1} + y_n}$$

and hence $x_i = y_i \frac{y_1+\ldots+y_{n-1}+y_n}{\max\{y_1,\ldots,y_{n-1},y_n\}}$ for all i.

Similarly, we can write down the inverse of the map θ given by (2). The map $\theta^{-1} : \lozenge^n \to \mathbb{R}^n$ is given by

$$\theta^{-1}(\mathbf{y}) = \left(y_1 \frac{s}{m}, \ldots, y_{n-1} \frac{s}{m}, \frac{1}{2}(y_n + 1)\frac{s}{m} \right) =: \mathbf{x}\,, \tag{4}$$

where $m = \max\{y_1, \ldots, y_{n-1}, |y_n|\}$ and $s = y_1 + \ldots + y_{n-1} + |y_n|$. If $\mathbf{y} = \mathbf{0}$, then we define $\mathbf{x} = (0, \ldots, 0, 1/2)$. We clearly have $\theta^{-1}(\lozenge^n) = \square^n$ by construction.

2.3 *Calculating Suprema and Infima*

In this subsection, we illustrate how to calculate suprema and infima of a set of points located in the double hypersimplex $\lozenge^n$. Generally, it should be noted that one could only define the supremum in the hypersimplex, but to make the calculation of a infimum straightforwardly, we decided to define the supremum in the double hypersimplex.

First, we illustrate how to calculate the supremum of a set of points $\mathscr{P}$ located in Δ^3. Let $\mathbf{x}^{(i)} := (x_1^{(i)}, x_2^{(i)}, x_3^{(i)})$ be the i-th point of the set $\mathscr{P}$. Note that each point is representing a pyramid with corners $E_1^{(i)} = (x_1^{(i)}, x_2^{(i)}, 0)$, $E_2^{(i)} = (x_1^{(i)} + h^{(i)}, x_2^{(i)}, 0)$, $E_3^{(i)} = (x_1^{(i)}, x_2^{(i)} + h^{(i)}, 0)$, and $E_4^{(i)} = (x_1^{(i)}, x_2^{(i)}, h^{(i)})$, where $h^{(i)} = x_3^{(i)}$. Obviously, each base of a pyramid is a triangle with vertices $(x_1^{(i)}, x_2^{(i)})$, $(x_1^{(i)} + h^{(i)}, x_2^{(i)})$, and $(x_1^{(i)}, x_2^{(i)} + h^{(i)})$. We call them base triangles. Note that we list only the first two coordinates, since the base triangles are located in the x_1–x_2–plane (the third coordinate is zero). In Fig. 5a, b, the point $(0.3, 0.4, 0.2)$ representing a pyramid inside Δ^3 (red) is shown in blue and its base triangle (green triangle with blue boundary) located in the $x_1 - x_2$-plane, respectively. Note that the point $(0.3, 0.4, 0.2)$ is the upper vertex of the blue pyramid and uniquely determines it.

The ordering we are considering is induced by the cone determined by the vertex $(0, 0, 0)$ and a base spanned by the three points $(0, 0, 1)$, $(0, 1, 1)$, and $(1, 0, 1)$. This means that the point $(0, 0, 1)$ is larger in this ordering than any other point of Δ^3. Hence, if one considers two points $a, b \in \Delta^3$, then the supremum $\sup(a, b)$ is the upper vertex of the smallest pyramid with triangular base in the $x_1 - x_2$-plane covering those two points. This smallest pyramid can be determined by the base triangles. To this end consider, for example, as above the point $(0.3, 0.4, 0.2)$ as the upper vertex of the small blue pyramid. Such a point determines a pyramid and its base triangle uniquely, and vice versa. We have a one-to-one correspondence between points in Δ^3 and the set of aforementioned base triangles. Now finding this smallest pyramid with triangular base in the $x_1 - x_2$-plane amounts to finding the smallest base triangle enclosing the two base triangles generated by the two points a and b. More general, to obtain the supremum of a set of points a_i with some index set $I \ni i$ boils down to determine the smallest base-triangle enclosing the base-triangles corresponding to the points a_i. In principle the same strategy is applicable in higher dimensions.

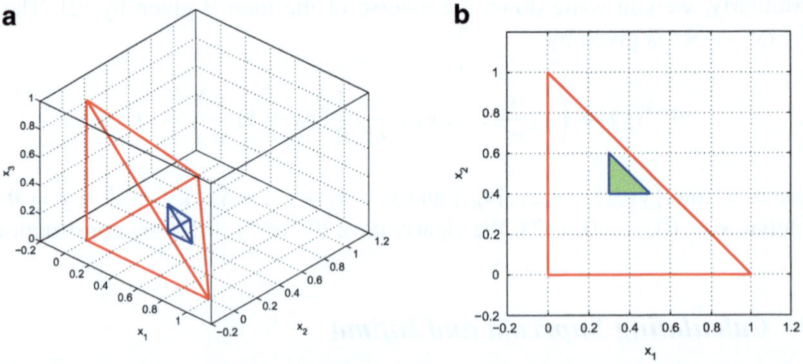

Fig. 5 Point (0.3, 0.4, 0.2) representing a pyramid (*blue*) inside Δ^3 (*red*) and its base triangle (*green triangle with blue boundary*) located in the $x_1 - x_2$-plane, respectively. (**a**) The point (0.3, 0.4, 0.2) representing a pyramid (*blue*) inside Δ^3 (*red*). (**b**) The base of the pyramid (*green triangle with blue boundary*) represented by the point (0.3, 0.4, 0.2) located in the $x_1 - x_2$-plane

Finding the smallest base triangle with vertices $v_{1,\max}$, $v_{2,\max}$, and $v_{3,\max}$ of a set of base triangles is an easy task, since it is an easy geometric problem. The first and second coordinate of $v_{1,\max}$ are given by $x_{1,\min} := \min_i x_1^{(i)}$ and $x_{2,\min} := \min_i x_2^{(i)}$, respectively. The length of the smallest base triangle is given by $l_2 := \max_i \sum_{j=1}^{3} x_j^{(i)} - x_{1,\min} - x_{2,\min}$. To see this, one has to calculate the i-th distance of the line spanned by the two points $(x_1^{(i)} + h^{(i)}, x_2^{(i)})$ and $(x_1^{(i)}, x_2^{(i)} + h^{(i)})$ by using the Hessian normal form and scale the distance by the factor $\sqrt{2}$. The maximum of the i-th distance is the length. Hence, the first and second coordinate of $v_{2,\max}$ are $x_{1,\min} + l_2$ and $x_{2,\min}$, respectively. The first and second coordinate of $v_{3,\max}$ are $x_{1,\min}$ and $x_{2,\min} + l_2$, respectively. It should be clear how to construct the corresponding pyramid. In sum, the supremum is given by the point $(x_{1,\min}, x_{2,\min}, l_2)$. In Fig. 6a, we show in blue the smallest enclosing base triangle given the three base triangles of the pyramids corresponding to the points (0.1, 0.4, 0.3), (0.2, 0.1, 0.1), and (0.6, 0.2, 0.1) in Δ^3. The supremum is given by the point (0.1, 0.1, 0.7). The triangle with edge color red is the maximum possible smallest triangle enclosing a set of triangles. In Fig. 6b, we show the smallest enclosing triangle for a set of five triangles which are represented by five randomly generated points.

Next, we illustrate how to calculate the supremum for a set of points located in Δ^4. Let $\mathbf{x}^{(i)} := (x_1^{(i)}, x_2^{(i)}, x_3^{(i)}, x_4^{(i)})$ be the i-th point of the set $\mathscr{P}$. The base of the i-th point is a pyramid with corners $(x_1^{(i)}, x_2^{(i)}, x_3^{(i)})$, $(x_1^{(i)} + h^{(i)}, x_2^{(i)}, x_3^{(i)})$, $(x_1^{(i)}, x_2^{(i)} + h^{(i)}, x_3^{(i)})$, and $(x_1^{(i)}, x_2^{(i)}, x_3^{(i)} + h^{(i)})$, where $h^{(i)} = x_4^{(i)}$. Note that we are again suppressing the last coordinate, since it zero. Finding the smallest base pyramid with vertices $v_{1,\max}$, $v_{2,\max}$, $v_{3,\max}$, and $v_{4,\max}$ of a set of pyramid works as follows. The first, second, and third coordinate of $v_{1,\max}$ are $x_{1,\min} := \min_i x_1^{(i)}$,

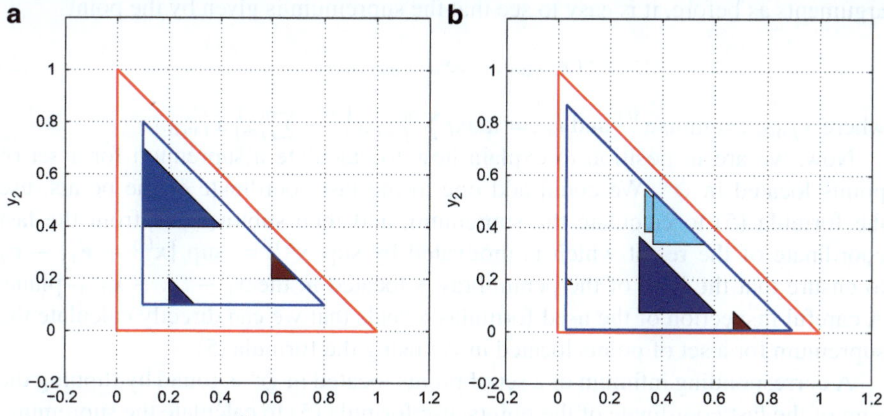

Fig. 6 Smallest enclosing base triangle given a set of base triangles. (**a**) Smallest enclosing base triangle given three base triangles. (**b**) Smallest enclosing base triangle given five base triangles (randomly generated)

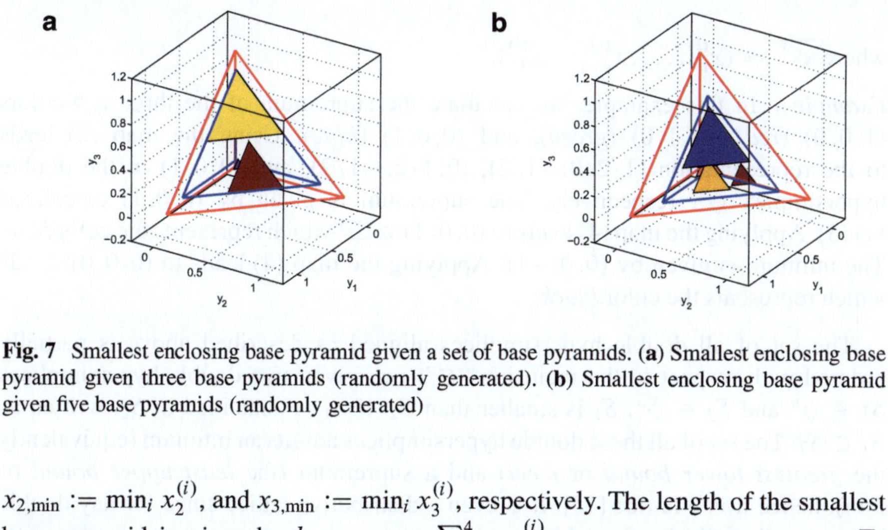

Fig. 7 Smallest enclosing base pyramid given a set of base pyramids. (**a**) Smallest enclosing base pyramid given three base pyramids (randomly generated). (**b**) Smallest enclosing base pyramid given five base pyramids (randomly generated)

$x_{2,\min} := \min_i x_2^{(i)}$, and $x_{3,\min} := \min_i x_3^{(i)}$, respectively. The length of the smallest base pyramid is given by $l_3 := \max_i \sum_{j=1}^{4} x_j^{(i)} - x_{1,\min} - x_{2,\min} - x_{3,\min}$. To verify this, one has to calculate the i-th distance of the plane spanned by the three points $(x_1^{(i)} + h^{(i)}, x_2^{(i)}, x_3^{(i)})$, $(x_1^{(i)}, x_2^{(i)} + h^{(i)}, x_3^{(i)})$, and $(x_1^{(i)}, x_2^{(i)}, x_3^{(i)} + h^{(i)})$ by using the Hessian normal form and scales the distance by the factor $\sqrt{3}$. The maximum of the i-th distance is the length. Hence, the supremum is given by the point $(x_{1,\min}, x_{2,\min}, x_{3,\min}, l_3)$. In Fig. 7a, b, we show the smallest base pyramid enclosing three and five base pyramids, respectively. The points representing the base pyramids are randomly generated.

Finally, we explain how to find the supremum given a set of points located in Δ^n. Let $\mathbf{x}^{(i)} := (x_1^{(i)}, \ldots, x_{n-1}^{(i)}, x_n^{(i)})$ be the i-th point of the set $\mathscr{P}$. Following the arguments as before, it is easy to see that the supremum is given by the point

$$(x_{1,\min}, \ldots, x_{n-1,\min}, l_n), \tag{5}$$

where $x_{j,\min} := \min_i x_j^{(i)}$ and $l_n := \max_i \sum_{j=1}^{n} x_j^{(i)} - \sum_{j=1}^{n-1} x_{j,\min}$.

Now, we are in position to explain how to calculate a supremum for a set of points located in $\lozenge^n$. We could add one to the last coordinate of the points, use the formula (5) to calculate the supremum, and then subtract one from the last coordinate of the result which is motivated by $\sup_i \mathbf{x}^{(i)} = \sup_i\{\mathbf{x}^{(i)} + \mathbf{e}_n\} - \mathbf{e}_n$ to ensure that the base of the penumbras is located in the $x_1 - \ldots - x_{n-1}$–plane. A careful inspection of the used formulas reveals that we can directly calculate the supremum for a set of points located in $\lozenge^n$ using the formula (5).

A corresponding infimum of a set of points located in $\lozenge^n$ is found by flipping the sign of the last coordinate of the points, use formula (5) to calculate the supremum, and then flip the sign of the last coordinate of the result, which is motivated through the use of $\inf_i a = -\sup_i\{-a\}$ for $a \in \mathbb{R}$. Precisely, we calculate

$$\inf_i \mathbf{x}^{(i)} := \overline{\sup_i\{\overline{\mathbf{x}^{(i)}}\}}$$

where $\overline{\mathbf{x}^{(i)}} := (x_1^{(i)}, \ldots, x_{n-1}^{(i)}, -x_n^{(i)})$.

Example 5 In this example, we calculate the supremum of the three *rgb*-colors $(1, 0, 0)$ (*red*), $(0, 1, 0)$ (*green*), and $(0, 0, 1)$ (*blue*). Using the map (2) leads to the representation $(1/2, 0, -1/2)$, $(0, 1/2, -1/2)$, and $(0, 0, 1)$ in the double hypersimplex $\lozenge^3$, respectively. The supremum is given by $(0, 0, 1)$ calculated via (5). Applying the map (4) leads to $(0, 0, 1) \in \square^3$ which represents the color *blue*. The infimum is given by $(0, 0, -1)$. Applying the map (4) leads to $(0, 0, 0) \in \square^3$ which represents the color *black*.

The set of all double hypersimplices aligned as described above is partially ordered with respect to the inclusion "$\subset$"; i.e., given two double hypersimplices $S_1 \in \lozenge^n$ and $S_2 \in \lozenge^n$, S_1 is smaller than S_2, if S_1 is contained in S_2, written as $S_1 \subset S_2$. The set of all these double hypersimplices admits an infimum (equivalently the *greatest lower bound* or *meet*) and a supremum (the *least upper bound* or *join*), hence it is a lattice [15]. It is even a distributive lattice since for any double hypersimplices $S_1, S_2, S_3 \in \lozenge^n$ the relations

$$\inf(S_1, \sup(S_2, S_3)) = \sup(\inf(S_1, S_2), \inf(S_1, S_3))$$

or equivalently

$$\sup(S_1, \inf(S_2, S_3)) = \inf(\sup(S_1, S_2), \sup(S_1, S_3))$$

are valid. To verify this, let $\mathbf{x}^{(1)}$, $\mathbf{x}^{(2)}$, and $\mathbf{x}^{(3)}$ be the point representing the double hypersimplices $S_1, S_2, S_3 \in \lozenge^n$, respectively. Let $i = 1, \ldots, n-1$, then the i-th component of $\inf(S_1, \sup(S_2, S_3))$ is given by $\min\{x_i^{(1)}, \min\{x_i^{(2)}, x_i^{(3)}\}\}$ which equals $\min\{\min\{x_i^{(1)}, x_i^{(2)}\}, \min\{x_i^{(1)}, x_i^{(3)}\}\}$ which represents the i-th component of the expression $\sup(\inf(S_1, S_2), \inf(S_1, S_3))$. The n-th component involves more steps of calculations and is left to the reader.

2.4 Difference of Two Points Located in the Double Hypersimplex

In this subsection, we explain how to subtract two points located in the double hypersimplex to make mathematical morphological operations such as top hats, gradients, and morphological Laplacian meaningful.

Before we define the new subtraction, we need the Einstein velocity addition $\oplus$ defined by

$$u \oplus v = \frac{1}{1 + u \cdot v} \left\{ u + \frac{1}{\gamma_u} v + \frac{\gamma_u}{1 + \gamma_u} (u \cdot v) \cdot u \right\}, \tag{6}$$

where

$$\gamma_u = \frac{1}{\sqrt{1 - u \cdot u}}.$$

Here, we used the constant $c = 1$ (see [23, p. 3]). Naturally, the Einstein subtraction $\ominus$ is defined by $u \oplus (-v)$.

Next, we will explain how to subtract two points located in $\Diamond^n$ ensuring that the result will be in $\Diamond^n$. The new subtraction will work as follows. Given two points $\mathbf{x}$ and $\mathbf{y}$ from $\Diamond^n$, we will transform the first $n-1$ coordinates of them by $\hat{x}_i = 2x_i - 1$ and $\hat{y}_i = 2y_i - 1$ $(i = 1, \ldots, n-1)$, respectively. Additionally, we set $\hat{x}_n = x_n$ and $\hat{y}_n = y_n$. Hence, it is ensured that each component of $\hat{\mathbf{x}}$ and $\hat{\mathbf{y}}$ is in $[-1, 1]$. Next, we apply componentwise the Einstein velocity subtraction using formula (6). Precisely, we calculate $\hat{z}_i := \hat{x}_i \oplus (-\hat{y}_i)$ for $i = 1, \ldots, n$. The result $\hat{z}_i$ will be in $[-1, 1]$ for each $i = 1, \ldots, n$. Finally, we convert the $n-1$ components of $\hat{\mathbf{z}}$ using $z_i = (\hat{z}_i + 1)/2$ ensuring that each z_i is in $[0, 1]$ for $i = 1, \ldots, n-1$. We set $z_n = \hat{z}_n$ which is $[-1, 1]$. Finally, we apply the map ψ given by (1) to $\mathbf{z}$. The result is in $\Diamond^n$ by construction. In sum, for $\mathbf{x}, \mathbf{y} \in \Diamond^n$, we define the difference operation $\boxminus$ as

$$\mathbf{x} \boxminus \mathbf{y} := \psi \left(\frac{[(2x_1 - 1) \oplus (-(2y_1 - 1))] + 1}{2}, \ldots, \right. \tag{7}$$
$$\left. \frac{[(2x_{n-1} - 1) \oplus (-(2y_{n-1} - 1))] + 1}{2}, x_n \oplus (-y_n) \right).$$

The result of $\mathbf{x} \boxminus \mathbf{y}$ is clearly an element of $\Diamond^n$ by construction.

3 A Glimpse at Scalar-Valued Morphology

In this section, we briefly review the definitions of some fundamental scalar-valued morphological operators that we will generalise to the multi-valued setting.

In gray-scale morphology an image is represented by a scalar function $f(x, y)$ with $(x, y) \in \mathbb{R}^2$. The so-called *structuring element* is a set $B \subset \mathbb{R}^2$ that determines the neighbourhood relation of pixels. Often convex sets such as disks, ellipses or squares are used as structuring elements.

Gray-scale *dilation* $\oplus$ replaces the gray-value of the image $f(x, y)$ by its supremum within a mask B. It is given by $(f \oplus B)(x, y) := \sup\{f(x - x', y - y') \mid (x', y') \in B\}$, while *erosion* $\ominus$ is determined by taking the infimum. It is given by $(f \ominus B)(x, y) := \inf\{f(x+x', y+y') \mid (x', y') \in B\}$. The *opening* operation, denoted by $\circ$, is defined as erosion followed by dilation: $f \circ B := (f \ominus B) \oplus B$. *Closing*, indicated by the symbol $\bullet$, consists of a dilation followed by an erosion: $f \bullet B := (f \oplus B) \ominus B$. Since erosion and dilation are antagonistic operations one can view opening and closing as an attempt to restore the image. A comparison with the original image by taking the difference to the opened or closed image leads to the so-called top-hats. The *white top-hat* which is the difference between the original image and its opening is defined as $\mathrm{WTH}(f) := f - (f \circ B)$. Its dual, the *black top-hat* is the difference between the closing and the original image; i.e., $\mathrm{BTH}(f) := (f \bullet B) - f$, while the *self-dual top-hat* is the difference between closing and opening; i.e., $\mathrm{SDTH}(f) := (f \bullet B) - (f \circ B)$. By construction these top-hats allow the detection of small details in an image.

In an image the boundaries or edges of objects are the loci of high gray-value variations. These variations can be detected by a derivative operator such as the gradient. Erosion and dilation are also the elementary building blocks of the basic morphological gradients: The so-called *Beucher gradient* is the difference between the dilation and the erosion: $\varrho_B f := (f \oplus B) - (f \ominus B)$. It is an analog to the Euclidean norm of the gradient $|\nabla f|$ if an image is regarded as a differentiable function. We also consider the *internal gradient* as the difference between the original image and its erosion given by $\varrho_B^- f := f - (f \ominus B)$, and the *external gradient* as the difference between the dilation and the original image given by $\varrho_B^+ f := (f \oplus B) - f$. The *morphological Laplacian* is defined by the difference of the external and internal gradient; i.e., $\Delta_B f := \varrho_B^+ f - \varrho_B^- f$.

All morphological operations can now be carried over for multispectral and hyperspectral images by using the supremum and infimum definition as explained on page 138, refer specifically to expression (5) and the next paragraph. The difference operation is replaced by the Einstein difference operation given by (8).

4 Numerical Results

In this section, we present the new proposed approach to process some data sets. First, we consider the case of a multispectral image with three bands; i.e., an *rgb*-image. We use the colored *peppers*-image of size 512×512 which is shown in Fig. 8a. As a first step, each pixel of the image is converted to a point located in the double hypersimplex using the map (2) with $n = 3$. The result is stored in a matrix field f. Next, we define a cross-shaped structuring element $\mathrm{SE}_{\mathrm{cross}}$ consisting of five pixels, B for short. Then, dilation $f \ominus B$ and erosion $f \oplus B$ are properly defined by using the supremum and infimum operation, respectively (refer to page 138, specifically to expression (5) and the next paragraph). Other morphological operations such as opening $f \circ B = (f \ominus B) \oplus B$, closing $f \bullet B = (f \oplus B) \ominus B$, white top-hat $f \boxminus (f \circ B)$, black top-hat $(f \bullet B) \boxminus f$, self-dual top-hat $(f \bullet B) \boxminus (f \circ B)$, Beucher gradient $(f \oplus B) \boxminus (f \ominus B)$, internal gradient $\varrho_B^- f = f \boxminus (f \ominus B)$, external gradient $\varrho_B^+ f = (f \oplus B) \boxminus f$, and the morphological Laplacian $\Delta_B f = \varrho_B^+ f \boxminus \varrho_B^- f$, where $\boxminus$ is the difference operation given by (8). After the application of a morphological operation the result is visualized by transforming back to a color image using the map (4) with $n = 3$. The results of the aforementioned morphological operations are presented in Fig. 8b–l. Now, we will show that it is possible to easily process a color image. The question of the interpretation of correct colors in the resulting color images is a completely different story. In our opinion, better results would be obtained if one can map the diagonal of the cube from $(0, 0, 0)$ and $(1, 1, 1)$ to the line of $\diamondsuit^3$ spanned by the two points $(0, 0, -1)$ and $(0, 0, 1)$ (in this instance the approach of gray-scale images is equivalent to basic gray-scale morphology), but unfortunately such a map has not yet been found. Note that our intention is not to give a better way of processing color images (this is just a byproduct of our new approach to process multispectral images for $n = 3$). As we can see the dilated *peppers*-image gets a *blue* touch. This happens because the supremum of the colors *red* and *green* is *blue* (refer to Example 5 and Fig. 8b). The minimum of those two colors is *black* as we can see in the eroded *peppers*-image (also refer to Example 5 and Fig. 8c). Opening and closing are the concatenation of the antagonistic operations dilation and erosion, hence in some way approximations to the identity map. Therefore the images in Fig. 8d, e are similar to the original, but not quite. The difference is obtained by the top hats as displayed in Fig. 8f–h. For large parts of the image there is hardly any difference (almost zero-difference) which accounts for the predominant color gray. Morphological derivatives imitate regular derivatives from calculus, hence help to detect locii of strong changes in the data as they appear, for example, at object boundaries. This explains the appearance of Fig. 8i–l where, to varying extent, contours are enhanced, while flat regions (almost zero-variance) are gray.

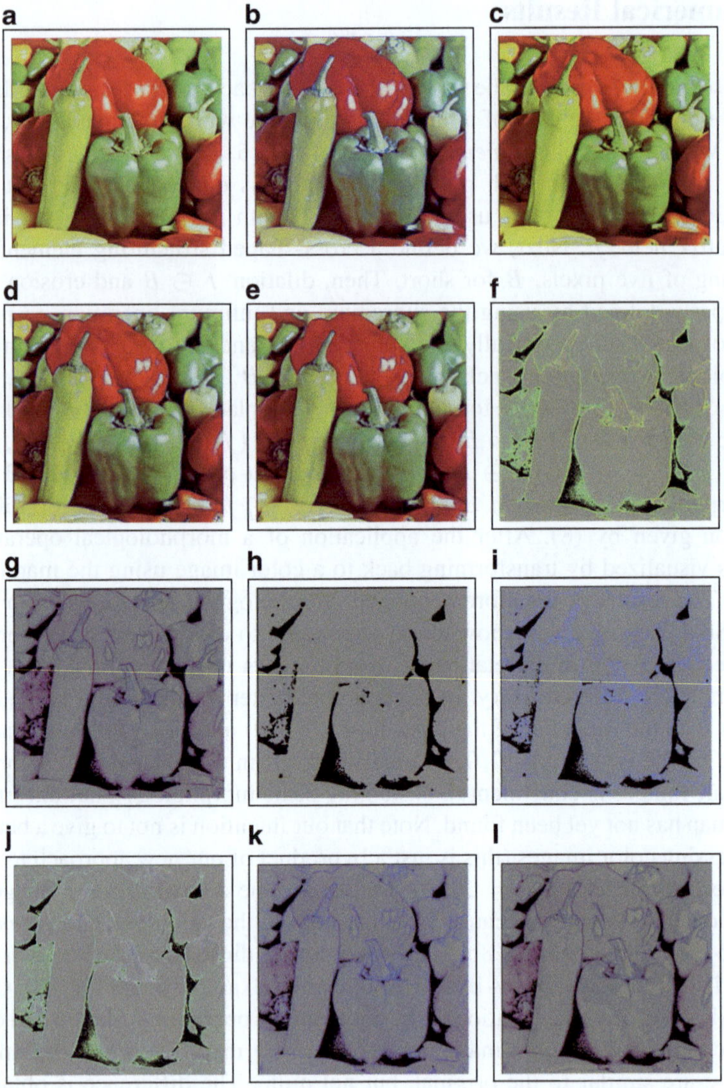

Fig. 8 Dilation, erosion, opening, closing, white top-hat, black top-hat, self-dual top-hat, Beucher gradient, internal and external gradient, and morphological Laplacian applied to *peppers* having resolution 512 × 512. (**a**) Original image of size 512 × 512. (**b**) Dilation, $\mathrm{SE}_{\mathrm{cross}}$. (**c**) Erosion, $\mathrm{SE}_{\mathrm{cross}}$. (**d**) Opening, $\mathrm{SE}_{\mathrm{cross}}$. (**e**) Closing, $\mathrm{SE}_{\mathrm{cross}}$. (**f**) White top-hat, $\mathrm{SE}_{\mathrm{cross}}$. (**g**) Black top-hat, $\mathrm{SE}_{\mathrm{cross}}$. (**h**) Self-dual top-hat, $\mathrm{SE}_{\mathrm{cross}}$. (**i**) Beucher gradient, $\mathrm{SE}_{\mathrm{cross}}$. (**j**) Internal gradient, $\mathrm{SE}_{\mathrm{cross}}$. (**k**) External gradient, $\mathrm{SE}_{\mathrm{cross}}$. (**l**) Morphological Laplacian, $\mathrm{SE}_{\mathrm{cross}}$

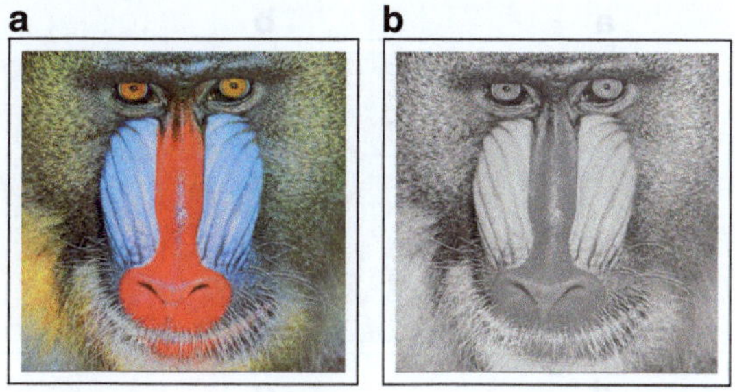

Fig. 9 Color image *baboon* and its gray-valued representation used as transparency. (**a**) Original color image *baboon* of size 512 × 512. (**b**) Gray-valued image *baboon* of size 512 × 512

We need 0.2, 0.7, 3, 11, and 36 s to dilate a *rgb*-image of size 32 × 32, 64 × 64, 128 × 128, 256 × 256, and 512 × 512, respectively. This is more than two times faster than our previous color processing approach (see [5]). Of course, it still very expensive if compared to the binary or gray-scale case.

Next, we will consider an *rgbα*-image; i.e., an *rgb*-image that contains an α-channel (transparency). We use the colored test image *baboon* of size 512 × 512 as shown in Fig. 9a and use as α-channel its gray-value representation (created with gimp using Image → Mode → Grayscale) as illustrated in Fig. 9b. We use this multispectral image of size 512 × 512 × 4 and apply opening, closing, white top-hat, external gradient, internal gradient, and morphological Laplacian to it. Figure 10a–l are displaying the processed outcomes of a 4-channel image. In Fig. 10a–c, g–i one can see the rgb-parts of the results of various morphological operators (first and third row of Fig. 10). Each of Fig. 10d–f, j–l shows the transparency as a gray-scale image (second and fourth row of Fig. 10).

Essentially, what has been said about Fig. 8 holds true for those images in Fig. 10 as well. However, since the original image contains rich texture-like structures with high variations, the results display more details and texture of the original image "covered" with a dominant gray tinge.

Next, we will process a hyperspectral image with 12 channels having resolution 949 by 220. We use the image *STC* taken from the *MultiSpec* website.[1] The image has been taken in June 1966 by an aircraft scanner Flighline C1 and shows a portion of Southern Tippecanoe County in Indiana. The data are stored in band sequential format (BSQ-format). The wavelength of the 12 channels are 0.4–0.44, 0.44–0.46, 0.46–0.48, 0.48–0.5, 0.5–0.52, 0.52–0.55, 0.55–0.58, 0.58–0.62, 0.62–0.66, 0.66–0.72, 0.72–0.8, 0.8–1 μm, respectively. In Fig. 11a, the image is shown as a *rgb*-image, where we have extracted the bands 11, 9, and 7, since band 11, 9, and 7 represent red, green, and blue of the spectrum. Using the same approach as before with $n = 12$ and the structuring element SE_{cross} consisting of five pixels, we get the following results for dilation, erosion, opening, closing, white top-hat, black top-hat, internal gradient, external gradient, and morphological Laplacian, where the result is again shown in Fig. 11b–j as a *rgb*-images using bands 11, 9, and 7. The same

[1]https://engineering.purdue.edu/~biehl/MultiSpec/hyperspectral.html.

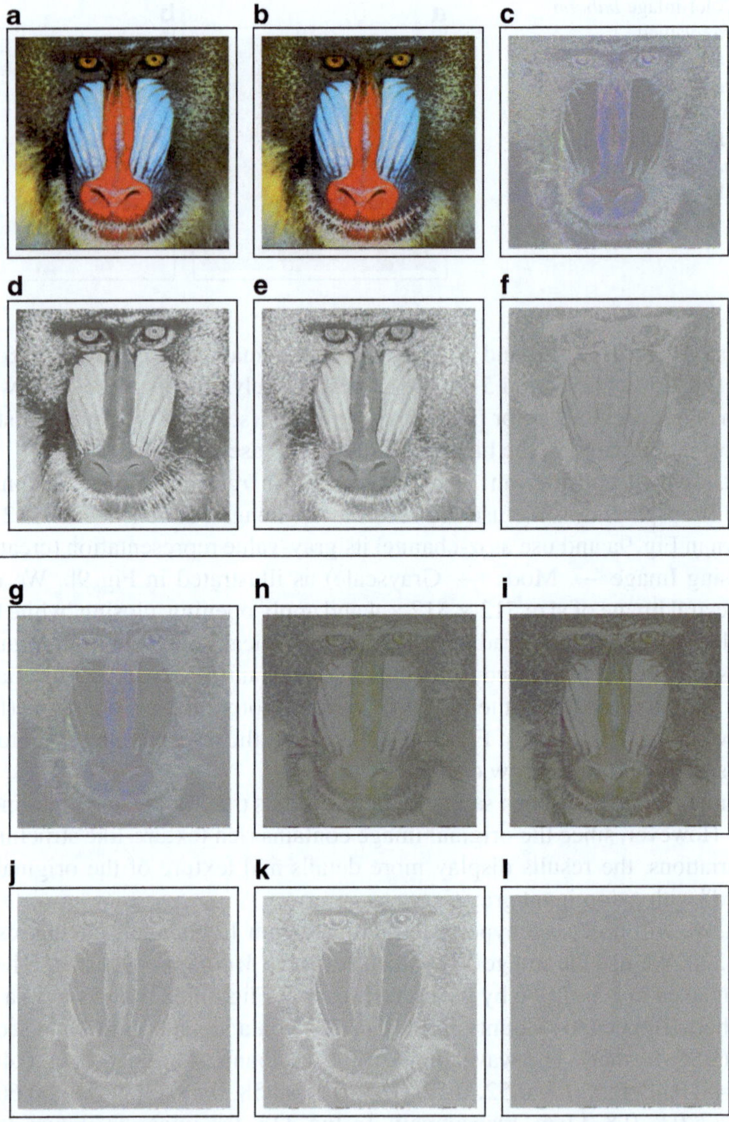

Fig. 10 Opening, closing, white top-hat, internal and external gradient, and morphological Laplacian applied to *rgbα*-image *baboon* having resolution 512 × 512. (**a**) Opening, SE_{cross}. (**b**) Closing, SE_{cross}. (**c**) White top-hat, SE_{cross}. (**d**) Opening, SE_{cross}. (**e**) Closing, SE_{cross}. (**f**) White top-hat, SE_{cross}. (**g**) External gradient, SE_{cross}. (**h**) Internal gradient, SE_{cross}. (**i**) Morphological Laplacian, SE_{cross}. (**j**) External gradient, SE_{cross}. (**k**) Internal gradient, SE_{cross}. (**l**) Morphological Laplacian, SE_{cross}

Fig. 11 *STC*-image and various morphological operations applied to it are shown as *rgb*-color image using bands 11, 9, and 7. (**a**) *STC*-image of size 949 × 220. (**b**) Dilation, SE_{cross}. (**c**) Erosion, SE_{cross}. (**d**) Opening, SE_{cross}. (**e**) Closing, SE_{cross}. (**f**) White top-hat, SE_{cross}. (**g**) Black top-hat, SE_{cross}. (**h**) Internal gradient, SE_{cross}. (**i**) External gradient, SE_{cross}. (**j**) Morphological Laplacian, SE_{cross}

Table 1 Time in seconds to dilate a multispectral/hyperspectral image of size $n \times 512 \times 512$ for various n ranging from 3 to 200

n	3	5	10	20	50	100	150	200
s	41.4	42.0	42.4	42.8	48.7	58.3	65.8	75.9

interpretation of the results carries over to the processed versions, see Fig. 11b–j, of the 12-channel *STC*-image as displayed in Fig. 11a. Note that the algorithm needs 30 s to dilate the image.

Finally, we dilate a multispectral/hyperspectral image of size $n \times 512 \times 512$ for various n using the structuring element $\mathrm{SE}_{\mathrm{cross}}$ to show that the computation time is quite low no matter how big n actually is. In Table 1 we list the time needed to dilate an image using the structuring element $\mathrm{SE}_{\mathrm{cross}}$ with the number of channels ranging from $n = 3$ to $n = 200$.

Hence, we are able to process quickly a hyperspectral image with 200 channels in 75.9 s.

5 Conclusion and Outlook

We have shown that it is a simple step to define mathematical morphological operations such as dilation and erosion for multispectral/hyperspectral images by first converting the data into a double hypersimplex using a simple geometric approach which is motivated by the Loewner ordering. Then, with the definition of a suitable difference operation motivated by the Einstein velocity addition we are able to define morphological operations such as top hats, gradients, and morphological Laplacian. Numerical results confirm that this approach has the potential to be applicable to multispectral images. Note that from the computational point of view the approach is fast in computation time and that we are not limited to multispectral images. As shown, the approach can also be used for hyperspectral images without any limitations. It remains to think about how to visualize a multispectral/hyperspectral image to extract meaningful information of a processed multispectral image such as the Beucher gradient or the morphological Laplacian.

References

1. Angula, J., Lefèvre, S., Lezoray, O.: Color representation and processing in polar color spaces. In: Fernandez-Maloigne, C., Robert-Inacio, F., Macaire, L. (eds.) Digital Color Imaging, pp. 1–40. Wiley-ISTE, Hoboken, New Jersey (2013)
2. Aptoula, E., Lefèvre, S.: A comparative study on multivariate mathematical morphology. Pattern Recognit. **40**(11), 2914–2929 (2007)

3. Banon, G.J.F., Barrera, J., Braga-Neto, U.d.M., Hirata, N.S.T. (eds.): Proceedings of the 8th International Symposium on Mathematical Morphology: Volume 1 - Full Papers. Computational Imaging and Vision, vol. 1. Instituto Nacional de Pesquisas Espaciais (INPE), São José dos Campos (2007)
4. Braun, K.M., Balasubramanian, R., Eschbach, R.: Development and evaluation of six gamut-mapping algorithms for pictorial images. In: Color Imaging Conference, pp. 144–148. IS&T - The Society for Imaging Science and Technology, Springfield (1999)
5. Burgeth, B., Kleefeld, A.: Morphology for color images via Loewner order for matrix fields. In: Luengo Hendriks, C.L., Borgefors, G., Strand, R. (eds.) Mathematical Morphology and Its Applications to Signal and Image Processing (Proceedings of the 11th International Symposium on Mathematical Morphology, Uppsala, 27–29 May). Lecture Notes in Computer Science, vol. 7883, pp. 243–254. Springer, Berlin (2013)
6. Burgeth, B., Kleefeld, A.: An approach to color-morphology based on einstein addition and loewner order. Pattern Recognit. Lett. **47**, 29–39 (2014)
7. Burgeth, B., Papenberg, N., Bruhn, A., Welk, M., Feddern, C., Weickert, J.: Mathematical morphology based on the loewner ordering for tensor data. In: Ronse, C., Najman, L., Decencière, E. (eds.) Mathematical Morphology: 40 Years On, Computational Imaging and Vision, vol. 30, pp. 407–418. Springer, Dordrecht (2005)
8. Burgeth, B., Welk, M., Feddern, C., Weickert, J.: Mathematical morphology on tensor data using the loewner ordering. In: Weickert, H.H.J. (ed.) Visualization and Processing of Tensor Fields. Springer, Berlin (2006)
9. Comer, M.L., Delp, E.J.: Morphological operations for color image processing. J. Electron. Imaging **8**(3), 279–289 (1999)
10. Heijmans, H.J.A.M.: Morphological Image Operators. Academic, Boston (1994)
11. Luengo Hendriks, C.L., Borgefors, G., Strand, R. (eds.): Mathematical morphology and its applications to image and signal processing. In: Proceedings of the 11th International Symposium on Mathematical Morphology. Lecture Notes in Computer Science, vol. 7883. Springer, Berlin (2013)
12. Matheron, G.: Eléments pour une théorie des milieux poreux. Masson, Paris (1967)
13. Matheron, G.: Random Sets and Integral Geometry. Wiley, New York (1975)
14. Ostwald, W.: Die Farbenfibel. Unesma, Leipzig (1916)
15. Ronse, C., Serra, J.: Algebraic foundations of morphology. In: Najman, L., Talbot, H. (eds.) Mathematical Morphology: From Theory to Applications, Chap. 2, pp. 35–80. ISTE/Wiley, London (2010)
16. Ronse, C., Najman, L., Decencière, E. (eds.): Mathematical Morphology: 40 Years On, Computational Imaging and Vision, vol. 30. Springer, Dordrecht (2005)
17. Serra, J.: Echantillonnage et estimation des phénomènes de transition minier. Ph.D. thesis, University of Nancy, France (1967)
18. Serra, J.: Image Analysis and Mathematical Morphology, vol. 1. Academic, London (1982)
19. Serra, J.: The false colour problem. In: Wilkinson, M.H.F., Roerdink, J.B.T.M. (eds.) Mathematical Morphology and Its Application to Signal and Image Processing. Proceedings of the 9th International Symposium on Mathematical Morphology. Lecture Notes in Computer Science, vol. 5720, Chap. 2, pp. 13–23. Springer, Heidelberg (2009)
20. Serra, J., Soille, P. (eds.): Mathematical Morphology and Its Applications to Image Processing. Computational Imaging and Vision, vol. 2. Kluwer, Dordrecht (1994)
21. Soille, P.: Morphological Image Analysis, 2nd edn. Springer, Berlin (2003)
22. Soille, P., Pesaresi, M., Ouzounis, G. (eds.): Mathematical Morphology and Its Applications to Image and Signal Processing. Proceedings of the 10th International Symposium on Mathematical Morphology. Lecture Notes in Computer Science, vol. 6671. Springer, Berlin (2011)
23. Ungar, A.A.: Einstein's special relativity: the hyperbolic geometric viewpoint. In: Conference on Mathematics, Physics and Philosophy on the Interpretations of Relativity, II. Budapest (2009)

24. van de Gronde, J.J., Roerdink, J.B.T.M.: Group-invariant frames for colour morphology. In: Luengo Hendriks, C.L., Borgefors, G., Strand, R. (eds.) Mathematical Morphology and Its Applications to Signal and Image Processing (Proceedings of the 11th International Symposium on Mathematical Morphology, Uppsala, 27–29 May). Lecture Notes in Computer Science, vol. 7883, pp. 267–278. Springer, Berlin (2013)
25. Wilkinson, M.H.F., Roerdink, J.B.T.M. (eds.): Mathematical Morphology and Its Application to Signal and Image Processing. Proceedings of the 9th International Symposium on Mathematical Morphology. Lecture Notes in Computer Science, vol. 5720. Springer, Heidelberg (2009)

Direction-Controlled DTI Interpolation

Luc Florack, Tom Dela Haije, and Andrea Fuster

Abstract Diffusion Tensor Imaging (DTI) is a popular model for representing diffusion weighted magnetic resonance images due to its simplicity and the fact that it strikes a good balance between signal fit and robustness. Nevertheless, problematic issues remain. One of these concerns the problem of interpolation. Because the DTI assumption forces Apparent Diffusion Coefficients (ADCs) to fit quadratic forms, destructive interference of diffusivity patterns tends to mask information on orientations. For some applications, notably tractography, one would like an interpolated DTI tensor to reflect not only some weighted average of its immediate grid neighbours, but also to preserve orientation information available at those points. This is possible if one declines from the quadratic restriction, considering general homogeneous functions of degree two instead. We show that one may interpret the interpolated ADC as a *family* of DTI tensors, parametrized by orientation. Any choice of a preferred direction—notably a stipulated fiber tangent—singles out a unique DTI tensor instance. Results are physically plausible and intuitive.

1 Introduction

Diffusion Weighted Magnetic Resonance Imaging (DWMRI) has become a standard MRI technique for in vivo imaging of apparent water diffusion processes in fibrous tissue, cf. Hagmann et al. [25]. Diffusion Tensor Imaging (DTI), pioneered by Basser et al. [5] and explored by many others, is an established model, based on a second order symmetric positive-definite tensor, or quadratic form. This so-called diffusion tensor relates the flux of diffusing particles to the concentration gradient in a solution under the assumption of steady state (Fick's first law [12]), and consequently enters as an attenuating factor in the DWMRI signal (cf. the Bloch-Torrey equation with diffusion term [45]). In this chapter we consider the

L. Florack (✉) • T. Dela Haije • A. Fuster
Eindhoven University of Technology, PO Box 513, NL-5600 MB Eindhoven, The Netherlands
e-mail: L.M.J.Florack@tue.nl; T.C.J.Dela.Haije@tue.nl; A.Fuster@tue.nl

I. Hotz, T. Schultz (eds.), *Visualization and Processing of Higher Order Descriptors for Multi-Valued Data*, Mathematics and Visualization,
DOI 10.1007/978-3-319-15090-1_8

classical problem of DTI interpolation, given a discrete image of diffusion tensors on a regular grid.

To alleviate shortcomings of component-wise linear interpolation in DTI, many alternatives have been proposed, such as geodesic [6, 14, 36], log-Euclidean [1], tensor spline [3], geodesic-loxodrome [30], linear invariant [21], dyadic-tensor [20], pde-based [48], and quaternion-based schemes [49]. None of these, however, is capable of preventing destructive interference when interpolating DTI tensors with conflicting orientations (the generic situation). To overcome this limitation, Schultz has proposed a higher order tensor inpainting method that is capable of handling ambiguous local orientations for the purpose of resolving fiber crossings [40]. In this chapter we revisit the problem of interpolation from an entirely novel vantage point so as to address this particular issue. The methodology we propose is in principle applicable to symmetric positive-definite second order tensors in a broader context than DTI.

Our aim is to construct an interpolation scheme that avoids destructive interference of orientational information provided by DTI evidence. A question that presents itself is whether this is possible at all. As a Gedanken experiment, consider two similar white matter fiber bundles arranged in a planar sheet and sampled on a square grid, crossing at right angles at the fiducial center point. Along the white matter tracts and away from the crossing the (level sets of the) inverse-DTI tensors are prolate spheroids, pointing towards the stipulated crossing. However, dictated by symmetry, any scheme that aims for an unambiguous DTI interpolation at the center must necessarily result in a profile that is isotropic in the crossing plane. Destructive interference of orientations seems unavoidable, confronting us with a paradox.

This state of affairs relates to a remark by Bernhard Riemann in his "Habilitation" [33, 37] in the context of Riemannian geometry, in which quadratic forms ("metric tensors") figure prominently. Riemann noted that the quadratic assumption is an unnecessary constraint for his theory, but declined to further investigate the general case. This study was taken up by Finsler [13] and subsequently by Cartan [7] and others. The key point is to replace quadratic forms by certain homogeneous functions of degree two. This idea has sparked new geometric developments in various fields of research, such as general relativity theory, cf. Gibbons et al. [22]. It is likewise of interest for extending the geometric approach in DWMRI [17], but here we will only discuss its application in the context of DTI interpolation.

Of course, ideally one would like to avoid destructive interference of orientations altogether. However, heterogeneity of orientations at the microscopic scale will always induce destructive interference at MRI observation scale (millimeter regime in a typical human scanner). More complex High Angular Resolution Diffusion Imaging (HARDI) or Diffusion Spectrum Imaging (DSI) schemes may avoid destructive interference at mesoscopic scales, but with a concomitant cost in terms of prolonged acquisition time and reduced robustness as compared to DTI [8–11, 15, 16, 19, 26–28, 31, 34, 35, 41, 46]. By virtue of its limited angular resolution the latter strikes a good balance between signal fit and robustness, is relatively fast, and is therefore often clinically preferred despite shortcomings. This reality has led us to consider the problem of how our recently proposed "Finsler-DTI" model [17] (another model in the HARDI/DSI category) could be fruitfully applied in the context of DTI. That is, we take it for granted that our input data are DTI images (or cannot be fit to any higher order model due to acquisition restrictions), and that we have no influence on the acquisition protocol. It might then seem inevitable that we have to take destructive interference due to the limitation of DTI to six degrees of freedom per point for granted, but this ignores the actual trade-off between spatial and angular resolution. There is no a priori reason why we would need to fix the number of degrees of freedom governing angular resolution if we *lower* spatial resolution. But this (lowering of spatial resolution) is precisely what happens in interpolation schemes, as these are invariably based on some weighted local neighbourhood averaging.

The paradox of DTI interpolation without destructive interference of orientations may be resolved if we (1) relax the quadratic assumption (and thus the restriction to six degrees of freedom) outside grid points, and (2) specify a projection back onto quadratic forms a posteriori given a context dependent bias in the form of a *preferred direction*. The methodology we propose in the next section can be seen as an application of Riemann-Finsler geometry [2, 7, 13, 17, 37–39] and exploits multiplicative calculus to implement positivity preserving "linear" operations [4, 18, 23, 24, 43, 47].

2 Theory

2.1 5D-DTI

The human brain consists mostly of water, but diffusion turns out to be anisotropic as a result of its fibrous architecture. The Stejskal-Tanner signal attenuation formula in the spin-echo experiment on spin diffusion in an isotropic medium [44] inspired Moseley, Basser, Le Bihan and others, [5] to capture this anisotropy in terms of a symmetric positive-definite second order diffusion tensor. This is the basis of DTI.

Associated with a DTI tensor is the nondegenerate positive-definite quadratic form[1]

$$F^2_{\text{DTI}}(x,\xi) = D_{ij}(x)\xi^i\xi^j \,, \tag{1}$$

with $F_{\text{DTI}}(x,\xi) \geq 0$. The coefficients $D_{ij}(x)$, $i,j = 1,2,3$, are the components of the *inverse* of the DTI tensor relative to a Cartesian coordinate basis; those pertaining to the DTI tensor itself are written as $D^{ij}(x)$. For our purpose we will consider a more general function, $F(x,\xi) \geq 0$, insisting on a quadratic scaling with

[1]Einstein summation convention applies to pairs of identical upper and lower indices.

respect to the *magnitude* of ξ by imposing the following homogeneity condition[2] in ξ-space:

$$F^2(x,\xi) = D_{ij}(x,\xi)\xi^i\xi^j \quad \text{with} \quad D_{ij}(x,\lambda\xi) = D_{ij}(x,\xi) \tag{2}$$

for all $\lambda \neq 0$. The function $F(x,\xi)$ is assumed to be smooth and positive for $\xi \neq 0$, and such that the ξ-Hessian of $F^2(x,\xi)$ is a nondegenerate positive-definite matrix. Using Euler's theorem for homogeneous functions, recall Footnote 2, it can be shown that

$$D_{ij}(x,\xi) = \frac{1}{2}\frac{\partial^2 F^2(x,\xi)}{\partial\xi^i\partial\xi^j}. \tag{3}$$

Equations (2) and (3) are in fact equivalent (i.e. the ξ-Hessian operator is invertible). Details can be found in Bao et al. [2].

The projection operator alluded to in the introduction—Π_ϑ, say—enables us to associate a quadratic form to the function $F(x,\xi)$ in (2) after specification of a preferred direction $\vartheta \in \mathbb{S}^2$, viz.

$$\Pi_\vartheta(F) = F_\vartheta\,, \tag{4}$$

in which $F_\vartheta(x,\xi) \geq 0$ is defined by

$$F_\vartheta^2(x,\xi) = D_{ij}(x,\vartheta)\xi^i\xi^j\,. \tag{5}$$

The tensor field $D_{ij}(x,\vartheta)$ lives on a 5D domain, $(x,\vartheta) \in \mathbb{R}^3 \times \mathbb{S}^2$, providing the desired flexibility for propagating orientation information in a spatial DTI interpolation without destructive interference, cf. Jonasson et al. [29] for a similar approach in the context of white matter segmentation and Duits and Franken [11] and Florack and Fuster [17] for a generic approach. The (convex) unit level set

$$F(x,\xi) = 1 \tag{6}$$

is called the *indicatrix* of F, and the (ellipsoidal) unit level set

$$F_\vartheta(x,\xi) = 1 \tag{7}$$

the *osculating indicatrix* of F associated with direction $\vartheta \in \mathbb{S}^2$. In DTI all osculating indicatrices equal the indicatrix, recall (1) and (5).

The corresponding level sets of the associated function $H(x,q) = D^{ij}(x,q)q_iq_j$, related to $F(x,\xi) = D_{ij}(x,\xi)\xi^i\xi^j$ via a Legendre transform, are referred to as "figuratrix", respectively "osculating figuratrix", cf. [17] for details.

[2]A function $f(z)$ is homogeneous of degree r if $f(\lambda z) = \lambda^r f(z)$ for all $\lambda > 0$.

(The distinction between indicatrices and figuratrices is not relevant for the present discussion.)

2.2 DTI Interpolation

We may consider a DTI image as the "vertical" projection of a hypothetical tensor field on $\mathbb{R}^3 \times \mathbb{S}^2$ in the sense of the previous section, recall (2) and (3):

$$D_{ij}(x) = \int_{\vartheta \in \mathbb{S}^2} D_{ij}(x, \vartheta)\mu(d\vartheta) . \tag{8}$$

The ϑ-integration[3]—with spherical measure $\mu(d\vartheta)$—expresses destructive interference of high angular resolution information due to the quadratic restriction. However, if we are given two DTI samples we may construct a weighted average by first lifting them to $\mathbb{R}^3 \times \mathbb{S}^2$, followed by averaging. The result may be backprojected to a quadratic form on $\mathbb{R}^3$ with the help of a suitably chosen parameter $\vartheta \in \mathbb{S}^2$. The details are as follows.

Let $F_g^2(x,\xi) = g_{ij}(x)\xi^i\xi^j$ and $F_h^2(x,\xi) = h_{ij}(x)\xi^i\xi^j$ be two inverse-DTI quadratic forms, and $\alpha \in [0, 1]$ a weighing parameter, then set

$$F(x, \xi; \alpha) = F_g^{1-\alpha}(x, \xi)\, F_h^{\alpha}(x, \xi) . \tag{9}$$

By construction $F^2(x, \xi; \alpha)$ is homogeneous of degree two in ξ, but typically not a quadratic form, unless $\alpha = 0$ or 1. We may thus apply (3) so as to obtain its indicatrix and associated family of osculating indicatrices. A tedious but straightforward computation reveals the explicit result in terms of $g_{ij}(x)$ and $h_{ij}(x)$ for any given value of α (this result holds in any spatial dimension):

$$D_{ij}(x, \xi; \alpha) = F^2(x, \xi; \alpha)\, \Delta_{ij}(x, \xi; \alpha) \quad \text{with} \tag{10}$$

$$\Delta_{ij} = (1-\alpha)\left[\frac{g_{ij}}{F_g^2} - 2\alpha\frac{\gamma_i}{F_g^2}\left(\frac{\gamma_j}{F_g^2} - \frac{\eta_j}{F_h^2}\right)\right] + \alpha\left[\frac{h_{ij}}{F_h^2} - 2(1-\alpha)\frac{\eta_i}{F_h^2}\left(\frac{\eta_j}{F_h^2} - \frac{\gamma_j}{F_g^2}\right)\right], \tag{11}$$

in which $\gamma_i(x,\xi) = g_{ik}(x,\xi)\xi^k$ and $\eta_i(x,\xi) = h_{ik}(x,\xi)\xi^k$. It is straightforward to apply this one-parameter weighted averaging formula to n-dimensional spatial interpolation for any $n = 1, 2, 3, \ldots$, cf. Fig. 1 for a sketch of the 2D procedure.

[3] In reality the r.h.s. should also entail a spatial integration over a voxel cell if the integrand is interpreted in distributional sense as a Riesz representant of an infinite-resolution source field, cf. the effective medium theory by Novikov and Kiselev [32].

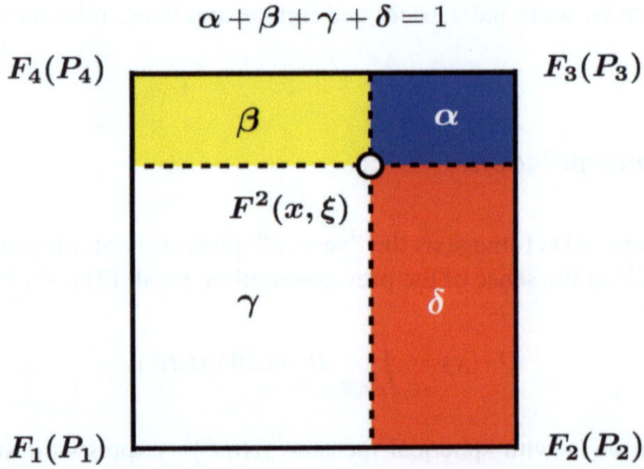

Fig. 1 Spatial interpolation in 2D: $F(x,\xi) = F_1(P_1,\xi)^\alpha F_2(P_2,\xi)^\beta F_3(P_3,\xi)^\gamma F_4(P_4,\xi)^\delta$, in which $F_k^2(P_k,\xi)$ are quadratic forms in ξ given at the grid points P_k, and $x \in \mathbb{R}^3$ is an interpolation point. The corresponding expression for the interpolated indicatrix follows by a computation similar to that in the one-parameter case, (9)–(11), now involving the relative area weights $\alpha, \beta, \gamma, \delta$ (all determined by the two components of $x \in \mathbb{R}^2$). The 3D case is similar

3 Results

Figure 2 shows the results of one-parameter interpolation between two 2D quadratic forms according to (9)–(11), both for the unambiguous indicatrix (left), as well as for the induced osculating indicatrices for a few orientations $\vartheta \in \mathbb{S}$ (middle and right). Looking at the results for two inverse-DTI tensors $g_{ij}(x)$ and $h_{ij}(x)$ we observe the following. The interpolated indicatrix is always convex, and ellipsoidal if $g_{ij}(x)$ and $h_{ij}(x)$ are proportional. Indeed, (10) and (11) imply that if $h_{ij} \propto g_{ij}$, then interpolation boils down to isotropic scaling:

$$D_{ij}(x,\xi;\alpha) = c(x,\xi;\alpha)\, g_{ij}(x) \quad \text{with} \quad c(x,\xi;\alpha) = \frac{F^2(x,\xi;\alpha)}{F_g^2(x,\xi)}. \tag{12}$$

Generically, however, the interpolated indicatrix is non-ellipsoidal. This effect is most pronounced for perpendicular orientations (main eigenaxes). The corresponding osculating indicatrices, on the other hand, are always ellipsoidal by construction. Given a pair $g_{ij}(x)$ and $h_{ij}(x)$ and a fixed weight $\alpha \in (0,1)$, there are, however, infinitely many such interpolated osculating indicatrices. A unique one is only singled out after providing an orientation $\vartheta \in \mathbb{S}$. If this orientation happens to be aligned with $g_{ij}(x)$ or $h_{ij}(x)$, then this alignment biases the interpolation to the extent that the corresponding orientation tends to be emphasized. To appreciate this phenomenon better, cf. the horizontal spoke in the middle subfigure of Fig. 2, and observe the occurrence of an isotropic transition for α *strictly less than* $\frac{1}{2}$; for the vertical spoke in the rightmost subfigure this is exactly opposite. The

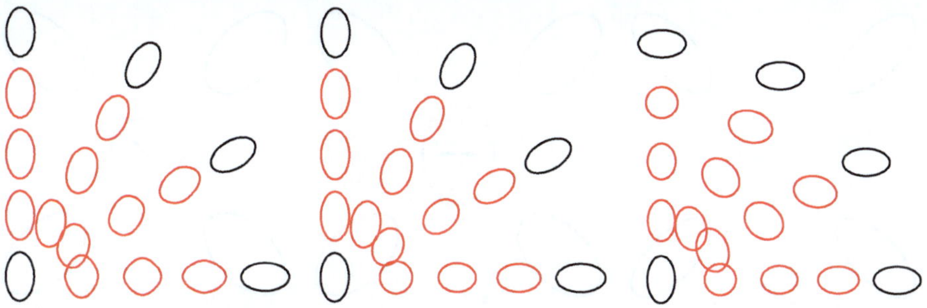

Fig. 2 Interpolated indicatrices and osculating indicatrices for two inverse-DTI tensors g_{ij} and h_{ij} shown as *black ellipses* at either end of a spoke, with g_{ij} in the lower left corner. Radial distance along a spoke reflects the value of $\alpha \in [0, 1]$, with $\alpha = 0$ at the center, recall (9)–(11). *Left*: Indicatrices. Polar angle reflects relative rotation of the outer ellipse. *Middle and right*: Osculating indicatrices. Polar angle now corresponds to the angular parameter $\vartheta \in \mathbb{S}$ of the osculating indicatrix, recall (3)

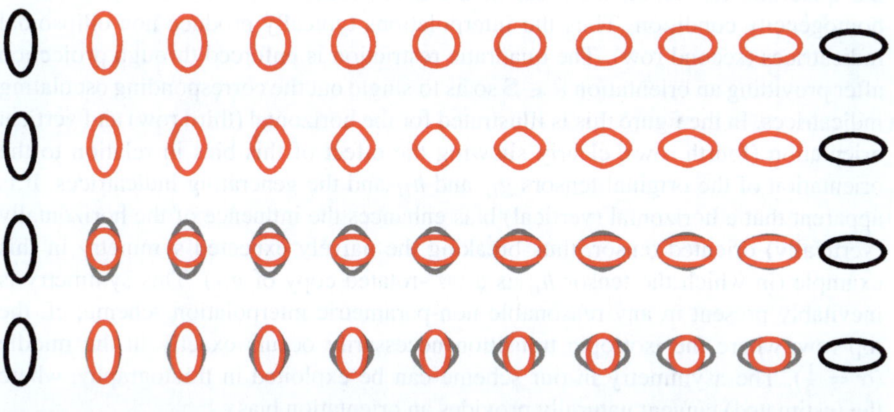

Fig. 3 Comparison of one-parameter interpolation schemes on 2D tensors for linearly increasing weighing parameters $\alpha \in [0, 1]$ between the tensors g_{ij} (*leftmost black ellipse*, $\alpha = 0$) and h_{ij} (*rightmost black ellipse*, $\alpha = 1$). Interpolations are shown in *red*. *Top row*: Log-Euclidean interpolation. *Second row*: The newly proposed scheme for indicatrices, based on (9)–(11). *Third and fourth rows*: Corresponding osculating indicatrices for a horizontal, respectively vertical bias $\vartheta \in \mathbb{S}$, overlayed onto their corresponding indicatrices (replicated in *grey*)

difference is explained by the a posteriori bias incorporated through the choice of ϑ, corresponding to a horizontal, respectively vertical orientation. In tractography this bias may be exploited in a spatial interpolation scheme to express one's expectation about the direction of fiber tracking, if available. In particular, crossing fibers "see" different DTI interpolations at the same crossing point, depending on their tangents.

Figure 3 shows a comparison of the newly proposed scheme with log-Euclidean interpolation (first row, cf. Arsigny et al. for details [1]). Again, the major difference is the fact that we do not directly aim for an interpolation that is closed under

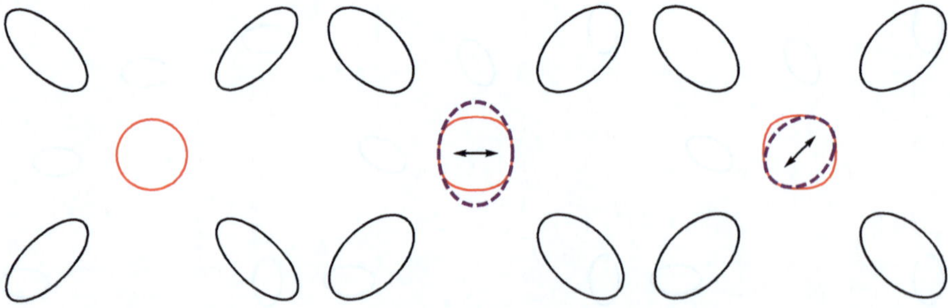

Fig. 4 Half-way 2D interpolation of a crossing pattern (*black ellipses*). *Left*: Log-Euclidean scheme. *Middle and right*: The newly proposed scheme for the indicatrix (*red convex shape*) and two instances of direction-controlled osculating indicatrices (*dashed ellipses*), for horizontal and diagonal orientations, respectively, as indicated by the *arrows*

the quadratic restriction manifest in the DTI model, but under a much weaker homogeneity condition. Thus the interpolations typically produce non-ellipsoidal indicatrices (second row). The quadratic restriction is enforced through projection after providing an orientation $\vartheta \in \mathbb{S}$ so as to single out the corresponding osculating indicatrices. In the figure this is illustrated for the horizontal (third row) and vertical orientation (fourth row), clearly showing the effect of this bias in relation to the orientation of the original tensors g_{ij} and h_{ij} and the generating indicatrices. It is apparent that a horizontal (vertical) bias enhances the influence of the horizontally (vertically) oriented tensor, thus breaking the naively expected symmetry in this example (in which the tensor h_{ij} is a 90°-rotated copy of g_{ij}). This symmetry is inevitably present in any reasonable non-parametric interpolation scheme, cf. the top row, where the isotropic transition necessarily occurs exactly in the middle ($\alpha = \frac{1}{2}$). The asymmetry in our scheme can be exploited in tractography, where the (estimated) tangent naturally provides an orientation bias.

Figure 4 illustrates the effect of 2D spatial interpolation, again in comparison to the log-Euclidean scheme. The latter produces an isotropic interpolation, as expected by symmetry. Our scheme produces genuine ellipsoidal shapes after providing a preferred orientation $\vartheta \in \mathbb{S}$, in this case $\vartheta \sim 0°$ and $\vartheta \sim 45°$. The osculating indicatrix with control parameter $\vartheta \sim 45°$ happens to be consistent with that of the end points, in the sense that its main axis has the same orientation, in contrast to the osculating indicatrix with control parameter $\vartheta \sim 0°$.

Figure 5 illustrates grid interpolation according to the "multiplicatively bilinear" interpolation scheme sketched in Fig. 1. The experiment shown on the left shows a discrete tensor field sampled from a smooth "single fiber orientation" region of interest, simulating a mildly curved underlying fiber bundle (black ellipses), together with its interpolation to a refined grid. The right subfigure illustrates the case of a sampled, randomly oriented tensor field.

It can be observed in the left subfigure of Fig. 5 that the interpolating gauge figures are almost, though not exactly, elliptical, and thus have a fairly well-defined, though not entirely crisp, preferred orientation. Although orientation is (by

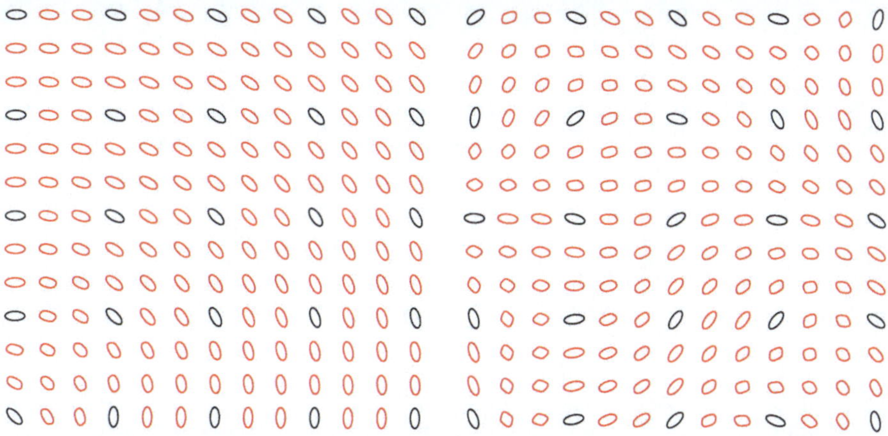

Fig. 5 Two-dimensional grid refinement through interpolation according to the scheme of Fig. 1. The *black ellipses* are grid samples, the *red* ones are interpolations. *Left*: Interpolation of a smoothly varying discrete tensor field. *Right*: Interpolation of a randomly varying discrete tensor field

construction) likewise unambiguous in each sample point in the right subfigure of Fig. 5, it is much more fuzzy and less anisotropic at typical interpolation points, and the non-elliptical shape of the (convex) gauge figures is also more apparent. In both cases one needs a "preferred orientation" field to single out an unambiguous member from the osculating family of ellipses so as to re-obtain a symmetric positive-definite second order tensor at each point. Such a field may be relatively straightforward and inspired by the data in the former case, but less trivial in the latter, recall the discussion on the results of Fig. 4.

Figure 6 illustrates one of the many possible projections of the interpolated image on the left of Fig. 5 back into the space of symmetric positive-definite second order tensors. The vertical has been chosen here as the preferred orientation for *all* interpolation points (a crude extrapolation of the preferred orientation apparent in the lower part of the discrete data). This might be a legitimate choice if one assumes that a bundle of fibers enters the region of interest from below in the upward direction. Clearly one could easily refine this crude global orientation selection pointwise, keeping tabs on the dominant direction as one progresses along the fiber tracks, but notice that by virtue of the almost elliptical nature of the indicatrices the back-projection is relatively insensitive to the choice of orientation parameter. (This is *not* the case in complex regions, recall the fiber crossing simulation in Fig. 4.)

The result of Fig. 6 may be compared with a standard DTI interpolation scheme, such as the log-Euclidean scheme illustrated for the same data in Fig. 7. But note that there are infinitely many ways to (pointwise) back-project indicatrices. The "right way" will depend on the application.

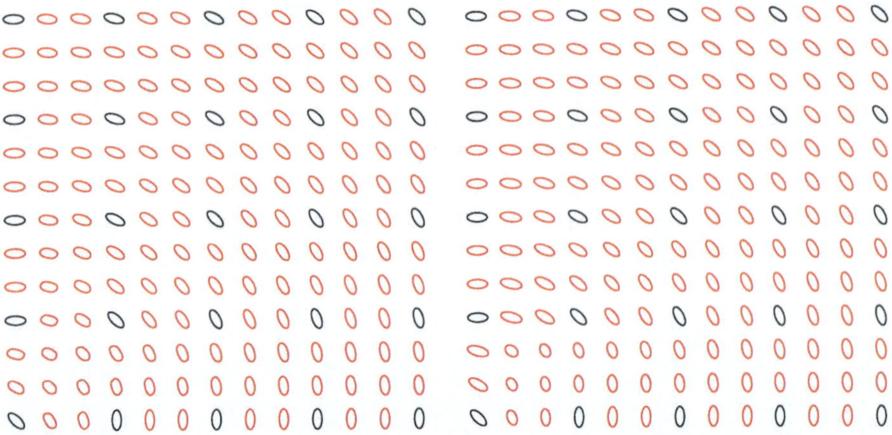

Fig. 6 Back-projection of the interpolated image on the left of Fig. 5 (replicated for ease of comparison) through specification of an orientation at each interpolation point so as to select an unambiguous osculating indicatrix, i.e. ellipse. The *black ellipses* are grid samples, the *red* ones are (original, respectively back-projected) interpolations. *Left*: Replication of the left subimage of Fig. 5. *Right*: Back-projection for a homogeneous vertical orientation parameter field

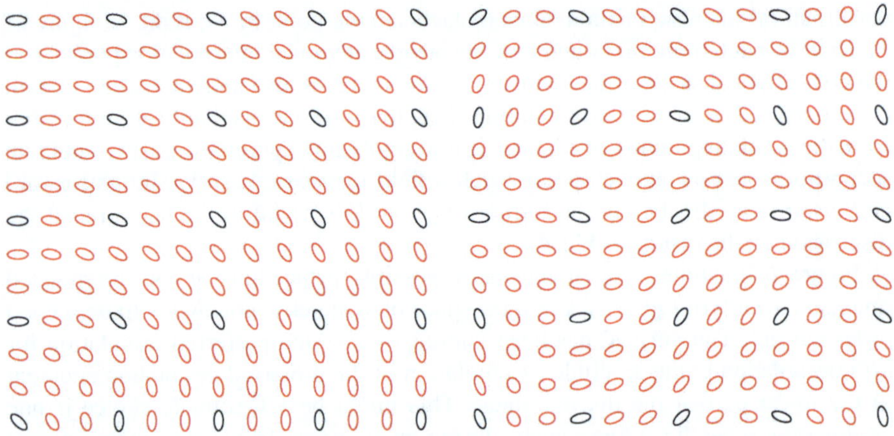

Fig. 7 Two-dimensional grid refinement through log-Euclidean interpolation. The *black ellipses* are grid samples, the *red* ones are interpolations. *Left*: Interpolation of a smoothly varying discrete tensor field. *Right*: Interpolation of a randomly sampled discrete tensor field. Cf. Figs. 5 and 6

Extension to three dimensions is straightforward. There are many more scenarios for the relative configurations of two ellipsoids and many more possible preferred directions to be specified in order to single out an osculating indicatrix/figuratrix. Figure 8 shows two examples.

Fig. 8 Three-dimensional DTI interpolation. The figures on either end are genuine ellipsoids, the ones in-between are convex indicatrices/figuratrices. A preferred direction $\vartheta \in \mathbb{S}^2$ will be needed in order to project each interpolated shape back to an ellipsoid (not shown here). *Top*: Interpolation between an oblate and a prolate spheroid. In this configuration, the minor axis of the former is aligned with the major axis of the latter. *Bottom*: Interpolation between two oblate spheroids with a 90° relative turn around one of the major axes

4 Summary and Conclusion

We have proposed a novel scheme for direction-controlled interpolation of DTI tensors (or other symmetric positive-definite second order tensors). It consists of two steps. In step 1 an unbiased interpolation is obtained that (generically) violates the quadratic restriction manifest in the 3D-DTI model. The resulting tensor can be interpreted as a 5D-DTI (an instance of a HARDI) tensor, i.e. a tensor living in the five-dimensional product $\mathbb{R}^3 \times \mathbb{S}^2$. The additional dimensions provided by $\mathbb{S}^2$ embody a "memory" for recollection of the orientation information available in the original tensor data, which would inevitably be destroyed in a parameter-free interpolation scheme. Step 2 entails a projection along $\mathbb{S}^2$ so as to reobtain a genuine 3D-DTI tensor on $\mathbb{R}^3$. This step requires specification of a preferred orientation $\vartheta \in \mathbb{S}^2$. Such an orientation may naturally present itself in certain applications, such as tractography (tangent directions).

Uncertainty may be accounted for by specifying a suitable distribution of orientation parameters in our interpolation scheme, rather than a crisp choice. Such a distribution of orientations naturally induces a probabilistic distribution of DTI tensors, akin to the model by Jian et al. [28], which may be useful for probabilistic tractography, cf. also Schultz et al. [42].

The methodology has been explained in the simplified context of one-parameter interpolation for the sake of clarity. We have indicated how to generalize the results, in particular to 1D, 2D and 3D spatial interpolation of sampled 3D tensor fields.

We have provided several illustrations to elucidate the closed-form interpolation formula, Eqs. (9)–(11).

All in all, the proposed scheme is more versatile than parameter-free schemes, and provides explicit control over *how* to average symmetric positive-definite second order tensors in a particular application context, such as tractography, and is not limited to DTI. It exploits the trade-off between spatial and angular resolution.

Acknowledgements The Netherlands Organisation for Scientific Research (NWO) is gratefully acknowledged for financial support.

References

1. Arsigny, V., Fillard, P., Pennec, X., Ayache, N.: Log-Euclidean metrics for fast and simple calculus on diffusion tensors. Magn. Reson. Med. **56**(2), 411–421 (2006)
2. Bao, D., Chern, S.S., Shen, Z.: An Introduction to Riemann-Finsler Geometry. Graduate Texts in Mathematics, vol. 200. Springer, New York (2000)
3. Barmpoutis, A., Vemuri, B.C., Shepherd, T.M., Forder, J.R.: Tensor splines for interpolation and approximation of DT-MRI with applications to segmentation of isolated rat hippocampi. IEEE Trans. Med. Imaging **26**(11), 1537–1546 (2007)
4. Bashirov, A.E., Kurpinar, E.M., Özyapici, A.: Multiplicative calculus and its applications. J. Math. Anal. Appl. **337**, 36–48 (2008)
5. Basser, P.J., Mattiello, J., Le Bihan, D.: Estimation of the effective self-diffusion tensor from the NMR spin echo. J. Magn. Reson. **103**, 247–254 (1994)
6. Batchelor, P.G., Moakher, M., Atkinson, D., Calamante, F., Connelly, A.: A rigorous framework for diffusion tensor calculus. Magn. Reson. Med. **53**, 221–225 (2005)
7. Cartan, E.: Les Espaces de Finsler. Hermann, Paris (1934)
8. Descoteaux, M., Angelino, E., Fitzgibbons, S., Deriche, R.: Apparent diffusion coefficients from high angular resolution diffusion imaging: estimation and applications. Magn. Reson. Med. **56**(2), 395–410 (2006)
9. Descoteaux, M., Angelino, E., Fitzgibbons, S., Deriche, R.: Regularized, fast, and robust analytical Q-ball imaging. Magn. Reson. Med. **58**(3), 497–510 (2007)
10. Duits, R., Franken, E.M.: Left invariant parabolic evolution equations on $SE(2)$ and contour enhancement via invertible orientation scores, part II: nonlinear left-invariant diffusion equations on invertible orientation scores. Q. Appl. Math. **68**(2), 293–331 (2010)
11. Duits, R., Franken, E.: Left-invariant diffusions on the space of positions and orientations and their application to crossing-preserving smoothing of HARDI images. Int. J. Comput. Vis. **12**(3), 231–264 (2011). doi:10.1007/s11263-010-0332-z
12. Fick, A.: Ueber diffusion. Ann. Phys. **170**(1), 59–86 (1855). doi:10.1002/andp.18551700105
13. Finsler, P.: Ueber kurven und Flächen in allgemeinen Räumen. Ph.D. thesis, University of Göttingen, Göttingen, Germany (1918)
14. Fletcher, P.T., Joshi, S.: Riemannian geometry for the statistical analysis of diffusion tensor data. Signal Process. **87**(2), 250–262 (2007)
15. Florack, L.M.J.: Codomain scale space and regularization for high angular resolution diffusion imaging. In: Aja Fernández, S., de Luis Garcia, R. (eds.) CVPR Workshop on Tensors in Image Processing and Computer Vision, Anchorage, AK, 24–26 June 2008. IEEE, New York (2008). Digital Proceedings
16. Florack, L., Balmashnova, E.: Two canonical representations for regularized high angular resolution diffusion imaging. In: Alexander, D., Gee, J., Whitaker, R. (eds.) MICCAI Workshop on Computational Diffusion MRI, New York, 10 Sept 2008, pp. 85–96

17. Florack, L.M.J., Fuster, A.: Riemann-Finsler geometry for diffusion weighted magnetic resonance imaging. In: Westin, C.F., et al. (eds.) Visualization and Processing of Tensors and Higher Order Descriptors for Multi-Valued Data. Mathematics and Visualization, pp. 189–208. Springer, New York (2014). doi:10.1007/978-3-642-54301-2
18. Florack, L., van Assen, H.: Multiplicative calculus in biomedical image analysis. J. Math. Imaging Vision **42**(1), 64–75 (2012)
19. Florack, L., Balmashnova, E., Astola, L., Brunenberg, E.: A new tensorial framework for single-shell high angular resolution diffusion imaging. J. Math. Imaging Vision **3**(38), 171–181 (2010). doi:10.1007/s10851-010-0217-3
20. Gahm, J.K., Ennis, D.B.: Dyadic tensor-based interpolation of tensor orientation: application to cardiac DT-MRI. In: Camara, O., Mansi, T., Pop, M., Rhode, K., Sermesant, M., Young, A. (eds.) Fourth International Workshop on Statistical Atlases and Computational Models of the Heart: Imaging and Modelling Challenges (STACOM 2013), Nagoya, 26 Sept 2013. Lecture Notes in Computer Science, vol. 8330, pp. 135–142. Springer, Berlin (2014)
21. Gahm, J.K., Wisniewski, N., Kindlmann, G., Kung, G.L., Klug, W.S., Garfinkel, A., Ennis, D.B.: Linear invariant tensor interpolation applied to cardiac diffusion tensor MRI. In: Ayache, N., Delingette, H., Golland, P., Mori, K. (eds.) Proceedings of the 15th International Conference on Medical Image Computing and Computer Assisted Intervention—MICCAI 2012, Nice, 1–5 Oct 2012. Lecture Notes in Computer Science, vol. 7510–7512, pp. 494–501. Springer, Berlin (2012)
22. Gibbons, G.W., Gomis, J., Pope, C.N.: General very special relativity is Finsler geometry. Phys. Rev. D **76**(8), 081701 (2007). doi:10.1103/PhysRevD.76.081701. http://www.link.aps.org/doi/10.1103/PhysRevD.76.081701
23. Grossman, M., Katz, R.: Non-Newtonian Calculus. Lee Press, Pigeon Cove (1972)
24. Guenther, R.A.: Product integrals and sum integrals. Int. J. Math. Educ. Sci. Technol. **14**(2), 243–249 (1983)
25. Hagmann, P., Jonasson, L., Maeder, P., Thiran, J.P., Wedeen, V.J., Meuli, R.: Understanding diffusion MR imaging techniques: from scalar diffusion-weighted imaging to diffusion tensor imaging and beyond. RadioGraphics **26**, S205–S223 (2006)
26. Hess, C.P., Mukherjee, P., Tan, E.T., Xu, D., Vigneron, D.B.: Q-ball reconstruction of multimodal fiber orientations using the spherical harmonic basis. Magn. Reson. Med. **56**, 104–117 (2006)
27. Jensen, J.H., Helpern, J.A., Ramani, A., Lu, H., Kaczynski, K.: Diffusional kurtosis imaging: the quantification of non-Gaussian water diffusion by means of magnetic resonance imaging. Magn. Reson. Med. **53**(6), 1432–1440 (2005)
28. Jian, B., Vemuri, B.C., Özarslan, E., Carney, P.R., Mareci, T.H.: A novel tensor distribution model for the diffusion-weighted MR signal. NeuroImage **37**(1), 164–176 (2007)
29. Jonasson, L., Hagmann, P., Bresson, X., Thiran, J.P., Van Wedeen, J.: Representing diffusion MRI in 5D for segmentation of white matter tracts with a level set method. In: Christensen, G.E., Sonka, M. (eds.) Proceedings of the 18th International Conference on Information Processing in Medical Imaging–IPMI 2005, Glenwood Springs. Lecture Notes in Computer Science, vol. 3565, pp. 311–320. Springer, Berlin (2005)
30. Kindlmann, G., San Jose Estepar, R., Niethammer, M., Haker, S., Westin, C.F.: Geodesic-loxodromes for diffusion tensor interpolation and difference measurement. In: Ayache, N., Ourselin, S., Maeder, A. (eds.) Proceedings of the 10th International Conference on Medical Image Computing and Computer-Assisted Intervention—MICCAI 2007, Brisbane, 29 Oct–2 Nov 2007. Lecture Notes in Computer Science, vols. 4791–4792, pp. 1–9. Springer, Berlin (2007)
31. Liu, C., Bammer, R., Acar, B., Moseley, M.E.: Characterizing non-Gaussian diffusion by using generalized diffusion tensors. Magn. Reson. Med. **51**(5), 924–937 (2004)
32. Novikov, D.S., Kiselev, V.: Effective medium theory of a diffusion-weighted signal. NMR Biomed. **23**, 682–697 (2010)
33. Riemann, B. (ed.): On the hypotheses which lie at the bases of geometry. Nature **8**(183, 184), 14–17, 36–37 (1873). Translated by William Kingdon Clifford

34. Özarslan, E., Mareci, T.H.: Generalized diffusion tensor imaging and analytical relationships between diffusion tensor imaging and high angular resolution imaging. Magn. Reson. Med. **50**(5), 955–965 (2003)
35. Özarslan, E., Shepherd, T.M., Vemuri, B.C., Blackband, S.J., Mareci, T.H.: Resolution of complex tissue microarchitecture using the diffusion orientation transform (DOT). NeuroImage **31**, 1086–1103 (2006)
36. Pennec, X., Fillard, P., Ayache, N.: A Riemannian framework for tensor computing. Int. J. Comput. Vis. **66**(1), 41–66 (2006)
37. Riemann, B.: Über die Hypothesen, welche der Geometrie zu Grunde liegen. In: Weber, H. (ed.) Gesammelte Mathematische Werke, pp. 272–287. Teubner, Leibzig (1892)
38. Rund, H.: The Differential Geometry of Finsler Spaces. Springer, Berlin (1959)
39. Rund, H.: The Hamilton-Jacobi Theory in the Calculus of Variations. Robert E. Krieger Publishing Company, Huntington (1973)
40. Schultz, T.: Towards resolving fiber crossings with higher order tensor inpainting. In: Laidlaw, D.H., Vilanova, A. (eds.) New Developments in the Visualization and Processing of Tensors Fields. Mathematics and Visualization, pp. 253–265. Springer, New York (2012)
41. Schultz, T., Fuster, A., Ghosh, A., Deriche, R., Florack, L., Lim, L.H.: Higher order tensors in diffusion imaging. In: Westin, C.F., et al. (eds.) Visualization and Processing of Tensors and Higher Order Descriptors for Multi-Valued Data. Mathematics and Visualization, pp. 129–162. Springer, New York (2014). doi:10.1007/978-3-642-54301-2
42. Schultz, T., Vilanova, A., Brecheisen, R., Kindlmann, G.: Fuzzy fibers: uncertainty in dMRI tractography. In: Hansen, C., Chen, M., Johnson, C., Kaufman, A., Hagen, H. (eds.) Scientific Visualization: Uncertainty, Multifield, Biomedical, and Scalable Visualization, pp. 79–92. Springer, New York (2014)
43. Stanley, D.: A multiplicative calculus. PRIMUS: Probl. Resour. Issues Math. Undergrad. Stud. **IX**(4), 310–326 (1999)
44. Stejskal, E.O., Tanner, J.E.: Spin diffusion measurements: spin echoes in the presence of a time-dependent field gradient. J. Chem. Phys. **42**(1), 288–292 (1965)
45. Torrey, H.C.: Bloch equations with diffusion terms. Phys. Rev. D **104**, 563–565 (1956)
46. Tuch, D.S.: Q-ball imaging. Magn. Reson. Med. **52**, 1358–1372 (2004)
47. Volterra, V.: Sulle equazioni differenziali lineari. Rendiconti Accademia dei Lincei (Series 4) **3**, 393–396 (1887)
48. Weickert, J., Welk, M.: Tensor field interpolation with PDEs. In: Weickert, J., Hagen, H. (eds.) Visualization and Processing of Tensor Fields. Mathematics and Visualization, Chap. 19, pp. 315–325. Springer, Berlin (2006)
49. Yang, F., Zhu, Y.M., Magnin, I.E., Luo, J.H., Croisille, P., Kingsley, P.B.: Feature-based interpolation of diffusion tensor fields and application to human cardiac DT-MRI. Med. Image Anal. **16**(2), 459–481 (2012)

Tensor Voting: Current State, Challenges and New Trends in the Context of Medical Image Analysis

Daniel Jörgens and Rodrigo Moreno

Abstract Perceptual organisation techniques aim at mimicking the human visual system for extracting salient information from noisy images. Tensor voting has been one of the most versatile of those methods, with many different applications both in computer vision and medical image analysis. Its strategy consists in propagating local information encoded through tensors by means of perception-inspired rules. Although it has been used for more than a decade, there are still many unsolved theoretical issues that have made it challenging to apply it to more problems, especially in analysis of medical images.

The main aim of this chapter is to review the current state of the research in tensor voting, to summarise its present challenges, and to describe the new trends that we foresee will drive the research in this field in the next few years. Also, we discuss extensions of tensor voting that could lead to potential performance improvements and that could make it suitable for further medical applications.

1 Introduction

Applications in the field of medical image analysis aim at extracting any kind of information from images that can be used within a medical context. These images are obtained by a broad range of different imaging modalities. Depending on their distinct characteristics concerning the resulting output images, each such modality is suitable for specific medical applications. In fact, in some cases a combination of several modalities is required.

A common problem of many modalities is the tradeoff between the achievable image quality and a limiting parameter inherent to the particular imaging device. For example, in the case of acquiring images in vivo through Computed Tomography (CT), regarding safety considerations, the radiation dose is a limiting factor for a higher contrast. At the same time, low doses result in a lower signal to noise

D. Jörgens • R. Moreno (✉)
School of Technology and Health, KTH Flemingsberg, Alfred Nobels Allé 10, 14152 Huddinge, Sweden
e-mail: daniel.jorgens@sth.kth.se; rodrigo.moreno@sth.kth.se

I. Hotz, T. Schultz (eds.), *Visualization and Processing of Higher Order Descriptors for Multi-Valued Data*, Mathematics and Visualization,
DOI 10.1007/978-3-319-15090-1_9

ratio (SNR) and therefore in decreased image quality. As another example, in Magnetic Resonance Imaging (MRI) higher resolutions usually result in longer acquisition times. In turn this makes the gathered signal more prone to artifacts due to movements of the patient like breathing and thereby leads to lower image quality as well.

In order to be able to retrieve as much reliable information as possible from images even in the case of low resolution and the presence of noise, the applied analysis methods must provide a robust behaviour in those contexts. In the past, perceptual organisation methods like tensor voting have proven to be suitable in such cases, as they have successfully been used within noisy scenarios in related applications like colour image segmentation [27], image restoration, curvature estimation [6, 14, 39, 41], edge detection [26, 31] or colour image denoising [28]. The strength of perceptual algorithms is that they usually do not rely on any prior knowledge about the image contents. Instead, they follow certain predefined rules that try to imitate the behaviour of the human visual system (HVS). As an example the authors of [3] developed a model based on the architecture of the visual cortex, which they show is capable of perceptual completion of structures. Referring to the structure of the visual cortex as well, stochastic completion fields are used in [44] for the inference of illusory contours. This technique was in fact shown to be strongly related to tensor voting (cf. [43]). Also founded—similarly to tensor voting—on the principles of Gestalt psychology (cf. [11]) the work on structural saliency in [38] aims at detecting global structures from local (image) features. An overview on further perceptual organisation methods subdivided in different categories can be found in [24].

Being one of the most versatile of these techniques, tensor voting has been applied to a variety of problems in the field of computer vision and has been successfully utilised in problems within medical image analysis as well. These include the correction of small segmentation errors of segmented blood vessels [34], the enhancement of ultrasound images [9], detection of adherens junctions in microscope images [16] (see also Fig. 1), surface inference for dental CAD/CAM [40] as well as cell tracking in [13]. Therefore we believe that it might also be beneficial in further applications like blood vessel segmentation, detection of bifurcations, detection of separation points and vortices in blood flow images, tractography or identification of nodes within the mesh-like structure of trabecular bone. The method retrieves salient information from the input data by basically following three steps. First, local features are encoded as tensors. After that, these are propagated in a neighborhood following the principles of proximity, good continuation and similarity borrowed from Gestalt psychology as mentioned above. Thereby, the underlying idea is that compatible features should reinforce each other whereas those which are contrary are disregarded. Finally, the resulting tensor field is processed in order to gather the desired information.

Tensor voting is on the one hand flexible for any kind of adaptions, as it constitutes a theoretical, mathematical framework. On the other hand, this flexibility comes at the cost of requiring extra efforts to adapt it to a particular application. Thereby, the key tasks for any adaption are, first, to find a way to encode the

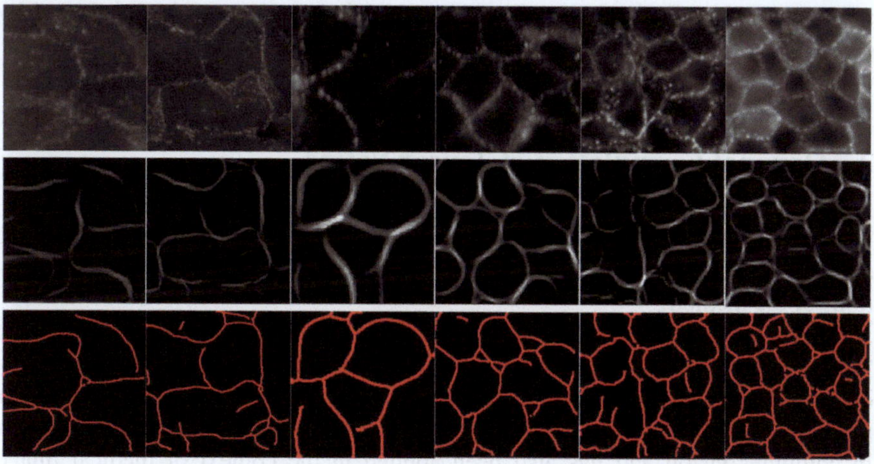

Fig. 1 In this example from [16] tensor voting is used to extract the curvilinear structures of *adherens junctions* from microscope images. The *first row* shows the original data in chosen regions, the *second* contains the resulting structures after iterative tensor voting. The inferred adherens junctions obtained as binary data after morphological thinning are depicted in the *third row*

features of interest by means of tensors, second, to extend the method for a particular type of input data and, in the end, to find a formulation that allows for an efficient implementation.

As described in [18], the original formulation of tensor voting in [20] provided promising results in different applications. Nevertheless, some issues, as for example its significantly high execution time due to several steps of numerical integration, limited its usability in many cases. Subsequent reformulations of the original equations, e.g. in [8, 29] or [25], proposed efficient modifications which led to a considerably reduced execution time of the algorithm. Although there have been even more approaches that dealt with further open problems, tensor voting still requires some more extensions in order to use it in applications like those mentioned above. Extended formulations for different types of input data like greyscale, vector- or tensor-valued images as well as modifications for the use of higher-order tensors and the handling of multiscale data can help to broaden the range of possible applications for this method. Even though there exist proposed solutions for some of these problems, there is still work to do on generalising underlying assumptions and further improving the performance.

Therefore, this chapter aims at further describing the current challenges in tensor voting and depicting possible approaches that, in our opinion, will be followed in the next few years in this field of research.

In the next section, an introduction to the original formulation of tensor voting is given at first, followed by a brief review on efficiency-related extensions. After that, we depict open issues inherent to tensor voting and briefly review attempts to overcome these in combination with a description of further possible solution strategies. In the final section, the main arguments are summarised followed by concluding remarks on the general implementation of these extensions.

2 Original Formulation of Tensor Voting

This section deals with the ideas that drove the original formulation of tensor voting and the modifications proposed to increase the efficiency of possible implementations of the method.

The first formulation of tensor voting (cf. [10, 20]), now referred to as the "original" or "classical" formulation, was applied to sparse and noisy three-dimensional data sets in order to extract surfaces, 3D curves and junctions from them. Since its introduction, the framework has been adopted to different problems in computer vision and has also been applied in the context of medical image analysis recently (cf. [9, 13, 16, 34] already mentioned above). Despite its success, applying tensor voting to specific medical applications requires further extensions of the method like the ones discussed in the next section.

In the following, we focus on the formulation of tensor voting in the case of three dimensions. Indeed, it can be formulated for spaces of arbitrary dimension $n \geq 2$. A description of the general formulation in the n-dimensional case is given in [18].

As mentioned above, tensor voting consists of three basic steps: Encoding of certain image features as tensors, propagation of these within the area of interest and, finally, analysis of the resulting tensor map to extract salient structures like those mentioned above.

The first step of encoding features requires the definition of these beforehand. Usually, estimated normals or tangents to structures found at certain positions in the input data, e.g. by utilising the image gradient, are encoded by tensorisation.[1] In case no prior information about a feature is available at a given point, an identity tensor is chosen for representation. In general, the number of these so-called *input tokens* that are initialised in the first step might be lower than the total number of output tensors of tensor voting. It must be noted, that the original formulation utilises only second-order tensors or, more precisely, symmetric positive semidefinite matrices throughout the method. Thereby, the basic idea is the observation that there exists an eigendecomposition for such a three-dimensional tensor T of second order, given by

$$\mathsf{T} = \sum_{i=1}^{3} \lambda_i \, \mathbf{e}_i \mathbf{e}_i^T \ , \tag{1}$$

[1]In this case tensorisation denotes the mapping $t : \mathbb{R}^3 \to \mathbb{R}^{3\times 3}$ with $t(\mathbf{n}) = \mathbf{n}\mathbf{n}^T \ \forall_{\mathbf{n}\in\mathbb{R}^3}$, also referred to as the *dyadic product* or the *outer product*.

λ_i and $\mathbf{e}_i$ being the ordered eigenvalues and their corresponding eigenvectors, respectively. By rewriting (1), one can derive three so-called *extreme cases*, that is tensors that encode only one of the mentioned features, as there are *curves*, *surfaces* and *junctions*. Based on their definitions as

$$\begin{aligned} \mathsf{S} &= (\lambda_1 - \lambda_2)\,\mathbf{e}_1\mathbf{e}_1^T\ , \\ \mathsf{P} &= (\lambda_2 - \lambda_3)\,(\mathbf{e}_1\mathbf{e}_1^T + \mathbf{e}_2\mathbf{e}_2^T)\ , \\ \mathsf{B} &= \lambda_3\,(\mathbf{e}_1\mathbf{e}_1^T + \mathbf{e}_2\mathbf{e}_2^T + \mathbf{e}_3\mathbf{e}_3^T)\ , \end{aligned} \tag{2}$$

the tensor T can then be expressed as

$$\mathsf{T} = \mathsf{S} + \mathsf{P} + \mathsf{B}\,, \tag{3}$$

where S, P and B are called the *stick*, *plate* and *ball* components of T. These usually indicate the *normal space* of the particular encoded features. Figure 2 illustrates this

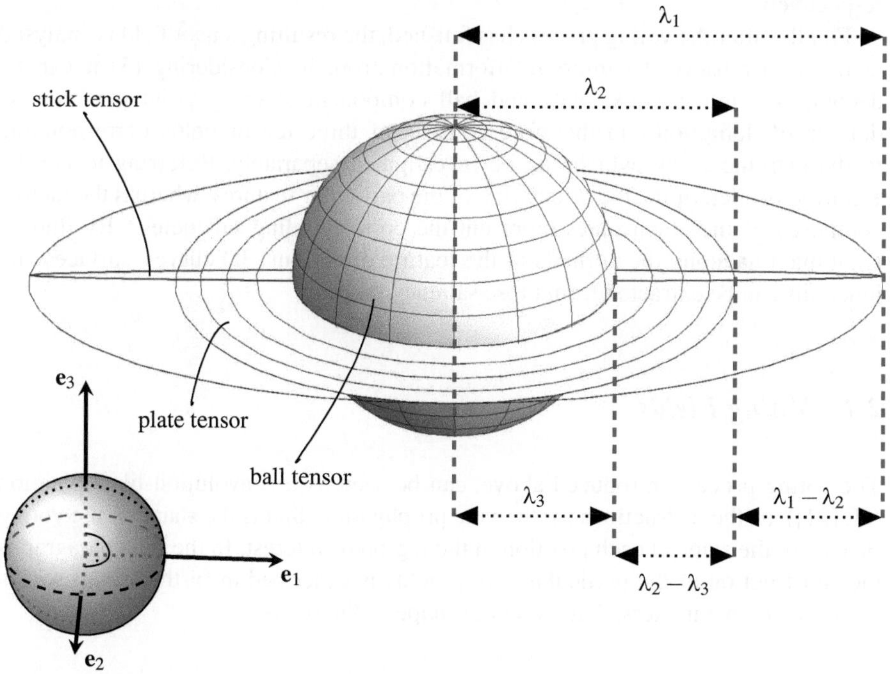

Fig. 2 Following (3), a three-dimensional second-order tensor T can be decomposed into a stick, plate and ball component. Their orientation is defined by the eigenvectors $\mathbf{e}_i$ of T and their saliency, i.e. their size, is determined by the combination of the eigenvalues λ_i as described in (2). This figure is based on [42]

decomposition using the ellipsoid-based representation of tensors of second order and provides additional interpretations of the factors in (2).

Following up, it is the step of propagation or *voting* during which each tensor-encoded feature induces a tensor (or voting) field describing its most likely continuation at all points within a given neighbourhood. These voting fields contain the implementation of the aforementioned perceptual rules of proximity, good continuation and similarity. In fact, the stick, plate and ball components of a specific tensor T generate their own respective tensor fields which, after summing them up, represent the overall voting field induced by T. Regarding that, the tensor vote TV received at a particular point $\mathbf{p}$ is then retrieved by integrating the votes collected from all points in its neighbourhood $\Omega(\mathbf{p})$. In accordance with [30] this yields

$$TV(\mathbf{p}) = \sum_{\substack{\mathbf{q} \in \Omega(\mathbf{p}),\\ \mathbf{v} = \mathbf{p} - \mathbf{q}}} SV(\mathbf{v}, \mathsf{S}_\mathbf{q}) + PV(\mathbf{v}, \mathsf{P}_\mathbf{q}) + BV(\mathbf{v}, \mathsf{B}_\mathbf{q}) \,, \tag{4}$$

where SV, PV and BV denote the particular stick, plate and ball voting field, and $\mathsf{S}_\mathbf{q}$, $\mathsf{P}_\mathbf{q}$ and $\mathsf{B}_\mathbf{q}$ the stick, plate and ball component of the tensor located at $\mathbf{q}$, respectively.

Finally, after the voting process has finished, the resulting tensor field is analysed in order to retrieve the inferred information from it. Considering (3) it can be decomposed into a stick, plate and ball component at every point $\mathbf{p}$ within its domain of definition and therefore consists of three tensor fields corresponding to the extreme cases, which can be investigated separately. Referring to (2) the eigenvectors determine the orientation of the particular features, whereas the factors composed of the eigenvalues represent the corresponding saliencies.[2] By finding local maxima along the normals of the feature directions, 3D curves, surfaces and junctions can be extracted from those saliency maps.

2.1 Voting Fields

The voting process introduced above, can be seen as a convolution-like operation (cf. [21]), as the instructions for the vote propagation, that is the shape of the voting fields, are the same at each position in the region of interest. In the next paragraphs the construction of the particular voting fields is described in further detail with a focus on key parameters, that affect the shape of the fields.

[2]The saliencies are also referred to as *curveness*, *surfaceness* and *junctionness*.

2.1.1 Stick Voting

The design of the stick voting field specially aims at implementing the principles of good continuation and proximity. It is postulated that the most likely continuation of a surface is that with the least curvature, which means straight lines are preferred over sharp edges, and a constant curvature, i.e. the connection of the points that cast and receive the vote, referred to as *voter* and *receiver*, is depicted by the osculating circle, as illustrated in Fig. 3. Thereby, the direction of the voting tensor, that is e.g. the normal to the most likely continuation of the surface, is aligned with the radius vector of the receiver position with respect to the centre of the circle. To further model proximity within the formulation, a Gaussian decaying function is introduced that decreases along with a larger distance between voter and receiver and with the curvature of the connecting circular arc. Thereby, the order of decay can be adjusted by the standard deviation parameter of the Gaussian, which mainly determines whether preferably large or small structures will be inferred within the voting step.

In summary, the stick voting consists of two parts: On the one hand, the tensorial function $R_{\mathbf{t}_2}$[3] performs a change of basis which is basically achieved through a rotation of the system of eigenvectors of the voting tensor by a given angle around the rotation axis $\mathbf{t}_2 = \mathbf{v} \times \mathbf{e}_1$ as depicted in Fig. 3. In this way the *orientation* of the tensor vote is determined. Second, the scalar decay function d controls the *strength*

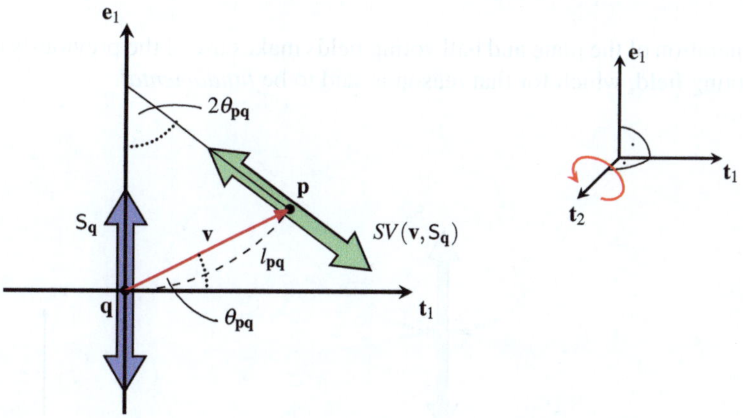

Fig. 3 Based on the explanations of Sect. 2.1.1, the construction of a stick vote $SV(\mathbf{v}, \mathsf{S}_\mathbf{q})$ at position $\mathbf{p}$ induced by the stick tensor $\mathsf{S}_\mathbf{q}$ at position $\mathbf{q}$ is visualised within its *local* coordinate system. This is defined by the principal eigenvector of $\mathsf{S}_\mathbf{q}$, denoted $\mathbf{e}_1$, and the tangent space to the encoded surface at $\mathbf{q}$ spanned by $\mathbf{t}_1$ and $\mathbf{t}_2$, whereby the latter is oriented perpendicularly to the image plane

[3]Function $R_\mathbf{t}$ is defined as $R_\mathbf{t}(\alpha, \cdot) : \mathbb{R}^{3\times3} \rightarrow \mathbb{R}^{3\times3}$ with $R_\mathbf{t}(\alpha, \mathsf{S}) = Q_{\alpha,\mathbf{t}}\ \mathsf{S}\ Q_{\alpha,\mathbf{t}}^T\ \forall_{\mathsf{S}\in\mathbb{R}^{3\times3}}$, where $Q_{\alpha,\mathbf{t}} \in SO(3)$ performs a rotation of angle α around axis $\mathbf{t}$.

of the propagated feature. This leads to

$$SV(\mathbf{v}, \mathsf{S}_\mathbf{q}) = d(\sigma_l, b_\kappa; l_\mathbf{pq}, \kappa_\mathbf{pq})\, R_{\mathbf{t}_2}(2\theta_\mathbf{pq}, \mathsf{S}_\mathbf{q})\,, \tag{5}$$

where $l_\mathbf{pq} \equiv l(\mathbf{v}, \theta_\mathbf{pq})$ denotes the arc length of the osculating circle between $\mathbf{p}$ and $\mathbf{q}$, $\kappa_\mathbf{pq} \equiv \kappa(\mathbf{v}, \theta_\mathbf{pq})$ the corresponding curvature and $\theta_\mathbf{pq}$ the angle enclosed by the direct connection $\mathbf{v} = \mathbf{p} - \mathbf{q}$ between voter and receiver and the tangent space of the surface encoded by the voter $\mathbf{q}$ (cf. Fig. 3). Following [30], d might be chosen as

$$d(\sigma_l, b_\kappa; l_\mathbf{pq}, \kappa_\mathbf{pq}) = \begin{cases} e^{-\frac{l^2}{2\sigma_l^2} - b_\kappa \kappa^2} & \text{if } \theta_\mathbf{pq} \in [-\frac{\pi}{4}, \frac{\pi}{4}] \\ 0 & \text{otherwise} \end{cases}. \tag{6}$$

It is obvious that the only degrees of freedom for the construction of the stick voting field are the scale parameter σ_l and the weighting parameter b_κ for adjusting the influence of the curvature within the exponent of the decay function d.

As large values of $\theta_\mathbf{pq}$ do not produce likely continuations of an encoded feature, the angle is limited to the range of $[-\frac{\pi}{4}, \frac{\pi}{4}]$ in (6) which is equivalent to restricting the voting field SV to a cone. Figure 4 shows a two-dimensional slice of the saliencies induced by the stick voting field.

2.1.2 Plate and Ball Voting

The generation of the plate and ball voting fields makes use of the previously defined stick voting field, which for that reason is said to be *fundamental*.

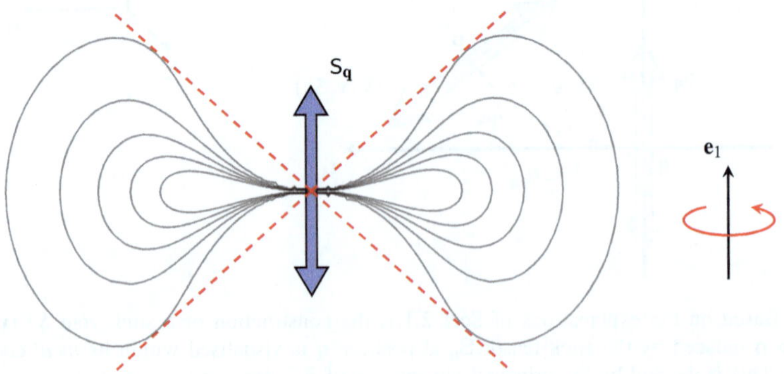

Fig. 4 A slice of the voting field induced by the stick tensor $\mathsf{S}_\mathbf{q}$ is schematically depicted by a number of *contour lines* of its saliencies. As the angle $\theta_\mathbf{pq}$ is restricted to the range of $[-\frac{\pi}{4}, \frac{\pi}{4}]$ the saliencies are equal to zero outside the cone that is indicated by the *dashed lines*. Actually, the three-dimensional field is retrieved by a rotation around the axis defined by the principal eigenvector $\mathbf{e}_1$ of $\mathsf{S}_\mathbf{q}$, as indicated in the figure

Fig. 5 The voting field of the plate tensor P is constructed by integrating the voting fields of all possible stick tensors $R_{\mathbf{t}}(\beta, \mathsf{S_P})$ within the plate. These are retrieved by rotating a particular stick tensor $\mathsf{S_P}$ contained in P by an angle β around the rotation axis **t**, which depicts the tangent space to the encoded curve, i.e. it is perpendicular to the eigenvectors $\mathbf{e}_1$ and $\mathbf{e}_2$ of P. In this figure only a finite number of possible stick tensors within P is exemplary depicted

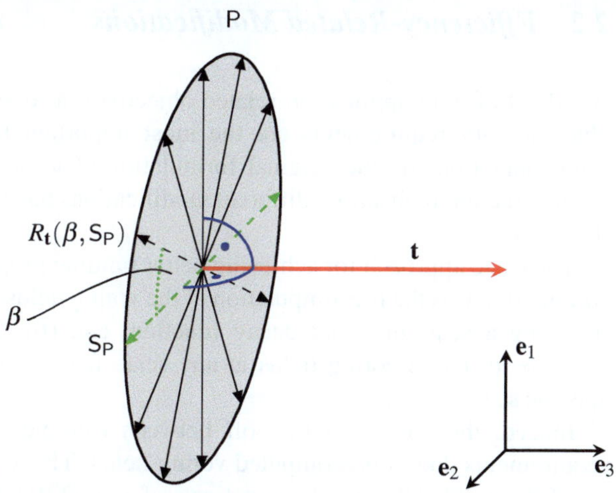

The relationship between plate and stick voting is based on the observation that a plate tensor can always "be decomposed into all possible *stick* tensors inside the *plate*" (cf. [30]). These are defined by means of rotations $R_{\mathbf{t}}(\beta, \mathsf{S_P})$, where $\mathsf{S_P}$ is an arbitrary stick tensor within the plate P and β the rotation angle within the normal space of the curve encoded by P (cf. Fig. 5). In fact, the plate vote is constructed by integrating the induced voting fields of all these stick tensors and therefore directly depends on the definition of the fundamental stick voting field, which can be formulated as

$$PV(\mathbf{v}, \mathsf{P_q}) \equiv PV(SV; \mathbf{v}, \mathsf{P_q}) = \frac{1}{\pi} \int_0^{2\pi} SV(\mathbf{v}, R_{\mathbf{t}}(\beta, \mathsf{S}_{\mathsf{P_q}}))\, \mathrm{d}\beta\,, \tag{7}$$

where $\mathbf{v} = \mathbf{p} - \mathbf{q}$.

In case of the ball voting the derivation is similar to that of plate voting with the only difference, that a second degree of freedom is added by means of an additional rotation angle. The decomposition of a ball tensor into all possible stick tensors within its defined sphere, can be seen as a subsequent application of two rotations of a specific stick within the ball tensor around two axes $\mathbf{r}_1$ and $\mathbf{r}_2$ that are perpendicular to each other as well as to that stick. Given the angles β_1 and β_2 with respect to these axes, the possible stick tensors can be written as $R_{\mathbf{r}_2}(\beta_2, R_{\mathbf{r}_1}(\beta_1, \mathsf{S_B}))$, where $\mathsf{S_B}$ denotes a random stick tensor within B. Integration of the induced voting fields of all these stick tensors yields the ball voting field, which again directly depends on the fundamental stick voting field, modelled as

$$\begin{aligned} BV(\mathbf{v}, \mathsf{B_q}) &\equiv BV(SV; \mathbf{v}, \mathsf{B_q}) \\ &= \frac{3}{4\pi} \int_0^{2\pi} \int_0^{\pi} SV(\mathbf{v}, R_{\mathbf{r}_2}(\beta_2, R_{\mathbf{r}_1}(\beta_1, \mathsf{S}_{\mathsf{B_q}}))) \sin(\beta_1)\, \mathrm{d}\beta_1 \mathrm{d}\beta_2\,, \end{aligned} \tag{8}$$

considering $\mathbf{v} = \mathbf{p} - \mathbf{q}$ again.

2.2 Efficiency-Related Modifications

Besides fulfilling application-related objectives, a reasonable runtime and reasonable memory requirements are the most important factors for the usage of an implementation. As the original formulation of tensor voting does not allow for a time efficient realisation, different modifications have been proposed to overcome this issue.

The basic approach for achieving better runtime properties in case of the original formulation was the precomputation of the voting fields. This issue can be addressed by using a scale-invariant decay function, e.g. (6). This modification makes it possible to derive voting fields at any scale from a single precomputed one at a specific scale.

Indeed, there exists a tradeoff between runtime considerations and memory requirements due to precomputed voting fields. This especially plays an important role for possible GPU implementations (cf. e.g. [22]) where "on-device" memory is a rather limited resource.

Focussing on the runtime of tensor voting implementations, further propositions mainly aim at avoiding the numerical integrations for the calculations of the plate and ball voting fields, since these are the most time consuming parts of the implementation.

Two alternative formulations are given in [29]. The first is based on numerical approximations derived from an analysis of the original voting processes. The second arrives at simplifications by taking "the perceptual meaning of the original tensor voting" into account. Both modifications reduce the complexity of the calculation of a *single* plate or ball vote (cf. (7) and (8) respectively) to *constant* time.

The authors of [8] rewrite the 2D stick voting field in order to derive a formulation of tensor voting in terms of an ordinary convolution employing steerable filters. Taking advantage of the convolution theorem this algorithm shows a rather good performance in the case of stick voting. In [33] a generalisation of that concept to three dimensions is proposed. It is further mentioned in [8] that due to their good performance concerning *convolutions* and the *fast fourier transform*, GPUs might lead to further runtime improvements of the proposed algorithm.

Further, there are two approaches also addressing n-dimensional tensor voting. The first one mainly consists of finding a certain orthonormal basis of the normal space of the feature encoded by the voting tensor [25]. Ensuring that this basis contains a component parallel to the projection $\mathbf{v}_n$ of the connection $\mathbf{v}$ between voter and receiver onto the normal subspace amounts in an optimised voting process consisting of separate stick votes for each component. Thereby, the complete stick voting procedure is only needed for the basis vector parallel to $\mathbf{v}_n$, whereas the others are just scaled by decay depending on the distance $||\mathbf{v}||$. This leads to a complexity of $\mathcal{O}(d\gamma^2)$ per vote (cf. [45]) where d is the dimension of the considered space and the term γ^2 stems from the applied *Gram-Schmidt* algorithm for finding the orthonormal basis. For a specific application case, i.e. the dimension d is fixed, this amounts to *constant* time as above.

A closed-form solution was also proposed in [45]. However, it was proven incorrect in [17].

3 Current Challenges and Proposed Solution Strategies

This section provides the description of further issues that appear within the framework of tensor voting, related to input data and additional objectives for the method. The need for the research to address these issues is derived from particular application cases, followed by summarising the current state of research and the proposition of possible solution strategies to be further investigated in the future.

3.1 Contextual Voting

The voting procedure in its classic form only relies on the angular and spatial relation between voter and receiver. In some cases that might not be enough information to determine an appropriate vote. For example in the context of colour image denoising in [28] where a stick vote can be interpreted as voting for a specific colour, the decision if voter or receiver carry either regular or noisy information is made based on their corresponding neighbourhoods (cf. Fig. 6 for further detail). If the voter is noisy its information should be suppressed. On the other hand, a noisy receiver needs more votes in order to refine its encoded information.

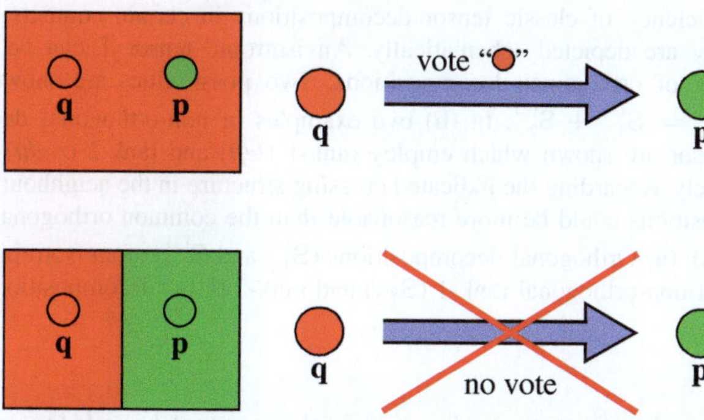

Fig. 6 This example of *contextual voting* is based on [28]. The two situations depicted in the *top* and *bottom row* only differ in the context at receiver site **p**. In case the information at **p** is different from that in its neighbourhood (*top*) the voter **q** should cast a vote whereas in case of consistency between **p** and its neighbourhood (*bottom*) no vote takes place

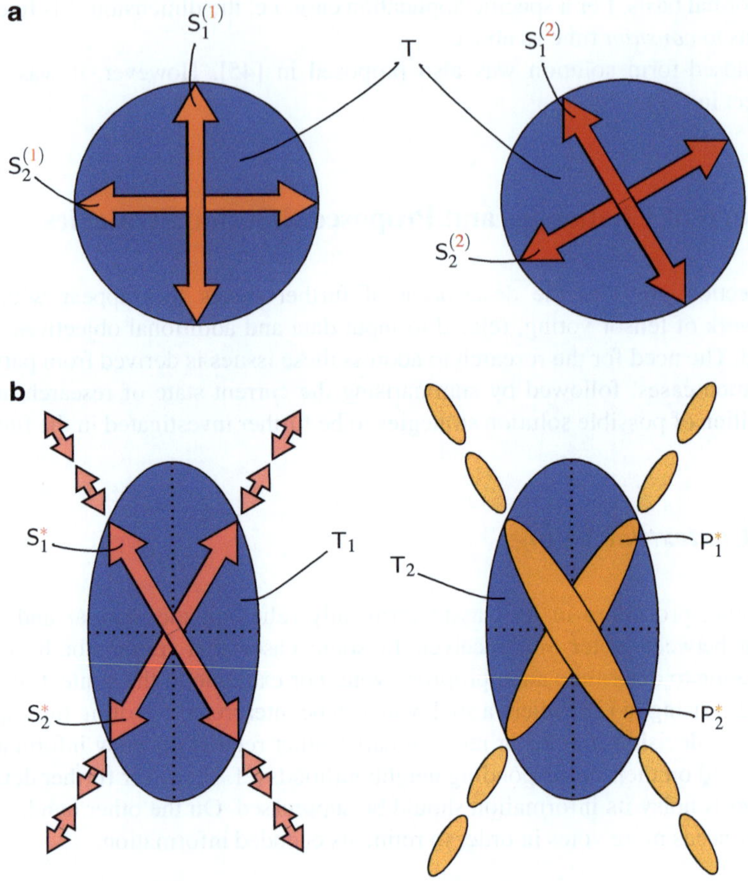

Fig. 7 The insufficiency of classic tensor decompositions in certain contexts and a proposed alternative strategy are depicted schematically. An isotropic tensor T can be decomposed in an infinite number of orthogonal decompositions. Two possibilities are shown in (**a**), where $T = S_1^{(1)} + S_2^{(1)} = S_1^{(2)} + S_2^{(2)}$. In (**b**) two examples of non-orthogonal decompositions of an anisotropic tensor are shown which employ rank-1 (*left*) and rank-2 (*right*) tensors for that purpose, respectively. Regarding the indicated crossing structure in the neighbourhoods of T_1 and T_2 these decompositions could be more reasonable than the common orthogonal one (along the *dotted* (⋯⋯) *lines*). (**a**) Orthogonal decompositions ($S_{1,2}^{(1)}$ and $S_{1,2}^{(2)}$) of an isotropic tensor (T). (**b**) Context-dependent non-orthogonal rank-1 ($S_{1,2}^*$) and rank-2 ($P_{1,2}^*$) decompositions of anisotropic tensors (T_1 and T_2)

Another example of *contextual voting* is the *context-based decomposition*. The classical eigendecomposition is not the only choice in the isotropic case. By rotation of the eigensystem one can retrieve several equivalent representations of the same isotropic tensor (cf. Fig. 7a). Also in the anisotropic case, the context, i.e. the neighbourhood, might favour a decomposition different from the one obtained from the eigenvectors. In particular, it must not necessarily be orthogonal or rank-1.

Regarding the indicated crossing structure in the neighbourhoods of T_1 and T_2 in Fig. 7b the decompositions based on $\mathsf{S}^*_{1,2}$ or $\mathsf{P}^*_{1,2}$ could be more reasonable than the common orthogonal one (along the dotted (······) lines).

One alternative for retrieving such a decomposition is to cluster the received votes at a certain position instead of adding them up. In the case of stick votes the direction of a particular stick, i.e. of the corresponding surface normal, is the base parameter for the clustering. For plate votes the direction of the encoded curve, i.e. the tangent to the feature, is used for that purpose. The particular components of the decomposition are then given by either the cluster centers or the sum of all cluster members. In view of the example in Fig. 7b, this choice might describe a tensor more appropriately in a specific context.

Generally, including such additional information from a broader neighbourhood can be seen as the implementation of additional perceptual rules. This principle can be adapted to different situations. Related examples are given in Sects. 3.3 and 3.4.

3.2 *Higher-Order Tensors*

As described in the previous section, the original formulation of tensor voting comprises only second-order tensors for encoding and propagating local features. However, in some applications second-order tensors are not appropriate to represent the desired characteristics of image structures. Figure 8 illustrates the summation of three orthogonal tensors in the second- and sixth-order case. It is obvious that in the first case it is not possible to extract the input tensors from the resulting sum, whereas the sixth-order result can still be decomposed into the three input components. This is in accordance with [35].

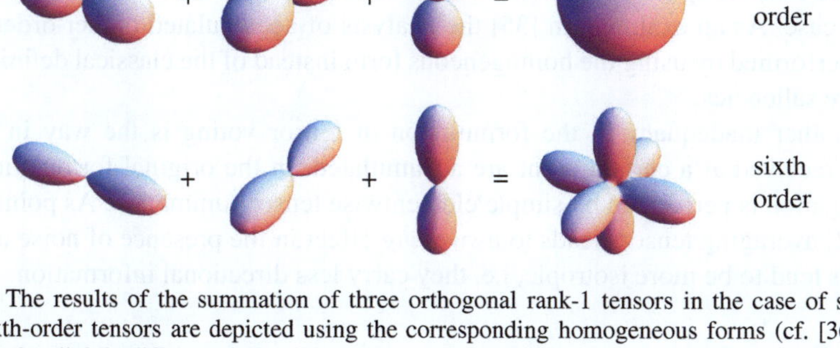

Fig. 8 The results of the summation of three orthogonal rank-1 tensors in the case of second- and sixth-order tensors are depicted using the corresponding homogeneous forms (cf. [36, 37]). Despite the slightly different shape of the rank-1 tensors, the loss of directional information in the second-order tensor sum is obvious. In that case, the decomposition of the sum into three rank-1 summands is not unique, as opposed to the sixth-order case, where the input components can still be extracted from the resulting tensor

Some applications, including analysis of blood vessels, extraction of neural tracts or detection of separation points in blood flow, involve modelling features like bifurcations[4] and crossroads.[5] As pointed out above, second-order tensors might not be appropriate in these cases because of their lack of directional information. Concerning those applications, the extension of tensor voting to higher-order tensors aims at getting improved results at Y- and X-shaped junctions.

In [35], such an extension of tensor voting is proposed as a preprocessing step for fiber tracking. Using higher-order tensor voting should overcome the lack of directional information at fiber crossings in an *inpainting* manner. Furthermore, an approach for the distinction of Y- and X-shaped crossings based on tensors of odd order, in particular of third order, is suggested. Although, both propositions led to improved fiber tracking results in the concerned areas, in the end, they only focus on inferring curve structures and do not take advantage of the whole framework, since only the stick tensor voting is applied.

A possible approach to adapt tensor voting to higher-order tensors is to decompose the latter into lower order components. Especially first- and second-order components are of interest, as classical tensor voting could be applied to these in a subsequent step. For this strategy, it is important to find efficient and appropriate higher-order tensor decompositions, since the concept of the eigendecomposition does not directly apply to higher-order tensors. Some approaches have already been suggested, like the *higher-order singular value decomposition* (HOSVD) proposed in [4] or the *lower-rank*[6] *approximation* of higher-order tensors in [36], which was actually used in [35]. However, a thorough evaluation of these two approaches must be performed in order to determine which of them is more appropriate for tensor voting.

An alternative option is to define extreme cases for higher-order tensors, in the same fashion as the ones mentioned in Sect. 2. That means, it is desired to find a decomposition into "basic" tensors that encode just one certain feature which permits to design the corresponding voting fields separately. By integrating these, the overall voting field could be obtained as before. At the same time, it must be ensured that the spirit of the classical tensor voting is still preserved in the higher-order case. As an example, in [35] the analysis of accumulated higher-order votes was performed by using the homogeneous form instead of the classical definition of feature saliencies.

Another inadequacy in the formulation of tensor voting is the way in which votes received at a certain point are accumulated. In the original formulation this aggregation is performed by simple elementwise tensor summation. As pointed out in [32] averaging tensors leads to a *swelling* effect in the presence of noise and the results tend to be more isotropic, i.e. they carry less directional information.

---

[4]That is Y-crossings.

[5]That is X-crossings.

[6]The rank of a tensor is defined as the minimal number of first-order tensors, i.e. vectors, which is needed to represent it as a sum of outer products of these.

To overcome this effect different aggregation schemes for accumulation of the tensor votes could be explored. There have been proposed different Riemannian metrics like *Euclidean*, *geometric* or *Log-Euclidean* [2] metrics, that led to different averaging methods for symmetric positive-definite matrices (cf. [23]). Including different metrics in the accumulation step of higher-order tensor votes might provide better characteristics than the former summing process.

Other aggregation schemes, for instance grouping the received votes in clusters and performing the summation only for votes within the same cluster (cf. Sect. 3.1), could be beneficial for the results in bifurcations and crossings. In fact, the grouping would require to choose an appropriate metric to measure the distance between two particular tensors of higher order.

Observing that the approximation of the angular distribution of the received votes is the central argument for employing tensors of higher order gives also rise for another approach. The propositions in [3, 7, 44] directly consider the full space of angular distributions in their framework. Using group convolution or diffusion equations the orientation information is being propagated. The method in [7] was shown to have good properties concerning the results at crossing in 2D images. Furthermore, the framework was adapted in order to apply crossing-preserving smoothing to HARDI data (cf. [5]). Since relationships between these methods and the original tensor voting have been established in the past, it is interesting to investigate how extensions of tensor voting such as the ones described in this chapter can be translated into those methods.

3.3 Multiscale Image Analysis

According to the description of tensor voting in Sect. 2, the scale parameter σ_l has strong influence on the results of tensor voting. Being the standard deviation parameter in the Gaussian decay from (5), it directly affects the shape of the voting fields. By adjusting σ_l, features of a certain size are preferred over others.

As an example, for blood vessel segmentation it is not only of interest to investigate the input data at one fixed scale, but instead over a number of different ones. This stems from the observation that thickness of blood vessels might gradually vary along their extent. There have been proposed different methods for performing multiscale analysis based on tensor voting.

The strategy in [15] is based on multiple thresholding steps, which are illustrated in Fig. 9. Initially, tensor voting is performed over several scales $\sigma_l^{(i)}$, $1 \leq i \leq I$, providing different saliency values for each particular input token, i.e. for each token there exists a mapping between scale and saliency, called the *saliency curve* (Fig. 9a). In a next step, this curve is subjected to a first thresholding procedure, which is depicted in Fig. 9b. During that procedure, the number of scales, denoted by N_k, at which the saliency lies above a given iteration-specific threshold T_s is determined for every input token, enumerated by k. Finally, as shown in Fig. 9c,

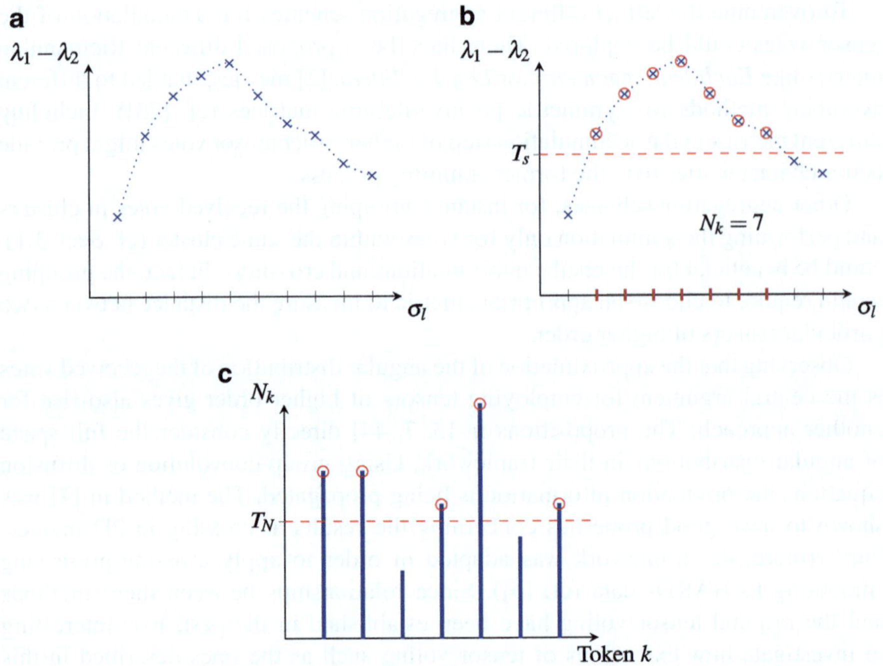

Fig. 9 Steps of one repetition of the iterative algorithm for multiscale analysis presented in [15]. (**a**) depicts the salience curve of the k-th token that basically relates the stick saliency to the scale parameter σ_l. Based on the threshold T_s the number of scales, N_k, for which the retrieved saliency of the token lies above T_s is extracted in (**b**). In the end, such tokens that satisfy $N_k > T_N$, are identified in (**c**) and kept for the next iteration starting with (**a**), whereas those below of the threshold T_N are discarded. (**a**) Saliency curve of the k-th token. (**b**) Thresholded saliency curve of the k-th token. (**c**) Thresholded quantities N_k of all tokens

another threshold T_N is applied to the extracted quantities N_k, obtaining only those tokens, which showed a sufficiently high saliency over a certain minimal number (T_N) of different scales $\sigma_l^{(i)}$. By repeating the three described steps, the authors aim at removing noise-like artifacts in each iteration based on the assumption that "non-salient segments do not exhibit consistent stability over multiple scales" which is motivated by scale space theory. At the same time this leads to less voting support for the remaining noise in the next repetitions. Thereby, the actual threshold T_s is slightly increased at each iteration in order to compensate for the strengthening of the structure saliencies due to the discarded background tokens. It is important to remark, that the number of different scales, I, as well as the number of repetitions are fixed throughout the algorithm.

Another approach, presented in [42], expands the tensor voting framework by a "first order augmentation". The authors observed, that even though the use of second-order tensors provides the capability to both measure the uncertainty and the direction of the votes themselves at the same time, the information from where the votes are received is not preserved. By using the first order augmentation through the so-called *polarity vector*, which is subsequently propagated after the usual voting steps, it is possible to determine the main direction from which the votes were received. As this vector is weighted with saliency, tokens within a particular structure should receive contrary polarity votes that cancel out. This gives the opportunity to distinguish between points located at the border of a certain feature, which means their polarity saliency is significantly high, and points that lie inside that feature, i.e. their polarity is supposed to be low (cf. Fig. 10). Considering the combination of *junction* and *polarity*, i.e. so-called *end-curve*, saliency leads to the definition of four distinct key situations which are listed in Table 1 together with the authors' interpretations within the particular medical applications described in [42]. Based on their aim to extract smooth and continuous surfaces they claim, that only the case of both low junction and polarity values at the same time describes a desired, that is smooth and continuous, feature.

By introducing an iterative and adaptive algorithm, one can now analyse the image data regarding several scales. After tensor voting was applied using a

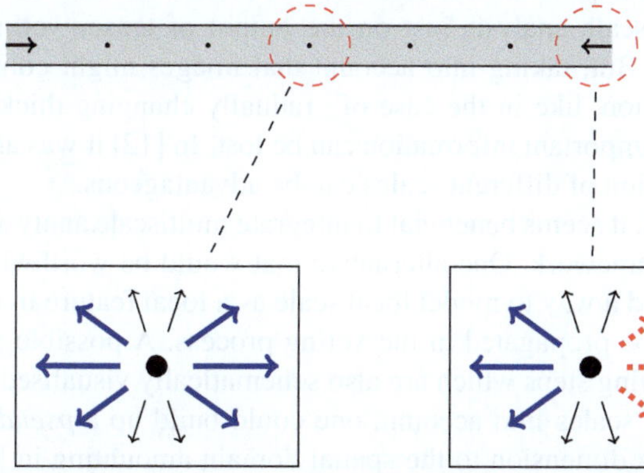

Fig. 10 By using the example of a *straight line* structure with two endpoints this figure schematically illustrates the concept of the polarity vector. In case of tokens within the interior of the structure the received polarity vectors, which are in fact weighted with the specific structure saliency, cancel out. Opposed to that, the endpoints obtain a non-null vector pointing to the interior of the line, which indicates the presence of a structure border in their neighbourhood

Table 1 This table lists the key situations for the classification of a token's neighbourhood based on the tuple of junction saliency and end-curve saliency defined in [42]

Junction saliency	End-curve saliency	Interpretation
High	High	Highly convoluted surfaces; neither smooth nor continuous
High	Low	Salient separated point; probably due to missing data or noise
Low	High	Discontinuous border of a structure without junction
Low	Low	Smoothness and continuity constraints fulfilled

The proposed iterative algorithm for multiscale analysis is based on the particular interpretations of these four combinations, which are based on the objective to infer mainly smooth surfaces certain scale parameter, the neighbourhood of each point is investigated regarding junctionness and polarity. If both values are low, a particular token is assumed to be valid and is therefore disregarded during the next iterations. In the other three cases the scale parameter is increased to promote smoothness through features of larger scale and the previous steps of voting and investigation are repeated again. As the presented medical applications in [42] only comprise the extraction of relatively smooth surfaces of organs or other tissue layers, the authors assume the particular case of high junctionness and low polarity, which indicates a separated point junction, to be unlikely to occur in their experiments. Instead they assume that it was induced by either noise or missing data. In that case, before approaching to the next iteration, all tokens in the investigated neighbourhood are removed, so that the thereby created gap can be closed by tensor voting at a larger scale.

Although the two modifications show good performance, it is questionable if discarding information in the analysis process of noisy data is a suitable strategy in a medical context. Instead, an approach based on the direct integration of scales in the formulation, like—though in a different context—in [12], appears to be more appropriate for medical images. Furthermore, both methods introduced above do not take into account possible interactions between different scales, since they perform their multiscale analysis just on the output of tensor voting at different independent scales. But, taking into account that images might convey important *inter-scale* information, like in the case of gradually changing thickness of blood vessels, potentially important information can be lost. In [12] it was also shown that combining information of different scales can be advantageous.

For these reasons, it seems beneficial to integrate multiscale analysis directly into the tensor voting framework. One alternative that would be worthwhile to explore for doing so is to find a way to model local scale as a local feature in the input data. Then it could also be propagated in the voting process. A possible strategy could consist of the following steps which are also schematically visualised in Fig. 11.

To take different scales into account, one could build up a *pseudo-scale-space* by adding scale as a dimension to the spatial domain amounting in $\mathbb{R}^n \times \mathbb{R}$ as the considered space. Starting with an n-dimensional tensor field whose tokens encode local features as before, the voting procedure is performed at several scales $\sigma_l^{(i)}$, $1 \leq i \leq I$, which can be seen as a pure spatial or *intra-scale* voting (Fig. 11a). By this, features of different scales are inferred separately.

Subsequently, information at different scales can be propagated between those in an *inter-scale* voting procedure. Again, this can be decomposed into further substeps. On the one hand a *global* inter-scale voting is performed (Fig. 11b), i.e. the receiver in $\mathbb{R}^n \times \mathbb{R}$ has both, different scale and a different position in $\mathbb{R}^n$ than the voter. On the other hand *local* inter-scale voting along the scale axis only (i.e. fixed position in $\mathbb{R}^n$, cf. Fig. 11c) could enforce the scale information at a certain site.

In the end, the analysis step can consist of two strategies. Either global features are extracted at each scale first and are combined afterwards, or tensors are locally combined first and then global structures are extracted.

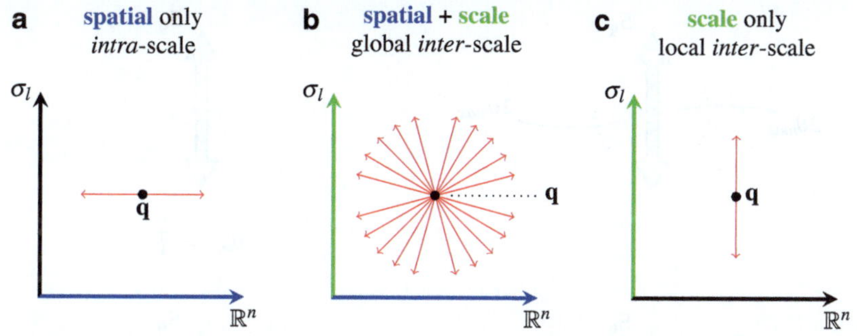

Fig. 11 The proposed voting strategy that takes interactions between multiple scales into account consists of three steps. In (**a**), the common voting is performed in the spatial domain separately for different scales $\sigma_l^{(i)}$. Second, voting is performed between the tensor fields (cf. (**b**)) created in the first step. At last, the tensors corresponding to different scales at the same spatial position are refined in a last (voting) step depicted in (**c**). (**a**) *Intra*-scale voting; (**b**) global *inter*-scale voting; (**c**) local *inter*-scale voting

The current description needs to be refined in order to be more efficient, especially concerning its memory consumption. Further, the voting fields used in the three steps have to be modified in a way so that they regard scale also as a feature. This could be handled within the decay function.

3.4 Greyscale Images

The original formulation of tensor voting was not intended to be directly applied to greyscale images and therefore lacks the handling of some issues appearing in that context. Besides the question of how to define the input tokens, the propagation of information from structured to non-structured regions constitutes the main problem. Propositions in literature primarily focussed on postprocessing the output of tensor voting.

In [30], the image gradient is used to define the input tokens as already mentioned in Sect. 2. By following this approach the input tensors are created as pure stick tensors. Therefore, the other components, i.e. the plate and the ball component, are equal to zero and the subsequent tensor voting steps only comprise stick voting.

An iterative application of tensor voting with thresholding steps at the beginning of each repetition depicts the basis of the method presented in [16] (cf. Fig. 1 for exemplary results), which especially aims at extracting curves from two-dimensional image data. Tokens that are found to provide low saliency are disregarded in the next iteration. As all tokens are initialised as unoriented tensors, the first voting step is performed as ball voting to infer preferred directions of the tokens. The following iterations only focus on stick voting, as the desired structures are represented by curves. Thereby, the essential idea of the iterative approach is

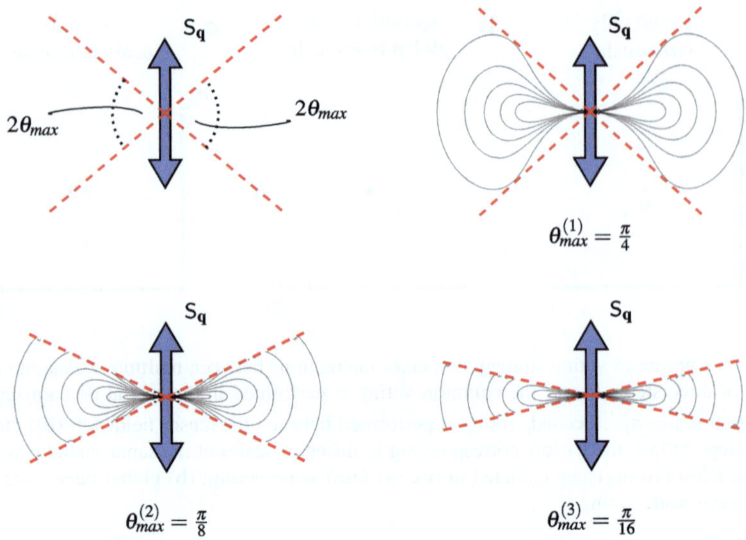

Fig. 12 As shown in the *upper left part* of the figure, the aperture angle θ_{max} determines the shape of the cone of the stick voting field. Using the example of three particular aperture angles, the other figures illustrate how decreasing θ_{max} affects the voting field induced by the stick tensor $\mathbf{S_q}$ in the method proposed in [16]

to decrease the limiting range for the angle $\theta_{\mathbf{pq}}$, that is the *aperture angle* θ_{max} of the stick voting field (cf. Fig. 12). This strategy reduces the *diffusion* of the votes and enforces the most promising structures in the image. Decreasing the aperture angle gradually over the iterations is described by the authors as "shifting from an exploratory to an exploitative mode". However, validation of this method showed that, even though good results for the particular application case of extracting adherens junctions were achieved, the approach still suffers from low accuracy at T- and X-shaped junctions.

Another strategy is described in [19], which is based on the introduction of additional voting fields that, rather than promoting their vote inside a given cone, aim at inhibiting votes at positions outside of that. The authors propose to utilise Gabor filters for the acquisition of the input tokens as a first step, as these filters are known for their optimal time-bandwidth product. Following up, there are two subsequent steps of, first, stick voting to infer directional information on curve structures and, second, ball voting to refine the regions around occurring junctions, i.e. close gaps between curve structures. Thereby, each of those steps involves an inhibitory voting performed prior to the classical (excitatory) tensor voting procedure. By using inhibitory voting fields, tokens with non-maximal saliency are expected to be suppressed so that their disturbing influence on the excitatory voting is minimised. Doing this, a better localisation of the features is expected.

Specific to both propositions, that were described in the previous two paragraphs, is the fact that they try to compensate the issues of tensor voting by adjusting either the input or the output of the classical method. Opposed to that, it would be an interesting approach to add modifications within the formulation of tensor voting itself to solve the problem in the source.

As we identified the propagation of information to unstructured regions to be the main problem in the context of greyscale images, it might be promising if tensor voting was aware of local *structuredness* of a region. Therefore we suggest to use a measure of *structuredness* that influences the voting steps. As a simple choice the gradient magnitude could be employed for that purpose. Alternatively, such a measure could be estimated by utilising both structure information defined in the classical tensor voting formulation and additional statistical methods. By using a kind of *contextual voting* scheme (cf. Sect. 3.1) *structuredness* could steer the propagation process of tensor voting. For example, based on the comparison of the structuredness of voter and receiver (and possibly their neighbourhoods), the decay function could be adjusted. Similar to the local scale in Sect. 3.3, the new parameter might also be propagated while other features steer that process.

3.5 Vector- and Tensor-Valued Input Data

Related to the challenge of processing greyscale, i.e. scalar, image data is the application of tensor voting to vector- or tensor-valued data. Some medical images obtained by different modalities inherently depict vector-, second-order tensor- or fourth-order tensor-valued images. For instance Phase-Contrast Cardiovascular Magnetic Resonance (PC-CMR) used for blood flow analysis produces flow maps that can be seen as time-resolved *vector fields*. Moreover Diffusion Tensor Imaging (DTI) and Diffusion Kurtosis Imaging (DKI) estimate the diffusion along different directions within every voxel and encode them in *second-* or *fourth-order tensors*.

Some possibilities to extend the tensor voting formulation to vector- and tensor-valued input data in the context of image structure estimation have been suggested in [30]. In particular the application to colour, i.e. vector-valued, images and higher-order tensor-valued data was explored. Recalling that the entries of a higher-order tensor can be associated with the components of a vector, basically two approaches are proposed which focus on processing vector-valued data. In case of both, initialisation is performed independently for each channel, providing d tensor fields of input tokens $\mathsf{T}^k_{\mathbf{q}}$, given by

$$\mathsf{T}^k_{\mathbf{q}} = \mathsf{S}^k_{\mathbf{q}} + \mathsf{P}^k_{\mathbf{q}} + \mathsf{B}^k_{\mathbf{q}} \,, \tag{9}$$

where $1 \leq k \leq d$ and the number of channels is denoted by d.

Following the first approach, the tensor voting formulation in (4) is applied to each channel separately, i.e.

$$TV_k(\mathbf{p}) = \sum_{\substack{\mathbf{q} \in \Omega(\mathbf{p}),\\ \mathbf{v}=\mathbf{p}-\mathbf{q}}} SV(\mathbf{v}, \mathsf{S}^k_{\mathbf{q}}) + PV(\mathbf{v}, \mathsf{P}^k_{\mathbf{q}}) + BV(\mathbf{v}, \mathsf{B}^k_{\mathbf{q}}) . \tag{10}$$

Summing up the specific intermediate results yields

$$TV(\mathbf{p}) = \sum_{k=1}^{d} w_k \, TV_k(\mathbf{p}) \tag{11}$$

with w_k being a weighting parameter depending on the particular channel. In fact, in case the image gradient ∇u^k is utilised for initialisation of the k-th channel like in [30], i.e. $\mathsf{T}^k_{\mathbf{q}} = \nabla u^k_{\mathbf{q}} (\nabla u^k_{\mathbf{q}})^T = \mathsf{S}^k_{\mathbf{q}}$, plate and ball voting can be omitted as for both components $\mathsf{P}^k_{\mathbf{q}} = \mathsf{B}^k_{\mathbf{q}} = 0$ holds true.

The second option proposed in [30] is to combine the initialised input tokens of all channels following

$$\mathsf{T}_{\mathbf{q}} = \sum_{k=1}^{d} w_k \, \mathsf{T}^k_{\mathbf{q}} , \tag{12}$$

where w_k again determines the weighting between the channels. Finally the formulation of tensor voting from (4) is applied to $\mathsf{T}_{\mathbf{q}}$ based on the decomposition into its stick, plate and ball components defined in (3).

The proposed adaptions of the tensor voting method have shown to lead to better results for image structure estimation compared to the concept of the structure tensor. Nevertheless, the modifications are based on simplifications and assumptions that are not fulfilled in all possible cases. Mainly the precondition that the different channels are independent from each other restricts the number of possible applications. Therefore, it is relevant to investigate how to process vector fields with correlated components.

Regarding metric spaces provides the opportunity to measure distances between elements of such a space. In particular, one is able to define the input tokens directly from the elements of a vector field. Metric partial derivatives (cf. [1]) could be used to calculate derivatives with respect to the metric, which one could use to define features in the same way as e.g. the image gradient in the scalar valued case. Higher-order tensor fields could be processed by either associating the tensors with a vector like in [30] and subsequently applying the extension of tensor voting to vector fields, or by defining the input tokens directly from the tensor elements.

4 Concluding Remarks

After providing a brief insight into the problems faced in the context of medical image analysis, we have presented the essential ideas of the original formulation of tensor voting. Based on that, selected challenges concerning the application of tensor voting in certain situations in medical image analysis were discussed in further detail. We described inherent issues related to contextual voting, utilising higher-order tensors within the method's formulation, performing multiscale analysis and to handling greyscale and vector- or tensor-valued input data. An overview of the current research which addresses these issues was given and possible strategies that are worthwhile to explore were described. Especially, the latter might potentially improve the results of tensor voting in order to make it applicable to further applications.

In the end, it must be remarked that along with the particular main objectives of the presented extensions there still exist secondary objectives. As already mentioned earlier, an efficient implementation is essential to provide a feasible extension of the method. This must be regarded when proposing the formulation of a mathematical solution. Furthermore, medical applications often require several of the extensions at the same time, which induces a relationship between those. For that reason, one has to consider that possible extensions solving different issues must still be compatible with each other, so that they can be used within the same framework.

To conclude, although tensor voting has been proven effective in many applications, it is far from being a mature technique and, as described in the previous sections, still comprises several issues to be solved. We think this fact will foster research in this field in the next few years.

References

1. Ambrosio, L., Gigli, N., Savaré, G.: Gradient Flows: In Metric Spaces and in the Space of Probability Measures. Springer, Berlin (2006)
2. Arsigny, V., Fillard, P., Pennec, X., Ayache, N.: Fast and simple calculus on tensors in the log-Euclidean framework. In: Medical Image Computing and Computer-Assisted Intervention—MICCAI 2005. Lecture Notes in Computer Science, pp. 115–122. Springer, Heidelberg (2005)
3. Citti, G., Sarti, A.: A cortical based model of perceptual completion in the roto-translation space. J. Math. Imaging Vision **24**(3), 307–326 (2006)
4. De Lathauwer, L., De Moor, B., Vandewalle, J.: A multilinear singular value decomposition. SIAM J. Matrix Anal. Appl. **21**(4), 1253–1278 (2000)
5. Duits, R., Franken, E.: Left-invariant diffusions on the space of positions and orientations and their application to crossing-preserving smoothing of HARDI images. Int. J. Comput. Vision **92**(3), 231–264 (2011)
6. Fischer, S., Bayerl, P., Neumann, H., Redondo, R., Cristóbal, G.: Iterated tensor voting and curvature improvement. Signal Process. **87**(11), 2503–2515 (2007)
7. Franken, E., Duits, R.: Crossing-preserving coherence-enhancing diffusion on invertible orientation scores. Int. J. Comput. Vision **85**(3), 253–278 (2009)
8. Franken, E., van Almsick, M., Rongen, P., Florack, L., ter Haar Romeny, B.: An efficient method for tensor voting using steerable filters. In: Proceedings of the 9th European Conference on Computer Vision - Volume Part IV (ECCV'06), pp. 228–240. Springer, Berlin (2006)
9. Guo, S., Yanagida, H., Fan, H., Takahashi, T., Tamura, Y.: Image enhancement of ultrasound image based on tensor voting. Trans. Jpn. Soc. Med. Biol. Eng. **47**(5), 423–427 (2009)

10. Guy, G., Medioni, G.: Inference of surfaces, 3D curves, and junctions from sparse, noisy, 3D data. IEEE Trans. Pattern Anal. Mach. Intell. **19**(11), 1265–1277 (1997)
11. Koffka, K.: Principles of Gestalt Psychology. Harcourt, Brace and Company, New York (1935)
12. Läthén, G., Jonasson, J., Borga, M.: Blood vessel segmentation using multi-scale quadrature filtering. Pattern Recogn. Lett. **31**(8), 762–767 (2010)
13. Leng, Z., Korenberg, J., Roysam, B., Tasdizen, T.: A rapid 2-D centerline extraction method based on tensor voting. In: IEEE International Symposium on Biomedical Imaging: From Nano to Macro, 2011, pp. 1000–1003 (2011)
14. Lombardi, G., Casiraghi, E., Campadelli, P.: Curvature estimation and curve inference with tensor voting: a new approach. In: Advanced Concepts for Intelligent Vision Systems, pp. 613–624. Springer, Heidelberg (2008)
15. Loss, L.A., Bebis, G., Nicolescu, M., Skurikhin, A.: An iterative multi-scale tensor voting scheme for perceptual grouping of natural shapes in cluttered backgrounds. Comput. Vision Image Underst. **113**(1), 126–149 (2009)
16. Loss, L.A., Bebis, G., Parvin, B.: Iterative tensor voting for perceptual grouping of ill-defined curvilinear structures. IEEE Trans. Med. Imaging **30**(8), 1503–1513 (2011)
17. Maggiori, E., Lotito, P., Manterola, H., del Fresno, M.: Comments on "A closed-form solution to tensor voting: theory and applications". IEEE Trans. Pattern Anal. Mach. Intell. **36**(12), 2567–2568 (2014)
18. Maggiori, E., Manterola, H.L., del Fresno, M.: Perceptual grouping by tensor voting: a comparative survey of recent approaches. IET Comput. Vision **9**(2), 259–277 (2015)
19. Massad, A., Babós, M., Mertsching, B.: Application of the tensor voting technique for perceptual grouping to grey-level images. In: Van Gool, L. (ed.) Pattern Recognition. Lecture Notes in Computer Science, pp. 306–314. German Association for Pattern Recognition (DAGM). Springer, Berlin (2002)
20. Medioni, G., Lee, M.S., Tang, C.K.: A Computational Framework for Segmentation and Grouping. Elsevier, Amsterdam (2000)
21. Medioni, G., Tang, C.K., Lee, M.S.: Tensor voting: theory and applications. In: In Proceedings of RFIA. TELECOM Paris, Dept. Traitement du Signal et des Images (2000)
22. Min, C., Medioni, G.: Tensor voting accelerated by graphics processing units (GPU). In: 18th International Conference on Pattern Recognition, 2006 (ICPR 2006), vol. 3, pp. 1103–1106. IEEE, New York (2006)
23. Moakher, M.: On the averaging of symmetric positive-definite tensors. J. Elast. **82**(3), 273–296 (2006)
24. Mordohai, P., Medioni, G.: Tensor voting: a perceptual organization approach to computer vision and machine learning. Synth. Lect. Image Video Multimed. Process. **2**(1), 1–136 (2006)
25. Mordohai, P., Medioni, G.: Dimensionality estimation, manifold learning and function approximation using tensor voting. J. Mach. Learn. Res. **11**, 411–450 (2010)
26. Moreno, R., Garcia, M.A., Puig, D., Julia, C.: Robust color edge detection through tensor voting. In: 16th IEEE International Conference on Image Processing (ICIP), 2009, pp. 2153–2156 (2009)
27. Moreno, R., Garcia, M.A., Puig, D.: Robust color image segmentation through tensor voting. In: 2010 20th International Conference on Pattern Recognition (ICPR), pp. 3372–3375 (2010)
28. Moreno, R., Garcia, M.A., Puig, D., Julià, C.: Edge-preserving color image denoising through tensor voting. Comput. Vision Image Underst. **115**(11), 1536–1551 (2011)
29. Moreno, R., Garcia, M.A., Puig, D., Pizarro, L., Burgeth, B., Weickert, J.: On improving the efficiency of tensor voting. IEEE Trans. Pattern Anal. Mach. Intell. **33**(11), 2215–2228 (2011)
30. Moreno, R., Pizarro, L., Burgeth, B., Weickert, J., Garcia, M.A., Puig, D.: Adaption of tensor voting to image structure estimation. In: Laidlaw, D.H., Vilanova, A. (eds.) New Developments in the Visualization and Processing of Tensor Fields. Mathematics and Visualization, pp. 29–50. Springer, Berlin (2012)
31. Moreno, R., Garcia, M.A., Puig, D.: Tensor voting for robust color edge detection. In: Advances in Low-Level Color Image Processing. Lecture Notes in Computational Vision and Biomechanics, pp. 279–301. Springer, Berlin (2014)

32. Pasternak, O., Sochen, N., Basser, P.J.: Metric selection and diffusion tensor swelling. In: Laidlaw, D.H., Vilanova, A. (eds.) New Developments in the Visualization and Processing of Tensor Fields. Mathematics and Visualization, pp. 323–336. Springer, Heidelberg (2012)
33. Reisert, M., Burkhardt, H.: Efficient tensor voting with 3D tensorial harmonics. In: Conference on Computer Vision and Pattern Recognition Workshops, 2008 (CVPRW'08). IEEE Computer Society, pp. 1–7. IEEE, New York (2008)
34. Risser, L., Plouraboue, F., Descombes, X.: Gap filling of 3-D microvascular networks by tensor voting. IEEE Trans. Med. Imaging **27**(5), 674–687 (2008)
35. Schultz, T.: Towards resolving fiber crossings with higher order tensor inpainting. In: Laidlaw, D.H., Vilanova, A. (eds.) New Developments in the Visualization and Processing of Tensor Fields. Mathematics and Visualization, pp. 253–266. Springer, Berlin (2012)
36. Schultz, T., Seidel, H.P.: Estimating crossing fibers: a tensor decomposition approach. IEEE Trans. Vis. Comput. Graph. **14**(6), 1635–1642 (2008)
37. Schultz, T., Fuster, A., Ghosh, A., Deriche, R., Florack, L., Lim, L.H.: Higher-order tensors in diffusion imaging. In: Westin, C.F., Vilanova, A., Burgeth, B. (eds.) Visualization and Processing of Tensors and Higher Order Descriptors for Multi-Valued Data. Mathematics and Visualization, pp. 129–161. Springer, Berlin (2014)
38. Sha'ashua, A., Ullman, S.: Structural Saliency: The Detection of Globally Salient Structures Using a Locally Connected Network, pp. 321–327. IEEE, New York (1988)
39. Tang, C.K., Medioni, G.: Curvature-augmented tensor voting for shape inference from noisy 3D data. IEEE Trans. Pattern Anal. Mach. Intell. **24**(6), 858–864 (2002)
40. Tang, C.K., Medioni, G., Duret, F.: Automatic, accurate surface model inference for dental CAD/CAM. In: Wells, W., Colchester, A., Delp, S. (eds.) Medical Image Computing and Computer-Assisted Interventation — MICCAI'98. Lecture Notes in Computer Science, vol. 1496, pp. 732–742. Springer, Berlin (1998)
41. Tong, W.S., Tang, C.K.: Robust estimation of adaptive tensors of curvature by tensor voting. IEEE Trans. Pattern Anal. Mach. Intell. **27**(3), 434–449 (2005)
42. Tong, W.S., Tang, C.K., Mordohai, P., Medioni, G.: First order augmentation to tensor voting for boundary inference and multiscale analysis in 3D. IEEE Trans. Pattern Anal. Mach. Intell. **26**(5), 594–611 (2004)
43. van Almsick, M., Duits, R., Franken, E., ter Haar Romeny, B.: From stochastic completion fields to tensor voting. In: Deep Structure, Singularities, and Computer Vision, pp. 124–134. Springer, Berlin (2005)
44. Williams, L.R., Jacobs, D.W.: Stochastic completion fields: a neural model of illusory contour shape and salience. In: Proceedings of the Fifth International Conference on Computer Vision, 1995, pp. 408–415. IEEE, New York (1995)
45. Wu, T.P., Yeung, S.K., Jia, J., Tang, C.K., Medioni, G.: A closed-form solution to tensor voting: theory and applications. IEEE Trans. Pattern Anal. Mach. Intell. **34**(8), 1482–1495 (2012)

Part III
Visualization

Visualization of Diffusion Propagator and Multiple Parameter Diffusion Signal

Olivier Vaillancourt, Maxime Chamberland, Jean-Christophe Houde, and Maxime Descoteaux

Abstract New advances in MRI technology allow the acquisition of high resolution diffusion-weighted datasets for multiple parameters such as multiple q-values, multiple b-values, multiple orientations and multiple diffusion times. These new and demanding acquisitions go beyond classical diffusion tensor imaging (DTI) and single b-value high angular resolution diffusion imaging (HARDI) acquisitions. Recent studies show that such multiple parameter diffusion can be used to infer axonal diameter distribution and other biophysical features of the white matter, otherwise not possible. Hence, this calls for novel visualization techniques to interact with such complex high-dimensional and high-resolution datasets. To date, there are no existing visualization techniques to visualize full brain images or fields of diffusion signal profiles and diffusion propagators reconstructed from them. It is important to be able to scroll in these images beyond single voxels, just as one would navigate in a whole brain map of fractional anisotropy extracted from DTI. In this chapter, we give a review of the existing visualization techniques for the local diffusion phenomenon and propose alternative visualization techniques for fields of high-dimensional 3D diffusion profiles. We introduce: (i) a volume rendering approach and (ii) a diffusion propagator silhouette glyph as a complement to existing DTI and HARDI visualization techniques. We show that these visualization techniques allow the real-time exploration of high-dimensional multi-b-value and multi-direction data such as diffusion spectrum imaging (DSI). Our visualization technique therefore opens new perspectives for 3D diffusion MRI visualization and interaction.

1 Introduction

New advances in multi-band acquisition [1–3], in compressed sensing technology [1, 4], improvements in magnetic resonance (MR) hardware [1, 3], such as multi-channel head coil allowing fast parallel imaging, and novel diffusion imaging

O. Vaillancourt • M. Chamberland • J.-C. Houde • M. Descoteaux (✉)
Computer Science Department, Sherbrooke Connectivity Imaging Lab (SCIL), Sherbrooke Molecular Imaging Center (CIMS), Université de Sherbrooke, Sherbrooke, QC, Canada J1K 2R1
e-mail: m.descoteaux@usherbrooke.ca

I. Hotz, T. Schultz (eds.), *Visualization and Processing of Higher Order Descriptors for Multi-Valued Data*, Mathematics and Visualization,
DOI 10.1007/978-3-319-15090-1_10

sequences allow the acquisition of many diffusion measurements in reasonable time. The motivation for these demanding acquisitions is to obtain both the radial and angular part of the diffusion signal at potentially multiple diffusion times. Hence, going beyond classical diffusion tensor imaging (DTI) and high angular resolution diffusion imaging (HARDI) tractography applications. In fact, not only is it nowadays important to densely measure the radial and angular part, but it is important to do so at multiple echo times (TE) [5–7]. Recent studies show promising multi-parameter diffusion techniques for axonal diameter distribution estimation and to study demyelination and abnormal white-matter tissue [6, 8, 9].

To date, there are no available visualization techniques to efficiently visualize a 2D field or full brain of such 3D diffusion images and diffusion propagators. The diffusion propagator is the full 3D probability distribution function describing the probability that a water molecule starting at origin will have displaced to a certain location during the diffusion time. Visualization of these 3D diffusion profiles is crucial to explore, analyze and understand the data efficiently in order reinforce hypothesis building and understanding of the data acquisitions, noise and features present in the data.

In this chapter, we first revisit the theory of the 1D and 3D q-space imaging and give a brief overview of existing visualization techniques. We then introduce two alternative ways to visualize fields of 3D diffusion profiles. We propose (i) to use a direct volume rendering method and (ii) define a new diffusion propagator silhouette glyph. We show that these visualization techniques allow the real-time interaction with complex 3D diffusion datasets and open new perspectives in diffusion MRI exploration. All software solutions are open and publicly available to the community.

2 Theory

2.1 *Diffusion-Weighted Imaging and Q-Space Imaging*

It is well-known that the diffusion-weighted signal is tuned with two key parameters: the direction of the applied diffusion gradients and a b-value acting like an inverse zoom factor, e.g. the higher the b-value, the smaller the range of the displacement of molecules observed. While these parameters are still used in clinical applications, they are inadequate to describe the basis of possible displacements in porous media. In 1991, Callaghan introduced **q**, representing the wave vector of displacements and leading to the dual space of displacement vectors, called *q-space* [10, 11]: $\mathbf{q} = \frac{\delta \mathbf{G} \gamma}{2\pi}$. The **q** wave vector is linked to the width (δ), the magnitude ($|\mathbf{G}|$) and orientation of the diffusion gradient pulse (**G**) of the diffusion sequence. The b-value itself is linked to the square of the **q** vector and the diffusion time τ: $b = 4\pi^2|\mathbf{q}|^2\tau$ and $\tau = (\Delta - \frac{\delta}{3})$, where Δ is the time between the two diffusion gradient pulses. The diffusion time thus corresponds to the duration of the observation of the diffusion

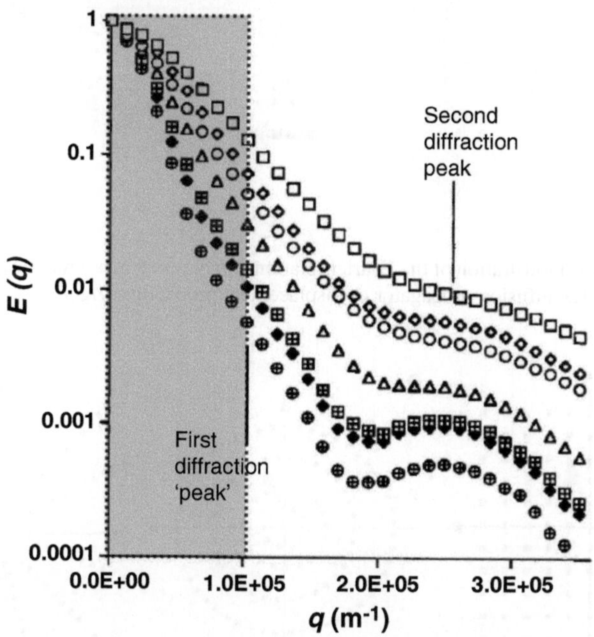

Fig. 1 Figure adapted from Kuchel et al. [12] representing the 1D q-space imaging diffusion signal attenuation decay curves for water in suspensions of human erythrocytes as a function of q-value at different hematocrit. The different *curves* on the graph represent the hematocrit values in decreasing order from the top of the figure, starting from a value of 93 %, followed by 83, 73, 63, 47, 42, and 25 % at the bottom of the figure. The *shaded area* represents the range of q values that are currently achievable on clinical MRI systems

process. This parameter is generally untuned but is the result of the optimization of the TE under the constraint of the chosen b-value, the hardware characteristics (maximum amplitude and slew-rate of the gradients and safety considerations). Yet, in recent *AxCaliber* and *ActiveAx* techniques [7, 9], diffusion acquisitions are done for multiple b-values, multiple gradient orientations as well as multiple TEs to be sensitive to different underlying microstructural sizes.

A very powerful illustration of the potential of these equations is Fig. 1. We see the evolution of the diffusion-weighted signal attenuation according to the magnitude of the **q** wave vector in a simple media composed of a network of impermeable cylinders filled with water. The attenuation is measured in the direction perpendicular to the axis of the cylinders and shows a typical diffraction phenomenon occurring because of water molecules hitting the boundaries of the cylinders. This simple 1D q-space imaging experiment illustrates how diffusion imaging could be used to probe microstructure of porous media and infer information about its compartments, shape and size.

One-dimensional q-space imaging was then generalized to 3D q-space imaging. Under the narrow pulse approximation [13], the relationship between the diffusion

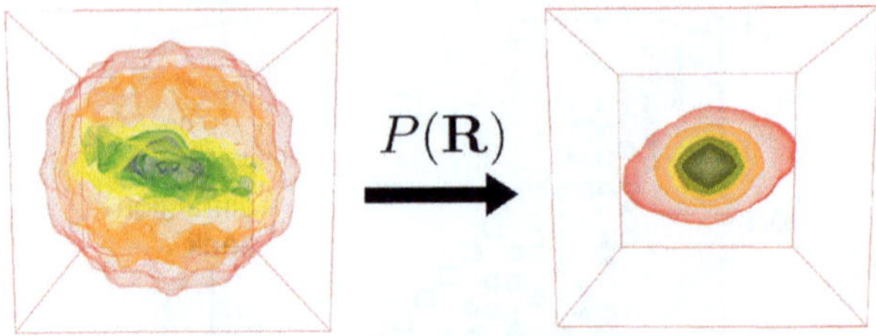

Fig. 2 Single voxel illustration of the Fourier relationship between measured 3D q-space diffusion signal (*left*) and 3D diffusion propagator or displacement probability (*right*)

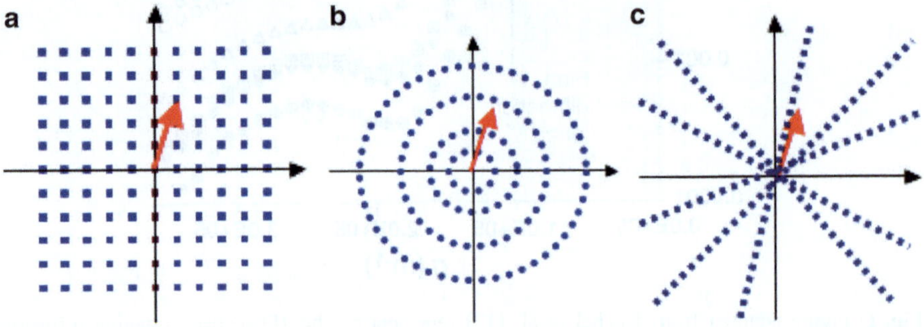

Fig. 3 2D slice representation of different q-space sampling schemes used to measure the diffusion signal and reconstruct the diffusion propagator. (**a**) Cartesian. (**b**) Spherical. (**c**) Radial

signal, $E(\mathbf{q})$, in q-space and the ensemble average diffusion propagator , $P(\mathbf{R})$, in real space, is given by a Fourier transform (FT) relationship [11] such that

$$P(\mathbf{R}) = \int_{\mathbf{q} \in \mathfrak{R}^3} E(\mathbf{q}) e^{-2\pi i \mathbf{q} \cdot \mathbf{R}} d\mathbf{q}, \tag{1}$$

where $E(\mathbf{q}) = S(\mathbf{q})/S_0$, $S(\mathbf{q})$ is the diffusion signal measured at position $\mathbf{q}$ in q-space related to the b-value and S_0 is the image without diffusion. We denote $q = |\mathbf{q}|$ and $\mathbf{q} = q\mathbf{u}$, $\mathbf{R} = r\mathbf{r}$, where $\mathbf{u}$ and $\mathbf{r}$ are 3D unit vectors. Equation (1) suggests a way to reconstruct the diffusion propagator; acquiring as many diffusion images $E(\mathbf{q})$, along as many q-vectors $\mathbf{q}$ as possible, before taking a Fourier transform to obtain the diffusion propagator $P(\mathbf{R})$. This is illustrated in Fig. 2 and is at the heart of Diffusion Spectrum Imaging (DSI) [14], which was recently used in several seminal papers exploring new theories of the grid organization of the brain [15] and connectomics studies [16, 17].

Equation (1) opens the way to several new acquisition schemes. Figure 3 highlights the three most used q-space sampling strategies: (i) sampling the Cartesian grid, (ii) multiple b-value shells in q-space and (iii) radial q-space lines. Every sampling scheme leads to a different reconstruction technique of the propagator and adapted mathematical tools for it.

2.2 Existing Visualization Techniques for Diffusion Imaging

To the best of our knowledge, there is no technique to visualize a field of diffusion propagators, or 3D diffusion signal profiles arising from DSI [14], multiple b-value diffusion-weighted imaging (DWI), and hybrid diffusion imaging (HYDI) [18]. As seen in Fig. 4, some techniques have been proposed at the voxel level, using (i) projections of isocontours along a certain view [14] (Fig. 4a), (ii) intersecting orthogonal planes [14] (Fig. 4b), or (iii) multiple concentric shells representing isosurfaces of the diffusion propagator [19] (Fig. 4c). It is true that a naive implementation of these techniques would be possible to make them visualize whole fields and volumes of diffusion data. However, making a useful tool to explore high-dimensional diffusion MRI data interactively is essential and non-existent at this time. Multiple-parameter diffusion MRI visualization will become important at the era of multiple b-value and multiple TE diffusion data acquisitions. This chapter sets the table for future visualization research.

It is also possible to visualize fields of isosurfaces of a certain radius (Fig. 4b) or fields of orientation distribution function (ODF) computed from the diffusion propagator [14, 18–22]. The ODF is computed with the equation:

$$\Psi(\theta, \varphi) = \int_{r=0}^{\infty} P(r\mathbf{r}, \theta, \varphi) r^2 dr, \tag{2}$$

which describes the *diffusion* ODF in unit direction $\mathbf{r} = (\theta, \varphi)$, because computed as the radial integral of the diffusion propagator . Many other HARDI angular profiles exist and a very popular one is the *fiber ODF* reconstructed using spherical deconvolution techniques [23]. High-order tensor glyphs for HARDI also exist [24–28] to generalize diffusion tensors from DTI and capture the apparent diffusion coefficient, the Kurtosis tensor [29] or a generalized tensor [30] from HARDI data. In all cases, the angular profile is visualized as a spherical function, where the radius of the sphere is scaled according to the spherical function at each vertex of the sphere mesh. This is illustrated in Fig. 5, where we see how each vertex point (θ, φ) of the spherical mesh is scaled to radius $r = \Psi(\theta, \varphi)$.

Data visualization for diffusion tensor imaging (DTI) is more developed. In DTI, classic ellipsoidal glyphs are used [31], superquadrics [32, 33], supertoroids [34], tensor glyphs [35] or more complex glyphs [36], highlighting the linear, planar or isotropic parts of the diffusion tensor, as seen in Fig. 6. New DTI glyph developments arise from the need to increase orientation and depth perception for single glyphs, and improve data inter-relation perception for glyph fields. As seen in Fig. 6, advances in DTI glyphs provide perceptual cues that reduce ambiguities caused by precedent models and thus must be considered in the creation

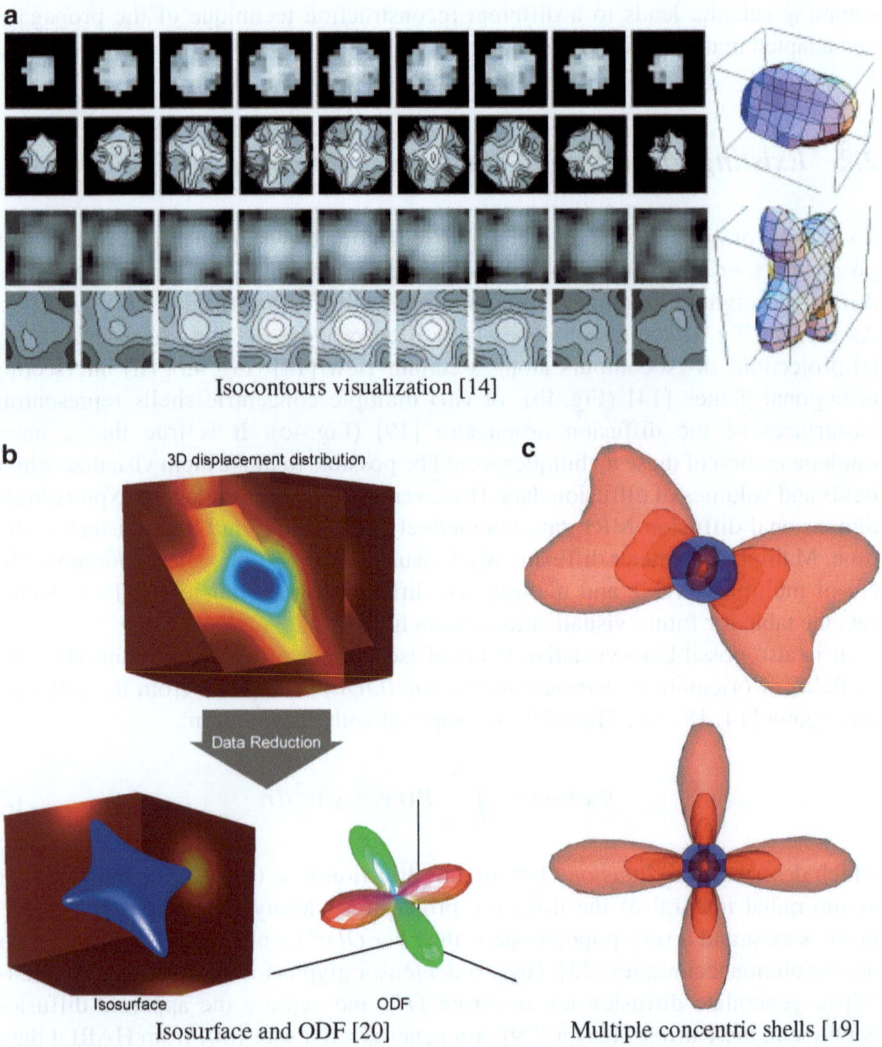

Fig. 4 Existing techniques to visualize the diffusion propagator and 3D q-space diffusion profile. All these techniques work only for a single voxel and do not exist to image a full brain or even a field of diffusion propagators. In (**a**), different 2D slices of the 3D diffusion signal profile (1st row) and propagator (2nd row) are seen, as well as an isosurface glyph through the volume. These images are adapted from [14, Fig. 2]. In (**b**), the diffusion propagator (also called 3D displacement distribution) is visualized using three orthogonal planes through its center. Below, an isosurface glyph at radius 5 of the propagator and a diffusion ODF integrating the propagator over radii. Images are adapted from [20, Fig. 6]. In (**c**), multiple concentric shells are seen on top of one another. Each shell represents an isosurface glyph overlaid with a different opacity and, a colormap from *blue* to *red* for low to high isosurface radii of the propagator. Images are adapted from [19, Fig. 6]

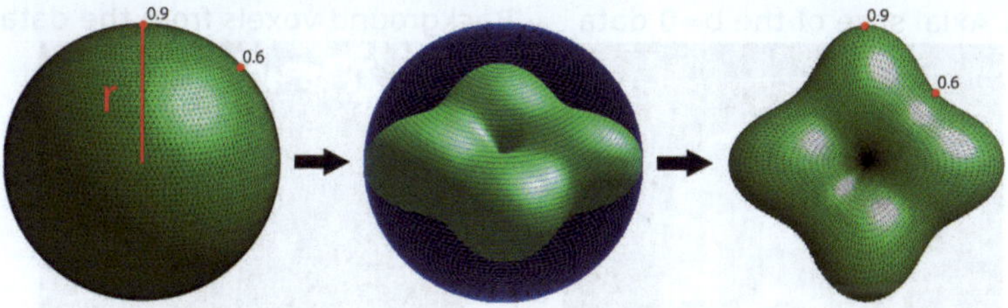

$$r = \psi(\theta, \phi) \mid \theta \in [0, \pi], \phi \in [0, 2\pi]$$

Fig. 5 Angular profile visualization with the orientation distribution function (ODF) glyph, $\Psi(\theta, \varphi)$. The radius of a spherical mesh is scaled according to the value of the sphere

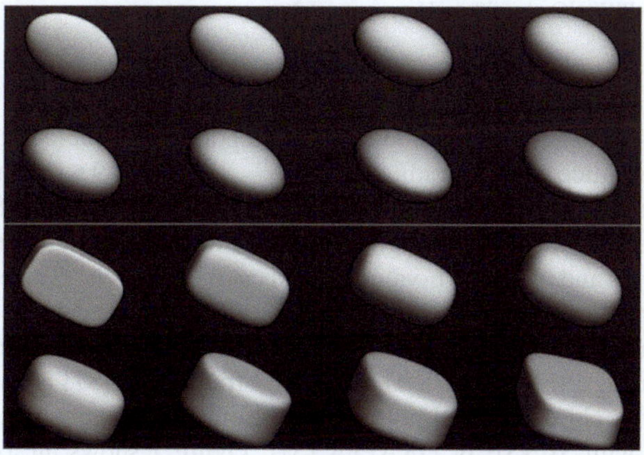

Fig. 6 Different prolate and oblate diffusion tensors visualized with classical ellipsoids (*top*) and superquadrics (*bottom*). Superquadrics are adapted from Kindlmann et al. [32] and provide better perceptual cues than ellipsoids

of new diffusion glyph models beyond DTI. In this regard, the proposed diffusion propagator silhouette glyphs in this chapter are inspired by Westin et al. [36] and superquadrics from [33] to allow the visualization of fields of propagators or 3D diffusion signal profiles.

3 Methods

3.1 Dataset for Visualization

A standard DSI acquisition mimicking the original DSI protocol [14] was done on a 3 T system (Philips Achieva X, Best, The Netherlands), equipped with a whole body gradient (40 mT/m and 200 T/m/s) and a 8-channel head coil. Single-shot spin-

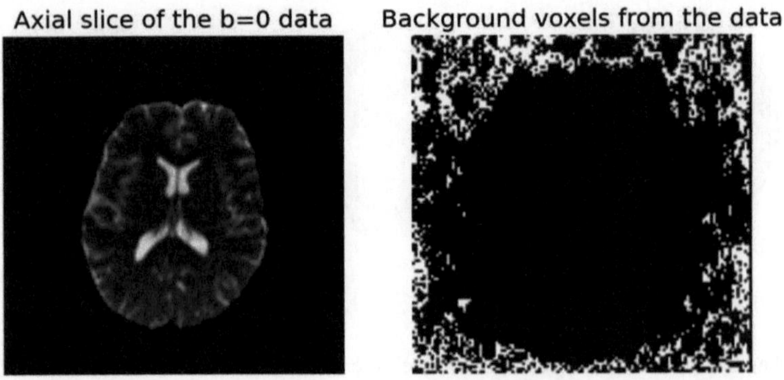

Fig. 7 Axial slice of the $b = 0$ image (S_0 image) and background voxels that are automatically estimated by the PIESNO algorithm. Here, $\sigma = 7.26$

echo EPI measurements with isotropic 2 mm spatial resolution and 515 diffusion measurements were acquired including q-space points of a cubic lattice within the sphere of five lattice units in radius. TE/TR = 116 ms/14.9 s (including time for dynamic B0 stabilization), bandwidth in EPI direction = 1,101 Hz, 128×128 matrix, 60 axial slices with a parallel imaging (SENSE) factor of 2, delta and Delta were 45.4 and 57.7 ms and maximal b-value of bmax = 6,000 s/mm^2. The estimated SNR in the white matter was approximately 38 for the $b = 0$ image, as computed by dividing the mean signal value in the white matter by the noise standard deviation σ estimated using PIESNO [37]. The PIESNO framework [37] is robust to Rician and non-Central chi distributed noise arising from multiple channel imaging data reconstructed with SENSE or GRAPPA [37]. Figure 7 shows an axial slice with background voxels automatically detected by PIESNO and then used to infer the noise standard deviations σ from these voxels. Note that the PIESNO technique is not meant to do a precise brain extraction or skull strip, but used to identify voxels that all have the same underlying noise statistics. Hence, it is expected that the background voxels form a mask with some holes, as shown in [37]. The PIESNO implementation is publicly available in the Diffusion in Python (www.dipy.org) library.

3.2 Visualization of a Field of Diffusion Propagators

Suppose we are given a 3D dataset with a 3D profile at each voxel. In our problem, this 3D profile is represented by discrete samples of $E(\mathbf{q})$ or $P(\mathbf{R})$ [Eq. (1)] for a certain sampling scheme of q-space or real-space. In practice, this means we have a four-dimensional dataset, $X \times Y \times Z \times N$, where at each voxel position (x, y, z), we have N diffusion measurements. For the rest of this chapter, we will assume that the N diffusion measurements are correspondent to a fixed Cartesian grid within each voxel (i, j, k). Hence, in the following equations, the voxel position (x, y, z) is omitted for simplicity. This is natural for diffusion images coming from a DSI acquisitions [14]. In the case of a multiple b-value or HYDI acquisitions, we assume that the data is first resampled to a fixed Cartesian grid. One can do this with trilinear interpolation [18] or more advanced techniques now exist using mathematically sound bases functions such as the 3D-SHORE basis [38], the BFOR basis [39] and novel interpolation strategies [40].

Given this representation, an isosurface glyph can be defined as the spherical function at a certain radius from the origin of the 3D profile. Hence, for the diffusion propagator, the isosurface glyph at radius r is defined in spherical coordinates as

$$P_r(\theta, \varphi) = P(\mathbf{R} = r, \theta, \varphi), \tag{3}$$

where $0 < r \leqslant \sqrt{i^2 + j^2 + k^2}$ is within each imaging voxel. Hence, r is a parameter of the isosurface glyph and can vary between 0 and the maximum radius prescribed by the Cartesian grid within the voxel. This corresponds to an isosurface glyph as seen in Fig. 4b[left]. Otherwise, from the diffusion propagator, we can reconstruct the ODF glyph using Eq. (2) and an upper bound prescribed by the acquisition grid. This corresponds to an ODF glyph seen in Fig. 4b[right]. Finally, the visualization of Fig. 4c is simply an extension of Eq. (3) for multiple radii and with different opacity and color for each isosurface. Multiple isosurfaces from the propagator are overlaid on top of one another. Isosurfaces at small radii look isotropic (blue) and isosurfaces at larger radii have a much sharper angular distribution (red) [19].

It is important to note that even though these visualization exists at the single voxel level, they are computationally heavy and demanding and, to our knowledge, none of these techniques can be visualized interactively in a 2D region or a full 3D brain. ODF-glyph visualization has only recently started to be efficiently visualized in fields using fields of spherical harmonics representations [41–43].

3.3 *Direct Volume Rendering*

Direct rendering proves to be an efficient way to convey the information contents of diffusion datasets since, in its simplest form, diffusion data can be seen as a N-dimensional scalar field, where N is the number of the diffusion measurements. Direct volume rendering methods differ from the previously mentioned visualization techniques by generating images of volumetric diffusion data without explicitly extracting geometric surfaces from it. The techniques instead use an optical model to render optical properties such as color and opacity. The general idea behind direct volume rendering is to accumulate optical properties along each viewing ray to obtain the final volume image [44].

In our case, the discretized volumetric diffusion data is stored as a single 3D texture in the graphics processing unit's (GPU) memory where each voxel of the

3D texture corresponds to a certain location in the data space of a diffusion volume. The data value at each voxel of the volume is used as the parameter of a transfer function that maps diffusion values to optical properties, which can then be used for rendering [45].

During rendering, any viewing ray going through the volume is sampled n times at regular intervals. In this work, we choose n to be two times the longest diagonal. Since the Cartesian grid is $16 \times 16 \times 16$, $n = 56$, $(ceil(2(16\sqrt{3})))$. The resulting fragment color C and opacity A for a viewing ray are respectively computed according to Eq. (4):

$$C = \sum_{i=1}^{n} C_i \prod_{j=1}^{i-1}(1 - A_j), \qquad A = 1 - \prod_{j=1}^{n}(1 - A_j), \tag{4}$$

where color C_i and opacity A_i are given by a transfer function that approximates the emission of optical properties between samples i and $i + 1$ from the data value at sample i.

Practically, the view-ray sampling scheme is constructed by generating n view-aligned, 3D texture-mapped slices that serve as sampling planes inside the bounding box defined by the DW image volume [46]. Equation (4) are evaluated by rendering the sampling planes in a back to front manner through an iterative compositing process performed via GPU alpha blending. The alpha blending is configured such that:

$$F = (1 - A_i)F_d + C_i, \tag{5}$$

where F is the final fragment color after an iteration and F_d the fragment color in the frame buffer at the beginning of the iteration.

For the rendering of diffusion dataset volumes, the transfer function is a simple 1D lookup texture that maps diffusion volume values to a certain color and alpha transparency component. The most significant visualization results were obtained by using a transfer function comprising of narrow color bands centered on low transparency color gradients. This provides a visualization similar to isosurfaces with supplementary inter-surface information provided by the color gradients. Moreover, the user can interactively edit the transfer function texture by adding, removing or displacing color bands, which provides precise interactive-time control over where and which isosurfaces are displayed.

Data-wise, direct volume rendering proves to be effective for the visualization of single propagators or small groups of propagator. It is also one of the few visualization techniques to allow precise rendering of raw diffusion data in the frequency domain [E(q)], less practicable with classic isosurface polygonal technique due to the high number of isosurfaces and extensive tessellation required to obtain a similar level of fidelity. Figure 8 presents examples of frequency domain diffusion profiles, $E(\mathbf{q})$, and their spatial domain diffusion propagators, $P(\mathbf{R})$.

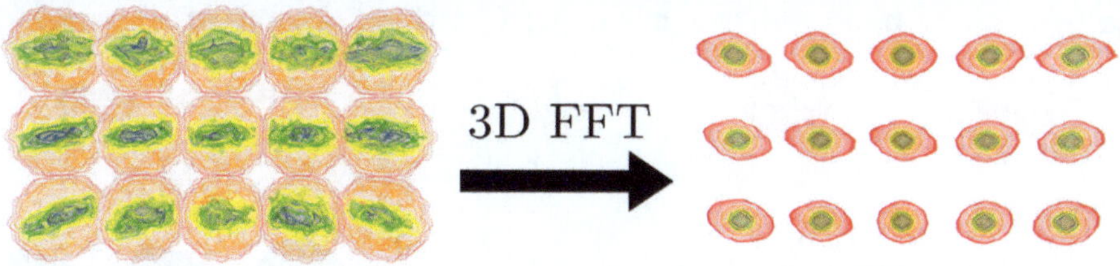

Fig. 8 Examples of glyphs obtained through direct volume rendering methods. [*Left*] Untransformed frequency domain DW profiles, $E(\mathbf{q})$. [*Right*] The corresponding spatial domain diffusion propagators. $P(\mathbf{R})$. Both datasets were normalized before display between their minimum and maximum values. The innermost *blue* regions represent values ranging from 1 to 0.65 and the outermost *red* region represent values ranging from 0.04 to 0.06

3.4 Diffusion Propagator Silhouette Glyphs

While effective for small datasets or precise data observation, direct volume rendering performs poorly for large fields of data. This can be explained by the focus of data field visualization, which concentrates on the observation of visual and structural tendencies rather than precise propagation information [47]. In this regard, propagator rendering through direct volume rendering provides poor direction and shape cues, which renders this visualization technique misadapted for large scale propagator field visualization. The shift in focus from a single propagator to propagator field visualization calls for visualization techniques that better describe data inter-relation.

As a solution to large scale diffusion field visualization, we introduce the diffusion propagator *silhouette* (DPS) glyph. The general objective of the silhouette glyph is to create a geometric glyph that preserves the unique propagator shapes obtained through diffusion acquisitions while sufficiently reducing the data to remain efficient in a large scale visualization context. The general glyph shape is built in such way as to emulate the visual cues provided by biaxial glyphs [32, 48], which were introduced for DTI visualization and have proven to perform well in terms of perception evaluation [49].

The glyph construction process is divided in five (5) steps presented in Fig. 9. Note that these steps were developed mainly for the diffusion propagator $P(\mathbf{R})$ visualization but could be adapted to any structure. Every step is performed independently for each propagator of the field:

1. **Principal axes extraction.** We begin by extracting the principal axes of the diffusion propagator glyph. The process is similar to eigen decomposition in DTI but is performed here with a principal component decomposition procedure on the full diffusion propagator. To do so, we apply a threshold to our propagator volume and only preserve values that are higher than the given threshold. The threshold can be modified to obtain different results and will be further detailed in the next subsection. Once the propagator has been thresholded, the data is stored

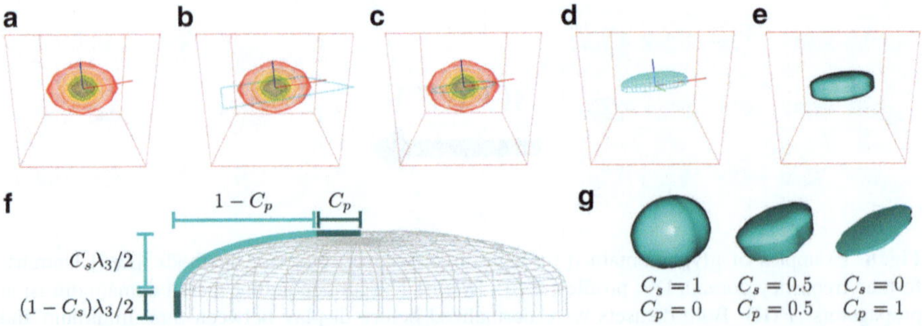

Fig. 9 Depiction of the silhouette glyph construction process. (**a**) The principal axes of the data volume are extracted by calculating the eigenvectors and eigenvalues of the thresholded volume. (**b**) A principal plane is formed parallel to the two eigenvectors corresponding to the longest eigenvalues. (**c**) The silhouette of the glyph is extracted at the precedently set threshold value. (**d**) The top half of the glyph geometry is extruded from the glyph silhouette. (**e**) The bottom half glyph is generated by mirroring the top half along the principal plane. (**f**) A side view of the glyph along with the contributions of measures C_s and C_p in the glyph construction. Note that $1 - C_p$ and C_p silhouette scaling factors. (**g**) Examples of various glyph profiles obtained from a given silhouette according to different C_s and C_p values

in an $M \times 3$ matrix X where each row corresponds to the location of a voxel in the data space of the diffusion propagator. The values contained in each column of X are centered on their mean, which creates a mean-subtracted $M \times 3$ data matrix B. The covariance matrix C is calculated from B such that $C = \frac{1}{M} \sum B^T B$. A matrix V of eigenvectors which diagonalizes the covariance matrix C is then computed such that $V^{-1} C V = D$ where D is the diagonal matrix of eigenvalues of C. The column vectors that make V are the three normalized eigenvectors of C. The elements of the diagonal of matrix D are the corresponding eigenvalues. The corresponding eigenvalues and eigenvectors are paired together and sorted in decreasing eigenvalue order.

2. **Principal plane construction** We create a plane parallel to the first two eigenvectors of our list which becomes the principal plane of the propagator. The plane will be used to generate our glyph silhouette in the next step.
3. **Silhouette extraction** The silhouette is extracted by circularly creating a number of rays initialized at the center of the propagator volume along the principal plane and pointing outside of the volume. Each ray is linearly sampled from the center to the exterior of the volume until the propagator volume value at that point falls below the threshold specified in step 1. The location of the point along the ray is kept. When all rays are sampled and all points obtained, the shape formed by linking the points together in order is the glyph's silhouette.
4. **Glyph geometry creation** The objective for this step is to produce a geometric shape that possesses visual cues akin to those provided by superquadric glyphs [32], while preserving the general propagation silhouette along the principal plane. To do so, we exploit two diffusion tensor metrics proposed by Westin et al. [36] in conjunction with the previously extracted silhouette to define

our silhouette glyph geometry. These metrics use the ordered eigenvalues λ_1, λ_2 and λ_3 obtained in step 1. such that $\lambda_1 \geqslant \lambda_2 \geqslant \lambda_3$. Westin et al. [36] defines three normalized metrics that can be calculated from the eigenvalues, the *Planar measure* C_p, the *Linear measure* C_l and the *Spherical measure* C_s, defined as

$$C_l = \frac{\lambda_1 - \lambda_2}{\lambda_1 + \lambda_2 + \lambda_3}, \qquad C_p = \frac{2(\lambda_2 - \lambda_3)}{\lambda_1 + \lambda_2 + \lambda_3}, \qquad C_s = \frac{3\lambda_3}{\lambda_1 + \lambda_2 + \lambda_3}. \tag{6}$$

For silhouette glyphs, only the C_p and C_s values are used. The linear measure C_l is left aside since the linearity of a given glyph is intrinsically defined in it's silhouette. In fact, C_l depends on λ_1 and λ_2, which correspond to the two eigenvectors used to create the principal plane on which the silhouette is defined. Once C_s and C_p have been computed, the glyph geometry is obtained by extruding the silhouette of the propagator on a straight extrusion path in the direction of the third eigenvector, on a distance corresponding to λ_3. The word extrusion comes from geology and basically means that we are carving the glyph's geometry using the diffusion propagator values. The extrusion is a process used to create objects of a fixed cross-sectional profile. This process has the ability to create very complex cross-sections. Each point on the extrusion path is defined by a position and a silhouette scale for that given position. The relation between the position t along the third eigenvector and the scale $s(t)$ at that position for the top half of a glyph is given by the piecewise function Eq. (7) and illustrated in Fig. 9f:

$$s(t) = \begin{cases} 1 : 0 \leqslant t \leqslant (1 - C_s)\lambda_3/2 \\ A : (1 - C_s)\lambda_3/2 < t \leqslant \lambda_3/2 \\ 0 : \text{otherwise}, \end{cases} \tag{7}$$

where

$$A = \sqrt{1 - (t/(\lambda_3(1 - (1 - C_s/2))))^2 (1 - C_p) + C_p}. \tag{8}$$

5. **Bottom half and shading** The bottom half of the glyph is a simple symmetry of the top half along the principal plane of the propagator. The resulting glyph is illuminated and colored using simple phong shading.

4 Results

Figure 10 presents three general propagator cases visualized through the classic ODF glyph, the direct volume rendering technique and the silhouette glyph. A single fiber voxel from the corpus callosum (row 1), a crossing fiber voxel between corpus callosum and corticospinal tract (row 2) and a voxel from the ventricles (row 3) were manually extracted from the DSI dataset.

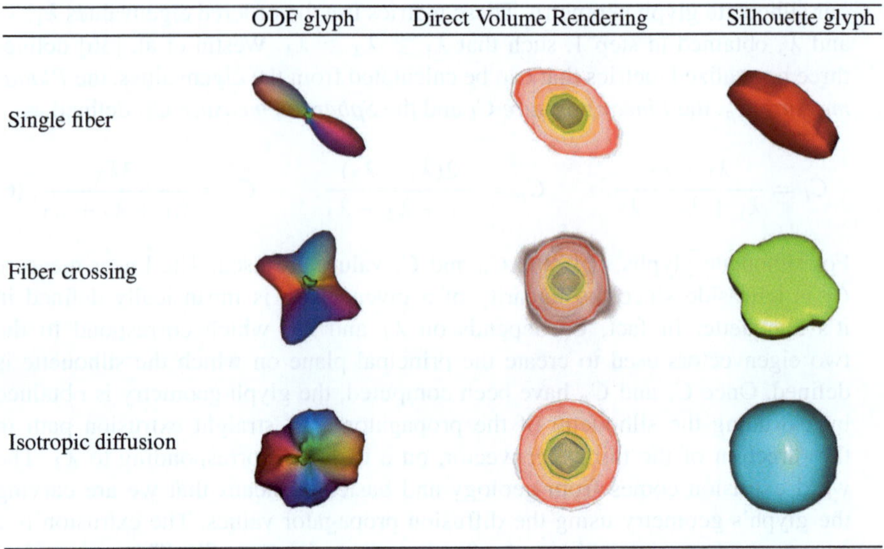

Fig. 10 Comparison of diffusion cases with an existing visualization method, ODF glyphs, and the proposed DWI visualization methods, direct volume rendering and diffusion propagator silhouette glyphs. The DPS glyphs are obtained with a threshold set at 0.035 and the ODF glyphs are obtained by integrating from a radius of 2 to a radius of 6 in data space

As previously stated, the distribution of diffusion around the propagator can be easily observed through direct volume rendering. However, it is obvious that the general shape perpendicular to the longest axis is better conveyed through the silhouette glyph, as shown with the "single fiber" case where the perpendicular plane around the primary axis is only apparent through the glyph visualization, and can hardly be seen with direct volume rendering of ODF rendering. This is a feature that resembles the superquadrics tensor visualization of Fig. 6. Looking at the ODF glyph only, one could be misled in believing that the signal is purely anisotropic and very focused around the principal direction of diffusion. However, it is quite clear from volume rendering and the silhouette glyph that there is diffusion propagation orthogonal to the principal direction.

The "crossing fibers" example (row 2) is also interesting in that the direct volume rendering clearly shows that the information relevant to the crossing visualization is located in the lower (outermost) propagator values. It can also be appreciated that the diffusivity along each of the fiber population is different and better appreciated in the volume rendering or silhouette glyph as opposed to the ODF glyph, where both fiber population seem to have the same diffusion properties. We have also highlighted with a black rectangle regions of 3-way crossings in Fig. 11. Here, the glyphs have a cuboid appearance with corners marking the 3-way crossings. Note the difference between the ODF glyph that shows a very different 3-way crossing between the corpus callosum (the red peak), corticospinal tract (the blue peak) and superior longitudinal fasciculus (the green peak coming out from the page).

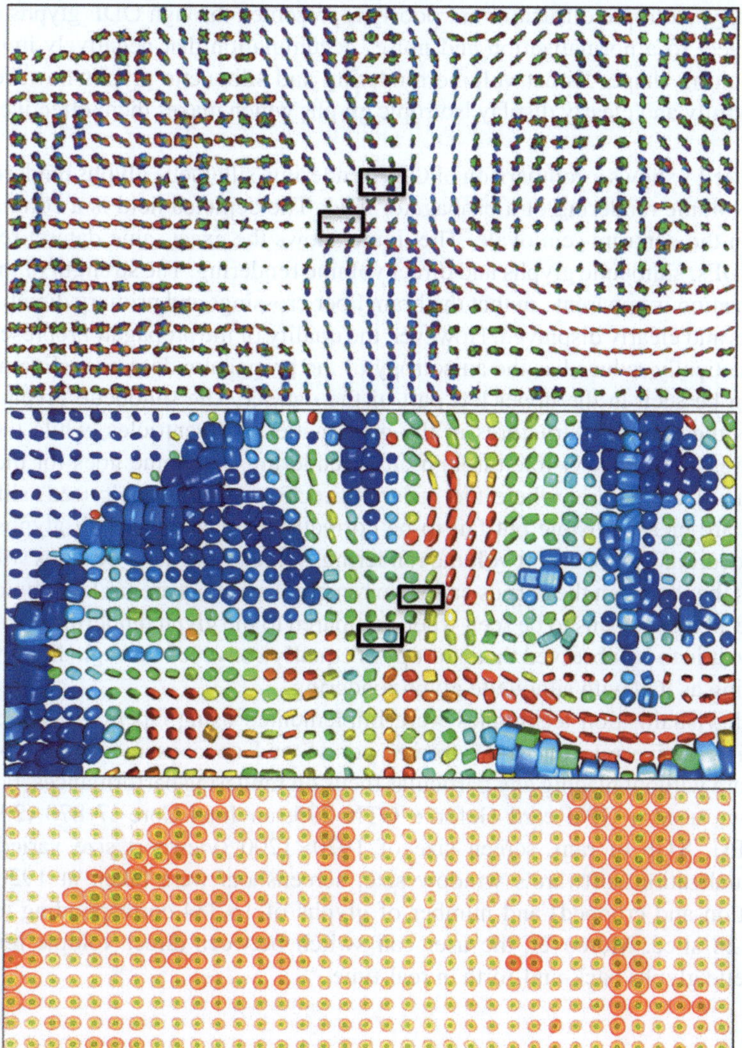

Fig. 11 Visualization of an experimental DWI dataset with an existing visualization techniques and the two techniques introduced in this paper. (*Top*) ODF glyph rendering of the DWI propagators. (*Middle*) DPS glyph rendering of the DWI propagator with a threshold set at 0.035. The glyph coloring is obtained from an FA map. (*Bottom*) The same dataset explored at large scale using direct volume rendering. In the *middle panel*, some glyphs with 3-way crossings have been identified with a *black rectangle*. These correspond to the crossings between corpus callosum, corticospinal tract and superior longitudinal fasciculus

Finally, isotropic diffusions are poorly represented through ODF glyphs as they are subject to a normalization and noise accumulation that negatively impact on the resulting shape. Direct volume rendering and silhouette glyphs however both successfully and clearly display isotropic glyphs which makes them effective in that regard.

Figure 11 shows a comparison of ODF rendering, silhouette glyphs rendering and direct volume rendering of a propagator field. The depicted field is a coronal slice taken in the centrum semiovale. The figure shows the exact same dataset rendered with ODFs, silhouette glyphs and direct volume rendering. The strength of the ODF visualization is apparent in that the brain fiber crossings and primary directions are obvious and clearly displayed. However, the quality of the propagator representation is dramatically reduced as the anisotropy of the propagator decreases. Examples of this can be seen in the upper left part of the dataset. The silhouette glyph clearly renders the isotropic structures in the gray matter and ventricles, while the same structures are hardly visible through ODF rendering. The same goes for the lower left part of the dataset where the general direction of anisotropic (red) silhouette glyphs are clearly visible but fail to be correctly displayed by ODF glyphs at the same location. As for direct volume rendering, some of the more obvious structures such as the voxels in the corpus callosum remain visible but are far less obvious and require a greater focus to be seen when compared to the glyph rendering techniques. Finally, it is to be observed that isotropy is equally conveyed through direct volume rendering as it is with silhouette glyph rendering.

Both visualization techniques were implemented using OpenGL and rendered at an interactive rate of 10–15 frames per second on consumer-grade computer hardware without placing any particular focus on rendering performances. For these tests, we used a GPU card Geforce GTX 295, an Intel Core i7 870 (2.93 GHz 4x) CPU, 8 Gb of ram, screen size of 1920×1200 (viewport size varied based on visualization software's window size, screenshots were taken at 1920×1200 resolution and cropped) and number of glyphs of 16,384 corresponding to a full 128×128 slice. Moreover, the user can switch between direct volume rendering silhouette glyphs back-and-forth in real-time.

5 Discussion

In this chapter, we have reviewed the existing visualization techniques and have proposed two alternatives to diffusion data visualization. Our presented methods are not proposed to replace current visualization techniques but can serve as a complement when combined. Creating scalar maps from derived information from the diffusion signal [18, 38] such as anisotropy indices, diffusivity, moments or higher-order features of the diffusion signal or propagator remain important to explore the full brain and can serve as scalar images to put behind ODFs, volumes or silhouette glyphs.

5.1 Preprocessing

During our development, data preprocessing had a clear impact on the obtained visualization. The influence of filtering windows, data normalization, zero padding and noise removal could be further investigated to see their impact on the final renderings. Several recent q-space interpolation techniques have also been proposed which could serve the visualization purpose [40].

Moreover, for the silhouette glyph, we have chosen to implement a PCA to determine the principal plane on which the data lived. Note that we have also tried to use a first DTI fit from the b-value images lower than $b = 1{,}500\,\text{s/mm}^2$, as used in the novel MAP-MRI technique [38]. Results were similar and we thus avoided adding a DTI fit in the procedure and preferred using the full q-space data available.

5.2 Threshold Adjustment for Silhouettes

The visualization of diffusion propagators through direct volume rendering shows that most of the directional information is contained within the lower valued parts of the propagator, that is, the propagation data located near the outermost parts of the volume. The innermost values of the diffusion propagator mostly appear isotropic and thus provide little visual information for the visualization of large propagator fields.

As a result, the selection of a correct threshold has a definite impact on the appearance of the generated glyphs. Lower thresholds will usually provide glyphs with greater directional cues and will expose more complex diffusion structures. In opposition, a higher valued threshold produces glyph fields that possess a greater visual uniformity but provide a smaller amount of visual directional cues. Complex diffusion structures can hardly be seen with a high threshold. Figure 12 displays

Fig. 12 Changing the threshold value has an observable effect on the resulting glyph. In the above images, the propagator volumes were independently normalized between their minimum and maximum values. [*Left*] Silhouette glyph field with a threshold value set at 2σ. [*Right*] The same DPS glyph field with a threshold value of σ

a comparison between two different threshold values resulting in a qualitatively visually different glyph fields.

In this work, we have set the threshold as a function of the dataset's noise standard deviation, σ, as automatically computed from the PIESNO framework [37]. In a sense, σ serves as a data-driven way to define the threshold. We found that a threshold of σ gave best visual cues for the silhouette glyphs. However, the σ remains a tunable parameter for users who require precise adjustment and other automatic selection could be investigated in the future.

5.3 *Beyond Fiber Crossings Visualization*

The "crossing fibers" problem has received a lot of attention and research since the invention of DTI. As such, visualization of crossing fibers is most visually appealing using sharp ODF glyphs or even just the extracted multiple directions from the ODFs, as seen in Fig. 13. Principal directions, also called peaks, are especially useful for fiber tractography and for orientation representation of the white matter architecture [43].

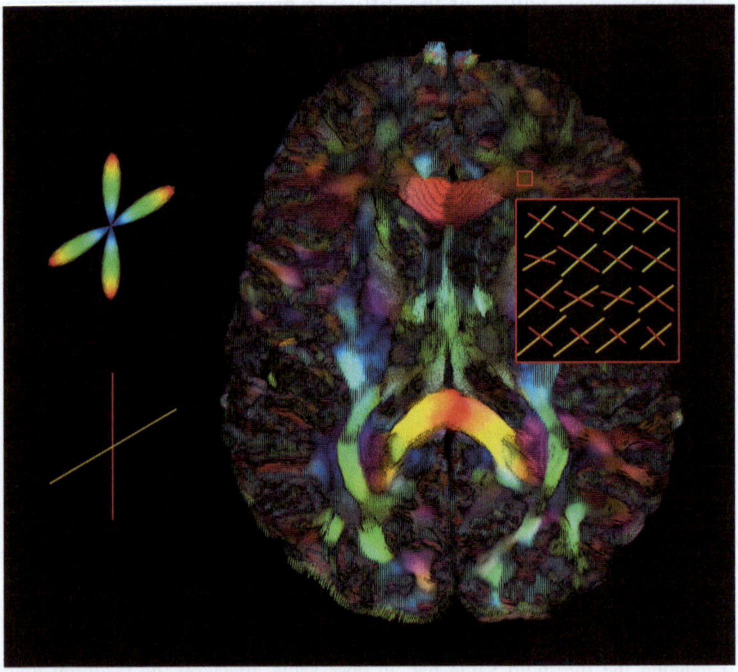

Fig. 13 Visualization a field of sharp fiber ODF glyphs using the extracted principal peaks, as powered by the FiberNavigator (http://scilus.github.io/fibernavigator/). Peaks are colored using the RGB (*red-green-blue*), left-right, anterior-posterior, inferior-superior convention [43]

The field of diffusion MRI is currently going beyond the crossing fiber problem. Models now incorporate dispersion of fibers [50, 51], branching and polarity of local representations [52], and for these, the volume rendering and silhouette glyphs could be a better alternative for visualization. As seen in Fig. 10, differences in diffusion properties of each fiber population and diffusion outside the principle axes diffusion can be better appreciated using a glyph or volume rendering than just the orientation.

5.4 *Volume Rendering Versus Glyphs*

Direct volume rendering and diffusion propagator silhouette glyphs both provide different visualization results in their own respect. In practice, a diffusion visualization system should allow the user to rapidly switch between direct volume rendering, diffusion propagator silhouette glyph rendering, ODFs and peaks to get the best of each approach. As previously explained, direct volume rendering is effective for the visualization of single propagators or small groups of propagators, when the study of the propagation distribution is the visualization focus. Conversely, diffusion propagator silhouette glyphs offer simplified visual information that is well suited for large dataset exploration as it places the emphasis on the general orientation features of the propagator while retaining its general shape.

5.5 *Multiple TEs*

The recent boom of diffusion microstructure techniques developed to estimate axonal diameter distributions and other such bio-physical features add the requirements of having multiple diffusion acquisitions at different TEs. This adds an extra dimension to diffusion signal profile visualization. Not only do we have a 3D profile at each voxel, but now, a 3D diffusion profile in time. Our current implementation only supports the direct volume rendering option. One can load multiple 3D profiles at different times and switch back and forth between them. In the near future, we want to add a continuation scroller to navigate through the time dimension. However, this remains a hard open question that will require a solution to interpolate q-space data acquired at different TE's. As the field of diffusion microstructure grows, this can serve as an important development to be made to go beyond 1D q-space plots as seen in Fig. 1 in different orientations to have a full 3D solution.

6 Conclusion

In this chapter, we have given a brief overview of visualization techniques from diffusion signal and diffusion propagator rendering. We have described two new alternative methods for diffusion visualization which allows large scale inspection of diffusion data in a real-time fashion. With respect to the volumic nature of diffusion data, we have proposed a direct volume rendering method for the visualization of diffusion profiles and a new diffusion propagator silhouette glyph adapted for diffusion propagator rendering. As multiple b-value imaging and imaging at different diffusion times has recently been propulsed by the many connectomics projects in the world and existing developments in diffusion microstructure imaging, the proposed diffusion signal profiles and diffusion propagator can serve visualization of all these novel multi-parameter diffusion datasets.

Acknowledgements We would like to thank our funding agencies, NSERC, MDEIE and CFI in Canada. Also, a special thanks to Michele Bosi for the open-source visualization library. Finally, thanks to Guillaume Gilbert from Philips Healthcare, MR Clinical Science, for the DSI datasets.

References

1. Setsompop, K., Kimmlingen, R., Eberlein, E., Witzel, T., Cohen-Adad, J., McNab, J., Keil, B., Tisdall, M., Hoecht, P., Dietz, P., Cauley, S., Tountcheva, V., Matschl, V., Lenz, V., Heberlein, K., Potthast, A., Thein, H., Horn, J.V., Toga, A., Schmitt, F., Al., E.: Pushing the limits of in vivo diffusion MRI for the Human Connectome Project. NeuroImage **80**, 220–233 (2013)
2. Sotiropoulos, S.N., Jbabdi, S., Xu, J., Andersson, J.L., Moeller, S., Auerbach, E.J., Glasser, M.F., Hernandez, M., Sapiro, G., Jenkinson, M., Feinberg, D.A., Yacoub, E., Lenglet, C., Essen, D.C.V., Ugurbil, K., Behrens, T.E.: Advances in diffusion MRI acquisition and processing in the Human Connectome Project. NeuroImage **80**, 125–143 (2013)
3. Uurbil, K., Xu, J., Auerbach, E.J., Moeller, S., Vu, A.T., Duarte-Carvajalino, J.M., Lenglet, C., Wu, X., Schmitter, S., de Moortele, P.F.V., Strupp, J., Sapiro, G., Martino, F.D., Wang, D., Harel, N., Garwood, M., Chen, L., Feinberg, D.A., Smith, S.M., Miller, K.L., et al.: Pushing spatial and temporal resolution for functional and diffusion MRI in the Human Connectome Project. NeuroImage **80**, 80–104 (2013)
4. Bilgic, B., Setsompop, K., Cohen-Adad, J., Yendiki, A., Wald, L.L., Adalsteinsson, E.: Accelerated diffusion spectrum imaging with compressed sensing using adaptive dictionaries. Magn. Reson. Med. Rapid Commun. **68**(6), 1747–1754 (2012)
5. Assaf, Y., Freidlin, R.Z., Rohde, G.K., Basser, P.J.: New modeling and experimental framework to characterize hindered and restricted water diffusion in brain white matter. Magn. Reson. Med. **52**, 965–978 (2004)
6. Assaf, Y., Basser, P.: Composite hindered and restricted model of diffusion (charmed) MR imaging of the human brain. NeuroImage **27**(1), 48–58 (2005)
7. Assaf, Y., Blumenfeld-Katzir, T., Yovel, Y., Basser, P.J.: Axcaliber: a method for measuring axon diameter distribution from diffusion MRI. Magn. Reson. Med. **59**(6), 1347–1354 (2008)
8. Cohen, Y., Assaf, Y.: High b-value q-space analyzed diffusion-weighted MRS and MRI in neuronal tissues - a technical review. NMR Biomed. **15**, 516–542 (2002)
9. Alexander, D.C., Hubbard, P.L., Hall, M.G., Moore, E.A., Ptito, M., Parker, G.J.M., Dyrby, T.B.: Orientationally invariant indices of axon diameter and density from diffusion MRI. NeuroImage **52**(4), 1374–1389 (2010)
10. Callaghan, P., Eccles, C., Xia, Y.: Rapid communication: NMR microscopy of dynamic displacements: k-space and q-space imaging. J. Phys. E Sci. Instrum. **21**, 820–822 (1988)
11. Callaghan, P.T.: Principles of Nuclear Magnetic Resonance Microscopy. Oxford University Press, Oxford (1991)

12. Kuchel, P., Coy, A., Stilbs, P.: Nmr 'diffusion-diffraction' of water revealing alignment of erythrocytes in a magnetic field and their dimensions and membrane transport characteristics. Magn. Reson. Med. **37**(5), 673–643 (1997)
13. Stejskal, E., Tanner, J.: Spin diffusion measurements: spin echoes in the presence of a time-dependent field gradient. J. Chem. Phys. **42**, 288–292 (1965)
14. Wedeen, V.J., Hagmann, P., Tseng, W.Y.I., Reese, T.G., Weisskoff, R.M.: Mapping complex tissue architecture with diffusion spectrum magnetic resonance imaging. Magn. Reson. Med. **54**(6), 1377–1386 (2005)
15. Wedeen, V.J., Rosene, D.L., Wang, R., Dai, G., Mortazavi, F., Hagmann, P., Kaas, J.H., Tseng, W.Y.I.: The geometric structure of the brain fiber pathways. Science (New York, NY) **335**(6076), 1628–1634 (2012)
16. Hagmann, P., Kurant, M., Gigandet, X., Thiran, P., Wedeen, V.J., Meuli, R., Thiran, J.P.: Mapping human whole-brain structural networks with diffusion MRI. PLoS ONE **2**(7), e597 (2007)
17. Honey, C.J., Sporns, O., Cammoun, L., Gigandet, X., Thiran, J.P., Meuli, R., Hagmann, P.: Predicting human resting-state functional connectivity from structural connectivity. Proc. Natl. Acad. Sci. USA **106**(6), 2035–2040 (2009)
18. Wu, Y.C., Alexander, A.L.: Hybrid diffusion imaging. NeuroImage **36**, 617–629 (2007)
19. Descoteaux, M., Deriche, R., Bihan, D.L., Mangin, J.F., Poupon, C.: Multiple q-shell diffusion propagator imaging. Med. Image Anal. **15**, 603–621 (2011)
20. Hagmann, P., Jonasson, L., Maeder, P., Thiran, J.P., Wedeen, V.J., Meuli, R.: Understanding diffusion mr imaging techniques: from scalar diffusion-weighted imaging to diffusion tensor imaging and beyond. RadioGraphics **26**, S205–S223 (2006)
21. Assemlal, H.E., Tschumperlé, D., Brun, L.: Efficient and robust computation of pdf features from diffusion mr signal. Med. Image Anal. **13**, 715–729 (2009)
22. Özarslan, E., Koay, C.G., Shepherd, T., Blackband, S., Basser, P.J.: Simple harmonic oscillator based reconstruction and estimation for three-dimensional q-space MRI. In: International Society for Magnetic Resonance in Medicine (ISMRM), p. 1396 (2009)
23. Descoteaux, M., Deriche, R., Knösche, T.R., Anwander, A.: Deterministic and probabilistic tractography based on complex fibre orientation distributions. IEEE Trans. Med. Imaging **28**(2), 269–286 (2009)
24. Özarslan, E., Mareci, T.H.: Generalized diffusion tensor imaging and analytical relationships between diffusion tensor imaging and high angular resolution imaging. Magn. Reson. Med. **50**, 955–965 (2003)
25. Descoteaux, M., Angelino, E., Fitzgibbons, S., Deriche, R.: Apparent diffusion coefficients from high angular resolution diffusion imaging: estimation and applications. Magn. Reson. Med. **56**, 395–410 (2006)
26. Barmpoutis, A., Vemuri, B.C., Shepherd, T.M., Forder, J.R.: Tensor splines for interpolation & approximation of DT-MRI with applications to segmentation of isolated rat hippocampi. IEEE Trans. Med. Imaging **26**(11), 1537–1546 (2007)
27. Basser, P.J., Pajevic, S.: Spectral decomposition of a 4th-order covariance tensor: applications to diffusion tensor MRI. Signal Process. **87**, 220–236 (2007)
28. Schultz, T., Kindlmann, G.L.: A maximum enhancing higher-order tensor glyph. Comput. Graphics Forum **29**(3), 1143–1152 (2010)
29. Jensen, J.H., Helpern, J.A., Ramani, A., Lu, H., Kaczynski, K.: Diffusional kurtosis imaging: the quantification of non-gaussian water diffusion by means of magnetic resonance imaging. Magn. Reson. Med. **53**, 1432–1440 (2005)
30. Liu, C., Bammer, R., Acar, B., Moseley, M.E.: Characterizing non-gaussian diffusion by using generalized diffusion tensors. Magn. Reson. Med. **51**, 924–937 (2004)
31. Pierpaoli, C., Jezzard, P., Basser, P., Barnett, A., Chiro, G.D.: Diffusion tensor MR imaging of human brain. Radiology **201**, 637–648 (1996)
32. Kindlmann, G.: Superquadric tensor glyphs. In: Proceedings of IEEE TVCG/EG Symposium on Visualization 2004, pp. 147–154 (2004)

33. Ennis, D.B., Kindlman, G., Rodriguez, I., Helm, P.A., McVeigh, E.R.: Visualization of tensor fields using superquadric glyphs. Magn. Reson. Med. **53**(1), 169–176 (2005)
34. Choukri, M., Mekkaoui, C., Jackowski, M., Martuzzi, R., SINUSAS, A.J.: Supertoroid-based fusion of cardiac DT-MRI with molecular and physiological information. In: 18th Annual Meeting of the International Society for Magnetic Resonance in Medicine (ISMRM), p. 1592 (2010)
35. Schultz, T., Kindlmann, G.: Superquadric glyphs for symmetric second-order tensors. IEEE Trans. Vis. Comput. Graph. **16**(6), 1595–1604 (2010)
36. Westin, C., Maier, S., Mamata, H., Nabavi, A., Jolesz, F., Kikinis, R.: Processing and visualization for diffusion tensor MRI. Med. Image Anal. **6**(2), 93–108 (2002)
37. Koay, C.G., Ozarslan, E., Pierpaoli, C.: Probabilistic Identification and Estimation of Noise (PIESNO): a self-consistent approach and its applications in MRI. J. Magn. Reson. **199**(1), 94–103 (2009)
38. Özarslan, E., Koay, C.G., Shepherd, T.M., Komlosh, M.E., Irfanoglu, M.O., Pierpaoli, C., Basser, P.J.: Mean apparent propagator (MAP) MRI: a novel diffusion imaging method for mapping tissue microstructure. NeuroImage **78**, 16–32 (2013)
39. Hosseinbor, A.P., Chung, M.K., Wu, Y.C., Alexander, A.L.: Bessel fourier orientation reconstruction (BFOR): an analytical diffusion propagator reconstruction for hybrid diffusion imaging and computation of q-space indices. NeuroImage **64**, 650–670 (2013)
40. Tax, C., Vos, S., Viergever, M., Froeling, M., Leemans, A.: Transforming grids to shells and vice versa: an evaluation of interpolation methods in diffusion MRI q- and b-space. In: Proceedings on International Society for Magnetic Resonance in Medicine, vol. 22, p. 4485 (2014)
41. Tournier, J.D., Calamante, F., Connelly, A.: MRtrix: diffusion tractography in crossing fiber regions. Int. J. Imaging Syst. Technol. **22**(1), 53–66 (2012)
42. Vaillancourt, O., Boré, A., Girard, G., Descoteaux, M.: A fiber navigator for neurosurgical planning (neuroplanningnavigator). In: IEEE Visualization, Salt Lake City, UT (2010)
43. Chamberland, M., Whittingstall, K., Fortin, D., Mathieu, D., Descoteaux, M.: Real-time multi-peak tractography for instantaneous connectivity display. Front. Neuroinformatics **8**, 59 (2014)
44. Max, N.: Optical models for direct volume rendering. IEEE Trans. Vis. Comput. Graph. **1**(2), 99–108 (1995)
45. Kindlmann, G., Durkin, J.W.: Semi-automatic generation of transfer functions for direct volume rendering. In: IEEE Symposium on Volume Visualization, January, pp. 79–86 (1998)
46. Cullip, T.J., Neumann, U.: Accelerating volume reconstruction with 3d texture hardware. University of North Carolina at Chapel, pp. 1–6 (1994)
47. Ropinski, T., Preim, B.: Taxonomy and usage guidelines for glyph-based medical visualization. In: Proceedings of the 19th Conference on Simulation and Visualization (SimVis08), pp. 121–138 (2008)
48. Jankun-Kelly, T.J., Mehta, K.: Superellipsoid-based, real symmetric traceless tensor glyphs motivated by nematic liquid crystal alignment visualization. IEEE Trans. Vis. Comput. Graph. **12**(5), 1197–1204 (2006)
49. Jankun-Kelly, T., Lanka, Y.S., Swan, J.E., II: An evaluation of glyph perception for real symmetric traceless tensor properties. Comput. Graphics Forum: Int. J. Eurographics Assoc. **29**(3), 1133–1142 (2010) [Special Issue on EuroVis 2010]
50. Zhang, H., Schneider, T., Wheeler-Kingshott, C.A., Alexander, D.C.: NODDI: practical in vivo neurite orientation dispersion and density imaging of the human brain. NeuroImage **61**(4), 1000–1016 (2012)
51. Sotiropoulos, S.N., Behrens, T.E.J., Jbabdi, S.: Ball and rackets: inferring fiber fanning from diffusion-weighted MRI. NeuroImage **60**(2), 1412–1425 (2012)
52. Savadjiev, P., Campbell, J.S.W., Descoteaux, M., Deriche, R., Pike, G.B., Siddiqi, K.: Labeling of ambiguous sub-voxel fibre bundle configurations in high angular resolution diffusion MRI. NeuroImage **41**(1), 58–68 (2008)

Visual Knowledge Discovery for Diffusion Kurtosis Datasets of the Human Brain

Sujal Bista, Jiachen Zhuo, Rao P. Gullapalli, and Amitabh Varshney

Abstract Classification and visualization of structures in the human brain provide vital information to physicians who examine patients suffering from brain diseases and injuries. In particular, this information is used to recommend treatment to prevent further degeneration of the brain. Diffusion kurtosis imaging (DKI) is a new magnetic resonance imaging technique that is rapidly gaining broad interest in the medical imaging community, due to its ability to provide intricate details on the underlying microstructural characteristics of the whole brain. DKI produces a fourth-order tensor at every voxel of the imaged volume; unlike traditional diffusion tensor imaging (DTI), DKI measures the non-Gaussian property of water diffusion in biological tissues. It has shown promising results in studies on changes in grey matter and mild traumatic brain injury, a particularly difficult form of TBI to diagnose. In this paper, we use DKI imaging and report our results of the classification and visualization of various tissue types, diseases, and injuries. We evaluate segmentation performed using various clustering algorithms on different segmentation strategies including fusion of diffusion and kurtosis tensors. We compare our result to the well-known MRI segmentation technique based on Magnetization-Prepared Rapid Acquisition with Gradient Echo (MPRAGE) imaging.

S. Bista (✉)
Institute for Advanced Computer Studies, University of Maryland, College Park, MD 20742, USA
e-mail: sujal@umiacs.umd.edu; sujal@cs.umd.edu

J. Zhuo • R.P. Gullapalli
University of Maryland School of Medicine, Baltimore, MD 21201, USA
e-mail: jzhuo@umm.edu; rgullapalli@umm.edu

A. Varshney
Department of Computer Science and Institute for Advanced Computer Studies, University of Maryland, College Park, MD 20742, USA
e-mail: varshney@cs.umd.edu

I. Hotz, T. Schultz (eds.), *Visualization and Processing of Higher Order Descriptors for Multi-Valued Data*, Mathematics and Visualization,
DOI 10.1007/978-3-319-15090-1_11

1 Introduction

Traumatic brain injury (TBI), caused by blows to the head, is a leading cause of death and disabilities. In 2010, the United States alone, TBI resulted in 2.5 million hospitalizations and 50,000 deaths [5]; Survivors often face lifelong disabilities. Medical professionals examine, diagnose, and treat these injuries; Once injury occurs, a major focus is on how to prevent further extensive degeneration of the brain. The examination can significantly impact recovery, as subsequent diagnosis and treatment depend on it. Different types of medical imaging modalities, including magnetic resonance imaging (MRI), are used for examining TBI injuries.

MRI is a non-invasive imaging device that uses powerful magnetic fields to image the diffusion patterns in biological tissues. Diffusion Tensor Imaging (DTI) is an increasingly popular MRI technique that detects diffusion of water to infer underlying tissue microstructure. DTI assumes that the water diffusion patterns follow a Gaussian distribution; it can effectively measure the dominant direction of water diffusion in tissues, and is widely used in studying white matter tracts in the brain. However, the Gaussian distribution assumption of the DTI fails whenever diffusion is restricted by injury or diseases. To address this problem, Jensen and Helpern [15] introduced diffusion kurtosis imaging (DKI), which measures the degree of the diffusional non-Gaussianity of water molecules in biological tissues. DKI has gained attention in the medical imaging community because of its ability to show a more detailed structure of underlying tissues and because it shows promise in detecting micro-structural tissue changes caused by mild traumatic brain injuries and other neurological diseases [43]. In DKI, second-order diffusion tensors (DT) and fourth-order kurtosis tensors (KT) are calculated. These tensors are spatio-angular fields that characterize the underlying imaged tissue, as shown in Fig. 1. Both DT and KT capture properties of underlying tissues that can be used to classify the whole brain by different tissue types. In both of these datasets, each sample point can be represented by a unique shape defined by its directional data. The per-sample shape of a spatio-angular field in KT is highly irregular and complicated compared with the DT, because KT is capable of estimating finer properties of the imaged tissue.

Classification and visualization of structures in the human brain provides vital information to medical professionals examining patients who suffer from brain diseases and injuries. Detailed information on the imaged tissues can help these professionals decide what actions to take to prevent further degeneration of the brain. Tissue segmentation is also important in studying the structure and function of the brain. There are numerous medical literature reviews that detail the classification of the brain's structure in brain fusing data from either single or multiple imaging techniques, such as DTI and high-angular-resolution diffusion imaging (HARDI) [12, 24, 28, 29, 33, 34]. Most methods rely on utilizing a statistical summarization of the datasets, such as the mean value, by identifying appropriate

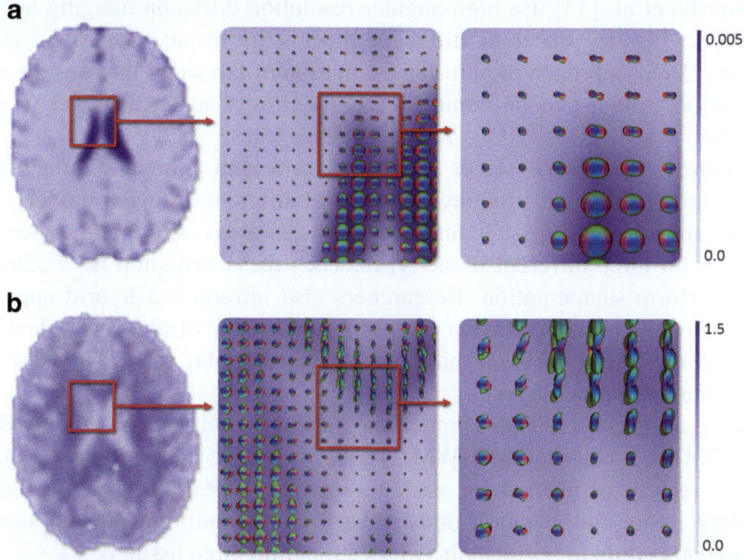

Fig. 1 The diffusion tensor (DT) and kurtosis tensors (KT) visualized using glyph overlays. *Each shape* shows properties of the underlying tissue. The diffusion values are high and isotropic in the cerebral spinal fluid (CSF) and gray matter (GM) regions, whereas they are low and anisotropic in the white matter (WM). The kurtosis values are high around the injury and the WM region. (**a**) Diffusion tensor. (**b**) Kurtosis tensor

ranges of various tissue types. This requires systematic domain knowledge and is error-prone because initial tissue selection used for training determines the quality of the output of classification. Also, it is difficult to find tissues for rare diseases. To our knowledge, no work has been done that performs segmentation by fusing the per-sample shapes of Gaussian and non-Gaussian diffusion estimated by diffusion and kurtosis tensors in DKI. In this paper, we report our classification and visualization results from DKI tensors based on tissue types, diseases, and injuries. We evaluate different segmentation strategies, and compare them to the latest MRI segmentation technique based on magnetization prepared rapid acquisition with gradient echo (MPRAGE) imaging. We also carry out efficient visualization of these segments using spherical harmonics lighting functions, to facilitate insights into the micro-structural properties of the imaged tissue volume.

2 Related Work

Numerous studies and literature reviews have been conducted on the segmentation and visualization of brain tissues using various types of MRI.

Prčkovska et al. [33] use high-angular-resolution diffusion imaging anisotropy measures to classify different diffusion models (isotropic, Gaussian, and non-Gaussian). Their approach requires an estimation of threshold intervals to perform classification, which can be complex and requires a advanced level of domain knowledge.

In another study, Prčkovska et al. [34] perform semi-automatic human-assisted classification of diffusion structures to separate different diffusion models, such as isotropic, anisotropic Gaussian, and non-Gaussian areas. A domain expert selects regions for all three different tissue types. Then the distribution is calculated and used to perform segmentation. Researchers also introduce a hybrid approach to visualize the structure of diffusion. Ellipsoids are used to display a simple diffusion shape, and ray-traced spherical harmonics glyphs display the complex structures based on the segmentation result.

Hasan et al. [12] use DTI to segment and partition cerebrospinal fluid (CSF), grey matter (GM) and white matter (WM). In their method, domain experts manually select 50 regions-of-interest for each tissue type. These regions are then used to create a tissue classification threshold used in a multidimensional supervised clustering procedure to segment the whole brain into three tissue types.

Liu et al. [28] use multiple domain-based attributes, such as the apparent diffusion coefficient and fractional anisotropy, to automatically segment CSF, GM, and WM. The apparent diffusion coefficient and eigenvectors from the diffusion tensor are used to separate CSF from other regions, such as GM and WM, then the fractional anisotropy value is employed to separate GM and WM. An expectation-maximization algorithm combined with a hidden Markov random field model is used to perform automatic segmentation.

Recently, constrained spherical deconvolution has been deployed on diffusion-weighted datasets to classify various tissue types and find fiber-track orientations [18, 35]. Jeurissen et al. [18] performed constrained spherical deconvolution on multi-shell diffusion weighted data with high angular resolution. Using a multi-shell multi-tissue model, they were directly able to classify CSF, GM, and WM. In this paper, we focus on DT and KT to perform classification. These datasets are acquired using significantly lower angular resolution readings compared to other diffusion imaging techniques such as high angular resolution diffusion imaging.

A few studies have looked into the classification and visualization aspects of DKI data. Lu et al. [29] use the spherical harmonics basis to analyze DKI datasets. Researchers limit the harmonic analysis to three bands (0, 2, and 4) and used coefficient summation ($C0$, $C2$, and $C4$) to describe the rotationally invariant property of each band. Then WM, GM, and fiber crossings are segmented based on the fractional anisotropy and $C0$ coefficient only, where $C0$ is a directionally-averaged apparent kurtosis coefficient equivalent to the mean kurtosis. In their paper, $C0$ values were 0.74 ± 0.03, 1.09 ± 0.01, and 0.84 ± 0.02 for GM, WM, and thalamus, respectively. It is interesting to note that, in their segmentation, they did not use $C2$ and $C4$, which are associated with a higher frequency signal in the rotationally invariant spherical harmonics coefficient. This higher frequency information has not yet been fully explored. In this paper, we classify brain tissues using the per-voxel shapes of DT and KT, which provides a better grouping of similar structures and also enhances the likelihood of detecting anomalies.

Volume rendering is widely used to visualize MRI datasets. A considerable amount of work has been done to improve visualization by incorporating advanced shading techniques, multiple depth cues, transfer functions, multiple lighting, and global illumination [6, 13, 14, 22, 23, 25–27, 30, 36, 39, 41, 42]. These studies on volume rendering contain significant ways to improve the visual quality of the volume being displayed. This work on visualization builds on our previous work [2], in which we used spherical harmonics lighting functions to facilitate a more meaningful visualization of dense spatio-angular datasets. In this work, we extend this method to support automatic segmentation and visualization of the entire brain.

3 Overview

The proposed method takes spatio-angular fields (such as DT and KT) as inputs and converts them into a spherical harmonics representation using spherical harmonics basis functions. Tissues are then classified using the spherical harmonics representation of both the DT and DK. Depending on the task and the complexity of the field, we choose to configure either single or multiple spherical harmonics lighting functions for visualization. Finally, by combining classified segments, the dynamic spherical harmonics lighting functions, and the input spatio-angular field, we render the image. We provide two modes to view the final output using either planar or volume rendering. An overview of our approach is shown in Fig. 2.

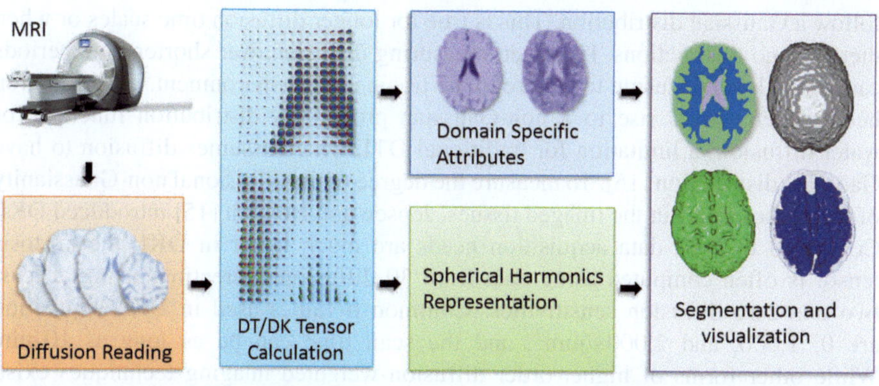

Fig. 2 An overview of our proposed method. First, a large number of diffusional readings are recorded by MRI. Then, we compute tensors and other domain-specific attributes. Next, the tensors are converted to a spherical harmonic form. After that, we use spherical harmonics approximation of DT and KT to classify various tissue types. Finally, by combining the dynamic spherical harmonics lighting functions and the segmented data, the image is rendered. The output is either a planar-rendered image, a volume-rendered image, or both

4 Background

4.1 Diffusion Tensor Imaging

DTI assumes a Gaussian diffusion process of water in the imaged tissue. A Taylor series expansion [19] is used to approximate the diffusion-weighted signal for each gradient direction, expressed by :

$$ln\left[S\left(g,b\right)\right] = ln\left[S_0\right] - bD_{app}\left(g\right) + O\left(b^2\right),$$

$$D_{app}\left(g\right) = \sum_{i=1}^{3}\sum_{j=1}^{3} g_i g_j D_{ij},$$

where g is the diffusion gradient, b is the MRI acquisition parameter b-value expressed in s/mm^2, S_0 is the signal without diffusion weighting, D_{ij} is the element of the diffusion tensor, and D_{app} is the apparent diffusion coefficient. The diffusion tensor, which is a second-order symmetric tensor with six independent elements, is calculated for each voxel. By using eigen-decomposition of the diffusion tensor we compute the dominant diffusion directions.

4.2 Diffusion Kurtosis Imaging

DKI measures the non-Gaussian property of water diffusion. The traditional DTI technique estimates the tensor, based on the assumption that water diffusion patterns follow a Gaussian distribution. This is true for longer diffusion time scales or when there are no obstructions. However, measuring diffusion over shorter time periods causes the local diffusion to adhere to the tissue micro-environment. This diffusion heterogeneity gives rise to a non-Gaussian probability distribution function for water diffusion; a limitation for traditional DTI, which assumes diffusion to have Gaussian distribution [16]. To measure the degree of the diffusional non-Gaussianity of water molecules in the imaged tissues, Jensen and Helpern [15] introduced DKI. Compared to DTI, data acquisition needs are much larger in DKI; the kurtosis tensor is often computed using data from 30 diffusional directions using at least two non-zero diffusion sensitivities. Common b-values used in DKI acquisition are 0, 1,000, and 2,000 s/mm^2, and the scan time can be as long as 10 min. While other forms of higher-order diffusion-weighted imaging techniques exist, such as high-angular-resolution diffusion imaging or diffusion spectrum imaging, they are less clinically practical because they take a considerably longer time to scan as they require a higher number of diffusional direction and b-values. The Taylor series equation in Sect. 4.1 is further expanded to measure the non-Gaussian property of the water diffusion [15, 16]. A fourth-order diffusion kurtosis tensor is calculated from the diffusional measurements in DKI using the equation described by Jensen and Halpern [16],

$$ln\left[S\left(g,b\right)\right] = ln\left[S_0\right] - bD_{app}\left(g\right) + \frac{1}{6}b^2 D_{app}\left(g\right)^2 K_{app}\left(g\right) + O\left(b^3\right),$$

$$K_{app}\left(g\right) = \frac{1}{D_{app}\left(g\right)^2}\sum_{i=1}^{3}\sum_{j=1}^{3}\sum_{k=1}^{3}\sum_{l=1}^{3} g_i g_j g_k g_l K_{ijkl},$$

$$K_{ijkl} = MD^2 W_{ijkl},$$

where MD is the mean diffusivity, K_{app} is the apparent kurtosis, and W_{ijkl} is the element of kurtosis tensor. The kurtosis tensor is a symmetric fourth-order tensor with 15 independent elements. In full form, the signal in each gradient direction is described by

$$ln\left[S\left(g,b\right)\right] = ln\left[S_0\right] - b\sum_{i=1}^{3}\sum_{j=1}^{3} g_i g_j D_{ij} + \frac{1}{6}b^2\sum_{i=1}^{3}\sum_{j=1}^{3}\sum_{k=1}^{3}\sum_{l=1}^{3} g_i g_j g_k g_l K_{ijkl},$$

4.3 Spherical Harmonics

We approximate DT and KT using spherical harmonics basis functions that are later used for classification and visualization. Spherical harmonics are basis functions used to represent and reconstruct any function on the surface of a unit sphere. Spherical harmonics are defined over the surface of a sphere in the same way Fourier functions are defined on a circle [32]. In computer graphics and visualization, spherical harmonics are used for lighting scenes with low frequency lights, for subsurface scattering and for global illumination, because they can inexpensively approximate a computationally-complex physical process [4, 10, 20, 36–38, 42].

Spherical harmonics are ortho-normal functions defined by

$$Y_l^m(\theta,\phi) = (-1)^m\sqrt{\frac{2l+1}{4\pi}\frac{(l-m)!}{(l+m)!}}P_l^m(\cos\theta)e^{im\phi},$$

where l is the band index, m is the order, P_l^m is an associated Legendre polynomial, and (θ,ϕ) is the representation of the direction vector in the spherical coordinate. We use real-valued spherical harmonics because the values used to define spatio-angular fields are positive and real.

To convert the function $f(\theta,\phi)$ into a spherical harmonics basis, spherical harmonics coefficients a_l^m are approximated using the equation

$$a_l^m = \int_s f(\theta,\phi)Y_l^m(\theta,\phi)ds,$$

A benefit of using spherical harmonics representation is that integrating two functions over the sphere can be estimated in an inexpensive way by performing a dot product of their spherical harmonics coefficients [3, 20].

$$\int U(s) \times V(s) ds = \sum_{i=0}^{l^2} u_i(s) \times v_i(s),$$

where U and V are two functions defined on the surface of a sphere, and $u(s)$ and $v(s)$ are their spherical harmonics coefficients.

5 Image Acquisition and Pre-processing

The 3T Siemens Tim Trio Scanner (Siemens Medical Solutions; Erlangen, Germany) was used to perform imaging. Diffusion weighted images were obtained with $b = 1{,}000, 2{,}000\,\text{s/mm}^2$ in 30 directions, together with 4 b_0 images, in-plane resolution $= 2.7\,\text{mm}^2$, echo time/time repetition $= 101\,\text{ms}/6{,}000\,\text{ms}$ at a slice thickness of 2.7 mm with two averages. DKI reconstruction was carried out on each voxel using a MATLAB program, as described by Zhuo et al. [43]. There are also newer alternative methods for computing kurtosis tensors by Ghosh et al. [9] and Tax et al. [40].

Once diffusion and kurtosis tensors are computed, we represent the shape of these tensors by using spherical harmonics approximation. From the diffusion and the kurtosis tensors, we use D_{app} and K_{app} to compute the shapes of Gaussian and non-Gaussian diffusion. Each shape is then represented in the spherical harmonics basis by computing spherical harmonics coefficients a_l^m. Based on the complexity of the shape, the number of coefficients used in spherical harmonics representation varies. The shape of the diffusion tensor is simpler than the kurtosis tensor. As described by Lu et al. [29], we used bands 1, 3, and 5 to represent the shape of the symmetric kurtosis tensor. This can be done using 15 spherical harmonics coefficients (there are 25 coefficients in total, but bands 2 and 4 are not used). Bands (>5) can be used too; however, high frequency data contains more noise, as discussed in [29]. These spherical harmonics coefficients capture the shape, magnitude, and direction of the tensors, which are used for segmentation and visualization.

5.1 Classification Reference Datasets

To compare various classification approaches, we perform tissue classification using $3D$ $T1$-weighted MPRAGE images, which is commonly used for brain tissue segmentation. These images were segmented to CSF, GM and WM using the SPM8 software package [1], and they served as the ground truth. The tissue masks were then aligned and under sampled to the DKI space through co-registration of the fractional anisotropy map and the WM tissue probability map, also through SPM8. As an initial step, we classify the DKI derived maps to different tissue types. It should be noted that T1-weighted MPRAGE images may not always be available, and that the image distortion inherent in diffusion weighted image may have an effect on co-registration, leading to inaccurate tissue classification. We demonstrate a method that can classify the tissue type reliably based on the DKI data.

6 Classification

Classifying spherical harmonics volume fields into smaller sub-regions is beneficial for both visualization and analysis. Local features can be enhanced or suppressed as desired, lighting functions can be optimized if the classification captures complexity of the spherical harmonics field, and grouping simplifies the analysis process because it can reflect domain-specific information. While there are several ways to accomplish segmentation in volume rendering, a popular method is to examine the intensity-gradient histogram to find the edge boundaries in order to segment different regions. In practice, there are different types of soft tissues in an image, and the boundaries may not be clearly defined. Instead of scalar values, our dataset contains irregular multi-dimensional geometric shapes. Furthermore, these datasets come with multiple attributes, which must be examined carefully in order to do the segmentation. This process can be very difficult. To classify the dataset, we examine two approaches: domain specific classification and shape based classification.

6.1 Domain Specific Classification

In DKI, several domain-specific attributes having biological relevance are computed. They are mean diffusion, fractional anisotropy, and mean kurtosis. To apply domain-specific classification, we apply the popular clustering algorithms K-means [11] and Gaussian mixture models (GMM) [31] on the mean kurtosis and mean diffusion. The relation between the mean diffusion and kurtosis has been explored by Jensen et al. [17]. They both capture properties of the imaged tissue. Here we use K-means and GMM to automatically cluster the dataset into segments. The results of GMM-based classification can be seen in Fig. 3.

6.2 Shape-Based Classification

We use the shape of the DT/KT tensor at each voxel to perform shape based classification across the entire volume. When comparing the shape of tensors, we

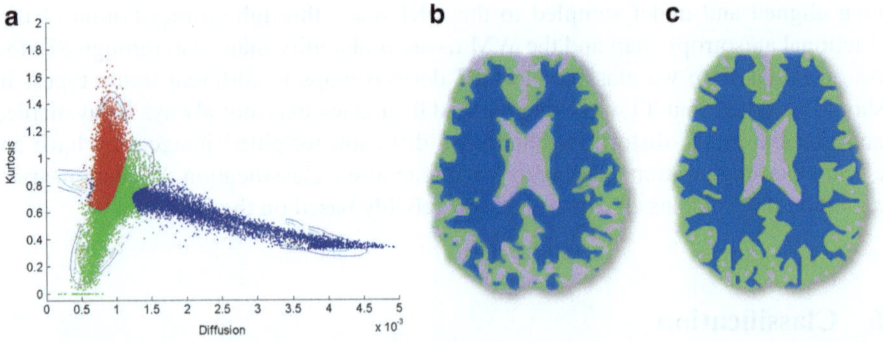

Fig. 3 The Gaussian mixture model is applied to mean diffusion and mean kurtosis data shown in (**a**). Three means are used to classify CSF, GM, and WM. We compare segmentation performed using MPRAGE (**b**) with GMM based segmentation performed using mean diffusion (MD) and mean kurtosis (MK) images (**c**)

consider two components: structure of the tensors and their orientation. Here we focus on just the shape by using the rotationally-invariant spherical harmonics form for classification. The rotationally invariant spherical harmonics form is computed by performing the summation of all the spherical harmonic coefficients within the same band as described by Funkhouser et al. [8] and Kazhdan et al. [21]. Coefficients of the spherical function become

$$R(f(\theta,\phi)) = \{\|f_0(\theta,\phi)\|, \|f_1(\theta,\phi)\|, \ldots, \|f_\infty(\theta,\phi)\|\}$$

where

$$f_l(\theta,\phi) = \sum_{m=-l}^{l} a_l^m Y_l^m(\theta,\phi)$$

Shape-based classification is a general approach that can be applied to any spatio-angular field. To categorize data into segments, clustering-based algorithms, K-means [11] and GMM are applied to the rotationally-invariant spherical harmonics attribute. The application of these clustering algorithms on shape-based attributes will group shapes together based on centroids or density. To apply these algorithms, we first adjust the rotationally invariant spherical harmonics representation of each dataset so that they are centered on the origin and have a unit standard deviation. $R_l' = \frac{R_l - \mu_{\{R_0,R_1,\ldots,R_L\}}}{\sigma_{\{R_0,R_1,\ldots,R_L\}}}$, where L is total number of bands used. This normalization is an important step as different datasets, such as rotationally-invariant spherical harmonics approximation of diffusion and kurtosis tensors, might have different data distribution. If one dataset is more compact than another, the properties of the compact dataset might not be well represented after segmentation. An extra

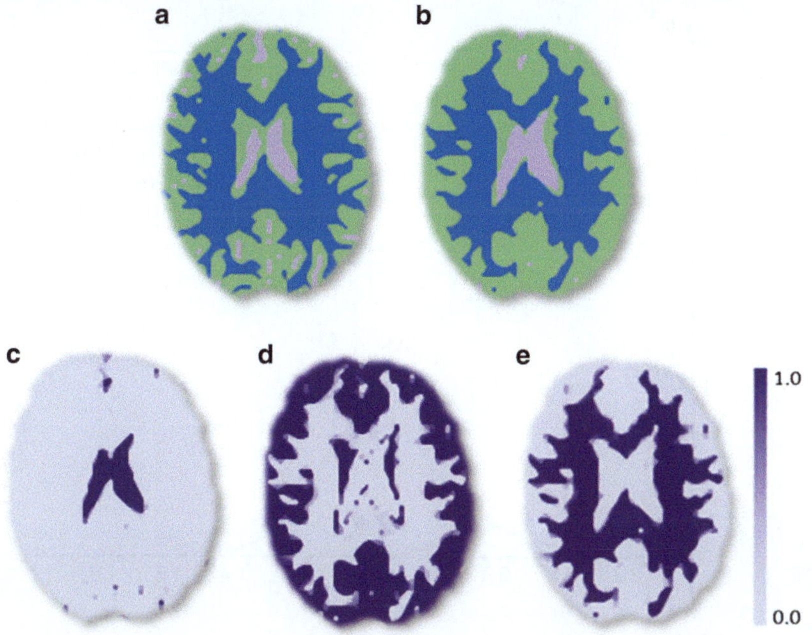

Fig. 4 We compare segmentation performed using MPRAGE image (**a**) with segmentation performed by applying K-means (with $k = 3$) on a rotationally-invariant spherical harmonics approximation of diffusion and kurtosis tensors (**b**). *Grey*, *green*, and *blue* represent CSF, GM, and WM respectively. In (**c**)–(**e**), we show the degree of membership of each pixel to different segments

weighting variable can also be applied depending on the need. Once the data is adjusted, we apply the clustering algorithms.

We have explored shape-based classification to segment various tissue types. For the KT dataset, we have three coefficients $(R'_{0_{dk}}, R'_{2_{dk}}, R'_{4_{dk}})$ for each voxel from the rotationally-invariant spherical harmonics attribute. For the DT and KT dataset we have six coefficients $(R'_{0_{dt}}, R'_{2_{dt}}, R'_{4_{dt}}, R'_{0_{dk}}, R'_{2_{dk}}, R'_{4_{dk}})$, three coefficients each for the diffusion and kurtosis tensors. The shape of the diffusion tensor characterizes the underlying Gaussian diffusion profile, whereas the kurtosis tensor describes the non-Gaussian diffusion profile. We cluster the dataset into three different segments and compare the result with the tissue classification performed on an MPRAGE image based on data from eight normal subjects. In Fig. 4, we show tissue classification performed on an MPRAGE image (Fig. 4a) along with segmentation performed by applying K-means (with $k = 3$) on the rotationally-invariant spherical harmonics approximation of the combined DT/KT dataset (Fig. 4b). In Fig. 4c–e we show degree of membership of each pixel with its segment.

In the study by Falangola et al. [7], three distinct peaks for CSF, GM, and WM were observed in the MK histogram around 0.45, 0.75, and 1.25 respectively in the frontal lobe white matter. We compare of segmentation result by apply the K-means

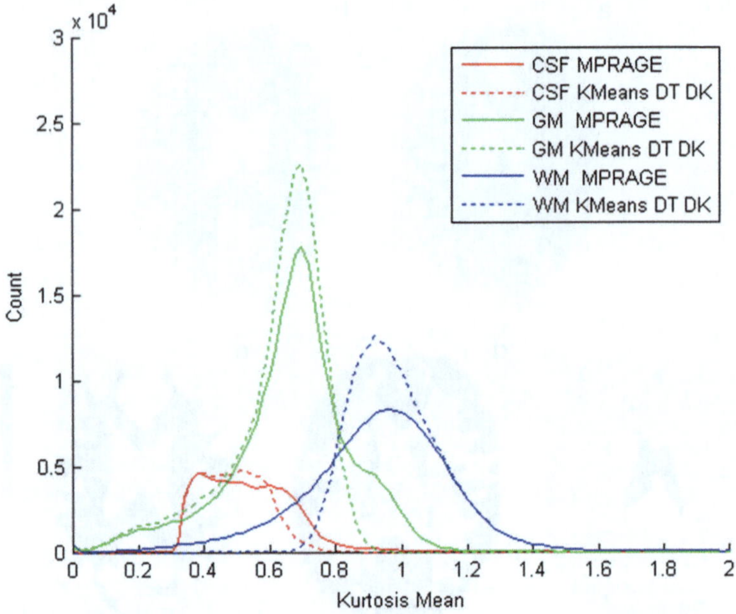

Fig. 5 Histogram plot of mean kurtosis (MK) for MPRAGE image and segmentation performed using K-means algorithm (with $k = 3$) on the rotationally invariant form of the spherical harmonics approximation of diffusion and kurtosis tensors. The peaks from both MPRAGE and the DT/KT segmentation are aligned with each other, which shows a good match between the two segmentation methods

algorithm (with $k = 3$) to the rotationally-invariant form of the spherical harmonics approximation of diffusion and kurtosis with MPRAGE-based segmentation.

We apply segmentation on 8 MRIs of healthy subjects and plot the combined histogram values of mean kurtosis (MK) for each segment, the results of which are shown in Fig. 5. The peaks of the histogram are aligned with each other. More interestingly, the MK histogram indicates a narrow distribution of MK values of all three tissue types, reflective of likely more accurate tissue classification using the shape based method.

The full result of classification is shown in Table 1. K-means applied on combined DT and KT performs best with 77 % match with the MPRAGE tissue classification. GMM produces good results when the CSF, GM, and WM have distinct density peaks. However, the distribution of each tissue type varies in each MRI and sometimes causes GMM to select a distribution that does not correspond to CSF, GM or WM. As the K-means algorithm searches for centroids and is geometric in nature, it provided better classification compared to the other techniques because the geometric properties of the tensors are closely tied to the underlying tissue types.

The volume ratio of a given tissue type is the ratio between the volume occupied by that tissue and the volume of the entire brain. We calculate the volume ratio for all three tissue types. The volume ratios of different tissue types using our

Table 1 Comparison of various segmentation methods on different data type with performance on MPRAGE image

Segmentation type	Classification type	Percentage match mean	Percentage match STD
GMM on MK	Domain	61.61	02.59
GMM on MD and MK	Domain	54.55	14.71
K-means on MD and MK	Domain	61.43	02.94
GMM on RI DT/KT	Shape	68.35	16.00
K-means on RI KT	Shape	64.92	03.09
K-means on RI DT/KT	Shape	77.50	01.32

Table 2 A comparison of various segmentation methods with MPRAGE based classification

	Volume ratio					
Segmentation type	CSF mean	CSF STD	GM mean	GM STD	WM mean	WM STD
GMM on MK	0.129	0.112	0.638	0.177	0.231	0.076
GMM on MD and MK	0.244	0.064	0.408	0.230	0.347	0.240
K-means on MD and MK	0.267	0.025	0.440	0.016	0.272	0.016
GMM on RI DT/KT	0.248	0.070	0.415	0.130	0.335	0.134
K-means on RI KT	0.251	0.029	0.448	0.015	0.286	0.020
K-means on RI DT/KT	0.126	0.035	0.508	0.028	0.363	0.017
MPRAGE	0.148	0.026	0.491	0.026	0.359	0.012

Volume ratio, which is a ratio between volume occupied by a tissue and the volume of the whole brain, is calculated for each tissue type

shape based segmentation and MPRAGE segmentation are shown in Table 2. We compare these with the volume ratios for MRI data from healthy subjects. K-means applied on combined DT/KT performs close to the MPRAGE tissue classification. As mentioned before, when there are no distinct density peaks, the output of the GMM algorithm degrades.

6.3 Representative Shape

After classifying various segments, we compute a representative shape for each segment for analysis and lighting in visualization. We determine a representative shape for lighting by using the mean value for each group based on attributes used for grouping. The voxel most closely representing the mean is chosen to represent the shape function. Figure 6 shows the segmentation performed on the DKI image of a patient with traumatic brain injury (TBI) using K-means (with k= 4) and the representative shape for each segment. The regions around the injury, as shown in red, have extreme kurtosis values, depicted by their elongated shapes.

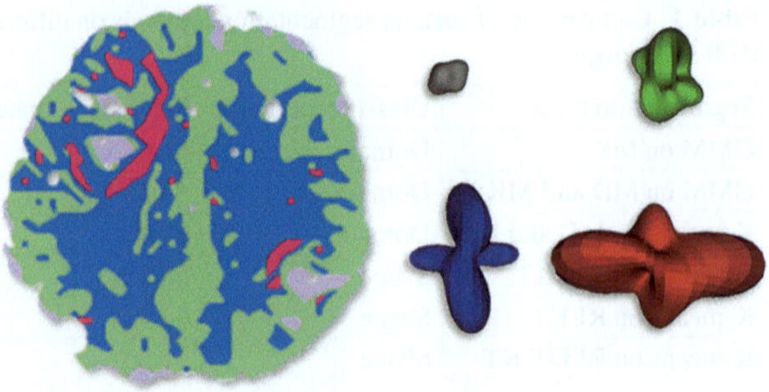

Fig. 6 Result of applying K-means (with k= 4) segmentation to differentiate CSF (*grey*), GM (*green*), and WM (*blue*) along with extreme kurtosis values (*red*). The area surrounding the injury site has very high kurtosis values. The representative DKI glyph for each segment is also shown

7 Visualization

In a previous paper [2], we used spherical harmonics lighting functions to analyze and visualize spatio-angular fields, such as diffusion and kurtosis tensors. Dynamic spherical harmonics lighting functions, which have unique directional shapes and sizes, are used as a query tool to illuminate the spatio-angular field and visualize the underlying structure. The output of the system is either a planar visualization or a volume rendering. In this work, our system uses the same tool with added support for visualizing segmented regions.

7.1 Planar Visualization

For planar visualization, we have several ways of visualizing the data. One direct way is to map segment identifiers to specific colors using a transfer function; this visualization mode allows easy identification of various segments. Although this method is straightforward, one needs to be careful in color assignment for different segment identifiers so that coloring across MRIs is consistent, as the segment identifiers from GMM or K-means can stochastically change for every run of the algorithm, because it is a stochastic process. An example visualization is shown in Fig. 7, comparing the MRI of a normal subject and an injured patient. For segmentation, shape-based classification using K-means segmentation (with three segments) on spherical harmonics approximations of the diffusion and kurtosis tensor is performed. Figure 7a shows the MRI segmentation of a normal subject, where segments relate to CSF (grey), WM (blue), and GM (green). In Fig. 7b, we use the MRI of a patient suffering from traumatic brain injury, in which segments show the injured region (red), WM (blue), and GM (green).

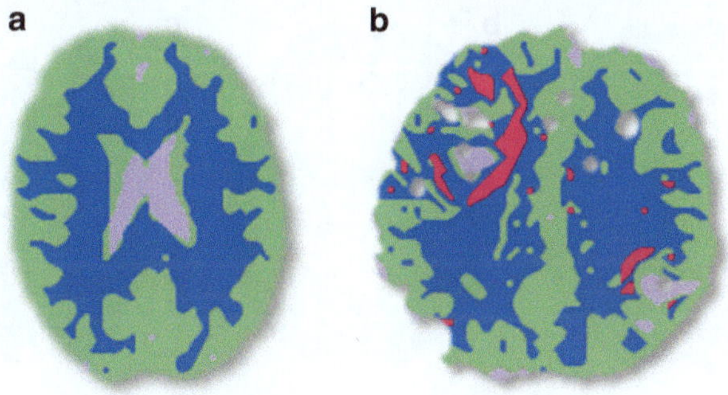

Fig. 7 Shape-based classification using K-means segmentation (with $k = 3$) on the spherical harmonics approximation of diffusion and kurtosis tensors. (**a**) shows segmentation into CSF (*grey*), WM (*blue*), and GM (*green*) on a normal subject. (**b**) We use the MRI of an injured patient. The segments show the injured region (*red*), WM (*blue*), and GM (*green*)

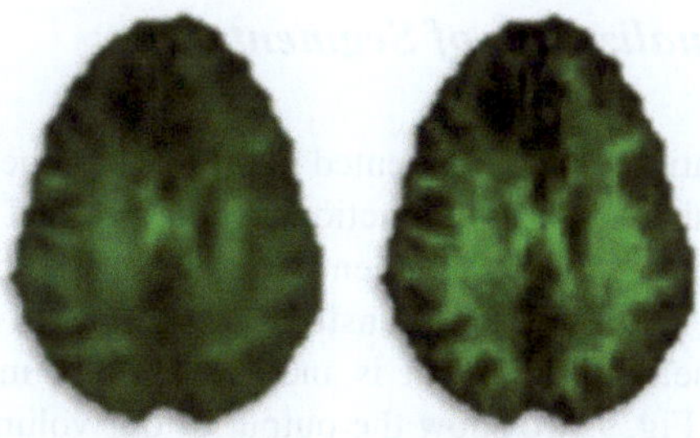

Fig. 8 The difference between lighting using a regular lighting function (*left*) and a local representative light (*right*). Using the representative glyph to light the volume field will emphasize exaggerate local differences, as seen in the second image

7.2 *Planar Visualization Using Representative Shapes*

The second form of planar visualization uses local shape-based lighting. In our previous work [2], lighting functions were used to illuminate spatio-angular fields to show the structural properties of the underlying tissues. The lighting functions can be modified or rotated to allow active exploration of the dataset. Most lighting functions used were pre-defined shapes. However, lighting functions do not have to be constrained to pre-defined shapes. In the previous section, we computed the representative shape for each cluster; using these shapes, each voxel can be lit by its group's representative shape, as shown in Fig. 8. With this lighting, a higher value characterizes the close approximation between the shape of the spherical harmonic voxel field and its representative shape, which is similar to the degree of membership used in segmentation.

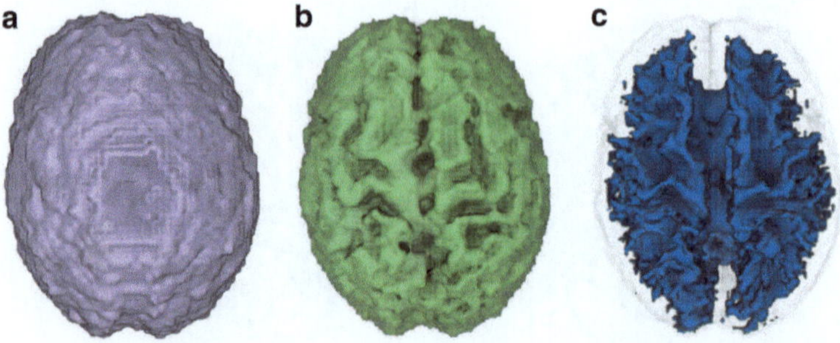

Fig. 9 The volume visualization of the segmented brain. The transfer function that maps the segment identifiers into color and opacity is automatically created based on segments selected by the user. In this example, the whole brain is classified into three segments (CSF, GM, and WM), as shown in the images

7.3 *Volume Visualization of Segments*

For volume visualization of the segmented data, we map the segment identifiers to color and opacity using a transfer function. After the MRI dataset is segmented, we create a scalar field using the segment identifiers. This field is used in volume rendering to perform a lookup of the transfer function. Based on the user preference, the opacity of the selected segment is increased while making other segments semi-transparent. In Fig. 9, we show the output of our volume visualization of the segmented MRI.

7.4 *Volume Visualization of Spatio-Angular Fields*

We use the framework, described in our previous work [2], to visualize spatio-angular fields and add support to display segmented data. In particular, we use two transfer functions. The first transfer function as described in [2], converts light response values to color and opacity. The second transfer function determines opacity based on the segments the user selects. By using both transfer functions at the same time, we allow the user to view the spatio-angular field of only the selected segments. In Fig. 10, the spatio-angular field of the segment related to the injury is visualized. By rotating the lighting function (shown on the right side of the figure) users can interact with the spatio-angular field.

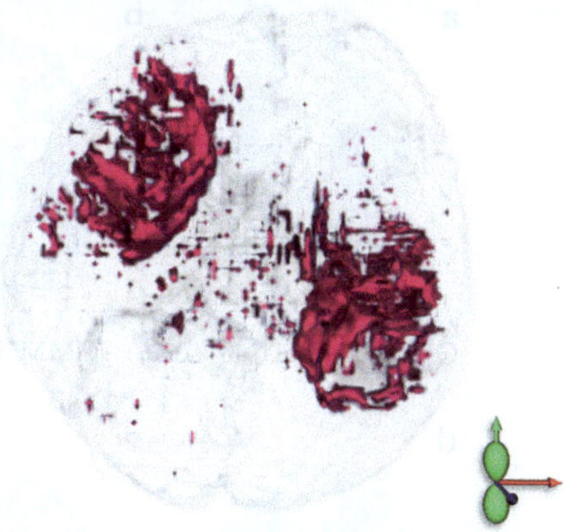

Fig. 10 The volume visualization of the spatio-angular field. Using an additional transfer function, only spatio-angular fields of selected segments are displayed. The lighting function, shown to the *right* of the image, is used to explore the directional strength of the spatio-angular field. As the user rotates the lighting function, a different direction is queried

8 Application

8.1 Visual Comparison

We visually compare results after applying different segmentation strategies. Since GMM performed well when segmentation was based on domain-specific attributes and K-means produced the best results when shape-based attributes were used, we visually compare these two results with each other. We apply segmentation to find three segments on the MRIs of both a normal patient and a patient with an injury, as shown in Fig. 11. The top row shows segmentation using domain-specific attributes, whereas the bottom row shows segmentation using shape-based attributes. Both segmentation strategies are able to distinguish basic segments, including injury. However, shape-based classification is able to capture the underlying properties of the tissues much better than segmentation done in with domain-specific attributes.

8.2 Segment Count Variation

In classification by shape-based attributes, we tested how increasing the number of segments affects classification. In Fig. 12, we show the output of segmentation using two datasets. In the MRI of the normal patient, which is shown in the top row of Fig. 12, CSF, GM, and WM are clearly segmented when the number of segments is 3. As the number is further increased, subdivision within GM and CSF, occurred as seen in Fig. 12c, d. In the case of the patient suffering from TBI, the region around the injury is clearly visible when the number of segments is greater than 3, as seen in Fig. 12g, h.

In most of the examples of an MRI of a normal subject, we classify the entire brain into CSF, GM, and WM. For these classifications $k \leq 3$ is used. CSF,

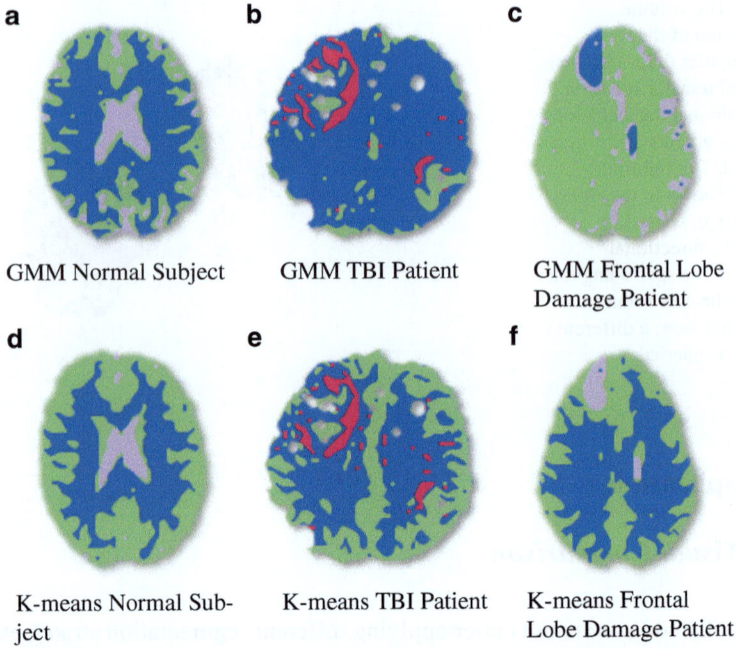

Fig. 11 A visual comparison between segmentation performed using domain-specific attributes and shape-based attributes. For segmentation based on domain-specific attributes, GMM is used, and for shape-based attributes, K-means is used. In the entire segmentation, we find three different segments. In (**a**) and (**d**), the MRI of a normal subject is used. In (**b**) and (**e**), the MRI of a patient suffering from TBI is shown. In (**c**) and (**f**), the MRI of a patient with frontal lobe damage is used. Shape-based classification captures the underlying properties of the tissues much better than segmentation created using domain-specific attributes

GM, and WM are structurally different; thus they have distinct diffusion profiles. In classifying a brain with an injury, we use $k \leq 4$ as we are dealing with four structurally distinct regions: CSF, GM, WM, and injury regions. If $k > 4$ is used, these regions are further classified. Additional study and evaluations are needed for these type of classifications.

8.3 *Traumatic Brain Injury*

We apply segmentation to the MRI of a patient suffering from traumatic brain injury. We used K-means segmentation with $k = 3$ on spherical harmonics approximations of diffusion and kurtosis tensors. In Fig. 13, the output of the segmentation is shown. The segmentation process is able to segment out the region around the injury (red) from other regions, such as WM (blue) and GM (green).

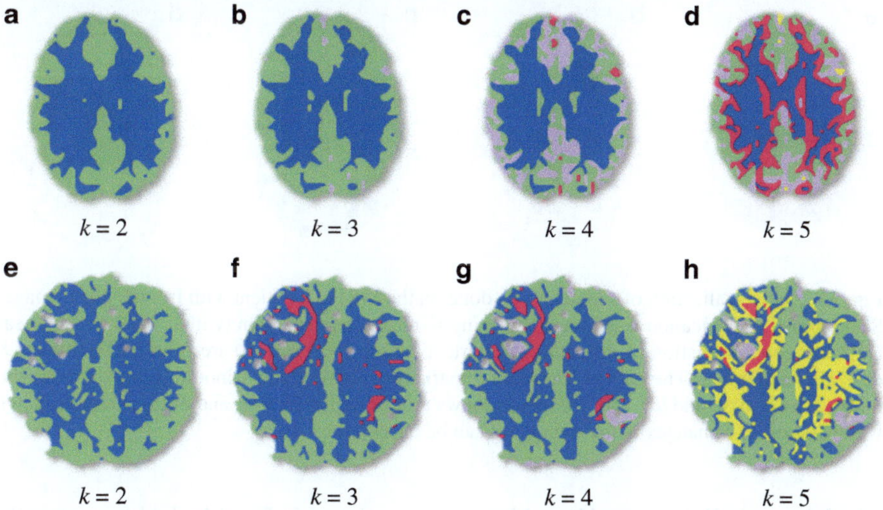

Fig. 12 A visual comparison of different segmentation strategies when the number of segments is varied for different MRI datasets. The *top row* uses the MRI of a normal subject and in the *bottom row* we use the MRI of a patient who is suffering from traumatic brain injury (TBI). Each segment in these images is colored differently. There is no relation between the coloring of segments for the normal subject and the patient suffering from traumatic brain injury. (**a**, **e**) $k = 2$. (**b**, **f**) $k = 3$. (**c**, **g**) $k = 4$. (**d**, **h**) $k = 5$

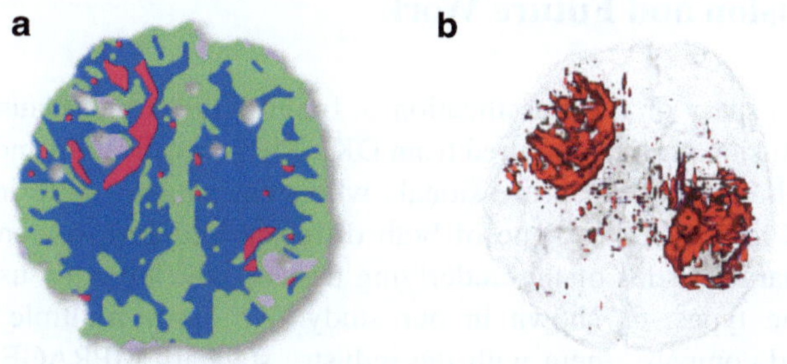

Fig. 13 The visualization of segmentation done on the MRI of a patient suffering from a traumatic brain injury. Shape-based classification was performed using K-means segmentation (with $k = 3$) on spherical harmonics approximations of diffusion and kurtosis tensors. The segment relating to the injury is shown in *red*. (**a**) Planar visualization. (**b**) Volume visualization

8.4 Frontal Lobe Injury

In the injury case shown in Fig. 14, the patient has sustained frontal lobe damage. We segment the MRI dataset using K-means segmentation (with $k = 4$) on spherical harmonics approximation of diffusion and kurtosis tensors. The region in blue is associated with white matter; the region in red is related to areas with high diffusion. Right after the injury, a high diffusion region was observed in the frontal lobe,

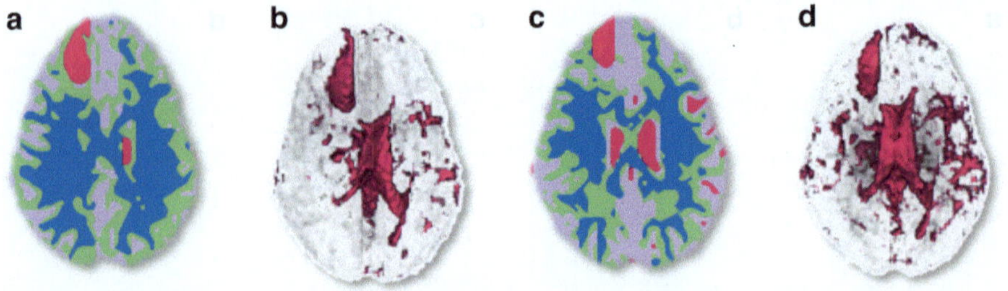

Fig. 14 The visualization of segmentation done on the MRI of a patient with frontal lobe damage. Shape-based classification was performed using K-means segmentation (with $k = 4$) on spherical harmonics approximation of diffusion and kurtosis tensors. (**a**) and (**b**) are from an MRI taken 8 days after the injury. The patient showed a remarkable recovery at a 1 month follow-up after the injury, shown in (**c**) and (**d**). The *red* region shows area of high diffusion, and the *blue* region shows white matter. The changes in the *red* region can be observed easily

which is normally occupied by white matter, as shown in Fig. 14a, b. After a month, some noticeable changes in the high diffusion region can be observed, as shown in Fig. 14c, d. This aligns with the clinical diagnosis, as the patient made a significant recovery within a month.

9 Conclusion and Future Work

We present- a study on the classification of brain tissues using Gaussian and non-Gaussian diffusion profiles acquired from DKI. MRI classification and visualization are vital tools for medical professionals who treat patients suffering from brain diseases and injuries. The shape of both diffusion and kurtosis tensors provides important characteristics of the underlying tissues, which can be used to classify various tissue types, as shown in our study. We apply multiple segmentation strategies and compare- them with the industry standard MPRAGE imaging. We also present a way to display the segmented data effectively in planar and in volume visualization modes.

In the future, we plan to extend the utility of our tool to automatically segment various disease biomarkers in the human brain to study inflammation and neurodegeneration. We also hope to include data from other forms of imaging to further improve classification.

Acknowledgements We are grateful to the anonymous reviewers whose constructive comments have greatly improved the presentation of our approach and results in this paper. We appreciate the support of the U.S. Army grant W81XWH-12-1-0098, NSF grants 09-59979 and 14-29404, the State of Maryland's MPower initiative, and the NVIDIA CUDA Center of Excellence. Any opinions, findings, conclusions, or recommendations expressed in this article are those of the authors and do not necessarily reflect the views of the research sponsors.

References

1. Ashburner, J., Friston, K.J.: Unified segmentation. Neuroimage **26**, 839–851 (2005)
2. Bista, S., Zhuo, J., Gullapalli, R.P., Varshney, A.: Visualization of brain microstructure through spherical harmonics illumination of spatio-angular fields. IEEE Trans. Vis. Comput. Graph. **20**(12), 2516–2525 (2014)
3. Cabral, B., Max, N.L., Springmeyer, R.: Bidirectional reflection functions from surface bump maps. ACM Siggraph Comput. Graph. **21**, 273–281 (1987)
4. Carr, N.A., Hall, J.D., Hart, J.C.: GPU algorithms for radiosity and subsurface scattering. In: Proceedings of the ACM SIGGRAPH/EUROGRAPHICS Conference on Graphics Hardware, pp. 51–59. Eurographics Association (2003)
5. Centers for Disease Control and Prevention: Traumatic Brain Injury in the United States: Fact Sheet. www.cdc.gov/TraumaticBrainInjury/get_the_facts.html (2014)
6. Correa, C., Ma , K.-L.: Size-based transfer functions: a new volume exploration technique. IEEE Trans. Vis. Comput. Graph. **14**(6), 1380–1387 (2008)
7. Falangola, M.F., Jensen, J.H., Babb, J.S., Hu, C., Castellanos, F.X., Martino, A.D., Ferris, S.H., Helpern , J.A.: Age-related non-Gaussian diffusion patterns in the prefrontal brain. J. Magn. Reson. Imaging **28**, 1345–1350 (2008)
8. Funkhouser, T., Min, P., Kazhdan, M., Chen, J., Halderman, A., Dobkin, D., Jacobs, D.: A search engine for 3D models. ACM Trans. Graph. **22**(1), 83–105 (2003)
9. Ghosh, A., Milne, T., Deriche, R.: Constrained diffusion kurtosis imaging using ternary quartics & mle. Magn. Reson. Med. **71**(4), 581–1591 (2014)
10. Hao, X., Baby, T., Varshney, A.: Interactive subsurface scattering for translucent meshes. In: ACM Symposium on Interactive 3D Graphics, 28–30 Apr, pp. 75–82 (2003)
11. Hartigan, J.A., Wong, M.A.: Algorithm AS 136: a k-means clustering algorithm. Appl. Stat. **28**(1), 100–108 (1979)
12. Hasan, K.M., Halphen, C., Sankar, A., Eluvathingal, T.J., Kramer, L., Stuebing, K.K., Ewing-Cobbs, L., Fletcher, J.M.: Diffusion tensor imaging-based tissue segmentation: validation and application to the developing child and adolescent brain. Neuroimage **34**(4),1497–1505 (2007)
13. Huang, R., Ma, K.-L.: RGVis: region growing based techniques for volume visualization. In: Proceedings of 11th Pacific Conference on Computer Graphics and Applications, pp. 355–363. IEEE, New York (2003)
14. Ip, C.Y., Varshney, A., JaJa, J.: Hierarchical exploration of volumes using multilevel segmentation of the intensity-gradient histograms. IEEE Trans. Vis. Comput. Graph. **18**(12), 2355–2363 (2012)
15. Jensen, J., Helpern, J.: Quantifying non-Gaussian water diffusion by means of pulsed-field-gradient MRI. In: Proceedings of the 11th Annual Meeting of ISMRM, vol. 2154 (2003)
16. Jensen, J.H., Helpern, J.A.: MRI quantification of non-Gaussian water diffusion by kurtosis analysis. NMR Biomed. **23**(7), 698–710 (2010)
17. Jensen, J.H., Helpern, J.A., Ramani, A., Lu, H., Kaczynski, K.: Diffusional kurtosis imaging: the quantification of non-Gaussian water diffusion by means of magnetic resonance imaging. Magn. Reson. Med. **53**, 1432–1440 (2005)
18. Jeurissen, B., Tournier, J.-D., Dhollander, T., Connelly, A., Sijbers, J.: Multi-tissue constrained spherical deconvolution for improved analysis of multi-shell diffusion {MRI} data. NeuroImage **103**, 411–426 (2014)
19. Jones, D.K.: Diffusion MRI Theory, Methods, and Applications. Oxford University Press, Oxford (2011)
20. Kautz, J., Snyder, J., Sloan, P.-P.J.: Fast arbitrary BRDF shading for low-frequency lighting using spherical harmonics. In: Eurographics Symposium on Rendering/Eurographics Workshop on Rendering Techniques, Pisa, pp. 291–296 (2002)
21. Kazhdan, M.M., Funkhouser, T.A., Rusinkiewicz, S.: Rotation invariant spherical harmonic representation of 3D shape descriptors. In: ACM International Conference Proceeding Series, pp. 156–165 (2003)

22. Kindlmann, G.L., Durkin, J.W.: Semi-automatic generation of transfer functions for direct volume rendering. In: Volume Visualization and Graphics, pp. 79–86 (1998)
23. Kniss, J., Kindlmann, G., Hansen, C.: Multidimensional transfer functions for interactive volume rendering. IEEE Trans. Vis. Comput. Graph. **8**(3), 270–285 (2002)
24. Lazar, M., Jensen, J.H., Xuan, L., Helpern, J.A.: Estimation of the orientation distribution function from diffusional kurtosis imaging. Magn. Reson. Med. **60**, 774–781 (2008)
25. Lee, C.H., Hao, X., Varshney, A.: Geometry-dependent lighting. IEEE Trans. Vis. Comput. Graph. **12**(2), 197–207 (2005)
26. Lee, C.H., Kim, Y., Varshney, A.: Saliency-guided lighting. IEICE Trans. Inf. Syst. **E92-D**(2), 369–373 (2009)
27. Levoy, M.: Display of surfaces from volume data. IEEE Comput. Graph. Appl. **8**, 29–37 (1988)
28. Liu, T., Li, H., Wong, K., Tarokh, A., Guo, L., Wong, S.T.: Brain tissue segmentation based on dti data. NeuroImage **38**(1), 114–123 (2007)
29. Lu, H., Jensen, J.H., Ramani, A., Helpern, J.A.: Three-dimensional characterization of non-gaussian water diffusion in humans using diffusion kurtosis imaging. NMR Biomed. **19**(2), 236–247 (2006)
30. Lum, E.B., Ma, K.-L.: Lighting transfer functions using gradient aligned sampling. In: Proceedings of the Conference on Visualization'04, pp. 289–296. IEEE Computer Society, Washington (2004)
31. McLachlan, G., Peel, D.: Finite Mixture Models. Wiley, New York (2004)
32. Mohlenkamp, M.J.: A fast transform for spherical harmonics. J. Fourier Anal. Appl. **5**(2–3), 159–184 (1997)
33. Prčkovska, V., Vilanova, A., Poupon, C., Haar Romeny, B., Descoteaux, M.: Fast classification scheme for HARDI data simplification. In: Davcev, D., Gómez, J. (eds.) ICT Innovations 2009, pp. 345–355. Springer, Heidelberg (2010)
34. Prčkovska, V., Peeters, T.H.J.M., Van Almsick, M., ter Haar Romeny, B., Vilanova i Bartroli, A.: Fused DTI/HARDI visualization. IEEE Trans. Vis. Comput. Graph. **17**(10), 1407–1419 (2011)
35. Roine, T., Jeurissen, B., Perrone, D., Aelterman, J., Leemans, A., Philips, W., Sijbers, J.: Isotropic non-white matter partial volume effects in constrained spherical deconvolution. Front. Neuroinform. **8**(28), 1–9 (2014)
36. Schlegel, P., Makhinya, M., Pajarola, R.: Extinction-based shading and illumination in GPU volume ray-casting. IEEE Trans. Vis. Comput. Graph. **17**(12), 1795–1802 (2011)
37. Sillion, F.X., Arvo, J.R., Westin, S.H., Greenberg, D.P.: A global illumination solution for general reflectance distributions. In: ACM SIGGRAPH Computer Graphics, vol. 25, pp. 187–196 (1991)
38. Sloan, P.-P., Kautz, J., Snyder, J.: Precomputed radiance transfer for real-time rendering in dynamic, low-frequency lighting environments. In: ACM Trans. Graph. **21**, 527–536 (2002)
39. Tao, Y., Lin, H., Bao, H., Dong, F., Clapworthy, G.: Structure-aware viewpoint selection for volume visualization. In: Visualization Symposium, Pacific Asia-Pacific, pp. 193–200 (2009)
40. Tax, C.M., Otte, W.M., Viergever, M.A., Dijkhuizen, R.M., Leemans, A.: Rekindle: robust extraction of kurtosis indices with linear estimation. Magn. Reson. Med. pp. 192–200 (2014). http://ieeexplore.ieee.org/xpls/abs_all.jsp?arnumber=4906856&tag=1
41. Tikhonova, A., Correa, C.D., Ma, K.-L.: An exploratory technique for coherent visualization of time-varying volume data. Comput. Graphics Forum **29**(3), 783–792 (2010)
42. Zhang, Y., Ma, K.-L.: Lighting design for globally illuminated volume rendering. IEEE Trans. Vis. Comput. Graph. **19**(12), 2946–2955 (2013)
43. Zhuo, J., Xu, S., Proctor, J.L., Mullins, R.J., Simon, J.Z., Fiskum, G., Gullapalli, R.P.: Diffusion kurtosis as an in vivo imaging marker for reactive astrogliosis in traumatic brain injury. NeuroImage **59**(1), 467–477 (2012)

A Survey of Illustrative Visualization Techniques for Diffusion-Weighted MRI Tractography

Tobias Isenberg

Abstract Fiber tracking is a common method for analyzing 3D tensor fields that arise from diffusion-weighted magnetic resonance imaging. This method can visualize, e.g., the structure of the brain's white matter or that of muscle tissue. Fiber tracking results in dense, line-based datasets that are often too large to understand when shown directly. This chapter provides a survey of recent illustrative visualization approaches that address this problem. We group this work into techniques that improve the depth perception of fiber tracts, techniques that visualize additional data about the tracts, techniques that employ focus+context visualization, visualizations of fiber tract bundles, representations of uncertainty in the context of probabilistic fiber tracking, and techniques that rely on a spatially abstracted visualization of connectivity.

1 Introduction

The visual representation of brain connectivity (e.g., Margulies et al. [43]) is an active research field within visualization. This work has lead to numerous techniques [51, 52] to explore and better understand the connections in the brain. Many of these approaches are based on diffusion-weighted magnetic resonance imaging (dwMRI) [37, 67] which yields estimates for the directional diffusion of water at each of the sampled locations. These datasets are typically represented as simplified 3D tensor fields (DTI; e.g., Kratz et al. [39]) and one fundamental visualization technique is the depiction of fiber tracts extracted from these tensor datasets (e.g., Behrens et al. [4] and Mori and van Zijl [45]). One sub-field within fiber tracking is deterministic fiber tracking. In this case a fiber tract is only extended along one direction—the diffusion tensor's principal eigenvector (e.g., see Fig. 1 as well as Mori and van Zijl [45] and Zhang et al. [66]). In contrast, probabilistic tracking approaches (e.g., Parker [48]) do not only follow a single direction when determining fiber tracts. Instead, they model the uncertainty of the tract direction at a location and

T. Isenberg (✉)
Inria-Saclay, Orsay, France
e-mail: tobias.isenberg@inria.fr

I. Hotz, T. Schultz (eds.), *Visualization and Processing of Higher Order Descriptors for Multi-Valued Data*, Mathematics and Visualization,
DOI 10.1007/978-3-319-15090-1_12

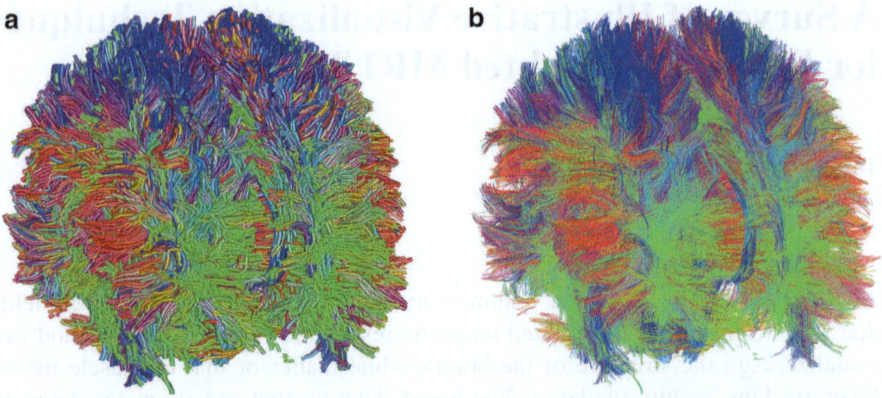

Fig. 1 A whole-brain HARDI dataset shown using (**a**) tubular and (**b**) a line-based visualization

follow several different directions at each sampled location. The results of this probabilistic tracking are typically presented visually as positional probabilities—rather than renderings of dense fiber tracts. Regardless of whether deterministic or probabilistic fiber tracking is used, however, the fiber tracking produces dense, line-based datasets[1] that are often difficult to understand due to their overall structure and spatial organization.

The fiber tracts are either computed for the whole brain or based on local seeding. Typically, the individual tracts are depicted using lines (e.g., Zhukov et al. [69] and Zöckler et al. [71]) or shaded cylindrical tubes (e.g., Zhang et al. [66]). An additional color map is often applied in which the color shades represent the local orientation of the tracts to better understand the spatial character of the data (see Fig. 1). It is important to understand that these extracted fiber tracts do not show actual brain fibers. Rather, each fiber tract is an abstracted representation of a likely direction of many fibers in the brain. As one can see in Fig. 1, however, these depictions can be quite overwhelming and it can be difficult to understand the spatial structure and other aspects of the connectivity data—even for small selections of fiber tracts.

For this reason researchers have begun to explore the use of illustrative visualization techniques to improve the visual representations of fiber tract data. Illustrative visualization [53] is a sub-field of visualization that is inspired, in particular, by the methods and techniques used in traditional illustration—which has a centuries-long history of creating understandable depictions of scientific subject matter. For example, illustrative visualization can be used to better depict the spatial structure of 3D datasets, to free up visual variables for the depiction of additional aspects of

[1] This is also true for probabilistic tracking [4], even if probabilistic tractography results—due to the size of the generated data—are typically visualized by displaying the scalar probabilities that different brain regions are connected to a seed region. In fact, these dataset sizes are one motivation to employ illustrative visualization as it promises to present the data in an understandable form.

the data, or to simultaneously show several layers of data. In this survey we review the different illustrative visualization approaches[2] that have been applied to DTI-based fiber tracking.[3] Specifically, we group the approaches according to whether they improve the depth perception of the dense fiber datasets, support the depiction of additional data such as uncertainty, enable focus+context visualizations, focus on abstracted fiber bundle representations, or use other forms of abstracted connectivity representations. This classification is not entirely unambiguous; however, it provides a useful structure and ambiguous cases are discussed appropriately.

2 Improving the Depth Perception of Fiber Tracts

In the majority of application scenarios, the visualizations generated based on the extracted fiber tracts are displayed as 2D projections using traditional PC-based workstations (in contrast to stereoscopic viewing environments such as CAVEs [14]). This means that, due to the denseness of the fiber tracts, it can become difficult to understand the spatial structure and the spatial relationship between different groups of fiber tracts. The problem can be alleviated somewhat by using shaded lines (e.g., Zöckler et al. [71] and Mallo et al. [42]) drawn with respect to a light source in the scene. When combined with line shadowing (e.g., Peeters et al. [49]), the results convey a much better sense of spatial structure. This basic approach or extensions of it have been used in cases when a complete, dense dataset of fibrous structure (such as muscle tissue) needs to be visualized [15, 49, 68].

A related way to address the spatial perception problem is to represent each tract with a shaded tube (Fig. 1a)—in contrast to a non-illuminated line rendering (Fig. 1b).[4] While the shading (combined with directional color coding) assists depth perception to some degree, it presents a challenge: Tubes with a larger diameter improve depth perception but also produce overlaps, reducing amount of visible detail. It is thus difficult to strike a good balance between the need for detail and the need for depth perception (even with additional measures such as tube halos [64]).

In a first attempt to address this issue, Klein et al. [38] proposed to remove the tube shading and, instead, to use distance-encoded contours and shadows for the tubes. This approach is realized using several rendering passes and enhances depth perception by visually grouping similar, neighboring fibers and using shadows as a visual cue for depth ordering. Inspired by this early work, Everts et al. [24]

[2]Surveys of the use of illustrative visualization techniques for domains other than brain connectivity have been presented for flow visualization [10] and as a general tutorial/overview [63].

[3]The chapter focuses on the visualization of brain connectivity. The discussed methods, however, can also be applied to other datasets that have similar characteristics, for example muscle fiber data.

[4]For a comparison of simple line rendering, shaded tubes, illuminated line rendering, and illuminated line rendering with shadowing see Figure 4 in Peeters et al.'s [49] paper.

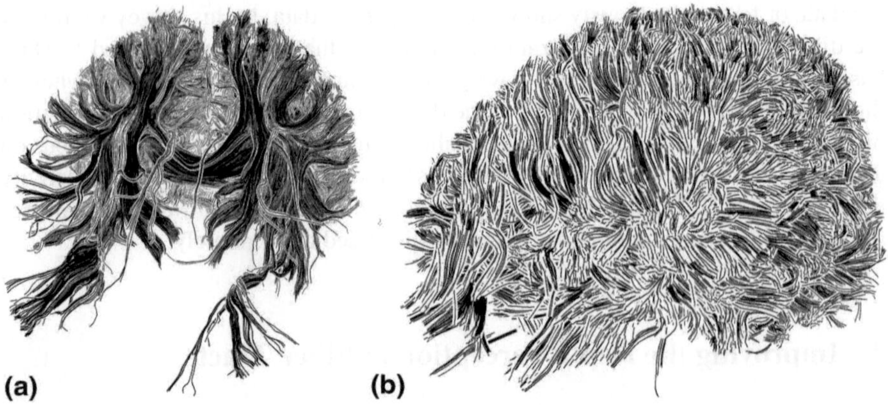

Fig. 2 Depth-dependent halos [24], applied to (**a**) a selection and (**b**) a whole-brain tractogram

presented a depth-dependent halo technique that used line-based triangle strips rather than tubes as the main primitive for depicting fiber tracts.[5] This decision was based on the observation that line-based techniques are able to show more detail than tube-based renderings (see the comparison in Fig. 1a, b). However, because using only lines does not provide good support for depth perception, they resort to using halos[6] to enhance depth perception. To prevent halos from overlapping neighboring fiber tracts, Everts et al. [24] apply them in a depth-dependent fashion by rendering view-aligned triangle strips (with separate line and halo components). These triangle strips are folded away from the viewer such that halos only appear for larger depth discontinuities. Everts et al. combine this approach with line width attenuation based on the distance of the line segment to the viewer.

As shown in Fig. 2,[7] the approach not only visually emphasizes fiber bundles but also shows a clear depth layering of the bundles—without the bundles ever being explicitly specified. As shown in Fig. 2, the depth-dependent halo technique works best for locally seeded selections of fiber tracts (Fig. 2a). In contrast, the spatial structure of whole-brain tractograms (Fig. 2b) is less clear due to the overlapping of the dense fiber tracts. The reduced co-linearity of the white matter fibers close to the gray matter of the brain further reduces clarity (filtering the dataset based on the fractional anisotropy (FA) value of the tensors can alleviate this problem somewhat).

This limitation of depth-dependent halos results from its very local approach for emphasizing spatial structure: it is only possible to provide halos and emphasize depth within the scope of individual line strips. Therefore, follow-up work has

[5]Everts et al.'s [24] approach could be viewed as an abstraction of line-based rendering with shadowing: it uses lines as the basic primitive, conveys occlusion, and does not rely on line shading.

[6]Halos had previously already been used in computer graphics [2] and visualization [13, 64].

[7]The example images in Fig. 2 were created with the depth-dependent halos demo; see the project website at http://tobias.isenberg.cc/VideosAndDemos/Everts2009DDH .

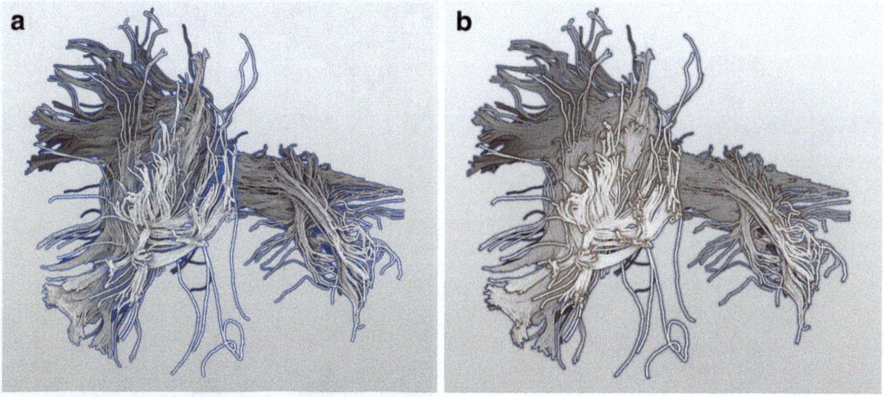

Fig. 3 The use of ambient occlusion halos to improve depth perception in fiber tract bundles [17, 18], using (**a**) screen-space ambient occlusion and (**b**) geometry-aware halos. Images © 2014 by Jesús Díaz-García and Pere-Pau Vázquez, used with permission

investigated more global approaches to assist people in perceiving the three-dimensional structure of fiber tract visualizations. These newer techniques are inspired by global illumination models in computer graphics, specifically ambient occlusion[8] [70].

A first technique in this group of approaches, presented by Díaz-García and Vázquez [17, 18], uses ambient occlusion halos around the fiber tracts. These halos are either computed in screen space (called SSAO halos) or in a geometry-aware fashion (called GA halos) and are combined with a depth-modulated line style (see the examples in Fig. 3). The screen-space ambient occlusion [44] essentially computes a local estimation of shadowing, while geometry-aware halos use a multi-pass approach that renders enlarged fiber tract geometries to generate halos in object space to simulate ambient occlusion. Díaz-García and Vázquez [17, 18] conclude that SSAO halos are best suited for large, dense datasets, while the GA halos are better for fiber tract selections which are more sparse.

Eichelbaum et al. [22] improved upon these approaches with line-based ambient occlusion technique called LineAO. Like Díaz-García and Vázquez's work [17, 18], this approach uses screen-space ambient occlusion. Based on an in-depth discussion of the sampling theory behind the technique, however, Eichelbaum et al. handle both local detail and global structures by separating the computation of the line-based ambient occlusion from that of the local illumination. Eichelbaum et al. thus not only produce grayscale visualizations (e.g., in Fig. 4a) but can also combine the method with illuminated lines or shaded tubes. These combinations can use,

[8]Ambient occlusion has also already been used in other sub-fields of visualization [62].

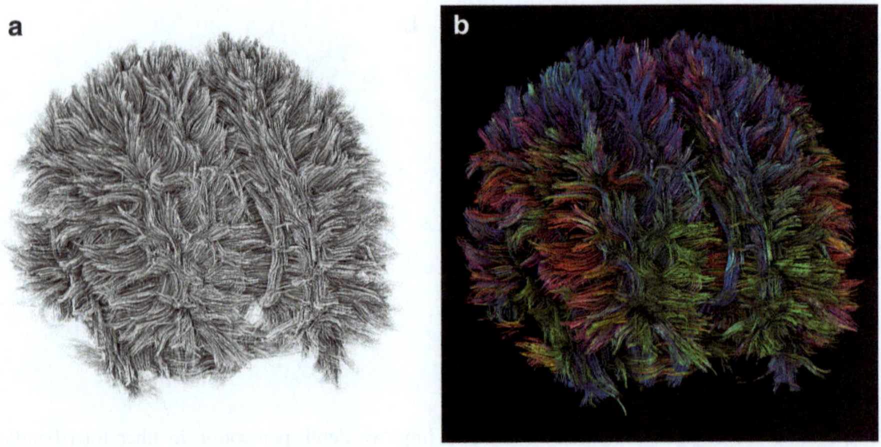

Fig. 4 The same dataset as shown in Fig. 1, but visualized using the LineAO technique [22]; in (**a**) grayscale and (**b**) with directional color-coding

for example, the established color coding based on the local line segment direction (e.g., Fig. 4b).[9]

Figure 4 demonstrates how the LineAO visualization is able to convey a whole-brain tractogram with an excellent support of spatial perception, in contrast to depth-dependent halos (Fig. 2b). Using LineAO, local individual fibers, fiber bundles, and the global spatial structure are clearly visible. However, the LineAO technique is less-suited for more coarse datasets which do not provide enough occlusion to benefit depth perception. Therefore, when visualizing subsets of fiber tracts, the approaches by Díaz-García and Vázquez [17, 18] or Everts et al. [24] may perform better.

3 Visualization of Additional Data About Fiber Tracts

In typical application scenarios for DTI-based fiber tract visualizations, researchers and health professionals are not only interested in understanding the spatial structure of a dataset but are also interested in other aspects. It is thus important to understand how different visualization techniques support the display of additional data dimensions. Some illustrative visualization techniques specifically make additional visual variables available that can be used for data display.[10]

[9]The example images in Figs. 1 and 4 were created with the tool OpenWalnut [19, 20]; see the website at http://www.openwalnut.org/ .

[10]In the context of the brain connectivity visualization, Laidlaw et al. [40] have used this principle to illustratively show slices of DTI data based on inspirations from oil painting.

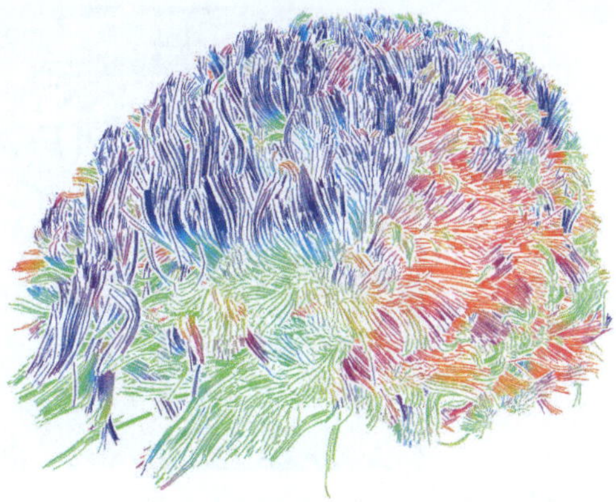

Fig. 5 Depth-dependent halos [24] with directional color coding

One common additional property is the local orientation of the fiber tracts. In traditional visualizations, local orientation is typically displayed using a directional color coding (as previously shown in Fig. 1). Also depth-dependent halos [24] and LineAO [22] can easily be combined with this encoding (Figs. 5 and 4b, respectively).

Going beyond this simple color mapping, Jianu et al. [35] explore encoding local properties computed based on the DTI data by using color pairs which are then mapped onto tubular fiber tract visualizations in horizontal, vertical, diagonal, or diamond patterns. This visual mapping is intended to allow viewers to easily compare properties of spatially co-located fiber tracts.[11] Hermosilla et al. [30] also used a similar texturing approach to indicate directionality of fiber tracts (see Fig. 7).

Bundle names are another important piece if information when interpreting fiber tract visualizations. Such annotations are a common element in traditional illustrations and, hence, are also important for illustrative visualization. Due to the complex nature of fiber tract datasets and the long length and intertwining nature of the tracts, however, it is not possible to employ existing external annotation placement methods created in illustrative visualization. Petrovic et al. [50] address this challenge by not only providing an impostor-based rendering technique that allows them to visualize the fiber tracts with a high visual quality but that also allows them to map the bundle names directly onto the respective fiber tract representations (see Fig. 6).

An essential property of fiber tract data is that there is always some degree of uncertainty about a tract's path. Depicting a tract using a discrete line or tube, however, suggests that it actually exists in that precise configuration. To encode several levels of uncertainty in fiber tract data, Hermosilla et al. [30] thus group fiber tracts based on each track's uncertainty value and render them using different colors (using lower saturations for higher uncertainties). They combine this approach with screen-space ambient occlusion, texture patterns, and unsharp masking to improve

[11] Everts et al. [25] extended this idea, encoding data properties in (colored) patterns for flow data.

Fig. 6 Visualization of a tract's bundle name on the tubular representation. Image © 2007 by James Fallon, used with permission

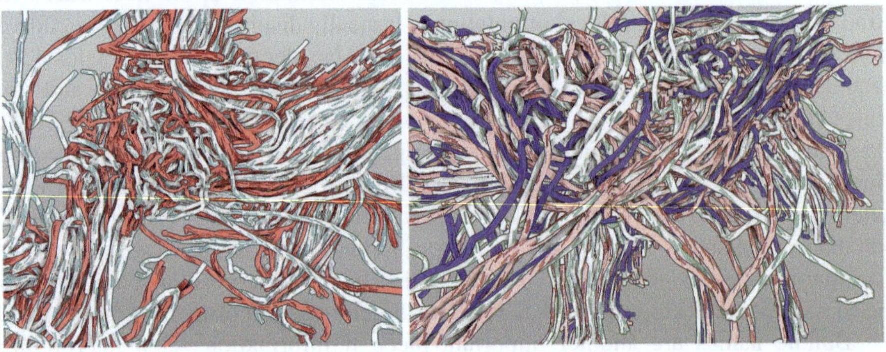

Fig. 7 Two examples of the uncertainty visualization by Hermosilla et al. [30]. Both images are by Pedro Hermosilla Casajus and are used under the CreativeCommonsAttribution-NonCommercial-NoDerivs3.0Unported(CCBY-NC-ND3.0)license

the visualization's visual quality. Two examples of this technique are shown in Fig. 7.

4 Focus+Context Visualization

In practical applications it is typically not sufficient to simply view and understand the structure of a fiber tract dataset by itself. Instead, viewers need to understand the spatial location of the fiber tracts with respect to anatomical landmarks such as the surface of the brain with its sulci and gyri. For this purpose researchers have developed a number of illustrative focus+context visualization techniques that combine fiber tract visualization with contextual rendering. Illustrative visualization

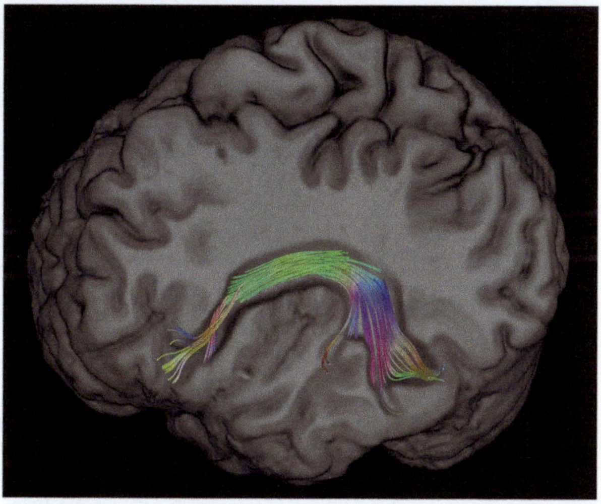

Fig. 8 Virtual Klingler dissection can serve as a way of showing the context for a traditionally visualized selection of fiber tracts. Image © 2014 by Thomas Schultz, used with permission

approaches are particularly well suited for this purpose because they can make use of the illustration principles of abstraction and emphasis.

A simple form of focus+context visualization is the combination of traditional fiber visualization with an illustratively rendered brain surface (e.g., Berres et al. [5, 6] and Eichelbaum et al. [21]). Such brain surface representations that wrap around the fiber tracts typically employ illustrative selective transparency such that sulci are shown in a more opaque form and gyri are more transparent. This rendering style allows most of the focus to be visible though the surrounding context, while at the same time also providing the necessary landmarks for reference. Another interesting alternative for generating focus+context is to render a subset of the brain volume in a fashion that resembles an existing dissection method as done by Anwander et al. [1] and Schultz et al. [59]. Here, the a virtual form of a Klingler dissection provides the context, while fiber tracts in focus are rendered in front of it as can be seen in Fig. 8.

Other illustrative rendering approaches attempt to combine the fiber tract visualization with the context visualization in such a way that the visual styles of both elements of the visualization match—as in traditional illustrations in medical textbooks. For example, Svetachov et al. [61] combine the black-and-white fiber tracts rendered using depth-dependent halos [24] with a similarly black-and-white hatched visualization of the brain surface (see Fig. 9),[12] and use cutting planes and stippling to show regions of gray matter. The hatched context visualization of the brain surface is stylized using screen-space ambient occlusion and halos are added around the fiber tracts to help the two visualization layers integrate appropriately.

---

[12]The example images in Fig. 9 were created with the project's demo; see the website at http://tobias.isenberg.cc/VideosAndDemos/Svetachov2010DCI .

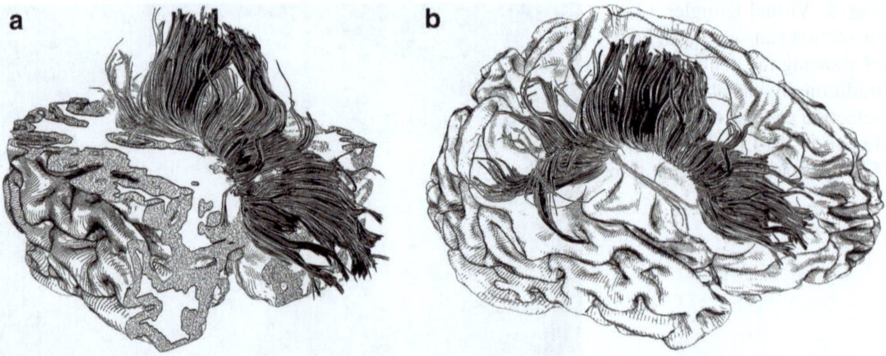

Fig. 9 Two examples of the *DTI in context* approach [61], using (**a**) cutting planes and (**b**) a focus halo to integrate the fiber tracts with the context representation. The images are in the public domain as declared in the original article's publication agreement

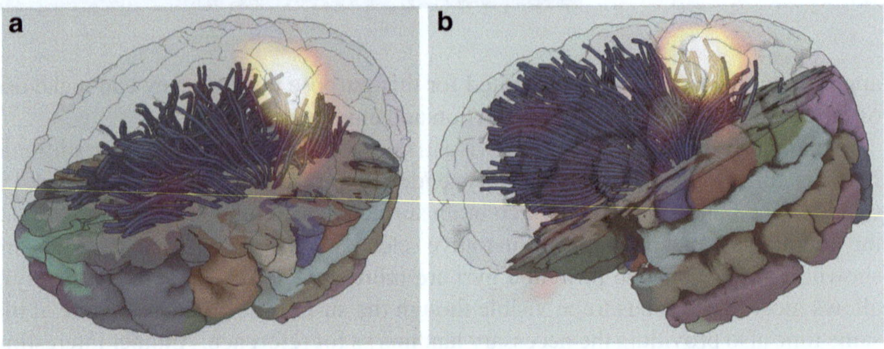

Fig. 10 (**a**) Born et al.'s [7] multi-modal combination of fiber tracts with the brain surface and functional MRI data in a focus+context visualization; (**b**) uses halftoning for sulci which resembles a hatching effect. Images © 2009 by Silvia Born, used with permission

Other illustrative visualization techniques specifically aim to combine fiber tract visualizations of the brain's structural connectivity with visualizations of other types of data to create illustrative multi-modal visualizations. For example, Born et al. [7] integrate tubular fiber tract visualizations with an existing illustrative visualization of functional brain connectivity data [33, 34]. The resulting visualization (see Fig. 10) uses cutting planes, color-coded regions of the gray matter in the opaquely visualized part of the brain and on the intersection surface, and a semi-transparent brain surface. This surface also shows brain activity obtained from functional MRI data. The semi-transparent context visualization employs illustrative techniques such as ambient occlusion, silhouettes, and halftoning.

Another approach by Schott et al. [57, 58] for illustrative multi-modal visualizations combines a direct volume rendering of MRI data as context with registered DTI fiber tracts as the focus. This visualization technique also uses ambient occlusion—not only within the volume data or within the fiber tracts individually, but between the two. This means that the volume generates occlusion for the tubular fiber tracts and vice versa. By using these illustrative effects together, Schott et al.'s [57, 58] technique nicely supports the perception of spatial depth of the depicted objects in their focus+context visualization (also see the approaches discussed in Sect. 2).

Rieder et al. [55] create a visualization that is visually similar to Born et al.'s [7]. They also combine the fiber tracts with an opaque and a semi-transparent rendering of the brain surface. The opaque rendering of the surface is produced using ambient occlusion and a texture-mapped cutting plane, while the transparent surface is rendered using silhouettes. Similar to Schott et al. [57, 58], they also ensure that the ambient occlusion effect of the fibers is not only applied to the fiber representation but also to the opaque volume rendering of the brain. Rieder et al.'s [55] visualization, however, is unique in that they combine their illustrative on-screen focus+context depiction with an interactive physical model that allows users to control the visualization. This physical model serves as a second layer of context that also provides an interaction proxy for exploring the visualization. In addition to adjusting the view on the visualization, users are also able to select specific fiber tract bundles (by touching one of the illuminated spots on the physical model).

Another recent focus+context visualization of brain anatomy and tractography data was presented by Reichenbach et al. [54]. In this visualization the authors focus on showing structural connectivity in the context of selected regions of interest and the general brain context. In both cases Reichenbach et al. provide context by emphasizing the cortex's parcellation (see Fig. 11). Using illustrative techniques such as depth-dependent outlines they highlight the relationships between different regions of the brain, assisted specifically by showing structural connectivity. It is

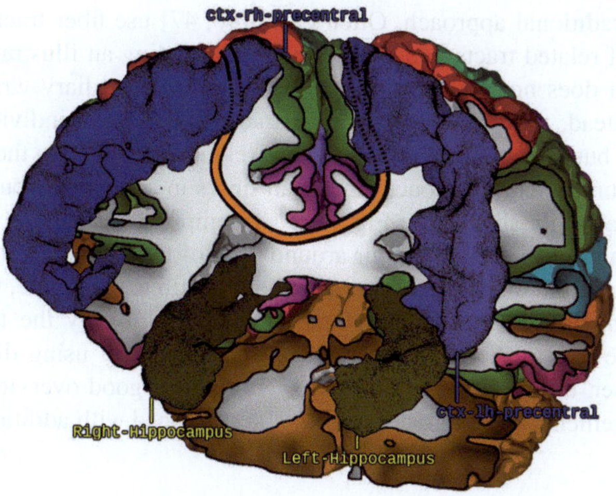

Fig. 11 Example of Reichenbach et al.'s [54] integrated focus+context visualization. Image © 2014 by André Reichenbach, Mathias Goldau, and Mario Hlawitschka, used with permission

particularly interesting that this approach and the previously discussed one do not show connectivity information by rendering individual fiber tracts, but instead show abstracted fiber bundles. Such bundling approaches are discussed in detail next.

5 Visualization of Fiber Tract Bundles

As the last two examples show, there are some cases in which it is not necessarily important to observe each extracted fiber tract individually. Instead, these only need to show meaningful fiber bundles. Some of the previously discussed techniques show fiber bundles implicitly, for example depth-dependent halos [24] and LineAO [22]. These methods, however, rely on a visual identification of a bundle. To make it easier for viewers to identify bundles, researchers have developed methods to explicitly represent the bundles, for example through wrapped representations [23]. Here, illustrative visualization can help to combine the bundle representation with that of the individual fibers, to indicate the fact that the bundles are only abstractions and do not physically exist as dedicated objects in the brain, or to provide additional visual variables that can be used to display bundle properties such as confidence.

The initial concept of extracting a dedicated bundle surface from a set of related fiber tracts was extended by Röttger et al. [56] in their BundleExplorer tool. BundleExplorer uses semi-transparent bundle surfaces with silhouettes and explicitly represented internal fiber tracts. The tool also uses marker-dependent cutaway views to see through the bundle surface as well as bundle intersection highlighting. In this combination of rendering techniques, the internal fiber tracts serve as focus objects while the bundle surface and the intersections represent context information.

In a less traditional approach, Otten et al. [46, 47] use fiber tract clustering to derive a set of related tracts which they then depict using an illustrative rendering technique that does not actually need to derive an intermediary wrapped surface geometry. Instead, they still process all extracted fiber tracts individually for the visualization, but render each of them with a relatively wide halo they color based on the fiber bundle. This approach causes all fibers in a particular bundle to merge *visually* into a single shape (see Fig. 12).[13] To emphasize this effect, Otten et al. [46, 47] add an image-space contour around each of these colored shapes. Finally, to better indicate the directionality of the fibers within each bundle, they add what they call "hint lines"—a fiber tract rendering in which only the top-most lines remain visible due to the occlusion by the tract halos. By using different colors for each cluster, this technique allows viewers to get a good overview of the fiber bundle arrangement. The approach can also be combined with additional focus and

[13] The example images in Fig. 12 were created with the vIST/e project's demo (using a test release); see the website at http://bmia.bmt.tue.nl/software/viste/ and the SourceForge repository at http://sourceforge.net/projects/viste/files/ .

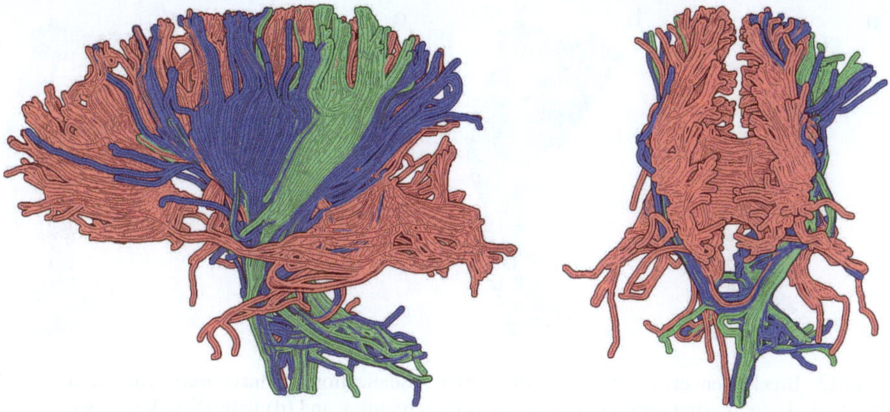

Fig. 12 Two views of Otten et al.'s [46, 47] fiber bundling approach using 2D line halos

context elements such as textured slices from volume data or explicit fiber tract visualizations.

6 Probabilistic Fiber Tracking and Uncertainty Visualization

Fiber tract bundles derived from DTI-based fiber tracking, however, are not clearly defined anatomical structures in the brain. This is due, in part, to limitations of the data acquisition process. Bundles are also uncertain, however, because the fiber tracking algorithm, in deterministic tractography, follows the direction of the tensor's primary eigenvector until the anisotropy falls below a pre-defined threshold. That means that, depending on whether or not a fiber tract is located in regions with a large fractional anisotropy, it can have varying degrees of confidence or uncertainty.

These different levels of confidence can be illustrated on a per-fiber-tract-basis as shown by Hermosilla et al. [30] (see Fig. 3). The uncertainty, however, can also be visualized within a fiber bundle as demonstrated by Brecheisen et al. [11, 12] (see Fig. 13).[14] Instead of representing the confidence of each individual fiber, Brecheisen et al. group the fiber tracts of a given bundle into intervals of the same confidence range, render them with a halo like in Otten et al.'s [46, 47] work. They then integrate the resulting layers into a single visualization. Similar to the other approaches discussed before, they emphasize some or all of the levels with silhouettes for a better visual perception of the layers. They also use different schemes to indicate the decreasing confidence levels. These levels include warm-to-cool shading [29] (Fig. 13a), increasing color saturation (Fig. 13b), decreasing

[14]The images in Fig. 13 were created with DTITool (provided by Ralph Brecheisen), an early version of the vIST/e software.

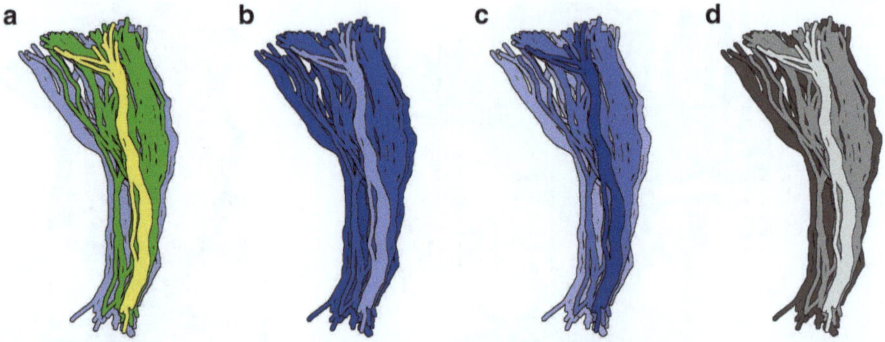

Fig. 13 Brecheisen et al.'s [11, 12] illustrative visualization of uncertainty: (**a**) warm-to-cool shading, (**b**) increasing saturation, (**c**) decreasing saturation, and (**d**) light-to-dark shading

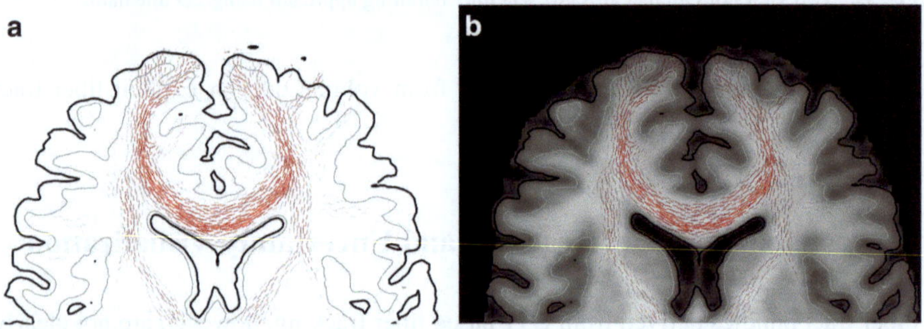

Fig. 14 Goldau et al.'s [28] fiber stippling technique for the visualization of probabilistic tractography data, (**a**) with context in form of silhouettes computed from T1-weighted MRI data and (**b**) as a multi-modal visualization with context shown in form of T1-weighted MRI data texture-mapped onto a cutting plane [27]. Images © 2014 by Mathias Goldau, used with permission

color saturation (Fig. 13c), decreasing opacity, light-to-dark shading (Fig. 13d), decreasing the amount of detail due to growing dilation around the fiber tracts, and an increasing amount of blur. The resulting bundle visualizations can also be combined with other data visualization techniques in focus+context views.

The problem of uncertainty in deterministic tractography can also be addressed by using probabilistic tractography (e.g., Parker [48]). As described at the beginning of the chapter, this technique does not follow a single, deterministic direction for each fiber tract integration step. Instead it follows several, depending on the probability that tracts follow a direction other than the tensor's primary eigenvector.

Using techniques from illustrative visualization and inspired by traditional illustrations of brain connectivity, Goldau et al. [28] created a slice-based fiber stippling technique that shows the distribution of the probability field (see Fig. 14). The stipples are small line segment that are oriented along the main diffusion direction. They are created by projecting the diffusion vector onto the slice such that long, narrow stipples represent diffusion parallel to the cutting plane and short, wide stipples represent diffusion at an angle to the cutting plane. The stipples representing different bundles are shown using different colors. Hlawitschka et al. [31] extend this approach by employing a Poisson-Disk sampling to distribute the fiber stipples in order to ensure an adequate perception of the pattern on the slices. In both cases, fiber stippling (i.e., small oriented dashes) as an illustrative visualization technique is well suited to the visualization of probability data because many people intuitively associate dashes to uncertainty [9]. Thus the visualization metaphor employed by Goldau et al. [28] and Hlawitschka et al. [31] is easily understood by viewers. Goldau and Hlawitschka [27] also demonstrated that such visualizations can be integrated in multi-modal depictions of brain data. These multi-modal visualizations can use a variety of other data modalities such as MRI data, functional MRI data, or CT data to provide the necessary context (e.g., see Fig. 14b).

7 Spatially Abstracted Visualization of Connectivity

All the illustrative visualization techniques discussed so far are intended to represent the fiber tracts and the resulting connectivity data as faithfully to the anatomical data as possible. However, sometimes people are interested in understanding connectivity in the brain at a higher level, one at which the visual representation can deviate to some degree from the anatomy. Illustrative visualization is naturally an ideal candidate for such spatially abstracting visualizations.

Jianu et al. [36] describe a visualization approach in this category in which they create two-dimensional neural map representations in which bundles of similar fiber tracts are grouped and visualized using simplified line-based primitives.[15] Specifically, these simplified representations are derived by clustering and selecting the tract with the smallest distance to all tracts in a bundle, favoring longer tracts over shorter ones. The results are then rendered in an illustrative fashion (using tract silhouettes and a schematic brain volume projection as context) on the sagittal, coronal, and transverse planes (see Fig. 15, bottom-left). These views are well established in medical practice and serve as a means to explore the correct locations of the corresponding fiber tracts in a linked 3D view. In an extension of this initial approach, Jianu et al. [36] also created an abstracted bundle representation of the clusters (see Fig. 15, right)—inspired by Holten's [32] edge bundling approach. This abstract representation can serve as an alternative way of exploring the 3D fiber tract visualization (Fig. 15, top-left). Both approaches together demonstrate that the use of abstraction and illustrative depiction styles can support interactive exploration of more traditional visualizations of 3D data.

[15] A downloadable demo of Jianu et al.'s [36] technique is available at http://graphics.cs.brown.edu/research/sciviz/newbraininteraction/tutorial.htm and an online demo can be found at http://graphics.cs.brown.edu/research/sciviz/newbraininteraction/BrainComplete/P3/gmap_brain.html .

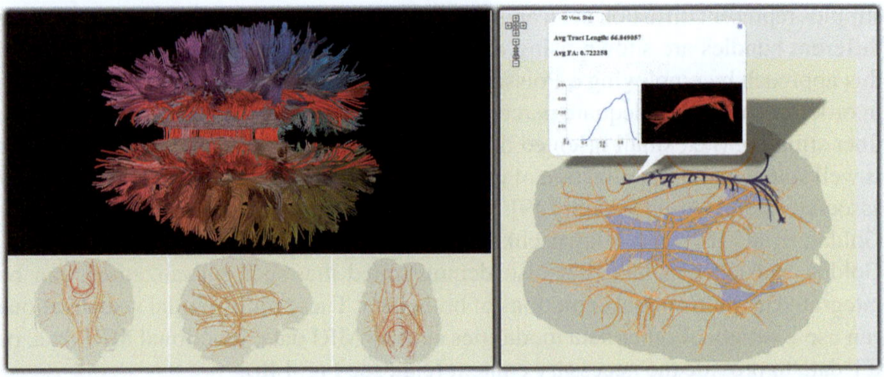

Fig. 15 Screenshot of Jianu et al.'s [36] interactive brain connectivity visualization system that makes use of a spatial abstraction of fiber bundles, in particular for the interactive selection of fiber tracts in a bundle. Image © 2014 by Radu Jianu, used with permission

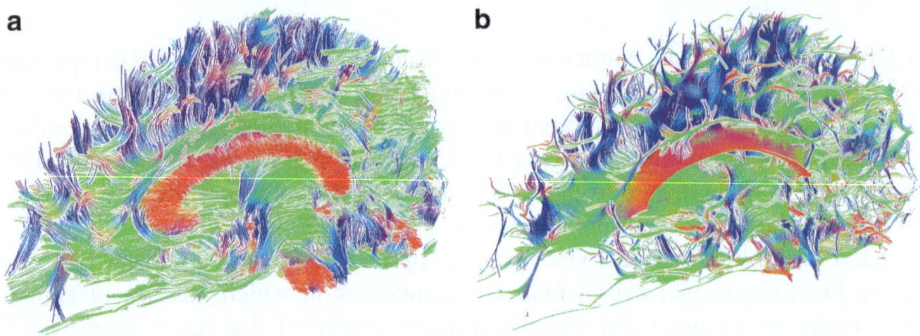

Fig. 16 Example of the effect of Everts et al.'s [26] fiber tract contraction at a scale of 4 mm

Jianu et al.'s [36] abstraction primarily supports the easy selection of previously computed fiber tract clusters or bundles since their approach removes the complexity of the dense line representations. Sometimes, however, it is also necessary to understand the inner structure of a complete set of fiber tracts (such as depicted in Fig. 1)—a goal that is not easily supported using traditional forms of depiction, even illustrative ones (e.g., Figs. 2b, 4, and 5). To address this problem, Everts et al. [26] describe an abstraction technique that spatially contracts full-brain fiber tract datasets based on their local similarity. This similarity is computed based on co-linearity within a neighborhood of a given scale level. The result of this processing is that fiber tracts locally contract perpendicularly to the tract direction, revealing the global structure of the brain's white matter by creating volumetric voids (e.g., see Fig. 16). For contractions at scale levels ≤ 2 mm, Everts et al. [26] show that the fiber tracts stay within the bounds of their corresponding fractional anisotropy areas. The authors also discuss how—when using larger displacements—the contraction can lead to anatomically incorrect depictions but show that these can reveal the brain's higher-level organization. As such their contraction relates to approaches that introduce significant distortions such as done by Correa et al. [16]. They used deformations of line data (including neurological fiber tracts) to provide insight into the complex structures by deforming their paths or visually separating subsets.

8 Summary and Conclusion

This chapter provided a survey of illustrative visualization techniques applied to tractography data that resulted from 3D tensor fields based on diffusion-weighted MRI. The survey showed that illustrative visualization is used, in particular, to improve the depth perception in complex visualizations, to facilitate the visualization of additional data such as tract names, tract confidence, or similarity between tracts, to combine fiber tract visualization with additional visual elements in focus+context visualizations, to visualize fiber tract bundles, and to facilitate further spatial abstraction, for example, for an interactive exploration of the data. Of course, such a classification is not necessarily exclusive: a technique that primarily aims for a focus+context visualization, for example, can also make use of techniques that improve the depth perception of the fiber tracts or of the whole visualization.

The techniques described in this paper, of course, closely relate to other illustrative visualization techniques in brain connectivity visualization and beyond. For instance, the structural connectivity that can be explored though deterministic and probabilistic tractography closely relates to functional connectivity that can be examined based on functional MRI data. An example for such a visualization of functional data was presented by Böttger et al. [8] who employ edge bundling and bundle-driven transparency as forms of illustrative visualization to create representations of whole-brain functional connectivity (e.g., Fig. 17).[16] Of course, the tract-based depictions of brain connectivity also closely relate to streamlines, pathlines, etc. in fluid mechanics which can use similar illustrative visualization methods (e.g., Brambilla et al. [10], Everts et al. [24, 25], Li and Shen [41], and Shafii et al. [60]).

Naturally, the work surveyed in this chapter only represents a subset of the field of illustrative visualization, with explores methods to highlight, provide emphasis, or introduce abstraction [53]. When using these methods, however, we need to be aware of the implications of the general approach. While an illustrative visualization style can be attractive and can thus positively affect viewers [3, 65], it can also have a high "suggestive" power: Illustrative visualization may suggest information that can sometimes be misleading precisely because they often provide precise depictions and clear representations. Everts et al. [24], for instance, reported that neurosurgeons mentioned that fiber tract visualizations that use depth-dependent halos appear to

[16]The example image in Fig. 17 was created with the braingl tool, see the webpage at http://code.google.com/p/braingl/ .

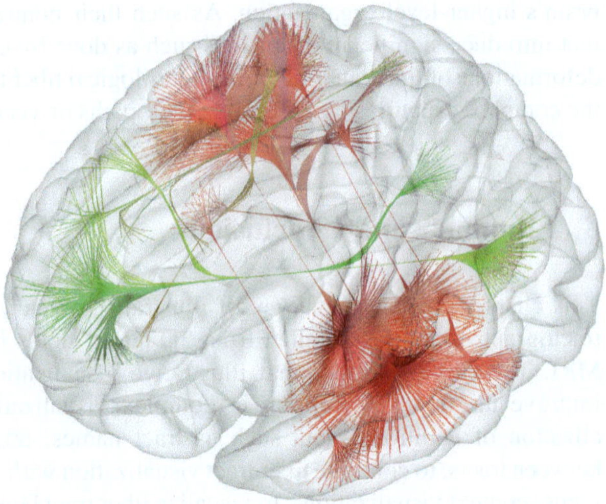

Fig. 17 Example of a bundling result generated by Böttger et al.'s [8] approach as a form of illustrative depiction of functional connections in the brain

depict the actual neuronal fibers (axons) of the brain—something that the data sources and the resulting visualizations do not provide. So while we should always be aware of these potential challenges for illustrative visualization, this example also illustrates the intriguing power of illustrative methods within visualization.

Acknowledgements I would like to thank all of the people who provided example images, tried to find material from old sources, or referred me to others—sometimes on very short notice. In particular, I would like to thank Joachim Böttger, Silvia Born, Ralph Brecheisen, Jesús Díaz-García, Mathias Goldau, Mario Hlawitschka, Radu Jianu, Daniel F. Keefe, Mathias Schott, Thomas Schultz, Pere-Pau Vázquez, and Anna Vilanova. I also specifically wish to thank those authors who provided demo applications of their publications that I could use to create my own example visualizations. Finally, I would like to thank Kai Lawonn and Wesley Willet as well as the anonymous reviewers for their valuable comments on drafts of this survey.

References

1. Anwander, A., Schurade, R., Hlawitschka, M., Scheuermann, G., Anderson, T.R., Knösche, T.R.: White matter imaging with virtual Klingler dissection. NeuroImage **47**(Suppl. 1), S105 (2009). doi:10.1016/S1053-8119(09)70916-4
2. Appel, A., Rohlf, F.J., Stein, A.J.: The haloed line effect for hidden line elimination. ACM SIGGRAPH Comput. Graphics **13**(3), 151–157 (1979). doi:10.1145/800249.807437
3. Bach, B., Dragicevic, P., Huron, S., Isenberg, P., Jansen, Y., Perin, C., Spritzer, A., Vuillemot, R., Willett, W., Isenberg, T.: Illustrative data graphics in 18th–19th century style: A case study. In: Posters at IEEE Visualization (2013). Extended abstract and poster
4. Behrens, T.E.J., Sotiropoulos, S.N., Jbabdi, S.: MR diffusion tractography. In: Johansen-Berg, H., Behrens, T.E.J. (eds.) Diffusion MRI: From Quantitative Measurement to In-vivo Neuroanatomy, 2nd edn., Chap. 19, pp. 429–451. Academic Press, San Diego (2014). doi:10.1016/B978-0-12-396460-1.00019-6

5. Berres, A.S., Goldau, M., Tittgemeyer, M., Scheuermann, G., Hagen, H.: Tractography in context: multimodal visualization of probabilistic tractograms in anatomical context. In: Proceedings of VCBM, pp. 9–16. Eurographics Association, Goslar (2012). doi:10.2312/VCBM/VCBM12/009-016
6. Berres, A.S., Goldau, M., Tittgemeyer, M., Scheuermann, G., Hagen, H.: Tractography in context: multimodal visualization of probabilistic tractograms in anatomical context. In: Posters of IEEE BioVis (2012)
7. Born, S., Jainek, W.M., Hlawitschka, M., Trantakis, C., Meixensberger, J., Bartz, D.: Multimodal visualization of DTI and fMRI data using illustrative methods. In: Bildverarbeitung für die Medizin, pp. 6–10. Springer, Berlin/Heidelberg (2009). doi:10.1007/978-3-540-93860-6_2
8. Böttger, J., Schäfer, A., Lohmann, G., Villringer, A., Margulies, D.S.: Three-dimensional mean-shift edge bundling for the visualization of functional connectivity in the brain. IEEE Trans. Vis. Comput. Graphics **20**(3), 471–480 (2014). doi:10.1109/TVCG.2013.114
9. Boukhelifa, N., Bezerianos, A., Isenberg, T., Fekete, J.D.: Evaluating sketchiness as a visual variable for the depiction of qualitative uncertainty. IEEE Trans. Vis. Comput. Graphics **18**(12), 2769–2778 (2012). doi:10.1109/TVCG.2012.220
10. Brambilla, A., Carnecky, R., Peikert, R., Viola, I., Hauser, H.: Illustrative flow visualization: state of the art, trends and challenges. In: Eurographics State of the Art Reports, pp. 75–94. Eurographics Association, Goslar (2012). doi:10.2312/conf/EG2012/stars/075-094
11. Brecheisen, R., Platel, B., ter Haar Romeny, B.M., Vilanova, A.: Illustrative uncertainty visualization of DTI fiber pathways. In: EuroVis Posters (2011)
12. Brecheisen, R., Platel, B., ter Haar Romeny, B.M., Vilanova, A.: Illustrative uncertainty visualization of DTI fiber pathways. Vis. Comput. **29**(4), 297–309 (2013). doi:10.1007/s00371-012-0733-9
13. Bruckner, S., Gröller, M.E.: Enhancing depth-perception with flexible volumetric halos. IEEE Trans. Vis. Comput. Graphics **13**(6), 1344–1351 (2007). doi:10.1109/TVCG.2007.70555
14. Bryson, S.: Virtual reality in scientific visualization. Commun. ACM **39**(5), 62–71 (1996). doi:10.1145/229459.229467
15. Chen, W., Yan, Z., Zhang, S., Crow, J.A., Ebert, D.S., McLaughlin, R.M., Mullins, K.B., Cooper, R., Ding, Z., Liao, J.: Volume illustration of muscle from diffusion tensor images. IEEE Trans. Vis. Comput. Graphics **15**(6), 1425–1432 (2009). doi:10.1109/TVCG.2009.203
16. Correa, C.D., Silver, D., Chen, M.: Illustrative deformation for data exploration. IEEE Trans. Vis. Comput. Graphics **13**(6), 1320–1327 (2007). doi:10.1109/TVCG.2007.70565
17. Díaz-García, J.: Rapid halo-based illustrative visualization of brain fiber tracts. Master's thesis, BarcelonaTech/Universitat Politècnica de Catalunya, Spain (2011). doi:2099.1/14366
18. Díaz-García, J., Vázquez, P.P.: Fast illustrative visualization of fiber tracts. In: Advances in Visual Computing. Lecture Notes in Computer Science, vol. 7431, pp. 698–707. Springer, Berlin/Heidelberg (2012). doi:10.1007/978-3-642-33179-4_66
19. Eichelbaum, S., Goldau, M., Philips, S., Reichenbach, A., Schurade, R., Wiebel, A.: OpenWalnut: A new tool for multi-modal visualization of the human brain. In: Posters of EG VCBM (2010)
20. Eichelbaum, S., Hlawitschka, M., Wiebel, A., Scheuermann, G.: OpenWalnut – An open-source visualization system. In: Proceedings of 6th High-End Visualization Workshop, pp. 67–78. Lehmanns Media—LOB.de, Berlin (2010)
21. Eichelbaum, S., Wiebel, A., Hlawitschka, M., Anwander, A., Knösche, T.R., Scheuermann, G.: Visualization of effective connectivity of the brain. In: Proceedings of VMV, pp. 155–162. Eurographics Association, Goslar (2010). doi:10.2312/PE/VMV/VMV10/155-162
22. Eichelbaum, S., Hlawitschka, M., Scheuermann, G.: LineAO—Improved three-dimensional line rendering. IEEE Trans. Vis. Comput. Graphics **19**(3), 433–445 (2013). doi:10.1109/TVCG.2012.142
23. Enders, F., Sauber, N., Merhof, D., Hastreiter, P., Nimsky, C., Stamminger, M.: Visualization of white matter tracts with wrapped streamlines. In: Proceedings of Visualization, IEEE Computer Society, Los Alamitos, pp. 51–58 (2005). doi:10.1109/VISUAL.2005.1532777

24. Everts, M.H., Bekker, H., Roerdink, J.B.T.M., Isenberg, T.: Depth-dependent halos: Illustrative rendering of dense line data. IEEE Trans. Vis. Comput. Graphics **15**(6), 1299–1306 (2009). doi:10.1109/TVCG.2009.138
25. Everts, M.H., Bekker, H., Roerdink, J.B.T.M., Isenberg, T.: Illustrative line styles for flow visualization. In: Proceedings of Pacific Graphics, pp. 105–110. Eurographics Association, Goslar (2011). doi:10.2312/PE/PG/PG2011short/105-110
26. Everts, M.H., Begue, E., Bekker, H., Roerdink, J.B.T.M., Isenberg, T.: Exploration of the brain's white matter structure through visual abstraction and multi-scale local fiber tract contraction. IEEE Trans. Vis. Comput. Graphics **21** (2015, to appear). doi:10.1109/TVCG.2015.2403323)
27. Goldau, M., Hlawitschka, M.: Multi-modal visualization of probabilistic tractography. In: Visualization in Medicine and Life Sciences III. Springer, Berlin (2015)
28. Goldau, M., Wiebel, A., Gorbach, N.S., Melzer, C., Hlawitschka, M., Scheuermann, G., Tittgemeyer, M.: Fiber stippling: An illustrative rendering for probabilistic diffusion tractography. In: Proceedings of BioVis, pp. 23–30. IEEE Computer Society, Los Alamitos (2011). doi:10.1109/BioVis.2011.6094044
29. Gooch, A.A., Gooch, B., Shirley, P., Cohen, E.: A non-photorealistic lighting model for automatic technical illustration. In: Proceedings of SIGGRAPH, pp. 447–452. ACM, New York (1998). doi:10.1145/280814.280950
30. Hermosilla, P., Brecheisen, R., Vázquez, P.P., Vilanova, A.: Uncertainty visualization of brain fibers. In: Proceedings of CEIG, pp. 31–40. Eurographics Association, Goslar (2012). doi:10.2312/LocalChapterEvents/CEIG/CEIG12/031-040
31. Hlawitschka, M., Goldau, M., Wiebel, A., Heine, C., Scheuermann, G.: Hierarchical Poisson-disk sampling for fiber stipples. In: Visualization in Medicine and Life Sciences, pp. 19–23. Eurographics Association, Goslar (2013). doi:10.2312/PE.VMLS.VMLS2013.019-023
32. Holten, D.: Hierarchical edge bundles: Visualization of adjacency relations in hierarchical data. IEEE Trans. Vis. Comput. Graphics **12**(5), 741–748 (2006). doi:10.1109/TVCG.2006.147
33. Jainek, W.: Illustrative visualization of brain structure and functional MRI data. Master's thesis, Eberhard-Karls-Universität Tübingen, Germany (2007)
34. Jainek, W.M., Born, S., Bartz, D., Straßer, W., Fischer, J.: Illustrative hybrid visualization and exploration of anatomical and functional brain data. Comput. Graphics Forum **27**(3), 855–862 (2008). doi:10.1111/j.1467-8659.2008.01217.x
35. Jianu, D., Zhou, W., Demiralp, Ç., Laidlaw, D.H.: Visualizing spatial relations between 3D-DTI integral curves using texture patterns. In: IEEE Visualization Poster Compendium (2007)
36. Jianu, R., Demiralp, Ç., Laidlaw, D.H.: Exploring brain connectivity with two-dimensional neural maps. IEEE Trans. Vis. Comput. Graphics **18**(6), 978–987 (2012). doi:10.1109/TVCG.2011.82
37. Johansen-Berg, H., Behrens, T.E.J. (eds.): Diffusion MRI: From Quantitative Measurement to In-vivo Neuroanatomy, 2nd edn. Academic Press, San Diego (2014). doi:10.1016/B978-0-12-396460-1.01001-5
38. Klein, J., Ritter, F., Hahn, H.K., Rexilius, J., Peitgen, H.O.: Brain structure visualization using spectral fiber clustering. In: ACM SIGGRAPH 2006 Research Posters, New York, p. 168(1) (2006). doi:10.1145/1179622.1179816
39. Kratz, A., Auer, C., Stommel, M., Hotz, I.: Visualization and analysis of second-order tensors: Moving beyond the symmetric positive-definite case. Comput. Graphics Forum **32**(1), 49–74 (2013). doi:10.1111/j.1467-8659.2012.03231.x
40. Laidlaw, D.H., Ahrens, E.T., Kremers, D., Avalos, M.J., Jacobs, R.E., Readhead, C.: Visualizing diffusion tensor images of the mouse spinal cord. In: Proceedings of Visualization, pp. 127–134. IEEE Computer Society, Los Alamitos (1998). doi:10.1109/VISUAL.1998.745294
41. Li, L., Shen, H.W.: Image-based streamline generation and rendering. IEEE Trans. Vis. Comput. Graphics **13**(3), 630–640 (2007). doi:10.1109/TVCG.2007.1009

42. Mallo, O., Peikert, R., Sigg, C., Sadlo, F.: Illuminated lines revisited. In: Proceedings of Visualization, pp. 19–26. IEEE Computer Society, Los Alamitos (2005). doi:10.1109/VISUAL.2005.1532772
43. Margulies, D.S., Böttger, J., Watanabe, A., Gorgolewski, K.J.: Visualizing the human connectome. NeuroImage **80**, 445–461 (2013). doi:10.1016/j.neuroimage.2013.04.111
44. Mittring, M.: Finding next gen: CryEngine 2. In: ACM SIGGRAPH Courses, New York, Chap. 28, pp. 97–121 (2007). doi:10.1145/1281500.1281671
45. Mori, S., van Zijl, P.C.: Fiber tracking: Principles and strategies – A technical review. NMR Biomed. **15**(7–8), 468–480 (2002). doi:10.1002/nbm.781
46. Otten, R.J.G.: Illustrative rendering of white matter fiber bundles. Master's thesis, TU Eindhoven, Department of Mathematics and Computer Science, the Netherlands (2009)
47. Otten, R., Vilanova, A., van de Wettering, H.M.M.: Illustrative white matter fiber bundles. Comput. Graphics Forum **29**(3), 1013–1022 (2010). doi:10.1111/j.1467-8659.2009.01688.x
48. Parker, G.J.M.: Probabilistic fiber tracking. In: Jones, D.K. (ed.) Diffusion MRI: Theory, Methods, and Applications, Chap. 23. Oxford University Press, Oxford (2010). doi:10.1093/med/9780195369779.003.0023
49. Peeters, T.H.J.M., Vilanova, A., Strijkers, G.J., ter Haar Romeny, B.M.: Visualization of the fibrous structure of the heart. In: Proceedings of VMV, pp. 309–316. Akademische Verlagsgesellschaft Aka GmbH, Berlin (2006)
50. Petrovic, V., Fallon, J., Kuester, F.: Visualizing whole-brain DTI tractography with GPU-based tuboids and LoD management. IEEE Trans. Vis. Comput. Graphics **13**(6), 1488–1495 (2007). doi:10.1109/TVCG.2007.70532
51. Pfister, H., Kaynig, V., Botha, C.P., Bruckner, S., Dercksen, V.J., Hege, H.C., Roerdink, J.B.T.M.: Visualization in connectomics. Technical Report 1206.1428 [cs.GR], arXiv.org (2012)
52. Pfister, H., Kaynig, V., Botha, C.P., Bruckner, S., Dercksen, V.J., Hege, H.C., Roerdink, J.B.T.M.: Visualization in connectomics. In: Hansen, C.D., Chen, M., Johnson, C.R., Kaufman, A.E., Hagen, H. (eds.) Scientific Visualization, Mathematics and Visualization, pp. 221–245. Springer, London (2014). doi:10.1007/978-1-4471-6497-5_21
53. Rautek, P., Bruckner, S., Gröller, E., Viola, I.: Illustrative visualization: New technology or useless tautology? ACM SIGGRAPH Comput. Graphics **42**(3), 4:1–4:8 (2008). doi:10.1145/1408626.1408633
54. Reichenbach, A., Goldau, M., Hlawitschka, M.: Combined three-dimensional visualization of structural connectivity and cortex parcellation. In: Proceedings of VMV, pp. 71–78. Eurographics Association, Goslar (2014). doi:10.2312/vmv.20141278
55. Rieder, C., Brachmann, C., Hofmann, B., Klein, J., Köhn, A., Ojdanic, D., Schumann, C., Weiler, F., Hahn, H.K.: Interactive visualization of neuroanatomical data for a hands-on multimedia exhibit. In: Proceedings of VMLS, pp. 37–41. Eurographics Association, Goslar (2013). doi:10.2312/PE.VMLS.VMLS2013.037-041
56. Röttger, D., Merhof, D., Müller, S.: The BundleExplorer: A focus and context rendering framework for complex fiber distributions. In: Proceedings of VCBM, pp. 1–8. Eurographics Association, Goslar (2012). doi:10.2312/VCBM/VCBM12/001-008
57. Schott, M., Martin, T., Grosset, A.V.P., Brownlee, C., Höllt, T., Brown, B.P., Smith, S.T., Hansen, C.D.: Combined surface and volumetric occlusion shading. In: Proceedings of PacificVis, pp. 169–176. IEEE, Los Alamitos (2012). doi:10.1109/PacificVis.2012.6183588
58. Schott, M., Martin, T., Grosset, A.V., Smith, S., Hansen, C.D.: Ambient occlusion effects for combined volumes and tubular geometry. IEEE Trans. Vis. Comput. Graphics **19**(6), 913–926 (2013). doi:10.1109/TVCG.2012.306
59. Schultz, T., Sauber, N., Anwander, A., Theisel, H., Seidel, H.P.: Virtual Klingler dissection: Putting fibers into context. In: Proceedings of EuroVis, pp. 1063–1070. Eurographics Association, Goslar (2008). doi:10.1111/j.1467-8659.2008.01243.x
60. Shafii, S., Obermaier, H., Kolář, V., Hlawitschka, M., Garth, C., Hamann, B., Joy, K.I.: Illustrative rendering of vortex cores. In: Proceedings of EuroVis, pp. 61–65. Eurographics Association, Goslar (2013). doi:10.2312/PE.EuroVisShort.EuroVisShort2013.061-065

61. Svetachov, P., Everts, M.H., Isenberg, T.: DTI in context: Illustrating brain fiber tracts in situ. Comput. Graphics Forum **29**(3), 1024–1032 (2010). doi:10.1111/j.1467-8659.2009.01692.x
62. Tarini, M., Cignoni, P., Montani, C.: Ambient occlusion and edge cueing for enhancing real time molecular visualization. IEEE Trans. Vis. Comput. Graphics **12**(5), 1237–1244 (2006). doi:10.1109/TVCG.2006.115
63. Viola, I., Gröller, M.E., Hadwiger, M., Bühler, K., Preim, B., Sousa, M.C., Ebert, D., Stredney, D.: Illustrative visualization. In: IEEE Visualization Tutorials, Chap. 4. IEEE Computer Society, Los Alamitos (2005)
64. Wenger, A., Keefe, D.F., Zhang, S., Laidlaw, D.H.: Interactive volume rendering of thin thread structures within multivalued scientific data sets. IEEE Trans. Vis. Comput. Graphics **10**(6), 664–672 (2004). doi:10.1109/TVCG.2004.46
65. Wood, J., Isenberg, P., Isenberg, T., Dykes, J., Boukhelifa, N., Slingsby, A.: Sketchy rendering for information visualization. IEEE Trans. Vis. Comput. Graphics **18**(12), 2749–2758 (2012). doi:10.1109/TVCG.2012.262
66. Zhang, S., Demiralp, Ç., Laidlaw, D.H.: Visualizing diffusion tensor MR images using streamtubes and streamsurfaces. IEEE Trans. Vis. Comput. Graphics **9**(4), 454–462 (2003). doi:10.1109/TVCG.2003.1260740
67. Zhang, S., Laidlaw, D.H., Kindlmann, G.: Diffusion tensor MRI visualization. In: Hansen, C.D., Johnson, C.R. (eds.) The Visualization Handbook, Chap. 16, pp. 327–340. Elsevier, Oxford (2004). doi:10.1016/B978-012387582-2/50017-4
68. Zhukov, L., Barr, A.H.: Heart-muscle fiber reconstruction from diffusion tensor MRI. In: Proceedings of Visualization, pp. 597–602. IEEE Computer Society, Los Alamitos (2003). doi:10.1109/VISUAL.2003.1250425
69. Zhukov, L., Barr, A.H.: Oriented tensor reconstruction. In: Hansen, C.D., Johnson, C.R. (eds.) The Visualization Handbook, Chap. 15, pp. 313–326. Elsevier, Oxford (2004). doi:10.1016/B978-012387582-2/50017-4
70. Zhukov, S., Iones, A., Kronin, G.: An ambient light illumination model. In: Rendering Techniques, pp. 45–55. Springer, Vienna (1998). doi:10.1007/978-3-7091-6453-2_5
71. Zöckler, M., Stalling, D., Hege, H.C.: Interactive visualization of 3D-vector fields using illuminated stream lines. In: Proceedings of Visualization, pp. 107–113. IEEE Computer Society, Los Alamitos (1996). doi:10.1109/VISUAL.1996.567777

Visualizing Symmetric Indefinite 2D Tensor Fields Using the Heat Kernel Signature

Valentin Zobel, Jan Reininghaus, and Ingrid Hotz

Abstract The Heat Kernel Signature (HKS) is a scalar quantity which is derived from the heat kernel of a given shape. Due to its robustness, isometry invariance, and multiscale nature, it has been successfully applied in many geometric applications. From a more general point of view, the HKS can be considered as a descriptor of the metric of a Riemannian manifold. Given a symmetric positive definite tensor field we may interpret it as the metric of some Riemannian manifold and thereby apply the HKS to visualize and analyze the given tensor data. In this paper, we propose a generalization of this approach that enables the treatment of indefinite tensor fields, like the stress tensor, by interpreting them as a generator of a positive definite tensor field. To investigate the usefulness of this approach we consider the stress tensor from the two-point-load model example and from a mechanical work piece.

1 Introduction

The Heat Kernel Signature (HKS) is a powerful shape signature and has been introduced by Sun et al. in [11]. They have shown that the HKS is an isometric invariant and contains almost all intrinsic information of a surface. Intuitively, the HKS can be considered as the curvature of the surface. Since the HKS is derived from the process of heat diffusion it is equipped with a time parameter which is a measure for the size of the neighborhood that influences the value of the HKS at a point. Common applications use the HKS to detect similarly shaped surfaces, see [2, 4, 7, 8].

V. Zobel (✉)
Leipzig University, Leipzig, Germany
e-mail: zobel@informatik.uni-leipzig.de

J. Reininghaus
Institute of Science and Technology Austria, Klosterneuburg, Austria
e-mail: jan.reininghaus@ist.ac.at

I. Hotz
Linköping University, Norrköping, Sweden
e-mail: ingrid.hotz@liu.se

I. Hotz, T. Schultz (eds.), *Visualization and Processing of Higher Order Descriptors for Multi-Valued Data*, Mathematics and Visualization,
DOI 10.1007/978-3-319-15090-1_13

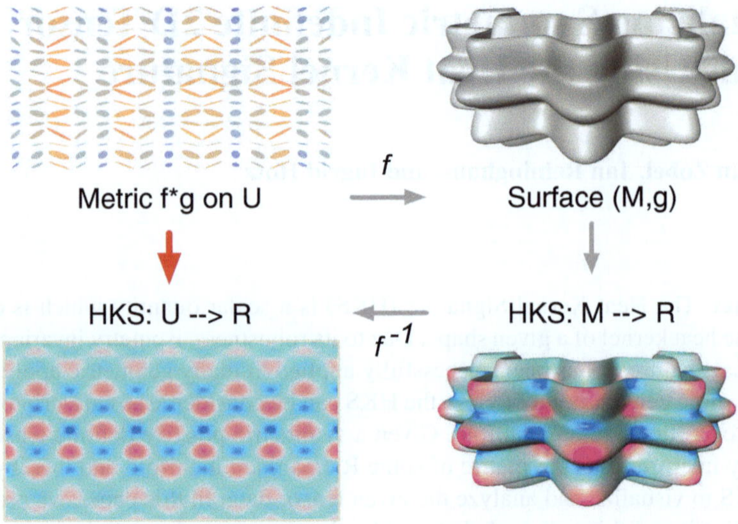

Fig. 1 Commutative diagram illustrating the relation between the HKS of a surface and a positive definite tensor field. Metric of the surface depicted as ellipses (*top left*), the parametrized surface (*top right*), HKS on the surface (*bottom right*) and the HKS on U (*bottom right*)

Motivated by these useful properties, the HKS has recently been proposed to visualize symmetric positive definite tensor fields [12]. The basic idea is to consider the HKS as a signature of the metric of the surface. By abstracting completely from the concept of an embedded surface, we can apply the HKS to tensor fields with the characteristics of a Riemannian metric, i.e. symmetric positive definite tensor fields.

The relation between the HKS of a two-dimensional surface M and a positive definite tensor field (i.e. the metric tensor field of the surface) is illustrated in Fig. 1. If g is the metric of the surface M and $f : \mathbb{R}^2 \supset U \mapsto \mathbb{R}^3$ a parametrization of M, i.e. $f(U) = M$, we can compute the pull back of the metric g on U by f, denoted by f^*g. The metric f^*g is a positive definite tensor field on U which is well characterized by the HKS of the surface. We can thereby compute the HKS for a positive definite tensor field defined on $U \subset \mathbb{R}^2$, by interpreting the tensor field as the metric of a surface.

Note that it is not necessary to compute an explicit embedding of the associated surface into some Euclidean space to compute the HKS of a given tensor field. This results in a significant difference for the computation of the HKS. While in case of surfaces the embedding is utilized to compute the HKS, in the case of general tensor fields all computations can be done using the tensor only. To do this efficiently, a realization employing a finite element method is described in [12]. If you are only interested in very short times scales there may be more accurate alternatives for the computation, see e.g. [10].

The concept described above has been successfully applied to positive define tensor fields [12]. For other tensor fields, e.g. stress tensor fields, this method is not directly applicable. In this paper, we propose to interpret such tensor fields as a generator of a time dependent deformation via a positive monotonic mapping, to obtain a field which describes a process close to a diffusion process [5]. This enables us to analyze these fields using the HKS.

A short introduction to the HKS and its application to tensor fields is given in Sect. 2.1. In Sect. 3 we motivate the use of the HKS to indefinite tensor fields and explain its generalization. Experiments and results from applying the method are shown in Sect. 4.

2 Fundamentals

In this section we recall the basic concepts and definitions of the heat kernel and its signature. It follows the original paper [12] closely.

2.1 Heat Kernel Signature

The Heat Kernel Signature (HKS) has been introduced in the field of visualization and computer graphics with the purpose of comparing surfaces. It is derived from the heat equation and assigns each point of the surface a time dependent function, which depends solely on the metric of the surface. The time parameter supports a multiscale comparison. It is used to control the size of the neighborhood of a point on the surface which is taken into account for the HKS. The definition of the HKS is applicable for arbitrary Riemannian manifolds, and thus can be used to visualize more general, positive definite tensor fields. A brief introduction to the HKS is given in this section. For details on the HKS we refer the reader to [11], while a formal treatment of the heat operator and the heat kernel can be found in [9].

Let (M, g) be a compact, oriented Riemannian manifold and Δ the Laplace-Beltrami operator on M which is a equivalent to the usual Laplacian in case of flat spaces. Given an initial heat distribution $h(x)$ on M, the heat distribution $h(t, x)$ at time t is governed by the *heat equation* $(\partial_t - \Delta)h(t, x) = 0$. The *heat kernel* $k(t, x, y)$ is satisfying $(\partial_t - \Delta_x)k(t, x, y) = 0$ with $\lim_{t\to 0} \int k(t, x, y)h(y)\,dy = h(x)$ where Δ_x denotes the Laplacian acting in the x variable. The heat kernel can be computed based on the eigenvalues λ_i and eigenfunctions ϕ_i of Δ by

$$k(t, x, y) = \sum_i e^{-\lambda_i t} \phi_i(x)\phi_i(y) \ . \tag{1}$$

The heat kernel signature (HKS) is defined in [11] as the function *HKS*

$$HKS(t, x) = k(t, x, x) \ . \tag{2}$$

Since the heat kernel is much more complex than the HKS, one might expect to lose a lot of information when regarding the HKS instead of the heat kernel. But, as shown in [11], the HKS of a surface contains almost all information of the metric of the surface itself and is much more informative than usual scalar quantities like the trace or the determinant. For small values of the time parameter t the HKS is strongly related to the curvature of the manifold. Intuitively, the heat is 'trapped' in regions with positive Gaussian curvature, while there is much 'space to escape' in regions with negative Gaussian curvature.

2.2 HKS for Symmetric Positive Definite Tensor Fields

The HKS introduced above is defined for any compact, oriented Riemannian manifold. Thus the HKS is not restricted to surfaces embedded in $\mathbb{R}^n$. If we have a metric tensor g, i.e. a symmetric positive definite tensor field, defined on a region $U \subset \mathbb{R}^n$, then (U, g) forms a Riemannian manifold. Since there is a Riemannian manifold associated with a positive definite tensor field in this way, we can compute the HKS for any positive definite tensor field. The relation of the HKS for surfaces and tensor fields is based on considering a parametrized surface and the pullback of its metric. This means, given a parametrized surface we can compute the HKS for the surface $f(U)$ or the HKS can be directly computed on U using the metric g even without knowing its embedding. This is equivalent to computing the HKS on the surface $f(U)$ and then pull it back to the parameter space U. This is nicely illustrated in the commuting diagram in Fig. 1. More details are given in the paper by Zobel et al. [12]

An example for the HKS for a symmetric positive definite tensor field is shown in Fig. 2, a diffusion tensor data set of a brain. Instead of using the diffusion tensor T itself we consider the metric $g = T^{-1}$. Large eigenvalues of the diffusion tensor correspond to high diffusivity in direction of the respective eigenvalue, whereas small eigenvalues correspond to low diffusivity. Since a high diffusivity should reflect small distances considering the inverse tensor is a natural way of assigning a metric to a diffusion tensor. For a detailed discussion see [6]. We evaluate the HKS for different time steps. Although the extraction of a single slice might discard valuable information, the structure of the brain becomes obvious by the HKS. The defined metric implies that blue regions (low values) reflect high diffusivity, whereas red regions (high values) reflect low diffusivity. Moreover, the time parameter t allows us to focus on small- as well as large-scale structures.

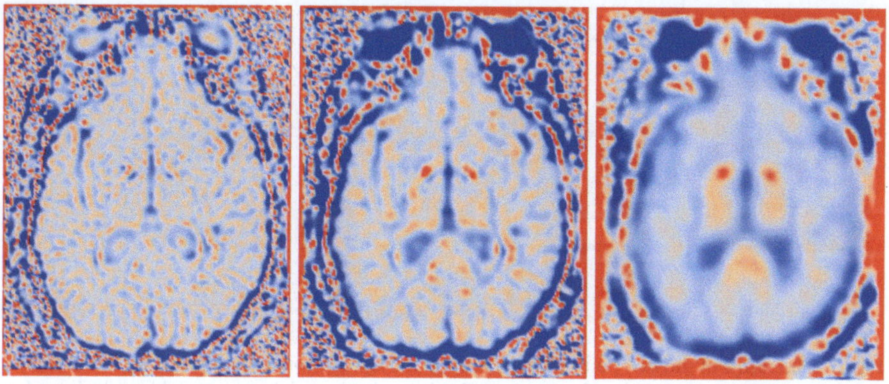

Fig. 2 HKS of a brain data set (Data set courtesy of Gordon Kindlmann at the Scientific Computing and Imaging Institute, University of Utah, and Andrew Alexander, W. M. Keck Laboratory for Functional Brain Imaging and Behavior, University of Wisconsin-Madison.) for different t. The inverse of the diffusion tensor is considered. The colormap ranges from the minimum (*blue*) to the maximum (*red*) of each individual image

2.3 Numerical Realization

The computation of the HKS for symmetric positive definite tensor fields has been described in [12] in detail. The method employs a finite element approach solving a generalized eigenvalue equation of the Laplacian. For the more general application of indefinite tensor fields this approach can be used without any changes. Also the choice of appropriate boundary conditions is discussed in this paper. Usual boundary conditions like Dirichlet or Neumann boundary conditions influence the HKS significantly. Neumann boundary conditions represent a perfectly insulated boundary. Dirichlet boundary conditions cause the HKS to have a fixed value at the boundary. To reduce these boundary artifacts we reflect a part of the field at the boundary such the heat at the boundary can diffuse outwards, see Fig. 3.

3 Using the HKS for Indefinite Stress Tensor Fields

While the relation of the heat diffusion process to other diffusion processes is obvious this is not the case for general indefinite tensors. For some applications however, there are indefinite tensors that can be considered as generator of a deformation process, which is described by a positive definite tensor. One example for such a tensor is the stress tensor, which is a central physical quantity for material modeling in mechanical engineering. In this section we propose an extension of the HKS method introduced by Zobel et al. [12] to a more general setting. The

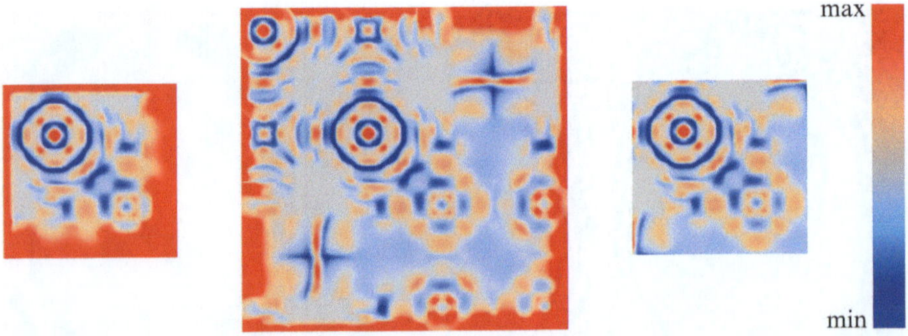

Fig. 3 The result of the two point load data set on the *left* is strongly influenced by the boundary. This effect can be reduced significantly by reflecting a portion of the tensor field on the boundary (*middle*) and cropping the result (*right*)

basic idea is to introduce a natural mapping of stress tensor fields to a positive definite tensor field which serves as input for the HKS computation. To motivate the mapping we will summarize some physical basics related to the stress tensor σ in the following. It should be noted, that this summary depicts a strongly simplified view on the much more complex topic of deformation and stress theory. The intention is mainly to justify the specific choices of mapping functions used for the visualization. We restrict the discussion to two-dimensional tensors to keep it simple. Our current implementation is also restricted to this case. For more details we refer to mechanical engineering textbooks, e.g. [1, 3]. The task of material modeling involves two essential tensor fields: the strain and the stress tensor field.

Considering the deformation of a continuous material, as a response to external forces, the deformation is essentially described by the displacement vector field. The *deformation gradient tensor* F measures all associated changes: stretches (local volume and shape changes) and local orientation changes due to rotations. Neglecting the rotational part, the stretch tensor is then derived from F employing the *polar decomposition*. It is a symmetric, positive definite tensor of second order. The physical quantity generally used for material modeling is *strain*, the relative stretch. In one dimension the uniaxial strain ϵ is defined as logarithm of the relative changes in length L

$$\epsilon = \int \frac{\mathrm{d}L}{L} = \ln \frac{L}{L_0} \quad . \tag{3}$$

Similarly, the multi-axial strain is defined as the logarithm of the rotation free part of the deformation tensor.

The *stress tensor* characterizes the local direction-dependent loads inside of a material. The sign of its eigenvalues are related to compressive respective tensile forces. Stress σ and strain ϵ tensor are linked through the constitutive relationship, which is close to linear for the large class of linear elastic materials (Hooke's law)

$$\sigma = C\,\epsilon\ ,$$

where C is a tensor of fourth order.

The naive intuition behind the mapping applied for the visualization of the stress tensor corresponds to a 'relaxation' of the material with respect to internal stresses. While the actual deformation process is much more complex, this mapping still gives an idea of a small scale transformation inside the material. In accordance to Eq. (3) the exponential mapping is the natural choice. For small scale changes it can be approximated linearly using the first term of the Taylor expansion. For visualization purposes we consider a larger variety of mappings with similar characteristics. They will be introduced in the following. A similar idea has also be used in [5] for the generation of texture visualizations.

Let T be a two-dimensional symmetric tensor field, $\lambda_1 \leq \lambda_2$ its eigenvalues and U the orthogonal matrix, such that

$$T = U^T \begin{pmatrix} \lambda_1 & \\ & \lambda_2 \end{pmatrix} U\ .$$

We define $\psi(T)$ for any positive, monotonic function $\psi : \mathbb{R} \rightarrow \mathbb{R}$ by

$$\psi(T) = U^T \begin{pmatrix} \psi(\lambda_1) & \\ & \psi(\lambda_2) \end{pmatrix} U\ .$$

The tensor $\psi(T)$ is now positive definite while its eigenvector fields remain unchanged. The selection and parametrization of the mapping ψ influences the HKS. Thus this choice has to be made carefully. We use the following mappings

- The exponential mapping $\psi(x) = exp(\alpha x)$, which is the most natural choice for stress tensor fields.
- Linear mapping $\psi(x) = c + \alpha x$, where α and c are constants such that we obtain a positive definite field.
- Arc tangent mapping $\psi(x) = \arctan(\alpha x) + \frac{\pi}{2}$. As for the exponential mapping, the range of this function is limited to $\mathbb{R}^+$. It further enhances changes for small absolute values of stresses while it is asymptotic for stresses with large absolute values.

Experiments using these mappings are shown in Sect. 4.

4 Results

In this section, we discuss some results of first experiments with indefinite stress tensor fields. We investigate the sensitivity of the HKS with respect to the three different mapping functions described in the previous section and the involved parameters. Therefore we consider two data sets which are both results from numerical finite element simulations of material stressing. Both data sets are originally three-dimensional, of which we have extracted two-dimensional slices. The first data set simulates two forces acting on a solid block, one pulling, one pushing the 'two point load'. This is a simulation with a very low resolution exhibiting some discretization artifacts, leading to a small scale structure. This data set is well-studied and therefore appropriate to evaluate our method. Throughout this section we use the colormap shown in Fig. 3, which ranges from the minimum to the maximum for the respective data set and setting.

We start with the exponential mapping, which is the most natural choice with respect to the physical interpretation. More precisely, we consider the tensor fields $\exp(T)$ (Fig. 4 first row) and $\exp(0.01T)$ (Fig. 4 second row), i.e. we use two different scalings of the original field. We can observe that the different scaling has hardly any influence on the result using this color map (from min blue to max red). Further, the HKS is evaluated for different time steps t. We see that for small time scales all details of discretization artifacts of the simulation are visible. Moving to larger times these details vanish and only the major features, the pushing and the pulling force remain.

In Fig. 5 the arc tangent mapping is applied to the tensor field. This mapping is symmetric with respect to positive and negative eigenvalues. It especially emphasizes changes for eigenvalues with small absolute value while the mapping

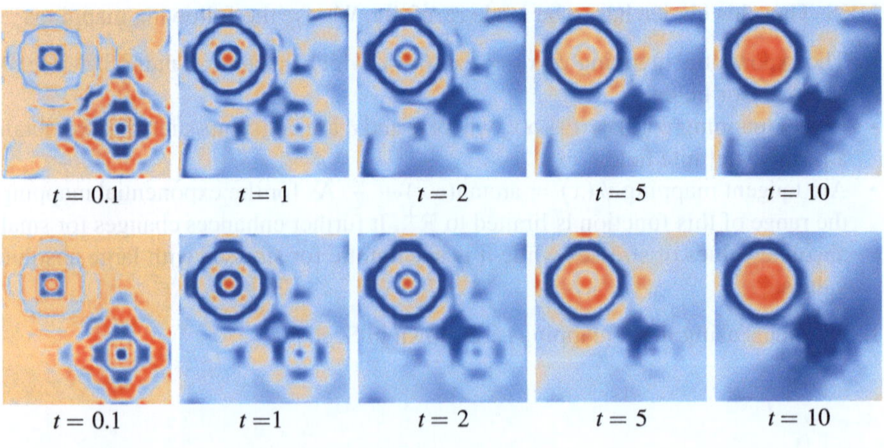

Fig. 4 HKS of a two point load data set (Data set courtesy of Boris Jeremić, University of California Davis.) using the exponential mapping. *First row* $\exp(T)$, *second row* $\exp(0.01T)$

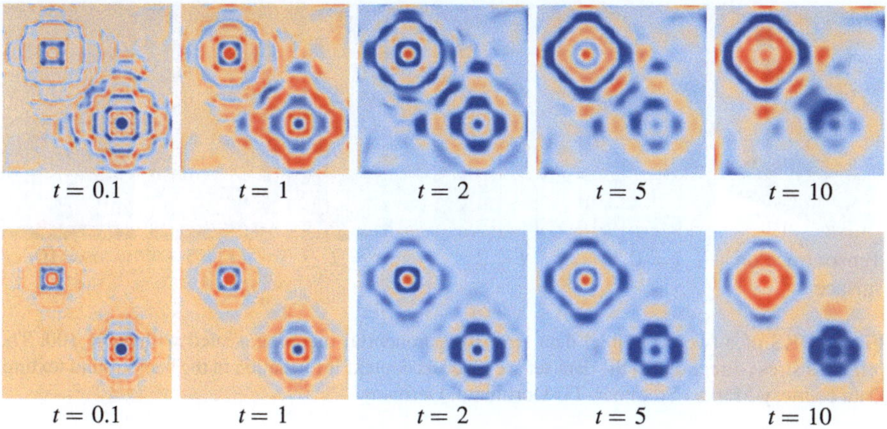

Fig. 5 HKS of a two point load data set mapped by arctan, scaled with $\alpha = 0.1$ for the *first row* and $\alpha = 0.001$ *second row*

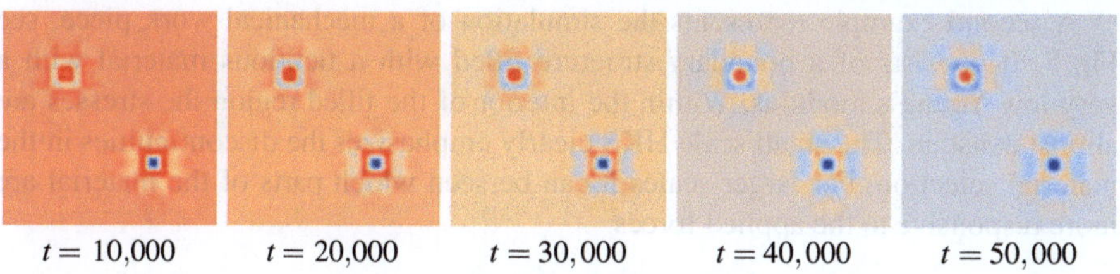

Fig. 6 HKS of a two point load data set using the linear mapping

is asymptotic for eigenvalues with large absolute value. In our example the tensor field is scaled with 0.1 and 0.001 in the first and second row, respectively, and shifted by $\frac{\pi}{2}$ to obtain a positive definite field, i.e. we consider $\arctan(0.1T) + \frac{\pi}{2}$ and $\arctan(0.001T) + \frac{\pi}{2}$. Using the arc tangent mapping, the scaling has more impact than for the exponential mapping. While the scaling of 0.1 also emphasizes regions with small eigenvalues of the tensor T, the scaling of 0.001 focuses on regions with more extremal eigenvalues, i.e. the points where the load is applied.

The simplest approach to obtain a positive definite tensor field is to shift the eigenvalues of T by a constant, such that the smallest eigenvalues occurring in the field are just above zero. Such a linear mapping is used in Fig. 6. Since the eigenvalues of our data set range from about −27,500 to 27,500, we add 27,500 to all eigenvalues. Therefore the area represented by the resulting metric is much larger than for the preceding mappings. As a consequence much larger time values have to be considered. However, the different time steps still show a very similar behavior. Shifting the eigenvalues by a constant causes the metric to represent a larger area at each point, thus the curvature decreases. Consequently there seem to arise some peaks of the curvature at points where the eigenvalues are close to the minimum, which dominate the behavior of the HKS for all time values. Thus, the

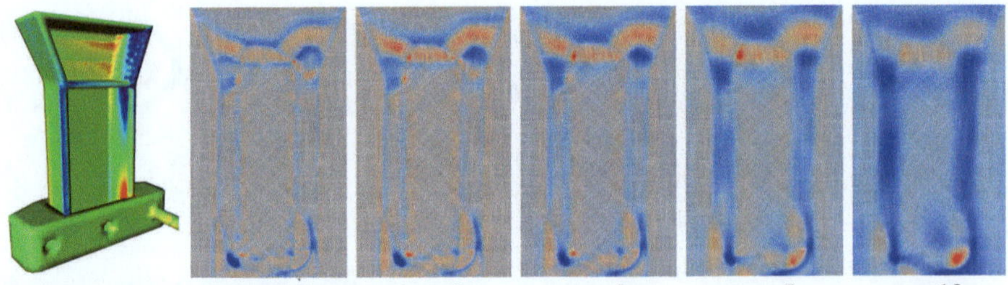

Fig. 7 HKS of a two point load data set with exponential mapping, scaled with $\alpha = 100$. The principal stress directions of the data set are given as context information in the background texture. Data courtesy Markus Stommel, TU Dortmund, Germany

linear mapping is maybe not the proper choice to obtain a positive definite tensor field, at least if there are strong negative peaks in the smallest eigenvalue field.

A second example represents the simulation of a mechanical work piece, see Fig. 7. It consists of a boundary structure filled with a fictitious material with a very low Young's modulus. Within the interior of the filled region the stresses are almost constant. The small scale HKS clearly emphasizes the discontinuities in the material selection. For larger scales it can be seen which parts of the material are more responsive to the applied forces.

5 Conclusion and Future Work

By applying a respective mapping to indefinite tensor fields it is possible to compute the heat kernel signature for such fields. This provides a new visualization method for stress tensors which differs strongly from common visualization methods. Due to its sensitivity with respect to the derivative of the tensor field it conveys additional information which is not visible in direct visualization. A special strength of the method is its inherent level of detail property. Thus, it is possible to emphasize smaller or larger structures. In contrast to basic Gaussian smoothing the scaling is directly driven by the tensor data itself. On the other side the interpretation of the results is not as easy and requires some effort. There are still many open questions in this respect. For the future we plan on further investigating the significance of the HKS for further applications. It might be of interest to compare the scaling properties to ideas of anisotropic diffusion.

From a theoretical point of view the method can be easily generalized to 3D tensor fields. With the exception of the formulas indicating the relation to Gaussian curvature, all statements are also valid in higher dimensions. The main obstacle is that the computation of the eigenvalues and eigenfunctions of the Laplacian is non-trivial. The computation of the first 500 eigenvalues for a data set with 256^2 points already takes a few minutes, thus the computation time for a data set with 256^3 points is going to be infeasible using standard approaches.

References

1. Brannon, R.M.: Functional and structured tensor analysis for engineers. UNM Book Draft (2003). http://www.mech.utah.edu/~brannon/
2. Bronstein, M., Kokkinos, I.: Scale-invariant heat kernel signatures for non-rigid shape recognition. In: 2010 IEEE Conference on Computer Vision and Pattern Recognition (CVPR), pp. 1704–1711 (2010)
3. Danielson, D.A.: Vectors and Tensors in Engineering and Physics, 2nd edn. Department of Mathematics, Naval Postgraduate School, Monterey, CA. Addison-Wesley, Reading, MA (1997) [Diss]
4. Dey, T., Li, K., Luo, C., Ranjan, P., Safa, I., Wang, Y.: Persistent heat signature for pose-oblivious matching of incomplete models. In: Computer Graphics Forum, vol. 29, pp. 1545–1554 (Wiley Online Library, 2010)
5. Hotz, I., Feng, L., Hagen, H., Hamann, B., Jeremic, B., Joy, K.I.: Physically based methods for tensor field visualization. In: VIS '04: Proceedings of IEEE Visualization 2004, pp. 123–130. IEEE Computer Society Press, Los Alamitos (2004)
6. O'Donnell, L., Haker, S., Westin, C.F.: New approaches to estimation of white matter connectivity in diffusion tensor mri: elliptic pdes and geodesics in a tensor-warped space. In: Medical Image Computing and Computer-Assisted Intervention MICCAI 2002, pp. 459–466. Springer, Berlin (2002)
7. Ovsjanikov, M., Bronstein, A., Bronstein, M., Guibas, L.: Shape Google: a computer vision approach to isometry invariant shape retrieval. In: 2009 IEEE 12th International Conference on Computer Vision Workshops (ICCV Workshops), pp. 320–327 (2009)
8. Raviv, D., Bronstein, M., Bronstein, A., Kimmel, R.: Volumetric heat kernel signatures. In: Proceedings of the ACM Workshop on 3D Object Retrieval, pp. 39–44 (2010)
9. Rosenberg, S.: The Laplacian on a Riemannian Manifold: An Introduction to Analysis on Manifolds. Cambridge University Press, Cambridge (1997)
10. Spira, A., Sochen, N., Kimmel, R.: A short time beltrami kernel for smoothing images and manifolds. IEEE Trans. Image Process. **16**(6), 1628–1636 (2007)
11. Sun, J., Ovsjanikov, M., Guibas, L.: A concise and provably informative multi-scale signature based on heat diffusion. In: Proceedings of Eurographics Symposium on Geometry Processing (SGP) (2009)
12. Zobel, V., Reininghaus, J., Hotz, I.: Visualization of two-dimensional symmetric positive definite tensor fields using the heat kernel signature. In: Topological Methods in Data Analysis and Visualization III, pp. 249–262. Springer, Berlin (2014)

Part IV
Statistical Analysis

A Framework for the Analysis of Diffusion Compartment Imaging (DCI)

Maxime Taquet, Benoit Scherrer, and Simon K. Warfield

Abstract The brain microstructure consists of the complex organization of cellular structures and extra-cellular space. Insights into this microstructure can be gained in vivo by means of diffusion-weighted imaging that is sensitive to the local patterns of diffusion of water molecules throughout the brain. Diffusion compartment imaging (DCI) provides a separate parameterization for the diffusion signal arising from each compartment of water molecules at each voxel. Their use in population studies and longitudinal monitoring of diseases hold promise for unraveling alterations of the brain microstructure in various disorders and conditions. Yet, to analyze multi-compartment models, high-level operations commonly used in scalar images need to be generalized. We present a framework that enables interpolation, averaging, filtering, spatial normalization and statistical analyses of multi-compartment data with a focus on multi-tensor representations. This framework is based on the generalization of linear combinations of voxel values through mixture simplification. We illustrate the impact of this framework in registration, atlas construction, tractography and population studies.

1 Introduction

The brain microstructure is the complex organization of cellular structures including the neurons, the axons, their myelin sheath, and glial cells. A key technique to probe the brain microstructure in vivo is to acquire diffusion-weighted images (DWI), the intensity of which depends on the local diffusion of water molecules through the microstructure. Diffusion tensor imaging (DTI) is the most widely used model to represent the brain microstructure from DWI. This model assumes that water molecules are diffusing in a single compartment, be it a single fascicle (in which the diffusion is anisotropic) or a compartment of free water (in which the diffusion is isotropic). This limits the validity of DTI to represent the brain

M. Taquet (✉) • B. Scherrer • S.K. Warfield
Computational Radiology Laboratory, Boston Children's Hospital, Harvard Medical School, Boston, MA, USA
e-mail: Maxime.Taquet@childrens.harvard.edu

I. Hotz, T. Schultz (eds.), *Visualization and Processing of Higher Order Descriptors for Multi-Valued Data*, Mathematics and Visualization,
DOI 10.1007/978-3-319-15090-1_14

microstructure which presents a variety of compartments in each voxel, including several crossing fascicles and partial volumes of CSF. The prevalence of voxels with crossing fascicles at typical resolutions has been shown to be around 60–90 % [6]. Hence any model of the microstructure that does not represent multiple fascicles is wrong in at least 60 % of the brain.

In contrast, diffusion compartment imaging (DCI) provides a separate representation for the diffusion of water molecules in each microstructural compartment in each voxel, thereby enabling a richer and more reliable characterization of the local brain microstructure. Properties of the diffusion compartment model can be directly related to properties of the underlying microstructure. For instance, the diffusion of water molecules in a direction that is orthogonal to a bundle of axons is facilitated if the axons have a larger diameter. The radial diffusivity of the corresponding compartment is therefore larger if axonal radii are larger. Typical examples of compartments are the intra-axonal space of a particular fascicle, the extra-axonal volume in the vicinity of the fascicle, the space within glial cells, and the extra-cellular space [10]. Various DCI models spanning a range of granularity and complexity have been proposed in the literature. These include the multi-tensor model with fixed diffusivities [23], full multi-tensor models with separate representations of fascicles [13], CHARMED [2], NODDI [27] and DIAMOND [14]. By providing a more reliable representation of the brain microstructure, DCI opens new opportunities to investigate the brain in disease and injury.

Analyzing DCI models has, however, proven challenging because of the absence of a one-to-one correspondence between compartments of different voxels (neighboring voxels or voxels from different subjects). This is because different voxels may have different numbers of fascicles crossing them (Fig. 1). Therefore, corresponding compartments cannot be identified and analyzed separately and a holistic approach is required. This chapter presents a holistic mathematical framework for the analysis of DCI.

Typical parameterizations of DCI models are introduced in Sect. 2.1. As we shall see, many operations on DCI models boil down to computing linear combinations of these models. Section 2.2 explains why computing linear combinations of diffusion compartment models is a mixture model simplification problem. Sections 2.3 and 2.4 presents two algorithms to compute linear combinations of DCI. Section 3 provides additional details about the implementation of these methods. Finally, Sect. 4 illustrates the far-reaching impact of this novel analysis framework on various applications, including registration, atlas construction, smoothing, tractography, population studies and model estimation.

2 Theory

In this section, we first introduce the typical parameterization of diffusion compartment models. We then explain why typical operations of image processing boil down to computing linear combinations of voxel values and how the concept of linear combinations can be generalized to diffusion compartment models. Finally,

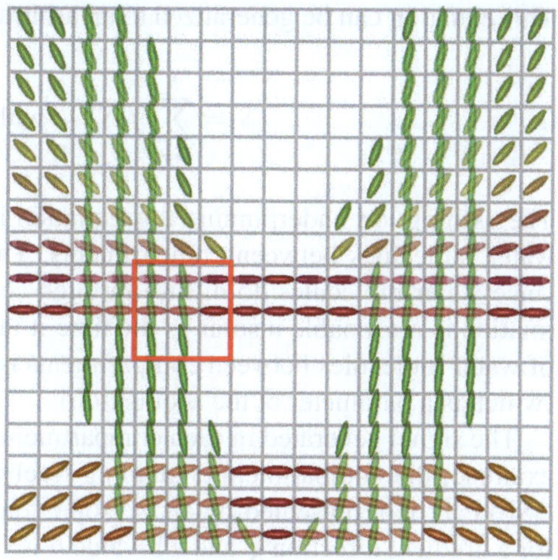

Fig. 1 Simple synthetic diffusion compartment model with fractions of occupancy encoded as the transparency of the tensors. The challenge of analyzing diffusion compartment imaging (DCI) stems from the presence of different numbers of compartments in each voxels. For example, in the *red area*, voxels containing one (isotropic), two and three compartments are present

we present two approaches to compute linear combinations of DCI. The choice between these two approaches depend on the quantity that one wants to preserve: the first approach preserves the *diffusion signal* that is generated by the model, whereas the second approach preserves the *microstructural properties* of the models. In summary, the theoretical contributions of this chapter are the following:

1. We propose a generic approach to generalize common operations to DCI, such as interpolation, averaging, smoothing and spatial normalization. We originally proposed this approach in [17] and applied it to registration, atlas construction and statistical analysis of DCI in [18] and [22].
2. We express two complementary approaches to computing linear combinations of DCI models: one that preserves the signal and one that preserves the microstructural properties. The former is similar (although slightly different) to the approach presented in our previous papers. The latter is a novel contribution of this chapter.

2.1 Diffusion Compartment Models

At each voxel, DCI models represent the diffusion signal arising from a set of several compartments as the weighted combination of the signals that would arise from each compartment individually. For a given set of acquisition parameters, if a compartment A is known to generate a signal S_A and a compartment B is known to generate a signal S_B, then the DCI model for the signal emanating from a voxel containing a fraction f_A of water molecules in A and a fraction $f_B = 1 - f_A$ of water molecules in B is:

$$S = f_A S_A + f_B S_B.$$

This equation can be generalized to an arbitrary number N of compartments as:

$$S = \sum_{i=1}^{N} f_i S_i, \quad \text{with} \sum_{i=1}^{N} f_i = 1.$$

The assumption underpinning this equation is that there is only slow exchange of water molecules between compartments. This is known to be an approximation since water molecules do diffuse through cell membranes. However, this approximation is reasonable insofar as the time it takes to observe a substantial exchange of water molecules between compartments is large compared to the diffusion time (which is a parameter of the acquisition).

The signal generated in each compartment, S_i, has its own parameterization. For example, the compartment containing freely diffusing water molecules (as in the CSF) can be represented by an isotropic tensor, i.e., its diffusion-weighted signal for a gradient direction $\boldsymbol{g}$ and a b-value b is:

$$S_{\text{iso}} = S_0 e^{-bd_{\text{free}}},$$

where $d_{\text{free}} = 3 \times 10^{-3}\,\text{mm}^2/\text{s}$ is the diffusivity of free water at 37 °C. In the sequel, we will focus on compartments whose signal decay can be represented by a multivariate Gaussian. This does not imply that we assume all fascicles to be represented by a single tensor. Indeed, the diffusion signal, S_j, arising from the j-th fascicle could be represented by two compartments, one for the intra-axonal water ($S_{j,\text{in}}$) and one for the hindered water molecules ($S_{j,\text{h}}$):

$$S_j = f_j^{\text{in}} S_j^{\text{in}} + (1 - f_j^{\text{in}}) S_j^{\text{h}} = f_j^{\text{in}} e^{-b\boldsymbol{g}^T \boldsymbol{D}_j^{\text{in}} \boldsymbol{g}} + (1 - f_j^{\text{in}}) e^{-b\boldsymbol{g}^T \boldsymbol{D}_j^{\text{h}} \boldsymbol{g}},$$

where f^{in} is the relative contribution of the intra-axonal space to the signal. In this representation, $\boldsymbol{D}^{\text{in}}$ and $\boldsymbol{D}^{\text{h}}$ have the same eigenvectors but the radial diffusivity of $\boldsymbol{D}^{\text{in}}$ is smaller than $\boldsymbol{D}^{\text{h}}$. This is similar to the CHARMED model in which the restricted diffusion would be approximated by a Gaussian diffusion tensor with small radial eigenvalues.

In summary, the class of models that we consider in the following sections is described as follows. Let N be the number of compartments including N_{iso} isotropic compartments and N_f fascicles, themselves represented by K compartments (usually $K = 1$ for a tensor representation or $K = 2$ for a bi-tensor representation). The signal decay S/S_0 is thus modeled as:

$$S/S_0 = \sum_{i=1}^{N} f_i e^{-b\boldsymbol{g}^T \boldsymbol{D}_i \boldsymbol{g}} = \sum_{l=1}^{N_{\text{iso}}} f_l^{\text{iso}} e^{-bd_l^{\text{iso}}} + \sum_{j=1}^{N_f} \sum_{k=1}^{K} f_j^k e^{-b\boldsymbol{g}^T \boldsymbol{D}_j^k \boldsymbol{g}}.$$

2.2 Weighted Combination of DCI: A Model Simplification Problem

Weighted combinations of voxel values are ubiquitous in medical image analysis. To name a few, interpolation consists in linearly combining the values of voxels on the grid to infer the value of the image at a non-grid location. Averaging (as used in atlas construction) consists in linearly combining the values of voxels in several subjects. Filtering consists in linearly combining the values of the voxels in an area with weights defined by a smoothing kernel.

Voxels in DCI contain compartment models and linearly combining these models is challenging. The challenge arises from the absence of a one-to-one correspondence between compartments of different voxels. There may be fascicles that are not present in all voxels (one-to-zero correspondence). There may be fascicles that are represented by different number of compartments in different voxels (one-to-many correspondence), for example in the case of fanning fascicles. For this reason, it is impossible to simply group compartments in pairs and compute linear combinations of single compartments in those pairs.

The goal of weighted combinations is to determine the parameters of a diffusion compartment model that can be interpreted as a weighted sum of other DCI models in a meaningful way. Let us define a virtual voxel containing all compartments from an original set of M compartment models, in proportion equal to some predefined weights $\alpha_m, m = 1, \ldots, M$. If S_{mi} is the signal arising from the i-th compartment in the m-th model ($i = 1, \ldots, N_m$) and f_{mi} is the fraction of its contribution to the total signal, then the diffusion signal arising from the microstructure in this virtual voxel would be accurately modeled as:

$$S = \sum_{m=1}^{M} \alpha_m \sum_{i=1}^{N_m} f_{mi} S_{mi} = \sum_{m=1}^{M} \sum_{i=1}^{N_m} \alpha_m f_{mi} S_{mi} \triangleq \sum_{k=1}^{N_c} w_k S_k.$$

A meaningful definition of the weighted combination of M diffusion compartment models could thus be the diffusion compartment model made of all these compartments with fractions equal to the original fractions multiplied by the weights of the combination. This *complete* model would, however, have an increased number of compartments (equal to $M\bar{N}$, where $\bar{N}$ is the average number of compartments in the original models) compared to all original models. This is not desirable for two reasons. First, the model complexity may become computationally intractable when linear combinations need to be recursively computed. Second, it may be that some compartments from the complete model actually represent the same microstructural environment and should therefore be merged in some way.

We want to simplify the *complete model* to obtain a *simplified model* that accurately represents the microstructure in the virtual voxel containing all compartments. To formalize this problem, let us introduce some notations. Let the complete model be denoted by $\mathcal{M}_c$ and let N_c be the number of its compartments. Let its k-th compartment be represented by a tensor $\boldsymbol{D}_k^c$ and generate a signal $S(\boldsymbol{D}_k^c, b, \boldsymbol{g}) = S_0 e^{-b\boldsymbol{g}^T \boldsymbol{D}_k^c \boldsymbol{g}}$ at a b-value b and for a gradient direction $\boldsymbol{g}$. Finally, let w_k be the relative contribution of the k-th compartment to the signal decay, so that the signal generated by the complete model at a b-value b and for a gradient direction $\boldsymbol{g}$ is modeled as:

$$S_{\mathcal{M}_c}(b, \boldsymbol{g}) = \sum_{k=1}^{N_c} w_k S(\boldsymbol{D}_k^c, b, \boldsymbol{g}).$$

Similarly, let $\mathcal{M}_s$ be the simplified model with $N_s \leq N_c$ compartments, with the j-th compartment described by a tensor $\boldsymbol{D}_j^s$ and generating a signal $S(\boldsymbol{D}_j^s, b, \boldsymbol{g})$ that contributes in a fraction f_j to the signal decay from the simplified model, so that the latter signal is modeled as:

$$S_{\mathcal{M}_s}(b, \boldsymbol{g}) = \sum_{j=1}^{N_s} f_j S(\boldsymbol{D}_j^s, b, \boldsymbol{g}). \tag{1}$$

The parameters of the simplified model are to minimize some discrepancy measure with respect to the complete model:

$$\mathcal{M}_s^* = \arg\min_{\mathcal{M}_s} d(\mathcal{M}_c, \mathcal{M}_s). \tag{2}$$

The discrepancy function depends on whether one is interested in preserving the signal or in preserving the microstructure. We detail the solutions to the mixture simplification problem in those two scenarios in the next two sections.

2.3 Signal-Preserving Model Simplification

If the model simplification should preserve, as much as possible, the signal that would arise from the complete model, then the discrepancy function reads:

$$\begin{aligned} d_S(\mathcal{M}_c, \mathcal{M}_s) &= \int \left(S_{\mathcal{M}_s}(b, \boldsymbol{g}) - S_{\mathcal{M}_c}(b, \boldsymbol{g})\right)^2 db\, d\boldsymbol{g} \\ &= \int \left(\sum_{j=1}^{N_s} f_j S(\boldsymbol{D}_j^s, b, \boldsymbol{g}) - \sum_{k=1}^{N_c} w_k S(\boldsymbol{D}_k^c, b, \boldsymbol{g})\right)^2 db\, d\boldsymbol{g}. \end{aligned}$$

In other words, we want the simplified model to generate a signal that is as close as possible to the signal generated by the complete model, throughout the q-space. Minimizing this function is challenging. However, Zhang and Kwok provided an efficient algorithm to compute the following upper bound to this function [25, 26]:

$$\begin{aligned}
\tilde{d}_S(\mathcal{M}_c, \mathcal{M}_s) &= N_s \sum_{j=1}^{N_s} \int \left(f_j S(\boldsymbol{D}_j^s, b, \boldsymbol{g}) - \sum_{k \in \Omega_j} w_k S(\boldsymbol{D}_k^c, b, \boldsymbol{g}) \right)^2 db\, dg \\
&\triangleq N_s \sum_{j=1}^{N_s} \tilde{d}_S^j \\
&\geq d_S(\mathcal{M}_c, \mathcal{M}_s),
\end{aligned} \quad (3)$$

where $\boldsymbol{\Omega} = \{\Omega_1, \ldots, \Omega_{N_s}\}$ is a partition of the N_c components of the complete model in N_s clusters (i.e., N_s groups of compartments that share some similarities) and $\tilde{d}_S^j$ is the discrepancy of the signal within each cluster. Minimizing $\tilde{d}_S$ is much easier than minimizing d_S because the optimization can be performed independently in each cluster, i.e., the terms $\tilde{d}_S^j$ can be independently optimized.

Zhang and Kwok proposed an iterative algorithm in which the partition variables (Ω_j) and the parameters of the simplified model in each cluster (f_j and $\boldsymbol{D}_j^s$) are alternatively optimized [25, 26]. Their method can be applied insofar as the components of the models can be expressed as kernel functions, $S_k = |\boldsymbol{H}_k|^{-1/2} K_{\boldsymbol{H}_k}(\boldsymbol{g} - \boldsymbol{g}_k)$, which is the case for Gaussian compartment models. The algorithm for the particular case of Gaussian DCI is presented in Algorithm 1 (with the step on Line 14 being detailed in Algorithm 2).

This algorithm deserves some comments regarding its interpretation and implementation.

1. The algorithm only involves simple matrix computations.
2. Both Algorithms 1 and 2 relies on iterative approaches. These approaches were empirically shown to converge to a local minimum [26] and this was also observed in all of our experiments.
3. The equations on Lines 10 and 12 of Algorithm 2 result from decoupling the minimization of $\tilde{d}_S^j$ among the parameters f_j and $\boldsymbol{D}_j^s$. Line 12 follows from the fact that $\tilde{d}_S^j$ is a quadratic form in f_j and Line 10 arises by computing the partial derivative of $\tilde{d}_S^j$ with respect to $(\boldsymbol{D}_j^s)^{-1}$.
4. The formulation of the distance on Line 20 of Algorithm 1 is a direct reformulation of the distance (3) for multivariate Gaussian.
5. Convergence checking for the clustering variables on Line 12 of Algorithm 1 simply consists in assessing whether any element k moved from one cluster to another during the last iteration.
6. Convergence for the tensor estimation on Line 5 of Algorithm 2 is verified by computing the Frobenius norm of the difference between two consecutive estimates and dividing it by the norm of the current estimate. When this ratio is sufficiently small, then convergence is claimed.
7. For computational efficiency, the inverse of all tensors should be computed a priori, given their frequent occurrence throughout the algorithm.

Algorithm 1 Signal-preserving weighted combinations of DCI

1: **Input:** (1) A set of M models $\{\mathcal{M}^m\}_{m=1,\ldots,M}$ so that $\mathcal{M}_m$ is described by fractions f_{mi} and tensors $\mathbf{D}_{mi}$ ($i = 1, \ldots, N_m$), (2) The weights α_m associated with each model, (3) The number N_s of compartments in the output.
2: **Output:** A multi-fascicle model: $\mathcal{M}_s$ such that $S_{\mathcal{M}_s}(b, \mathbf{g}) = \sum_{j=1}^{N_s} f_j e^{-b\mathbf{g}^T \mathbf{D}^s_j \mathbf{g}}$
3: $k \leftarrow 0$
4: **for** m in 1 to M **do** ▷ Construct the complete model $\mathcal{M}^c$
5:     **for** i in 1 to N_m **do**
6:         $k \leftarrow k + 1$
7:         $w_k \leftarrow \alpha_m f_{mi}$
8:         $\mathbf{D}^c_k \leftarrow \mathbf{D}_{mi}$
9:     **end for**
10: **end for**
11: $\boldsymbol{\Omega} \leftarrow$ Initialization($\{D^c_k, w_k\}, N_s$) ▷ Initialize clustering
12: **while** $\boldsymbol{\Omega}$ has not converged **do**
13:     **for** j in 1 to N_s **do** ▷ Representation
14:         $\mathbf{D}^s_j, f_j \leftarrow$ Update($\{\mathbf{D}^c_k, w_k\}_{k\in\Omega_j}$) (see **Algorithm 2**)
15:         $Z_j \leftarrow \sum_{k\in\Omega_j} w_k$
16:     **end for**
17:     Reset all Ω_j to empty sets
18:     **for** k in 1 to N_c **do** ▷ Clustering
19:         **for** j in 1 to N_s **do**
20:             Distance$(j,k) \leftarrow \left|2\mathbf{D}^c_k\right|^{1/2} + \frac{f_j^2}{Z_j^2}\left|2\mathbf{D}^s_j\right|^{1/2} - \frac{2f_j}{Z_j}\left|(\mathbf{D}^s_j)^{-1} + (\mathbf{D}^c_k)^{-1}\right|^{-1/2}$
21:         **end for**
22:         $c(k) \leftarrow \arg\min_x$ Distance(x, k)
23:         $\Omega_{c(k)} \leftarrow \Omega_{c(k)} \cup \{k\}$
24:     **end for**
25: **end while**

8. The initialization of clusters on Line 11 of Algorithm 1 can be done in several manners. We found empirically that the spectral clustering approach of Ng et al. [8] with the similarity between two tensors defined as the cosine of the scalar product between their principal orientation is both efficient and accurate. This approach involves a k-means algorithm on the projection of similarity vectors onto the basis of the first N_s eigenvectors of the Laplacian matrix. The efficiency of the algorithm enables us to run it multiple times with random initializations and select the best clustering (based on the within-cluster sum of distances).

The results of this algorithm in terms of interpolation (a prototype application of weighted combinations) are depicted in Fig. 2. As expected from the minimization of the discrepancy between signals, the interpolated models tend to preserve the signal of the original tensors. This preservation should be understood in the following way. When the two original models (at the extreme left and extreme right of Fig. 2a) generate the same signal (for a particular b-value and gradient direction), then all the interpolated models (in between original models) also generate the same signal, so that the signal is preserved along the interpolated line. When the original

Algorithm 2 Update step in signal-preserving weighted combinations of DCI

Input: The tensors and fractions that are in Cluster Ω_j : $\{\boldsymbol{D}^c_k, w_k\}_{k\in\Omega_j}$

Output: The updated tensor and fraction of the j-th compartment in the simplified model : $\boldsymbol{D}^s_j$ and f_j

$$(\boldsymbol{D}^s_j)^{-1} \leftarrow \frac{1}{\sum\limits_{k\in\Omega_j} w_k} \sum_{k\in\Omega_j} w_k (\boldsymbol{D}^c_k)^{-1}$$

$\delta \leftarrow \infty$

while $\boldsymbol{D}^s_j$ has not converged **do**

  for k in Ω_k **do**

$$\boldsymbol{W}_k \leftarrow \frac{\left((\boldsymbol{D}^s_j)^{-1} + (\boldsymbol{D}^c_k)^{-1}\right)^{-1}}{\left|(\boldsymbol{D}^s_j)^{-1} + (\boldsymbol{D}^c_k)^{-1}\right|^{1/2}} w_k$$

  end for

$$\boldsymbol{P}_j \leftarrow \sum_{k\in\Omega_j} \boldsymbol{W}_k$$

$$(\boldsymbol{D}^s_j)^{-1} \leftarrow \boldsymbol{P}_j^{-1} \sum_{k\in\Omega_j} \boldsymbol{W}_k (\boldsymbol{D}^c_k)^{-1}$$

end while

$$f_j \leftarrow \left|\tfrac{1}{2}\boldsymbol{D}^s_j\right|^{-1/2} \sum_{k\in\Omega_j} \frac{w_k}{\left|(\boldsymbol{D}^s_j)^{-1} + (\boldsymbol{D}^c_k)^{-1}\right|^{1/2}}$$

models generate different signals then the sequence of interpolated models should generate signals that monotonically evolve from one model to the other. This can be seen by looking at all the signals at any x-location on the graph of Fig. 2b.

As observed in the top row of Fig. 2a, the tensors tend to look inflated. This is confirmed by the plot of the radial and axial diffusivities and the fractional anisotropy which all present non-monotonic evolutions (Fig. 2c). By preserving the signal, we do not preserve the microstructure. In the next section, we will show that the opposite is also true.

To understand why we introduce microstructural artifacts by preserving the signal, let's picture the weighted combinations of many identical highly anisotropic compartments spanning a spectrum of orientations (Fig. 3). Preserving the microstructure would result in one such highly anisotropic compartment aligned with the mean direction in the spectrum. However, as a function of gradient direction, the signal decay generated by this average compartment would present a sharp peak in the orientation of the compartment. This sharp peak is not present in the signal decay of the original spectrum which presents moderate decay values for all gradient directions within the spectrum. To most accurately represent these moderate values spanning a larger area, an inflated tensor needs to be fit.

Another way to preserve the signal would be to compute weighted sums of the signal, i.e., the set of original scalar diffusion-weighted images (DWI). This is sometimes performed in the literature as a work-around to the problem of processing DCI. In this view, the signal-preserving weighted combination presented above can be interpreted as an acceleration technique to carrying all the processing in the space

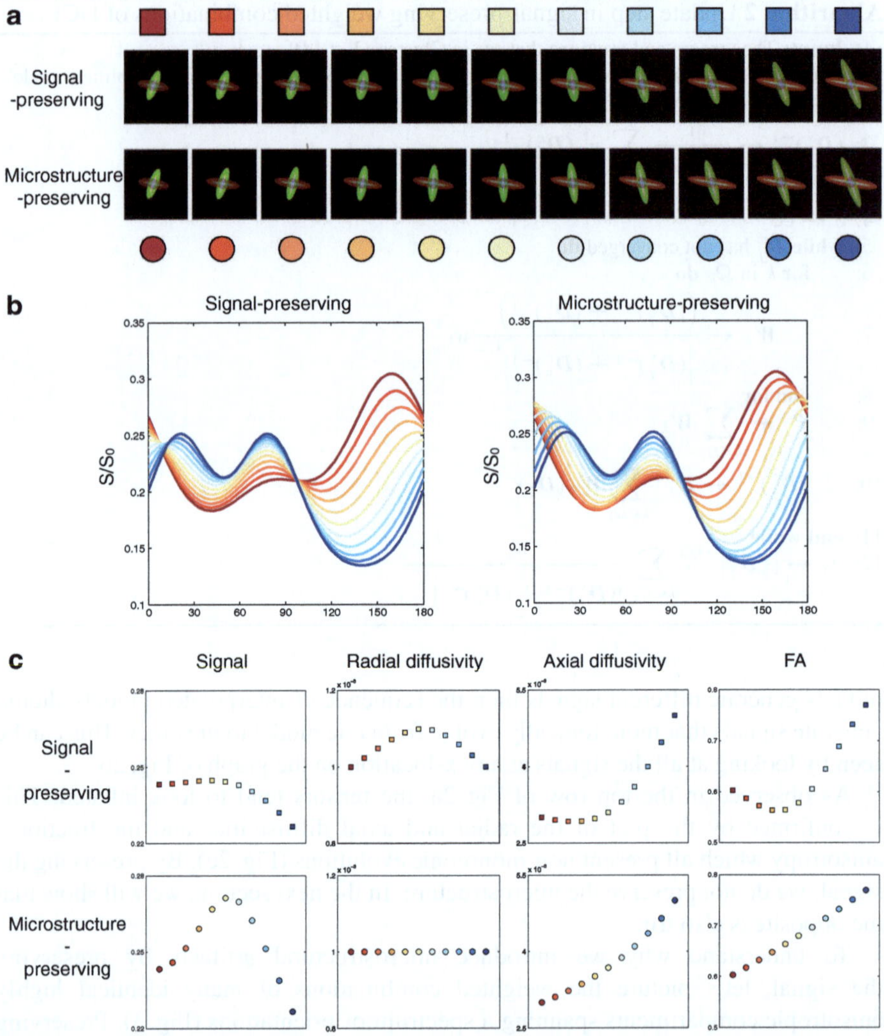

Fig. 2 Interpolation results using the signal-preserving and microstructure-preserving interpolations. Signal-preservation implies that the signal monotonically transits from one value to another, whereas microstructure-preservation implies that microstructural features monotonically transits from one value to another. When signal-preservation is achieved, microstructure is not preserved and vice versa. (**a**) The models at the left and right extremities are fixed and the others are interpolation results with weights increasing from 0.1 to 0.9 with steps of 0.1. The *colored circles* and *squares* match those in the graphs below. (**b**) Signals arising from each model for a b-value of 1,000 s/mm^2 and for in-plane gradients making an angle between 0 and π with the horizontal. (**c**) Signal generated in a direction of $\pi/25$ and various microstructural properties of the green tensor

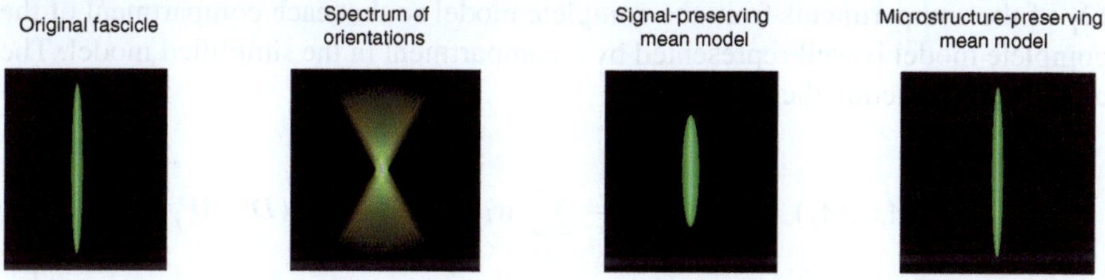

Fig. 3 Let an identical highly anisotropic original compartment be repeated many times (here 21) with various orientations. The signal-preserving mean of these compartments is an inflated tensor with higher radial diffusivity and lower axial diffusivity than the original fascicle. This inflated tensor better represents the signal generated by the spectrum of fascicles although it has microstructural features that do not correspond to original compartments. By contrast, the microstructure-preserving mean accurately represents the microstructure of the compartments in the spectrum but its generated signal departs from the signal generated by the spectrum

of DWI (since by processing DCI, one avoids the computational burden associated with re-estimating the DCI at each processing step). However, the discussion above also makes it clear that if the goal is to preserve properties of the microstructure then one should not process DWI directly.

Finally, alternatives to the distance in Eq. (3) can also be proposed. For instance, the original method proposed to compute weighted combinations of DCI was based on the minimization of the differential entropy between compartments [17]. The associated algorithm, presented in [22], is framed as an Expectation-Maximization clustering problem and is guaranteed to converge to a local minimum. The preserved quantity of this energy function is less clear. However, the simplified model simply consists in the weighted mean of covariance matrices in each cluster, which speaks more to the intuition than the equation on Line 10 of Algorithm 2. The two cost functions mostly differ theoretically and lead to very similar results in practice.

2.4 *Microstructure-Preserving Model Simplification*

In many applications, preserving the microstructural features is arguably more important than preserving the signal. For instance, when we align DCI from various subjects in order to compare their microstructure, it is important that the interpolation does not result in inflated tensors. This may jeopardize our capability to detect group differences because the amount of inflation depends on the weights of the interpolation (tensors further away form the extremities in Fig. 2 are more inflated).

If weighted combinations must preserve the microstructure, then the discrepancy function in Eq. (2) should express differences between the parameters of the DCI models rather than differences between the generated signals. The j-th compartment of the simplified model should accurately represent a subset

Ω_j of the compartments from the complete model so that each compartment of the complete model is well represented by a compartment in the simplified model. The discrepancy function therefore reads:

$$d_{\mathcal{M}}(\mathcal{M}_c, \mathcal{M}_s) = \sum_{j=1}^{N_s} \left(\left| f_j - \sum_{k \in \Omega_j} w_k \right| + \sum_{k \in \Omega_j} w_k d'(\boldsymbol{D}_k^c, \boldsymbol{D}_j^s) \right).$$

The first term expresses that the fraction f_j of the j-th compartment in the simplified model should be close to the total fraction of all the compartments that the j-th compartment represents. The second term expresses that $\boldsymbol{D}_j^s$ should be close to the tensors $\boldsymbol{D}_k^c$ that it represents in a proportion that is weighted by the fraction w_k of occupancy of those compartments. This discrepancy should be simultaneously minimized for all f_j, all $\boldsymbol{D}_j^s$ and all Ω_j. Given any clustering of compartments into N_s clusters ($\Omega_j, j = 1, \ldots, N_s$), one can always find a set of fractions ($f_j, j = 1, \ldots, N_s$) that globally minimizes the first term. Indeed, for $f_j = \sum_{k \in \Omega_j} w_k$, the first term is minimum and equal to zero and the constraints that $\sum_{j=1}^{N_s} f_j = 1$ with $0 \leq f_j \leq 1$ are respected. It is therefore sufficient to simultaneously optimize the tensors and the partition by minimizing the second term of the discrepancy:

$$\begin{aligned} \mathcal{M}_c^* &= \arg\min_{\mathcal{M}_c} d_{\mathcal{M}}(\mathcal{M}_c, \mathcal{M}_s) \\ &= \arg\min_{\mathcal{M}_c} \sum_{j=1}^{N_s} \sum_{k \in \Omega_j} w_k d'(\boldsymbol{D}_k^c, \boldsymbol{D}_j^s), \quad \text{with } f_j = \sum_{k \in \Omega_k} w_k. \end{aligned}$$

This is a weighted k-means clustering problem which can be solved by iterating the following two steps:

$$\boldsymbol{D}_j^s = \text{WeightedMean}\{(w_k, \boldsymbol{D}_k^c)\}_{k \in \Omega_j} \tag{4}$$

$$\Omega_j = \left\{ k \,\middle|\, d'\left(\boldsymbol{D}_k^c, \boldsymbol{D}_j^s\right) \leq d'\left(\boldsymbol{D}_k^c, \boldsymbol{D}_l^s\right), \forall l = 1, \ldots, N_s \right\}. \tag{5}$$

In order for the k-means algorithm to converge, the definitions of the distance function d' and the weighted mean must be interrelated so that the weighted mean is the tensor that is at the smallest (weighted) cumulative distance of all the elements in the cluster. Here, we choose the recently introduced anisotropy-preserving metric of Collard et al. [4] for it best preserves the microstructural features while being computationally tractable.

In this anisotropy-preserving framework, the weighted mean of two tensors, $\boldsymbol{D}_1$ with weight w and $\boldsymbol{D}_2$ with weight $1 - w$, is defined as follows (the generalization to

N tensors is straightforward). First, we compute the spectral decomposition of each tensor:

$$\boldsymbol{D}_i = \boldsymbol{U}_i \boldsymbol{\Lambda}_i \boldsymbol{U}_i^T.$$

Second, the eigenvector matrices $\boldsymbol{U}_i$ are transformed to their quaternion representations $\boldsymbol{q}_i = (a_i, \boldsymbol{v}_i)$ where a_i is a scalar and $\boldsymbol{v}_i$ is a vector in $\mathbb{R}^3$ such that $\|q_i\|^2 = 1$. The quaternion representation can be understood as the rotation required to align the canonical basis of $\mathbb{R}^3$ to the eigenvectors $\boldsymbol{U}_i$. The scalar a_i is related to the angle of the rotation and $\boldsymbol{v}_i$ is proportional to the invariant axis of rotation.

Averaging the quaternions directly would give equal importance to the orientation of all tensors regardless of their anisotropy. However, tensors with low anisotropy have less relevant orientational information than highly anisotropic ones. To reflect this relative importance, Collard et al. weight quaternions by a function of the anisotropy of the tensors. Formally, if we let $\text{HA}_i = \log \frac{\lambda_{\max}}{\lambda_{\min}}$ be the Hilbert anisotropy of the tensors, we compute the weighted mean of quaternions as follows (for details on the derivation, see [4]):

$$\bar{\boldsymbol{q}} = w^* \boldsymbol{q}_1 + (1 - w^*) \boldsymbol{q}_2$$
$$\text{with } w^* = \frac{w f\left(\min(\text{HA}_1, \overline{\text{HA}})\right)}{w f\left(\min(\text{HA}_1, \overline{\text{HA}})\right) + (1-w) f\left(\min(\text{HA}_2, \overline{\text{HA}})\right)},$$

where f is any sigmoid function. From the representation of $\boldsymbol{D}_i$ as $(\boldsymbol{q}_i, \boldsymbol{\Lambda}_i)$, the weighted mean of two tensors is then defined as:

$$\begin{aligned} &\text{WeightedMean}\{(w, \boldsymbol{q}_1, \boldsymbol{\Lambda}_1), (1-w, \boldsymbol{q}_2, \boldsymbol{\Lambda}_2)\} \\ &= \left(w^* \boldsymbol{q}_1 + (1-w^*) \boldsymbol{q}_2, \exp\left(w \log \boldsymbol{\Lambda}_1 + (1-w) \log \boldsymbol{\Lambda}_2\right)\right). \end{aligned} \tag{6}$$

In other words, in the anisotropy-preserving framework, the eigenvalues are independently averaged in the log-domain and the eigenvectors are averaged in their quaternion representations, with weights that depend on the anisotropy of tensors.

Since the sign of eigenvectors is undefined, there are 2^3 different representations of $\boldsymbol{U}_i$ as a basis of $\mathbb{R}^3$ and therefore eight different equivalent quaternions. Before summing quaternions, they must therefore be aligned. Collard et al. proposed to perform this alignment by first selecting one arbitrary tensor from the set of tensors to be averaged, computing one of its quaternion representation, calling it the *reference* quaternion and then selecting for each other tensor the quaternion representation that maximizes the scalar product with the reference quaternion [4].

The distance between two tensors in the anisotropy-preserving framework of [4] can be defined in terms of a metric compose of one term related to the quaternions (the chordal distance) and one term related to the eigenvalues. Similarly to the weighted mean, the term related to the quaternion is weighted by a function of the Hilbert anisotropy. The distance induced by this metric has no obvious closed-form and we therefore employ the following approximation which is an upper bound of the distance [4]:

$$d'(\boldsymbol{D}_1, \boldsymbol{D}_2) = f\,(\min(\mathrm{HA}_1, \mathrm{HA}_2))\,\|\boldsymbol{q}_1 - \boldsymbol{q}_2\|_2 + \sum_{i=1}^{3} \left| \log \frac{\Lambda_{1,ii}}{\Lambda_{2,ii}} \right|. \tag{7}$$

The weighted mean [Eq. (6)] and distance [Eq. (7)] in the anisotropy-preserving framework are ill-posed when the tensors are cylindrically symmetric, because each cylindrically symmetrical tensor can be decomposed in a quaternion in an infinite number of ways. Cylindrical symmetry is often enforced in DCI to reduce the number of parameters to estimate [10, 21]. We thus introduce the following weighted average and distance operators that can be used for cylindrically symmetric tensors whose representations only depend on the main orientation $\boldsymbol{e}_i$ of tensors and their eigenvalues:

$$\begin{aligned} &\mathrm{WeightedMean}_c\{(w, \boldsymbol{e}_1, \boldsymbol{\Lambda}_1), (1-w, \boldsymbol{e}_2, \boldsymbol{\Lambda}_2)\} \\ &= \left(\frac{w^*\boldsymbol{e}_1 + (1-w^*)\boldsymbol{e}_2}{\|w^*\boldsymbol{e}_1 + (1-w^*)\boldsymbol{e}_2\|_2}, \exp\left(w \log \boldsymbol{\Lambda}_1 + (1-w)\log \boldsymbol{\Lambda}_2\right) \right) \end{aligned} \tag{8}$$

and

$$d'_c(\boldsymbol{D}_1, \boldsymbol{D}_2) = f\,(\min(\mathrm{HA}_1, \mathrm{HA}_2))\,\mathrm{acos}(|\boldsymbol{e}_1 \cdot \boldsymbol{e}_2|) + \sum_{i=1}^{3} \left| \log \frac{\Lambda_{1,ii}}{\Lambda_{2,ii}} \right|. \tag{9}$$

We now have operators to compute the weighted mean and distance for non-symmetric tensors [Eqs. (6) and (7)] and for symmetric tensors [Eqs. (8) and (9)]. We can plug these single-tensor operators into the k-means algorithm [Eqs. (4) and (5)] to obtain a framework for the analysis of DCI that preserves the microstructure.

Expressing the weighted combinations of diffusion compartment models as a k-means clustering problem enables the incorporation of useful extensions developed in the general k-means framework. One particularly interesting extension is the so-called *constrained* k-means in which prior knowledge about the clusters can be incorporated [24]. This is useful if we know that some compartments definitely represent different microstructural environments despite the possible similarity between their parameters. For instance, one may want to force the free water compartments to remain separate from other compartments; or one may want to keep all compartments that were separate in one voxel to remain so in the simplified model. Conversely, if some tensors are known to be represent the same microstructural environments in advance, this can also be incorporated in the k-means algorithm. Both types of constraints can be expressed as a constraint matrix

Algorithm 3 Microstructure-preserving weighted combinations of DCI

Input: (1) A set of M models $\{\mathcal{M}^m\}_{m=1,\dots,M}$ so that $\mathcal{M}_m$ is described by fractions f_{mi} and tensors $\boldsymbol{D}_{mi}$ ($i = 1, \dots, N_m$), (2) The weights α_m associated with each model, (3) The number N_s of compartments in the output, (4) a constraint matrix $\boldsymbol{C}$.
Output: A multi-fascicle model: $\mathcal{M}_s$ such that $S_{\mathcal{M}_s}(b, \boldsymbol{g}) = \sum_{j=1}^{N_s} f_j e^{-b\boldsymbol{g}^T \boldsymbol{D}^s_j \boldsymbol{g}}$
$k \leftarrow 0$
for m in 1 to M **do** ▷ Construct the complete model $\mathcal{M}^c$
  for i in 1 to N_m **do**
    $k \leftarrow k + 1$
    $w_k \leftarrow \alpha_m f_{mi}$
    $\boldsymbol{D}^c_k \leftarrow \boldsymbol{D}_{mi}$
  end for
end for
$\boldsymbol{\Omega} \leftarrow$ Initialization$(\{D^c_k, w_k\}, , N_s, C)$ ▷ Initialize clustering
while $\boldsymbol{\Omega}$ has not converged **do**
  for j in 1 to N_s **do** ▷ Representation
    $\boldsymbol{D}^s_j \leftarrow$ WeightedMean$\{(w_k, \boldsymbol{D}^c_k)\}_{k\in\Omega_j}$ ▷ given by Equation (6) or (8)
    $f_j \leftarrow \sum_{k\in\Omega_j} w_k$
  end for
  Reset all Ω_j to empty sets
  for k in 1 to N_c **do** ▷ Clustering
    if $\exists k' < k : C(k, k') = 1$ **then** ▷ Check if k must be forced to some cluster
      $c(k) \leftarrow c(k')$
      $\Omega_{c(k)} \leftarrow \Omega_{c(k)} \cup \{k\}$ **else**
      ForbiddenClusters $\leftarrow \{j \,|\, \Omega_j \supset k' < k$ and $C(k, k') = -1\}$
      for j in 1 to N_s **do**
        if $j \in$ ForbiddenClusters **then**
          Distance$(j, k) \leftarrow \infty$ **else**
          Distance$(j, k) \leftarrow d'(\boldsymbol{D}^s_j, \boldsymbol{D}^c_k)$ ▷ d' given by Equation (7) or (9)
        end if
      end for
      $c(k) \leftarrow \arg\min_x$ Distance(x, k)
      $\Omega_{c(k)} \leftarrow \Omega_{c(k)} \cup \{k\}$
    end if
  end for
end while

C whose k_1k_2-entry defines the link between the k_1-th and k_2-th tensors in the complete model:

$$C(k_1, k_2) = \begin{cases} -1 & \text{if } k_1 \text{ and } k_2 \text{ must not belong to the same cluster} \\ 1 & \text{if } k_1 \text{ and } k_2 \text{ must belong to the same cluster} \\ 0 & \text{otherwise.} \end{cases}$$

The full algorithm to perform microstructure-preserving weighted combinations of DCI is presented in Algorithm 3. The results using this algorithm with constraints imposing that two tensors from a same voxel cannot be clustered together are depicted in Fig. 2. As expected, the microstructural features are preserved in this framework, whereas the signal is not preserved.

3 Details of Implementation

In this section, we provide additional details regarding the implementation of the weighted combinations of DCI.

3.1 *Selecting the Number of Compartments in the Output*

One aspect of the algorithm that we have not yet discussed is how the number of compartments of the output [N_s in Eq. (1)] is determined. We argue that any choice for N_s that is lower than the maximum number of non-empty (i.e., with a non-zero fraction) compartments in the input models is not self-consistent. In other words, we argue that if we have M input DCI models and if the m-th input model has N_m non-empty compartments, then we need:

$$N_s \geq \max_{m=1,\ldots,M} N_m.$$

To understand why other choices would not be self-consistent, imagine that we want to compute the average DCI model in the following two situations. In the first situation, the first model is a three-compartment model $S_1 = f_a S_a + f_b S_b + f_c S_c$ and the second model is a two-compartment model $S_2 = (1 - \epsilon) S_A + \epsilon S_B$ with a fraction ϵ that is arbitrarily close to zero. In the second situation, the two-compartment model is replaced by a one compartment model $S_2 = S_A$. There is essentially no difference between the two situations in terms of the microstructure being represented and in terms of the diffusion signal being generated (the signal generated by the compartment with an infinitesimal fraction will itself be infinitesimal). Yet, if the number of compartments of the output depends on the number of compartments in the second model (e.g., if we set N_s to be the average number of compartments in the input), then we may end up with two different average DCI models in the two situations.

Since there is no obvious reason to increase the number of compartments in the output DCI model compared to the input models, we set, in every applications, the number of output compartments to be equal to the maximum number of non-empty compartments in the input models, i.e.,

$$N_s = \max_{m=1,\ldots,M} N_m.$$

3.2 Computational Time

Several factors impact the computational time required to estimate weighted combinations of DCI models. In general, weighted combinations are computed at every voxel of an image, whether it is to apply a transformation field to a DCI image or to estimate an atlas from a set of spatially aligned image (see Sect. 4). This process is therefore completely parallelizable, which significantly decreases computational time. Furthermore, efficient initializations as described above makes the actual algorithm converge extremely fast. For 27 input voxels each containing a three-compartment model (which is a typical situation in the tridimensional interpolation of three-compartment models), less than ten iterations are usually required to achieve convergence of the K-means. As a typical illustration of the computational time, the registration of three-compartment DCI models at a resolution of 1 mm × 1 mm × 1 mm containing 220 × 220 × 176 voxels takes approximately 40 min on a 10 cores Linux machine. Amongst others, this process requires tens of millions of calls to the weighted combinations of DCI models (to interpolate the model at every iteration of the algorithm). This demonstrates that the proposed method can be used in practice for the analysis of DCI in a large number of subjects.

4 Applications

In this section, we demonstrate how the proposed framework for weighted combinations of diffusion compartment models can be used in various applications. Unless otherwise mentioned, we present results for the signal-preserving approach with the differential entropy as a cost function since it corresponds to the method used in most of our published results. When appropriate, we compare the obtained results with those of a heuristic approach described in [17]. The heuristic approach considers DCI as multi-channel images with a tensor in each channel and performs log-euclidean single-tensor processing in each channel [1]. Attribution of a tensor to a channel is based on the FA with the tensor with the highest FA occupying the first channel. When, in a particular voxel, a channel is missing a tensor (because the selection of the number of fascicles in that voxel resulted in fewer tensors than the maximum number allowed), then the channel with the highest fraction is split into two to enforce one-to-one correspondences.

4.1 Registration

Registration is pervasive in medical imaging. A subject's image can be registered to an atlas for further comparison between groups (we will discuss this application in

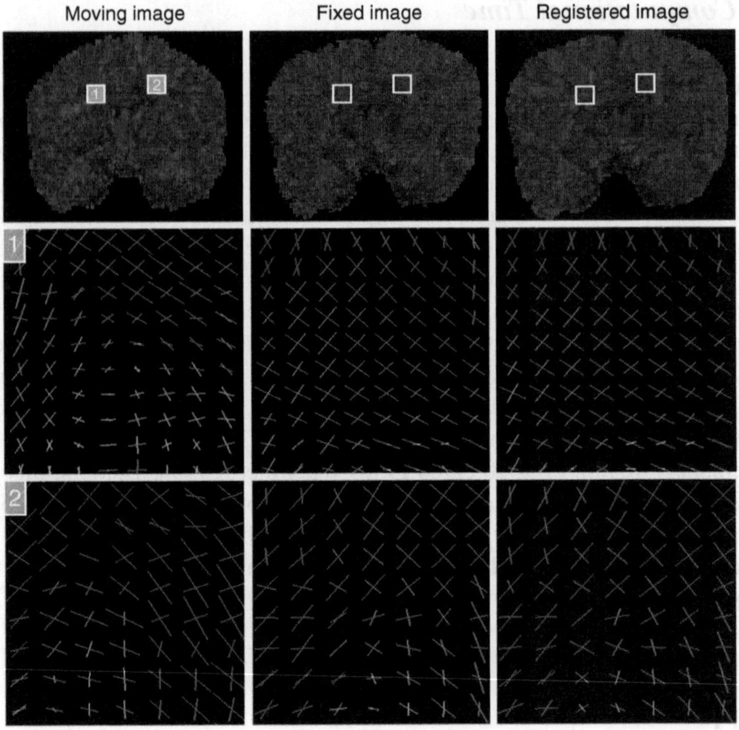

Fig. 4 The capability to compute weighted combinations of DCI combined with an appropriate similarity metric for block matching (here the generalized correlation coefficient for DCI) enables us to spatially align DCI

a following section) or to a previous image from the same subject for longitudinal analysis.

We developed a block matching approach to registration of DCI in [18, 22]. This approach requires the definition of a similarity metric specifically dedicated to DCI models and a method to interpolate DCI images. Our definition of a generalized correlation coefficient as a similarity metric can be found in [16] and the interpolation of DCI was performed using the framework presented in this chapter. Examples of registered images using this approach are depicted in Fig. 4. To assess the performance of this registration approach, ten random log-Euclidean polyaffine fields were applied to the DCI of 24 subjects and noise at six different levels was then added to both the original and the transformed DCI. Details of the experiment can be found in [22]. Registering DCI using the presented approach is significantly more accurate than performing it with the heuristic multi-channel approach: both the magnitude and the variance of the registration errors are significantly smaller (Fig. 5).

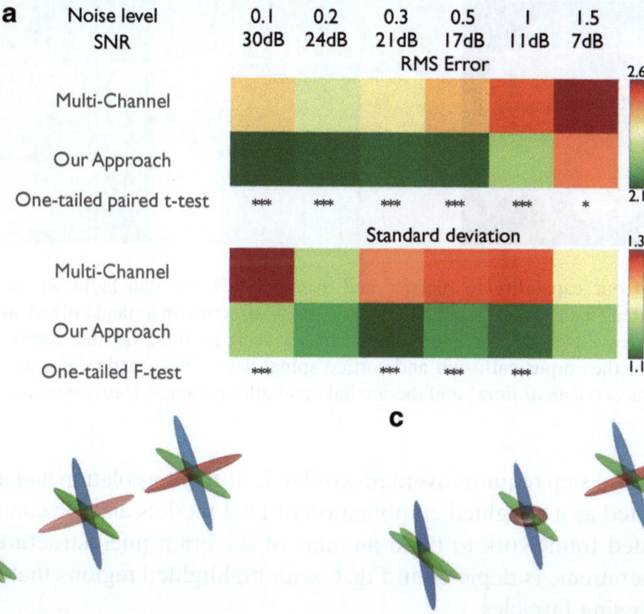

Fig. 5 (**a**) Registering the DCI with the proposed approach to weighted combinations is significantly more accurate than the multi-channel heuristic, with registration errors that have both smaller magnitude (*top table*) and smaller variance (*bottom table*) (*$p < 0.05$, **$p < 0.005$, ***$p < 0.001$). The poor performance of the heuristic approach to registration can be understood from its inability to deal with the absence of one-to-one comparisons between compartments. The model in the middle is the result of interpolating the models on the left and on the right with equal weights, (**b**) using the presented approach to interpolate DCI and (**c**) using the heuristic approach (figure reproduced from [22])

The poor performance of the multi-channel approach can be understood by its inability to deal with the absence of one-to-one correspondence. Figure 5b, c illustrates how the heuristic approach fails in those cases and results in aberrant tensors that may eventually mislead conclusions from population studies or longitudinal studies.

4.2 *Construction of an Atlas of the Brain Microstructure*

An atlas of the brain microstructure represents the average microstructure in a standardized anatomy. Such an atlas is an important asset to conduct population studies as we will see in a subsequent section. Constructing an atlas typically consists in iterating between the following three steps: registering all subjects to a common frame, averaging the aligned images, and applying the average inverse transform to the mean image [5]. The first and last steps require interpolation of DCI

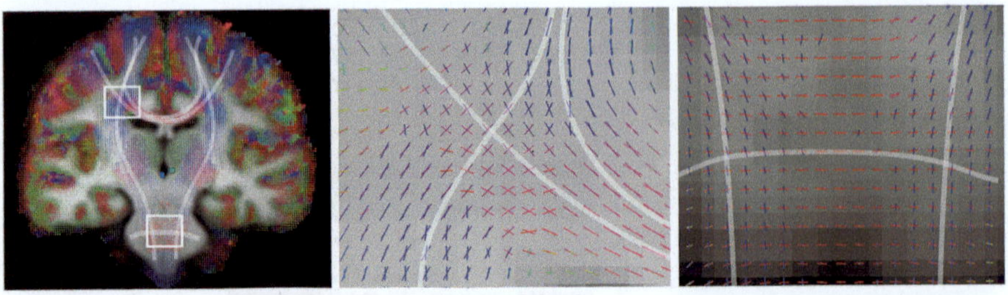

Fig. 6 From our capability to register and average DCI, we can build an atlas of the brain microstructure. This atlas represents the average microstructure in a standardized anatomy. Regions where fascicles cross are accurately represented, such as (*middle*) the corona radiata where projections of the corpus callosum and cortico-spinal tracts cross, and (*right*) a region where the pyramidal tracts (*vertical lines*) and the medial cerebellar peduncle (*horizontal lines*) cross

and the second step requires averaging of DCI. Both interpolation and averaging can be interpreted as a weighted combination of DCI models and we can therefore use the presented framework to build an atlas of the brain microstructure. The result, after ten iterations, is depicted in Fig. 6 with highlighted regions that are known to present crossing fascicles.

4.3 Filtering

Filtering DCI data proves very useful when the models have been estimated from noisy data. Filtering consists in replacing every voxels of the image by a weighted combination of voxels in a neighborhood. This is a simple application of the weighted combination operator for DCI. We illustrate this capability by generating a noisy version of the phantom presented in Fig. 1. To obtain a realistic noisy version of this phantom, we generated synthetic DWI in a three-shell HARDI sequence with 30 gradient directions on each shell and with b-values of 1,000, 2,000 and 3,000 s/mm^2. We added Rician noise to each DWI with a scale parameter of 40 ($= S_0/10$) and estimated the DCI model from these noisy DWI using the technique developed in [13]. The noisy result is depicted in Fig. 7. We then applied a 3×3 Gaussian filter with a standard deviation of 0.5 and a 5×5 Gaussian filter with a standard deviation of 1 to the noised DCI using the microstructure-preserving weighted combination operator. The result, also depicted in Fig. 7, shows that aberrant tensors are adequately filtered and brought closer to the source image.

Importantly, this filtering approach leverages information from both single-tensor and multi-tensor areas in an adaptive manner: the green fibers in the highlighted region of Fig. 7 are regularized in part by the single-tensor region located above the two-tensor region whereas the orange fibers are regularized by the single-tensor region located to the right.

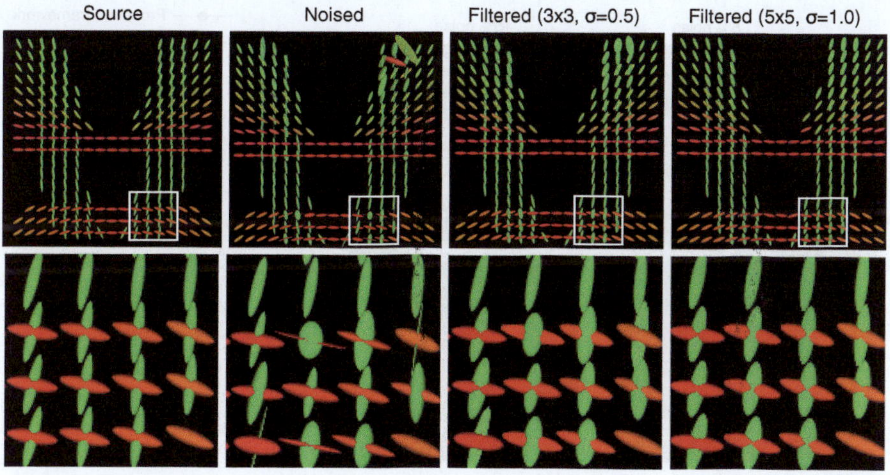

Fig. 7 Gaussian smoothing of DCI based on the weighted combination operator. The *highlighted area* presents voxels with very noisy tensor estimates that are being regularized by the filter

4.4 Application to Multi-Fiber Tractography

Single-tensor tractography typically consists in (a) shooting tracts from seeding voxels in the direction of the tensor, (b) making one step in that direction and (c) interpolating the DTI field at the new location and reiterating the process from there. The capability to interpolate DCI fields thus enables us to perform multi-fiber tractography that can disentangle crossing fibers. However, if interpolation leads to aberrant tensors, as with the heuristic approach, then spurious tracts may appear (false positives) and expected tracts may go missing (false negatives). In [17], this effect was demonstrated on a simple synthetic phantom consisting of two crossing fascicles (Fig. 8). With the proposed approach, however, tracts can reliably be drawn even in the region of crossing fibers and under the influence of noise.

4.5 Population Studies of the Microstructure

One of the most important applications of diffusion compartment imaging is to learn how diseases affect the brain microstructure. This can be achieved by conducting DCI-based population studies. Conducting population studies requires to construct an atlas of the brain microstructure and to register individual DCI to it. These techniques have been described in the above sections. Once all subjects have been aligned to a common coordinate frame, the parameters of the DCI models can be statistically analyzed. In [18, 22], two different statistical tools for DCI analysis were introduced: fascicle-based spatial statistics and isotropic diffusion analysis.

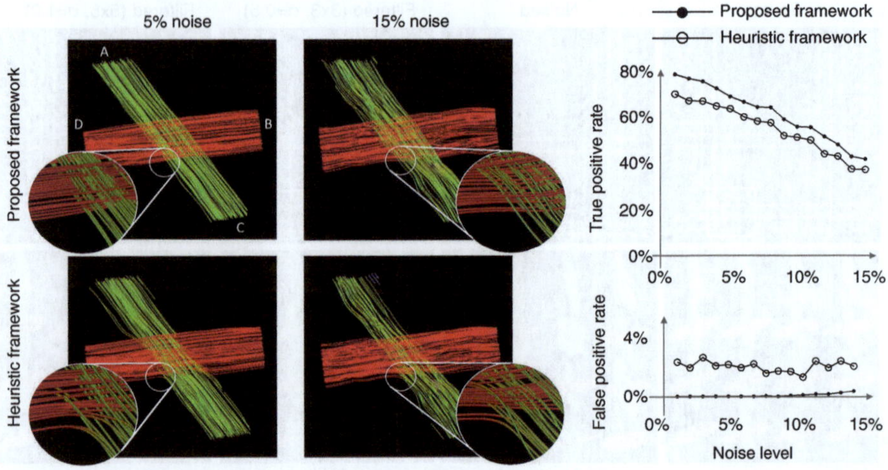

Fig. 8 The capability to compute weighted combinations enables us to perform multi-fiber tractography by interpolating the DCI field at each step. Using the heuristic multi-channel approach to interpolate DCI field results in more spurious tracts (higher false positive rate) and more missing tracts (lower true positive rate), mostly when the noise level increases

The idea behind these two tools is that some properties of DCI models pertain to individual fascicles whereas some other properties pertain to the surrounding extra-axonal volume. These tools therefore contrast with traditional DTI-based tools, such as tract-based spatial statistics (TBSS [15]) that investigate group differences in microstructure on a per-voxel basis.

Fascicle-based spatial statistics (FBSS) consists in comparing some property of the fascicle (e.g., FA, MD, radial diffusivity, axial diffusivity, etc.) along a particular fascicle of interest drawn on the atlas. Specifically, FBSS proceeds in the following steps:

1. Build a DCI atlas of the microstructure.
2. Perform multi-fiber tractography on the atlas to extract the fiber bundle of interest (tractography is only computed on the atlas).
3. Select a representative fiber tract from the bundle of interest.
4. Spatially align the DCI of all subjects to the atlas.
5. Interpolate the DCI field from each subject along the fascicle of interest.
6. Select, for each subject and at every location along the fascicle, the tensor most aligned with the fascicle.
7. Perform statistical analysis of the resulting vectors of microstructural features.

Many of the steps above (atlas construction, registration and multi-fiber tractography) require the presented framework. The last step was originally computed using cluster-based statistics [18, 22]. Recently, however, a Bayesian approach to FBSS was introduced to circumvent the caveats of p-values in population studies [21]. In the latter approach, a local model of the microstructure at the population level

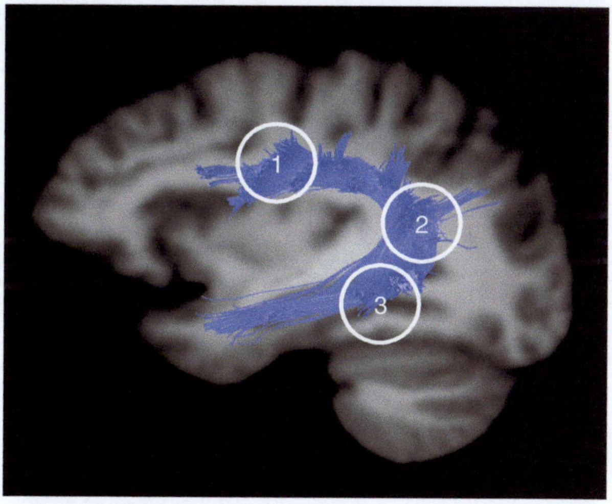

Fig. 9 The dorsal language circuit is a set of fascicles thought to connect Broca's area in the frontal lobe (Region 1), Geschwind's territory in the parietal lobe (Region 2), and Wernicke's area in the temporal lobe (Region 3)

is estimated and a Markov random field prior is assigned to its latent variables to express spatial coherence. As an illustration, FBSS was conducted in a population studies of Tuberous Sclerosis Complex (TSC) to compare the dorsal language circuit (Fig. 9) between patients with TSC and controls and, within patients with TSC, between those with autism spectrum disorder (ASD) and those without the disorder [12]. The results depicted in Fig. 10 demonstrate that FBSS reveals group differences that single-tensor DTI analysis fails to detect. FBSS also reveals specific clusters that are missing when the multi-channel heuristic processing of DCI is used. In particular, a cluster with significantly lower FA in TSC patients with autism (TSC+ASD) in the vicinity of Broca's area (Region 1 in Figs. 9 and 10) was only revealed with FBSS.

Isotropic diffusion analysis (IDA) is used to compare properties that pertain to the extra-axonal space. It was shown that a higher fraction of water molecules diffusing in this environment may be a surrogate to the presence of neuroinflammation [11]. IDA proceeds in the following steps:

1. Extract the scalar map of the fraction of isotropic diffusion (f_{iso}) from the aligned DCI of all subjects.
2. Transform the f_{iso} maps in l_{iso} maps with $l_{\text{iso}} = \text{logit}(f_{\text{iso}})$ to bring the distribution of the statistics close to a Gaussian.
3. Perform cluster-based statistics on the resulting l_{iso} maps as in [9].

Figure 11 presents the results of IDA in a population study comparing children with TSC with a comorbid diagnosis of autism (TSC+ASD) and children with TSC but without autism (TSC−ASD).

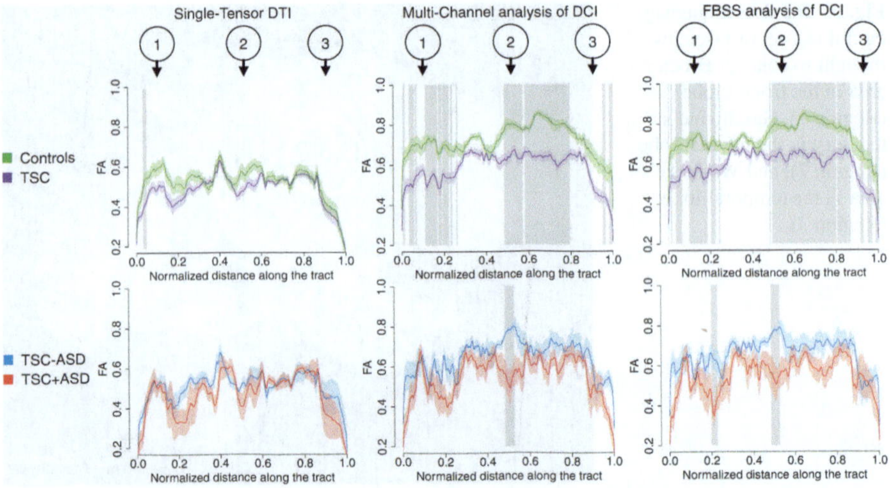

Fig. 10 Results of fascicle-based spatial statistics (FBSS) in a population studies comparing the dorsal language circuit in patients with tuberous sclerosis complex (TSC) and controls and comparing, within patients with TSC, those with autism spectrum disorder (TSC+ASD) and those without the disorder (TSC−ASD). The *dark curves* are the mean fractional anisotropy in the group and *shaded areas* around the curve represent two standard errors. *Grey rectangles* are locations where the FA is significantly different between the groups. FBSS reveals microstructural differences that the analysis of single-tensor DTI fails to detect. If a heuristic multi-channel approach is used to compute weighted combinations of DCI models, then some significant clusters go missing. The landmarks 1, 2 and 3 correspond to those in Fig. 9 (figure adapted from [22])

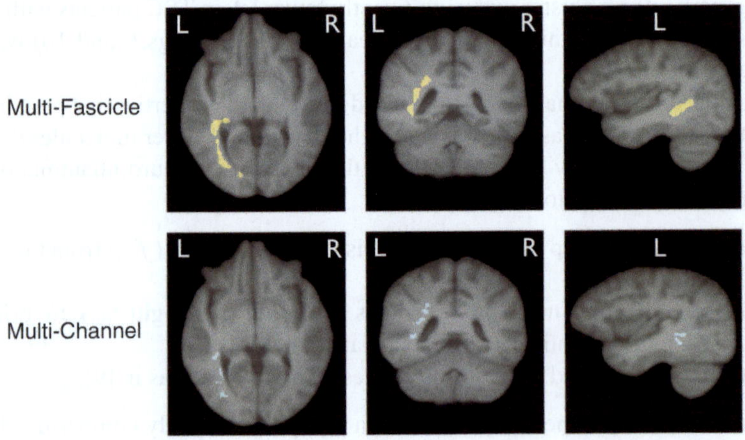

Fig. 11 IDA reveals clusters of significantly higher fraction of isotropic diffusion in children with TSC and a comorbid diagnosis of autism spectrum disorder than in children with TSC but without autism. The significant clusters are represented in the atlas space on an axial (*left*), coronal (*center*) and sagittal (*right*) slice. When the multi-channel heuristic approach is used (*bottom line*), significant clusters are located in the same areas but are less spatially coherent (figure adapted from [22])

4.6 *Estimation of Diffusion Compartment Models*

Estimating a DCI model is an ill-posed problem when data at only a single b-value are available [13, 19]. In that case, additional information from an external population of subjects can be incorporated in the estimation to regularize it [20]. This prior information is encoded in a probabilistic atlas of the brain microstructure which contains, in every voxel, a distribution over the model parameters. This distribution is used as a prior in the estimation of the DCI models in a new subject [19]. The distribution is spatially aligned to the subject's space using the proposed registration method. As a result, population studies of the brain microstructure can be conducted with single b-value data that are clinically widely available.

5 Conclusion and Discussion

This chapter introduced a framework for the analysis of diffusion compartment imaging data. At the heart of this framework is the capability to compute weighted combinations of DCI models. This is a challenging problem because of the absence of one-to-one correspondences between the compartments of different DCI voxels (such as adjacent voxels or voxels from different subjects).

By combining all the compartments from all the DCI models of the weighted combinations, one obtain a new DCI model. This complete model is, however, not practical due to its large number of compartments. For this reason, we want to estimate a simplified model that approaches the complete model. Two approaches were introduced to simplify the complete model: a *signal-preserving* approach and a *microstructure-preserving* approach. The former leads to a simplified model whose generated signal is close to that generated by the complete model, whereas the latter has microstructural properties that are close to the original models. Importantly, the microstructure-preserving framework preserves all eigenvalues and can work equally well for tensors that have a cylindrical symmetry and for tensors that have all three eigenvalues different. In particular, averaging, smoothing or interpolating ball-and-sticks models lead to a ball-and-sticks models with the microstructure-preserving operator. This is not the case with the signal-preserving operator for which combining ball-and-sticks models may lead to multi-tensor models. In that sense, the microstructure-preserving operator generalizes the approach of Cabeen et al. [3] for clustering ball-and-sticks, and defines a unique framework that can be used whether the radial diffusivities are assumed null or finite.

The proposed weighted combinations operator for DCI models has far-reaching applications in microstructure imaging. We have shown in the last section how it can be used to perform registration, tractography, smoothing, atlas construction, population studies and how it can help in estimating DCI. Importantly, the operator is general enough to be incorporated in various implementations of these applications. For instance, one may be interested in developing a more advanced filtering technique for DCI based on bilateral smoothing. Such an implementation would simply require to adapt the weights of a smoothing kernel and the proposed operator could still be used. Similarly, one may want to use another algorithm for tractography, such as multi-tensor filtered tractography [7]. There again, the weighted combination operator can be incorporated to better represent the DCI field at a non grid location.

Incorporating the novel operator in various applications enables us to fully leverage DCI models in population studies of the brain microstructure, from the estimation of the models to the statistical analysis of DCI models aligned to an atlas. This opens new opportunities for the in vivo analysis of the brain microstructure in the normal development and in diseases and injury.

References

1. Arsigny, V., Fillard, P., Pennec, X., Ayache, N.: Log-euclidean metrics for fast and simple calculus on diffusion tensors. Magn. Reson. Med. **56**(2), 411–421 (2006)
2. Assaf, Y., Basser, P.J.: Composite hindered and restricted model of diffusion (charmed) MR imaging of the human brain. NeuroImage **27**(1), 48–58 (2005)
3. Cabeen, R.P., Bastin, M.E., Laidlaw, D.H.: Estimating constrained multi-fiber diffusion MR volumes by orientation clustering. In: Medical Image Computing and Computer-Assisted Intervention–MICCAI 2013, pp. 82–89. Springer, Berlin (2013)
4. Collard, A., Bonnabel, S., Phillips, C., Sepulchre, R.: Anisotropy preserving DTI processing. Int. J. Comput. Vis. **107**(1), 58–74 (2014)
5. Guimond, A., Meunier, J., Thirion, J.P.: Average brain models: a convergence study. Comput. Vis. Image Underst. **77**(2), 192–210 (2000)
6. Jeurissen, B., Leemans, A., Tournier, J.D., Jones, D.K., Sijbers, J.: Investigating the prevalence of complex fiber configurations in white matter tissue with diffusion magnetic resonance imaging. Hum. Brain Mapp. **34**(11), 2747–2766 (2013)
7. Malcolm, J.G., Shenton, M.E., Rathi, Y.: Filtered multitensor tractography. IEEE Trans. Med. Imaging **29**(9), 1664–1675 (2010)
8. Ng, A.Y., Jordan, M.I., Weiss, Y., et al.: On spectral clustering: analysis and an algorithm. Adv. Neural Inf. Process. Syst. **2**, 849–856 (2002)
9. Nichols, T.E., Holmes, A.P.: Nonparametric permutation tests for functional neuroimaging: a primer with examples. Hum. Brain Mapp. **15**(1), 1–25 (2002)
10. Panagiotaki, E., Schneider, T., Siow, B., Hall, M.G., Lythgoe, M.F., Alexander, D.C.: Compartment models of the diffusion MR signal in brain white matter: a taxonomy and comparison. NeuroImage **59**(3), 2241–2254 (2012)
11. Pasternak, O., Westin, C.F., Bouix, S., Seidman, L.J., Goldstein, J.M., Woo, T.U.W., Petryshen, T.L., Mesholam-Gately, R.I., McCarley, R.W., Kikinis, R., et al.: Excessive extracellular volume reveals a neurodegenerative pattern in schizophrenia onset. J. Neurosci. **32**(48), 17365–17372 (2012)
12. Peters, J.M., Taquet, M., Prohl, A.K., Scherrer, B., van Eeghen, A.M., Prabhu, S.P., Sahin, M., Warfield, S.K.: Diffusion tensor imaging and related techniques in tuberous sclerosis complex: review and future directions. Future Neurol. **8**(5), 583–597 (2013)
13. Scherrer, B., Warfield, S.K.: Parametric representation of multiple white matter fascicles from cube and sphere diffusion mri. PLoS One **7**(11), e48232 (2012)

14. Scherrer, B., Schwartzman, A., Taquet, M., Prabhu, S.P., Sahin, M., Akhondi-Asl, A., Warfield, S.K.: Characterizing the distribution of anisotropic micro-structural environments with diffusion-weighted imaging (diamond). In: Medical Image Computing and Computer-Assisted Intervention–MICCAI 2013, pp. 518–526. Springer, Berlin (2013)
15. Smith, S.M., Jenkinson, M., Johansen-Berg, H., Rueckert, D., Nichols, T.E., Mackay, C.E., Watkins, K.E., Ciccarelli, O., Cader, M.Z., Matthews, P.M., et al.: Tract-based spatial statistics: voxelwise analysis of multi-subject diffusion data. NeuroImage **31**(4), 1487–1505 (2006)
16. Taquet, M., Macq, B., Warfield, S.K.: A generalized correlation coefficient: application to DTI and multi-fiber DTI. In: 2012 IEEE Workshop on Mathematical Methods in Biomedical Image Analysis (MMBIA), pp. 9–14. IEEE, Breckenridge (2012)
17. Taquet, M., Scherrer, B., Benjamin, C., Prabhu, S., Macq, B., Warfield, S.K.: Interpolating multi-fiber models by gaussian mixture simplification. In: 2012 9th IEEE International Symposium on Biomedical Imaging (ISBI), pp. 928–931. IEEE, Barcelona (2012)
18. Taquet, M., Scherrer, B., Commowick, O., Peters, J., Sahin, M., Macq, B., Warfield, S.K.: Registration and analysis of white matter group differences with a multi-fiber model. In: Medical Image Computing and Computer-Assisted Intervention–MICCAI 2012, pp. 313–320. Springer, Berlin (2012)
19. Taquet, M., Scherrer, B., Boumal, N., Macq, B., Warfield, S.K.: Estimation of a multi-fascicle model from single b-value data with a population-informed prior. In: Medical Image Computing and Computer-Assisted Intervention–MICCAI 2013, pp. 695–702. Springer, Berlin (2013)
20. Taquet, M., Scherrer, B., Macq, B., Warfield, S.K., et al.: Multi-fascicle model reconstruction from acquisitions at a single b-value with a population-informed prior. In: Proceedings of the 21st International Symposium on Magnetic Resonance in Medicine (ISMRM), vol. 30 (2013)
21. Taquet, M., Scherrer, B., Peters, J.M., Prabhu, S.P., Warfield, S.K.: A fully bayesian inference framework for population studies of the brain microstructure. In: Medical Image Computing and Computer-Assisted Intervention–MICCAI 2014. Springer, Berlin (2014)
22. Taquet, M., Scherrer, B., Commowick, O., Peters, J., Sahin, M., Macq, B., Warfield, S.: A mathematical framework for the registration and analysis of multi-fascicle models for population studies of the brain microstructure. IEEE Trans. Med. Imaging **33**(2), 504–517 (2014)
23. Tuch, D.S., Reese, T.G., Wiegell, M.R., Makris, N., Belliveau, J.W., Wedeen, V.J.: High angular resolution diffusion imaging reveals intravoxel white matter fiber heterogeneity. Magn. Reson. Med. **48**(4), 577–582 (2002)
24. Wagstaff, K., Cardie, C., Rogers, S., Schrödl, S., et al.: Constrained k-means clustering with background knowledge. In: ICML, vol. 1, pp. 577–584 (2001)
25. Zhang, K., Kwok, J.T.: Simplifying mixture models through function approximation. In: Advances in Neural Information Processing Systems, pp. 1577–1584 (2006)
26. Zhang, K., Kwok, J.T.: Simplifying mixture models through function approximation. IEEE Trans. Neural Netw. **21**(4), 644–658 (2010)
27. Zhang, H., Schneider, T., Wheeler-Kingshott, C.A., Alexander, D.C.: NODDI: practical in vivo neurite orientation dispersion and density imaging of the human brain. NeuroImage **61**(4), 1000–1016 (2012)

Statistical and Machine Learning Methods for Neuroimaging: Examples, Challenges, and Extensions to Diffusion Imaging Data

Lauren J. O'Donnell and Thomas Schultz

Abstract In neuroimaging research, a wide variety of quantitative computational methods enable inference of results regarding the brain's structure and function. In this chapter, we survey two broad families of approaches to quantitative analysis of neuroimaging data: statistical testing and machine learning. We discuss how methods developed for traditional scalar structural neuroimaging data have been extended to diffusion magnetic resonance imaging data. Diffusion MRI data have higher dimensionality and allow the study of the brain's connection structure. The intended audience of this chapter includes students or researchers in neuroimage analysis who are interested in a high-level overview of methods for analyzing their data.

1 Introduction

The study of the human brain was originally performed by expert dissection of fixed brains. Now, with the advent of structural and functional neuroimaging, we can apply quantitative computational analyses to study and model the brain in vivo. Neuroimaging analyses have important scientific and clinical applications that include the study or diagnosis of disease, the measurement of change, the detection of neural activation, and the modeling of anatomy. In this chapter, we aim to provide a general overview of analysis approaches for neuroimaging data, including some specific examples of neuroimaging studies.

Much of the research in the neuroimage analysis field has focused on the analysis of scalar data, such as structural magnetic resonance imaging (MRI) or computed tomography (CT), where a single scalar value is present at each voxel. Another large body of analysis research focuses on detection of neural activations using

L.J. O'Donnell (✉)
Harvard Medical School, Boston, MA, USA
e-mail: odonnell@bwh.harvard.edu

T. Schultz
University of Bonn, Bonn, Germany
e-mail: schultz@cs.uni-bonn.de

I. Hotz, T. Schultz (eds.), *Visualization and Processing of Higher Order Descriptors for Multi-Valued Data*, Mathematics and Visualization,
DOI 10.1007/978-3-319-15090-1_15

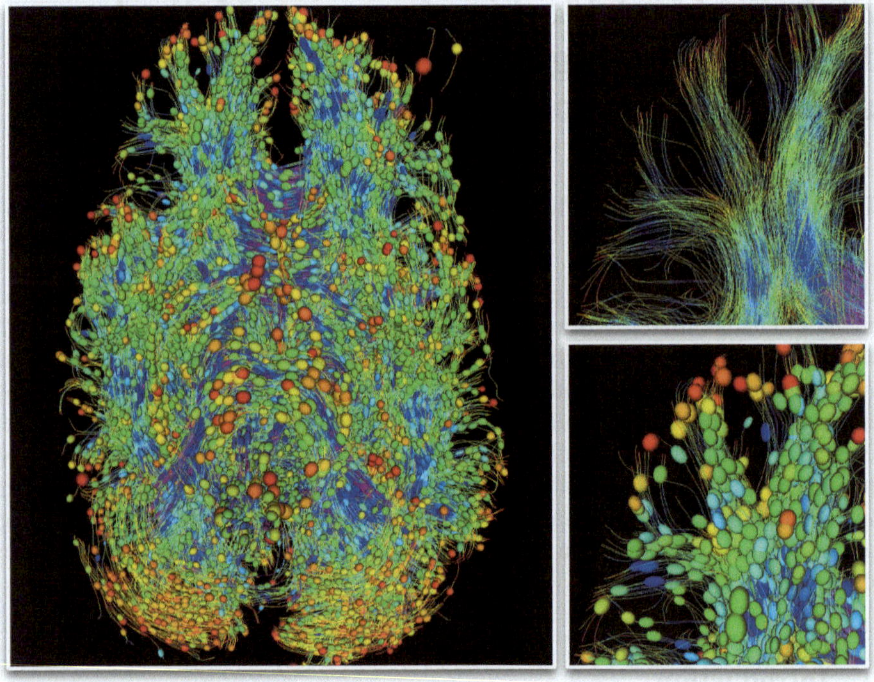

Fig. 1 Example diffusion MRI data, including fiber tract trajectories from tractography with selected (randomly sampled) ellipsoids to visualize diffusion tensors along the tracts. At *left*, the whole brain is shown in an inferior view. At *right*, zoomed images show fiber tract trajectories (*top*) plus ellipsoids (*bottom*). The tracts and ellipsoids are colored by fractional anisotropy (FA), a popular scalar measure for statistical analyses of diffusion MRI. *Blue* and *purple* are high FA, *green* and *yellow* are intermediate values, and *red* is low FA

timecourse data: the blood-oxygen-level dependent (BOLD) signal of functional magnetic resonance imaging (fMRI). Of particular interest in this chapter is the analysis of diffusion MRI, the only non-invasive scan for measurement of the brain's connectional structure. The traditional representation of diffusion MRI data is not scalar. Rather, it is a tensor (specifically, a 3×3 symmetric, positive-definite matrix) at each voxel. Data employing the tensor representation are called diffusion tensor MRI or DTI. In current diffusion MRI research, higher-order models (as well as connectivity data) may also be reconstructed from the scan [58]. Figure 1 shows both tensor and connectivity (fiber tract) data from a diffusion MRI scan of a healthy human brain.

Because different types of neuroimaging data have different data dimensionalities as well as vastly different interpretations in the context of the brain, and because neuroimaging studies have many possible designs, the analyses developed for neuroimaging data are manifold. Analyses have been developed for scalar data, for timecourse data, for tensor-valued data, and for many other types of data representations such as measurements from regions of interest or along image skeletons.

To organize this chapter, we categorize quantitative neuroimaging analyses into two groups according to their overall philosophy: statistical testing or machine learning. In statistical testing, the goal is to obtain a result that is statistically significant: unlikely to have arisen by chance. Most often, these approaches are applied to measure a result, such as a functional or structural difference, between groups of subjects. Statistical testing methods are also in regular clinical use to detect functional brain activations in individual patients. The supervised machine learning methods that are treated in the second part of this chapter learn computational models that estimate or predict the values of unobserved variables. During learning, they are given access to labeled training data, for which the value of the variable of interest is known, such as images categorized into healthy control and patient images, or annotated with subject age. In a second step, the respective quantity—such as disease state or age—is estimated based on other brain scans for which it is unknown. Statistical and machine learning methods for exploratory data analysis, such as clustering or Principal Component Analysis (PCA), are outside of our main focus, even though a use of PCA as part of a predictive model is discussed in Sect. 3.4.

In the rest of this chapter, we survey examples from the scalar neuroimaging field, and where possible we describe extensions or new methods developed for the analysis of diffusion MRI data. The chapter is divided into two parts: first, the more traditional statistical testing approaches, and second, the more recent machine learning approaches.

2 Methods for Neuroimaging Analysis That Use Statistical Tests

In neuroimaging research, statistical tests are used in many scenarios. Examples include: to find regions of significant difference between two populations in a clinical neuroimaging study, to find regions of neural activation in fMRI, or to detect abnormal regions that differ from a model of the healthy brain. In the rest of this section, we first describe basic concepts, then we give examples of popular methods that employ statistical testing in neuroimaging data, and finally we describe extensions of the statistical testing frameworks that have been proposed for analysis of diffusion MRI data.

2.1 Basic Concepts and Potential Problems

We begin this section with a simple example that motivates the vocabulary and the basic concepts used in statistical hypothesis testing. Readers familiar with this may wish to skip ahead to the overview of methods that have been developed for diffusion MRI.

In the statistical hypothesis testing framework, there is generally a *null hypothesis* H_0, such as "There is no difference between the two study groups, thus their data have the same mean." A corresponding *alternative hypothesis* H_a^1 could be, in this simple example, that the means of the data from the two groups are different. (Another H_a^2 could be, for example, that the mean of one group is larger than the other.)

To assess this possible difference, a *test statistic* is chosen. In our example, the test statistic should be a quantity related to the difference in means, such as the popular t-statistic [64]. The *null distribution* is a probability distribution that gives the probability, under the null hypothesis, of observing values of the test statistic. The null distribution can be known or estimated from the data. In our example, armed with the null distribution and an observed test statistic, the researcher will determine the conditional probability of observing the test statistic if both groups have the same mean (the null hypothesis).

If the observed test statistic is found to have low probability under the null hypothesis, the reasoning is that the observed test statistic is unlikely to have occurred by chance. Thus there may be an experimental finding: it may be possible to reject the null hypothesis in support of the alternative hypothesis. To decide whether to reject the null hypothesis, the *statistical significance* of the observed test statistic is determined by calculating a *p-value*, the probability of observing a statistic at least as extreme as the observed statistic (under the null distribution). Here, the word "extreme" refers to the tails of the null distribution, where the probability of observing the test statistic values is low: For H_a^1, the first alternative hypothesis mentioned above, both tails would be considered to be extreme ("two-tailed test"). For H_a^2, only the tail corresponding to larger values would be taken into account ("one-tailed test"). If the calculated p-value falls below a predetermined threshold or *alpha level*, such as 0.05, the result may be considered significant. Alternatively, for a given alpha level, the test statistic can be compared to a threshold for which 5 % of the area of the null distribution is located under the tail(s).

Potential problems in hypothesis testing have been widely discussed, for example in the book "The Cult of Statistical Significance" [76]. Issues include incorrect rejection of a true null hypothesis, called *type I error*, or false positive error. In the context of neuroimaging, this type of error would lead to publication of a false finding. Type I errors are typically controlled at an alpha level of 0.05, which means that statistical tests commonly used in neuroimaging have a 5 % chance of rejecting a true null hypothesis due to chance. A second issue is *type II error* or false negative error. This means that statistical significance of an effect, even though it is true, cannot be shown based on the acquired data.

It is clear that performing multiple tests (*multiple comparisons*) is dangerous: eventually, one of the tests will produce a significant value. If this is not correctly accounted for, the overall chance of a type I error can increase drastically. An infamous illustration of this was given by an fMRI experiment in which activation was found in the brain of a dead salmon [8]. Popular strategies for correcting this potential source of error are mentioned in the next section.

2.2 *Popular Neuroimaging Analyses*

Here we give a brief overview of two main approaches to data analysis: voxel-based, where data are measured and statistics are performed in a large number of voxels throughout the brain, and region-based, where data measurement and statistical analyses are restricted to neuroanatomical regions generated by image segmentation. We note that neuroimage analysis methods may also be categorized according to the number of subjects analyzed. Often, analyses employ a population, or a neuroimaging dataset that includes data from multiple subjects. However, some analyses are inherently single-subject, such as fMRI activation detection in neurosurgical patients.

Voxel-Based Statistics

There is a large and sophisticated body of literature on voxel-based morphometry (VBM) and statistical parametric mapping (SPM) in structural and functional imaging [26]. These approaches use the general linear model (GLM) framework, a linear regression model that incorporates covariates and any indicator variables reflecting study design [28]. The overall idea is that parameters of interest are calculated from the GLM, then a parametric statistical test is applied at each voxel, such as the t-test or F-test. In VBM, traditionally the gray matter is segmented and smoothed, giving a map of gray matter concentration that is compared across groups [4]. In fMRI analysis, where the per-voxel information is a vector of time-course data, traditionally the GLM approach uses regression to obtain a single scalar parameter for univariate statistical testing [29]. The voxel-based approach assumes that anatomy corresponds across subjects at the voxel level, and thus smoothing and image registration play important roles. Statistical analyses called deformation-based or tensor-based morphometry generally analyze the Jacobian determinants of the vector-valued deformation fields generated by image registration [5].

In voxel-based analyses, multiple comparisons arise naturally because the tests are performed at many anatomical locations within the brain. Several statistical methods may be employed to correct for multiple comparisons, including the stringent Bonferroni correction, where the threshold for statistical significance is adjusted to account for the multiple tests. The Bonferroni correction assumes tests are independent, which is not the case in spatially smooth image data, and leads to an overly conservative correction, reducing statistical power. Thus, the theory of Gaussian random fields is employed in SPM to correct for multiple comparisons [4]. An alternative that controls the expected proportion of false positives within a statistical map, rather than the probability that any part of the map includes a false positive, is the false discovery rate (FDR) [7, 30]. However, simple application of FDR does not take into account the fact that voxels are spatially contiguous and represent continuous data [11]. In another approach, a summary or maximal test statistic (such as maximum suprathreshold cluster size) may be used to summarize information from multiple statistical tests across voxels, and the null distribution may be estimated for this new, overall test statistic. This strategy may be used in combination with permutation testing for computation of the null distribution [47]. Permutation tests are increasingly used because they are powerful, non-parametric, and simple to perform by repeatedly randomizing the labels of the data. However, they can be computationally intensive.

Region-Based Statistics

In the case where there is a hypothesis about the likely region of an effect (for example, if the corpus callosum is hypothesized to differ between groups), a region of interest (ROI) can be created for measurement. This may be done via a manual or automated image segmentation procedure. Scalar measurements are made, such as the ROI's volume or the mean value of image voxels within the ROI. This approach can avoid the multiple comparisons problem, if only one ROI is measured, and only one type of information is measured from that ROI. More typically, data from more than one ROI are measured, and Bonferroni or FDR correction would be appropriate. Traditional t-tests and ANOVA are very commonly used in the neuroimaging literature to identify possible differences between groups in ROI-based studies, for example [62].

2.3 Extension of Analyses to Diffusion MRI

We give examples of analyses in the voxel-based and region-based frameworks, as well as methods where statistical tests have been developed to deal with unique types of data from diffusion MRI. We begin with voxel-based and region-based methods that operate on scalar values derived from diffusion MRI, most commonly the fractional anisotropy (FA). Next we describe statistical methods that have been developed for diffusion MRI tracts, followed by methods for vector and tensor data estimated from diffusion MRI. This is by no means an exhaustive list of references from the field; rather, we intend to provide examples illustrating the main concepts.

Voxel-Based Statistics Proposed for Diffusion MRI

At this point, standard VBM studies are not often performed on diffusion MRI data. It has been shown that results are highly sensitive to the size of the smoothing kernel [36] and that image registration often fails to match the high FA core of the white matter tracts [63]. Furthermore, there are issues with non-normally distributed residuals after fitting a GLM model [36].

The most popular voxel-based analysis of diffusion MRI data was designed to address these issues. Though it is a voxel-based method, it is called Tract-Based Spatial Statistics [63]. In this method, to ameliorate registration difficulties and to restrict analyses to the presumed core of the tract, locally high FA values are projected onto voxels of a group FA skeleton. After this procedure, the voxels of the groupwise skeleton have been attributed with data from every subject in the study, and standard GLM analyses may be used.

Methods have also been investigated for diagnostic analysis of diffusion MRI on the single subject level. Diffusion MRI is of particular interest as a sensitive marker for traumatic brain injury, where a quantitative marker is desired to help in diagnosis and prognosis. Initially, standard VBM techniques were applied to investigate brain changes by comparing an individual to controls [42]. Then alternative voxel-based analyses were designed to detect abnormal regions within the single subject, based on comparison to a model of normal diffusion that employs data from multiple control subjects. An FA-based method that employs bootstrap methods for estimating control population variance and corrects for covariates such as age and gender has been developed to assess departure from the normal model using z-scores [43].

Region-Based Statistics Proposed for Diffusion MRI

Existing image segmentation and measurement pipelines may be applied to any scalar data derived from diffusion MRI. Additionally, many diffusion-MRI specific methods exist for defining white matter tracts, including deterministic and probabilistic tractography methods for estimation of white matter connections. For a recent overview of tractography segmentation methods, see [49]. Many diffusion MRI analysis pipelines use atlases derived from tractography, such as the Mori atlas [69] to define regions of interest in individual subjects. Once tract ROIs have been defined, they can be used for measurement of quantities such as the average FA within the tract. This enables region-based statistical analyses.

After ROI definition, measurement and statistical analysis are the same as for any imaging ROI study, except for the fact that there are many scalar parameters that may be measured from one diffusion MRI scan. For a basic diffusion tensor reconstruction, scalars can include FA, mean diffusivity (MD), and more. Thus the multiple comparisons problem may be more severe in diffusion MRI studies.

Tract-Based Statistics Proposed for Diffusion MRI

In more sophisticated data analyses than average measurement within an entire tract region, fiber tracts have been used for structure-specific statistical mapping. Since tracts can be considered to have a linear structure (connecting brain region A to brain region B), one option is to analyze data along the tract. This style of analysis measures data versus arc length along a tract [13]. Methods for measurement and analysis have been proposed by many authors. Simple averaging of data at points along the tract and use of permutation testing found significant differences across hemispheres [48]. Authors have proposed more sophisticated machinery, such as using an extension of multivariate statistics called functional regression analysis [31, 75]. Fiber tractography in a DTI atlas was employed to define and parameterize tracts in conjunction with the Hotelling T^2 statistic to analyze both FA and tensor norm [31]. Analysis of data along tracts has been shown to have advantages over simple averaging of the data, which may mask differences [12, 48]. Related approaches have proposed analysis over the entire tract surface, representing it as a sheet, rather than attributing a single trajectory with data [74].

Eigenvector and Tensor Statistical Tests Proposed for Diffusion MRI

Some disagreement exists regarding an appropriate manifold for diffusion tensors. A Riemannian metric between diffusion tensors was proposed [3, 6, 23, 24, 41] in order to restrict analyses to the space of positive definite symmetric matrices. However, others believe that a Euclidean metric is more appropriate for actual diffusion MRI data [51]. Several groups have investigated geodesics for interpolation of diffusion tensors [22, 38]. However, recent work on smoothing may indicate that the metric between tensors has little practical effect for data analyses [67]. Each metric may be useful for certain computational tasks: in registration, the log-Euclidean metric may be used for reducing blurring when averaging, while the Euclidean metric performs well for the objective function [37].

Limited work exists on statistical testing for group differences in principal diffusion directions (major eigenvectors) and in entire diffusion tensors. A statistical method based on the bipolar Watson distribution was proposed to test whether the principal diffusion direction had the same mean in two groups of subjects [60]. This test was shown to detect differences that were invisible to a more standard FA analysis [60]. Additional work by the same author investigated tensor statistics [59] and gave further insight into FDR correction for the eigenvector testing [61]. Another group investigated the application of several multivariate statistical tests directly to the components of the full diffusion tensor [73].

3 Regression and Classification in Neuroimaging

Given samples from one or multiple populations, statistical hypothesis testing allow us to infer statements about parameters describing those populations. In the context of neuroimaging, frequent examples of populations in the statistical sense include groups of subjects, voxels, or time steps.

Recently, methods from machine learning are increasingly being used to make statements about individual samples, rather than populations. Example applications include supporting the diagnosis of disease based on examples of both healthy and diseased subjects [25], estimating a person's brain maturity [21, 27], detecting which class of object a person is currently looking at [14], whether or not he or she is telling the truth [19], or predicting behavior [32].

Building a system that facilitates such predictions requires selecting a suitable machine learning method, extracting mathematical descriptors ("features") on which further analysis can be based, and selecting features that are particularly relevant to the task. Obtaining a reliable estimate of a method's accuracy can pose serious and surprising pitfalls. Finally, it is desirable, though unfortunately difficult, to gain insight on how the machine learning method arrived at its final estimate.

In this section, we will elaborate on each of these steps. Since the field is young, new methods are evolving rapidly, and no widely used standards have been established so far. Therefore, we cannot hope to provide a final and exhaustive overview, but rather focus on general principles and examples of solutions that have been found to be effective on more than a single dataset and, ideally, by different groups. We are particularly interested in examples involving diffusion MRI and multimodal imaging, which have been excluded from an earlier, related overview [52].

3.1 *Methods for Classification*

In the context of neuroimaging, classification is the assignment of a subject or a cognitive state to a specific class, such as recognizing that a subject suffers from a specific disease, or is currently looking at an example from a certain class of objects. Mathematically, classification is performed by a function $f(x)$ that maps an instance $x \in \mathscr{X}$, the subject or cognitive state, to a discrete output variable ("label") y, which encodes the different classes. In practice, x is usually represented by an m-dimensional feature vector $\mathbf{x} \in \mathbb{R}^m$.

Training a classifier amounts to learning the function f from a training dataset $\{(\mathbf{x}_i, y_i)\}$, $i = 1, \ldots, n$ so that $f(\mathbf{x}_i) = y_i$ for as many training examples as possible. At the same time, f should be as "simple" as possible, in a sense that can be made mathematically precise [57], to maximize the chance that it will produce correct results also for novel inputs $\tilde{\mathbf{x}}$ which have not been part of the training data.

In neuroimaging applications, it is common to have a high-dimensional feature space, but relatively little training data ($m \gg n$). Support Vector Machines (SVM) are widely used as a classifier, since they are known to be able to deal with this situation relatively well [57]. They are based on finding a hypersurface in the feature space that correctly separates as many of the training samples as possible, while also maximizing the distance of the decision boundary to the samples that are correctly classified.

SVMs can be generalized to nonlinear classification by implicitly mapping the features into an abstract higher-dimensional space using the "kernel trick" [57]. While LaConte et al. [40], working with very high-dimensional feature vectors to begin with, do not find a clear benefit from mapping them to an even higher-dimensional space, Wee et al. [71] report a noticeable increase in accuracy when using nonlinear kernels with moderately sized feature vectors, and Rasmussen et al. [53] construct an example in which a nonlinear kernel aids classification even in high-dimensional space. Ultimately, no single kernel is optimal for all applications, and classification accuracy can often be increased by trying different alternatives.

Aside from support vector machines, the machine learning literature offers a wide range of classifiers that are occasionally used in neuroimaging, including Fisher Linear Discriminant Analysis (LDA) [25] and maximum uncertainty Linear Discriminant Analysis (MLDA) [18], naive Bayesian classifiers [45], k Nearest Neighbor classifiers [70], neural networks [2], and random forests [1]. For more detailed explanations of these methods and further pointers to the machine learning literature, we refer the reader to [9].

Sometimes, it is desirable to combine the results from multiple classifiers. For example, in multimodal imaging, a separate classifier might be created for each modality, and a single prediction y has to be derived from their outputs. In the simplest case, it can be based on a majority vote [35]. A natural improvement of this is to weight the impact of each classifier by its estimated accuracy [18]. In adaptive boosting (AdaBoost), this idea is combined with an iterative training of classifiers on re-weighted training samples, so that classifiers trained at later stages focus on examples misclassified previously [44].

3.2 Methods for Regression

Regression differs from classification mainly in the fact that the output variable y is continuous, such as age or brain maturity [21, 27], rather than discrete. Many methods for classification have a closely related variant that can be used for regression. An example is support vector regression [57] which, like support vector classification, produces a function f that can be written in terms of a subset of the training data, the so-called support vectors. Relevance vector regression, as it was used in [27], generally provides an even sparser representation of a similar form and at similar accuracy, at the cost of a more difficult and time consuming training process.

Many aspects of learning a function $f(x)$ that will be discussed in the remainder of this section are common to classification and regression. In this case, we will refer to methods that create such functions as "learning machines".

3.3 Feature Extraction

Feature extraction is the process of producing a feature vector $\mathbf{x}$ from the image data. It will subsequently represent a subject or cognitive state and contain information relevant to the classification or regression task. Initially, the individual images often undergo the same preprocessing that would be used for voxel-based statistical analysis, as it was explained in section "Voxel-Based Statistics". This includes normalization to a standard space, so that each voxel position (approximately) corresponds to the same anatomical structure, often followed by smoothing to reduce image noise and to compensate for residual misalignment.

At this point, each voxel could in principle be turned into an entry of the feature vector [40]. Often, a shorter feature vector is desired and is achieved by averaging values over larger blocks of voxels [19] or over predefined functional regions [18, 21], whose selection may be informed by prior knowledge on the regions involved in specific tasks or conditions [20].

In the context of diffusion MRI, feature extraction often makes use of the pipeline developed for Tract-Based Spatial Statistics (TBSS), which was explained in section "Voxel-Based Statistics Proposed for Diffusion MRI". In this case, the features are given by the values on the TBSS skeleton [33, 56], sometimes averaged over predefined white matter regions [15].

A more complex way of deriving feature vectors from diffusion MRI involves a brain connectivity graph constructed using tractography. To this end, Wee et al. [71] first parcellate the brain into anatomical regions of interest and detect which of them are connected by a deterministic full-brain tractography. The resulting graph is represented as an adjacency matrix, where edges are alternatively weighted by fiber count, Fractional Anisotropy, Mean Diffusivity, or any of the three diffusion tensor eigenvalues, and the resulting six matrices are vectorized and concatenated to form the final feature vector. In a follow-up work, these dMRI-based connectivity matrices have been combined with ones constructed from correlations in resting-state fMRI [72].

3.4 Feature Selection and Feature Weighting

Even though many learning machines are in principle able to operate on high-dimensional feature spaces, their effectiveness can be reduced when feature vectors include components whose variation does not carry any information about the desired output y, especially when, in addition, little training data is available. This is particularly relevant for some of the feature vectors described in the previous subsection, which can be very high-dimensional ($m \approx 10^6$), and often include information from all regions of the brain, even if only some small specialized area may be affected by a disease or relevant to a task.

Initially, one often attempts to give similar influence to all features ("feature normalization"), for example by subtracting the mean and dividing by the standard deviation, or by linearly rescaling all features to some fixed interval [15]. Subsequently, a crucial step in most applications of machine learning in neuroimaging is to reduce the impact of features which are less relevant to the task at hand.

Feature selection methods attempt to find a subset of features which is particularly well suited for building a learning machine. In order to arrive at an optimal solution, one would have to evaluate each possible combination of features, which is infeasible in most cases. Therefore, a frequently used strategy is to first rank features according to their expected utility, and to include the top k features in the final feature vector.

In neuroimaging, the Fisher score (as it would be used in an F-test [15]), the t score (as it would be used in Student's t-test [71]), and, in case of regression, the Pearson correlation coefficient [21], are particularly popular for ranking features, possibly due to their ubiquitous use for statistical testing on the same type of data. As an alternative to these straightforward methods, a family of heuristics known as Relief, ReliefF, and RReliefF [54] is occasionally used [33, 50], and offers the advantage of being able to detect nonlinear dependencies between features and labels, as well as providing a higher rating of features that are only useful when used in combination, whereas the simple methods rate each feature in isolation.

Once a ranking has been achieved, the number k of features that should be included can be found by cross-validation [71], which will be explained in greater detail in Sect. 3.5. As a computationally less demanding alternative, sometimes only features are used whose difference between labels is statistically significant [18], or the number of retained features is simply set to some constant value [21].

Traditional techniques for dimensionality reduction such as Principal Component Analysis have also been used [27], but have occasionally been found to perform worse than other feature selection schemes [71]. This might be explained by the fact that, unlike all methods described above, they only consider the feature vectors $\mathbf{x}_i$ in isolation, and do not account for their relationship to the labels y_i.

An alternative to feature selection is feature weighting, which assigns a greater influence to some features than to others, rather than eliminating features completely. For example, Schmidt-Wilcke et al. [56] and Schlaffke et al. [55] scale all features by their corresponding F score, which avoids the need to decide how many features to retain.

Related to the idea of feature weighting are multiple-kernel SVMs, which are based on a weighted sum of several distance measures (kernels) between the $\mathbf{x}_i$, each of which might depend only on a certain subset of features. A natural application of this concept is multimodal imaging, where each modality is represented by a separate kernel [72].

3.5 Validation and Parameter Tuning

Once a function $f(x)$ has been trained for classification or regression, its accuracy can be estimated by applying it to a set of test data $\{(\tilde{\mathbf{x}}_i, \tilde{y}_i)\}$, and measuring the difference between the predictions $f(\tilde{\mathbf{x}}_i)$ and the true $\tilde{y}_i$. In order to ensure that the resulting estimate is unbiased (i.e., not overly optimistic), it is essential that the test data may not overlap with the data that has been used for training.

When data is available from only relatively few subjects, as it is quite common in neuroimaging, setting part of it aside for testing only allows us to evaluate accuracy on very few examples, leading to estimates that may be unbiased (on average, we do not overestimate accuracy), but have high variance (individual estimates of accuracy are highly uncertain). This problem can be reduced by applying cross validation, in which the learning machine is trained repeatedly on part of the data. In particular, in n-fold cross validation, the data is distributed equally between n sets ("folds"). Based on these, the learning machine is trained n times, each time using data from $n - 1$ folds, and evaluating the result on the data from the remaining fold. The final estimate of accuracy is obtained by averaging the results from all n iterations. A special case of this is leave-one-out cross validation, in which the number of folds coincides with the size of the available training dataset, so that, in each iteration, only one sample $(\mathbf{x}_i, y_i)$ is left out of the training set.

Most learning machines have parameters that need to be set, such as choosing a kernel and setting a regularization parameter in support vector machines, or deciding how many features to retain in feature selection. While some authors simply use fixed default settings [21], results can often be improved greatly by evaluating alternative settings using cross-validation, and using the one that led to the highest estimated accuracy.

When cross-validation is used for parameter tuning, obtaining a reliable estimate of the final accuracy requires nested cross-validation, so that an outer cross-validation loop, which is responsible for estimating the overall accuracy, separates the data into a training and a testing set, and the inner loop, which performs the parameter tuning, may only access the training data set from the outer loop.

A subtle consequence of this, which is sometimes overlooked, is that in order for cross-validation to be effective, *only the training data* may be used for feature selection. A pragmatic safeguard against accidental double dipping is to attempt classification of the same data with randomly permuted labels y_i, to repeat this a large number of times, and to observe the resulting distribution of accuracies. Dosenbach et al. [21] perform this experiment as a part of validating their method; Schmidt-Wilcke et al. [56] use it as a permutation-based test to assess significance of their classification results.

Since a classifier cannot be expected to predict random labels with larger-than-random accuracy, an unbiased estimate should, on average, result in the same accuracy as a random guess. This is illustrated in Fig. 2: It is based on 100 iterations in which random group labels have been assigned to 18 healthy subjects, and a support vector machine with feature weighting has been trained to predict those

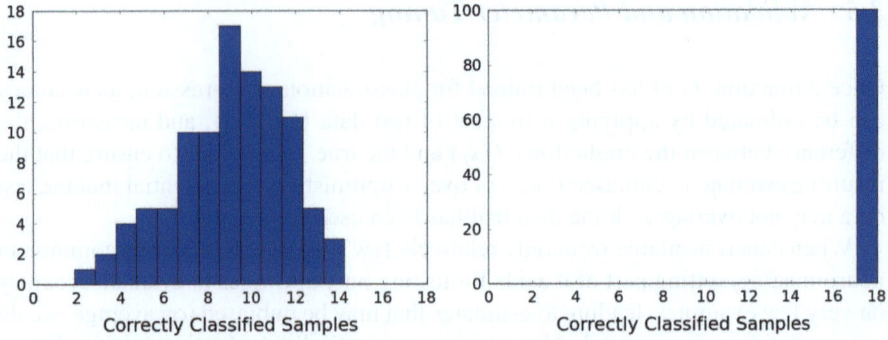

Fig. 2 On average, trying to predict 100 different sets of random labels cannot lead to better-than-random accuracy (*left*). However, if feature weighting is performed outside of the cross validation loop, the classifier is erroneously reported to achieve perfect results in each case (*right*)

random labels from MRI data. On the left, feature weighting is done correctly, within the leave-one-out cross validation. As expected, on average, the classifier does not achieve higher accuracy than a random guess. On the right, the same classification is attempted using the same method; the only difference is that, similar to [33, 50], feature weighting has now been performed as a pre-process on all data, including the test data. This leads to the misleading estimate that random labels can be predicted with perfect accuracy in all cases.

This surprising pitfall can be explained by the fact that we are given few data points with a huge number of features, many of which take on random values. This means that, given arbitrary class assignments, the feature vector includes features that happen to separate the data into those classes by pure chance. Performing feature selection on the whole dataset allows the classifier to operate on those features, without having any independent data left to check whether or not they actually contained legitimate information, or were correlated with the labels only by chance. This is similar to the fact that, after selecting a region of interest (ROI) based on correlations with another random variable, it is no longer meaningful to perform a statistical test on those correlations within that same ROI [68].

In summary, Fig. 2 illustrates that performing feature selection or feature weighting outside the cross validation loop can bias estimates so severely that they lose all meaning. While many works have avoided this problem by a correct setup [21, 27, 35, 71], some others merely acknowledge that performing feature selection as a pre-process on all data might lead to results that are "too optimistic, probably related to some degree of over-fitting" [33] or that "validation in an independent sample will be essential to determine how robust the current approach is when applied to a fully independent dataset" [50], which does not appreciate the full severity of the problem. Importantly, one should never attempt to compare accuracies from a correct setup with those reported after doing feature selection on the full dataset.

The fact that this section has illustrated one difficulty in correctly applying machine learning to neuroimaging data should not be taken as an indication that these techniques are fundamentally flawed: They rest on a solid statistical foundation [66] and, when applied correctly, they have already led to results that could be reproduced across different datasets [21, 35], and learning machines trained on one scanner have been tested successfully on data from other scanners [27].

3.6 *Interpretation and Visualization*

Since a fundamental goal in most neuroimaging studies is to better understand how the structure of the brain and its activity relate to specific functions, or to factors such as gender, age, and disease, it is desirable to obtain not only a classification or regression result from applying a learning machine to the data, but to gain at least some level of understanding of how it arrives at its prediction, e.g., which regions of the brain were most important for detecting a specific disease.

Most machine learning methods are designed to achieve the highest possible accuracy, whereas interpretability by a human operator is usually not a primary design goal. One common way to still glean some insight is to consider the results of feature selection. For example, if each feature corresponds to the average over some region of interest (ROI), the selected features indicate which regions were used to achieve the classification. When cross validation is used, different features might be selected in each iteration, and it is common to only report the features that are selected most frequently [15, 71] or even in all cases [18, 21].

Closely related to this, some authors compare the accuracies that can be achieved when making different parts of their data available to the classifier. For example, in the context of diffusion tensor MRI, this may indicate whether Fractional Anisotropy, Mean Diffusivity, or individual eigenvalues allow for more reliable detection of a certain disease [71].

As part of their training, linear classifiers, such as linear support vector machines or linear discriminant analysis, compute a weight with which each feature contributes to the final result. If features were appropriately normalized, this makes it natural to inspect the weight vectors as an indicator of feature importance. In fact, when features correspond to individual voxels or small ROIs, weight vectors can be visualized as spatial maps, similar to activation maps from mass-univariate statistical analysis [40]. Support vector machines have recently been extended to increase spatial regularity of the resulting maps, with the goal of making them more interpretable [17].

However, an important caveat in the interpretation of weight vectors is that classifiers may put significant weight on features that are unrelated to the given task or disease, and that the largest weights do not necessarily correspond to the features which are most strongly related to the label. In particular, Haufe et al. [34] provide examples in which features are only included to cancel out artifacts that are also present in truly informative features and might obtain an even greater weight.

Even though weight vectors have been found to agree with prior knowledge about abnormalities in Alzheimer's disease [17, 39] and have indicated neuroanatomically plausible regions in cases where mass univariate analysis failed to detect significant differences [16], Haufe et al. [34] conclude that the only truly firm conclusion that can be drawn from weight vectors of a successful classifier is that at least one of the features with non-zero weight is associated with the given condition or task.

When support vector machines are used with a nonlinear kernel, the weight vector is defined in an abstract higher-dimensional space, and generally cannot be mapped back to the original feature vector [57]. However, sensitivity analysis [40, 53] can still quantify how much impact each feature has on the classification. In the linear case, sensitivities amount to the squared feature weights [53], so they suffer from the same limitations with respect to their interpretability.

4 Main Challenges and Conclusions

As discussed in section "Voxel-Based Statistics", spatially contiguous regions play an important role in maintaining statistical power while correcting for multiple comparisons in mass-univariate statistical testing. In contrast, most learning machines act on abstract feature vectors, and are oblivious of the underlying spatial structure. Even though attempts have recently been made to increase accuracy and interpretability of classifiers by spatial regularization [17], it is still widely unexplored how to best account for spatial and anatomical structures when training learning machines, and how much is to be gained from it. Taken to the extreme, Honorio et al. [35] have demonstrated that, on several datasets with a limited number of subjects each, classification based on a single discriminative region of interest outperformed some widely used multivariate methods that were found to make use of a larger number of scattered voxels.

While there is hope that the multivariate analysis afforded by machine learning techniques will lead to an understanding of interactions and dependencies that would remain hidden to mass-univariate approaches, interpretation and visualization of what allows a learning machine to perform successful classification or regression remains a difficult task [34], and merits further work.

Applications of machine learning to diffusion MRI have so far mostly been based on features derived from the second-order diffusion tensor model. However, it is now common to acquire more complex diffusion MR data that requires higher-order models, including High Angular Resolution Diffusion Imaging (HARDI), Diffusion Spectrum Imaging (DSI), and multi-shell data. Only few initial works exist on extracting features suitable for machine learning from such models, based on spherical deconvolution [10], or a spherical harmonics expansion of apparent diffusivities [46]. There is still a need to explore alternative features based on such rich and complex data, and to evaluate their power and reliability in a range of applications.

Finally, training highly accurate learning machines and obtaining a realistic impression of their performance requires more data than is typically acquired for traditional statistical analysis. Currently, relatively few groups have the opportunity to apply machine learning to sufficiently uniform datasets that include hundreds of subjects [21, 27]. However, larger datasets, such as the ones from the Human Connectome Project [65], are currently becoming available to the general research community, and are about to open up new horizons for the development and evaluation of machine learning on neuroimaging data.

Acknowledgements This work has resulted from a series of breakout sessions at Dagstuhl seminar 14082. We thank Anna Vilanova (TU Delft, The Netherlands) for her collaboration in those sessions, and for her help in organizing the LaTeX structure of this chapter. Author LJO thanks NIH grant support R01MH074794, P41EB015902, R21CA156943, P41EB015898, and U01NS083223.

References

1. Anderson, A., Dinov, I.D., Sherin, J.E., Quintana, J., Yuille, A.L., Cohen, M.S.: Classification of spatially unaligned fMRI scans. NeuroImage **49**(3), 2509–2519 (2010)
2. Arribas, J.I., Calhoun, V.D., Adalı, T.: Automatic bayesian classification of healthy controls, bipolar disorder and schizophrenia using intrinsic connectivity maps from fMRI data. IEEE Trans. Biomed. Eng. **57**(12), 2850–2860 (2010)
3. Arsigny, V., Fillard, P., Pennec, X., Ayache, N.: Log-euclidean metrics for fast and simple calculus on diffusion tensors. Magn. Reson. Med. **56**(2), 411–421 (2006)
4. Ashburner, J., Friston, K.J.: Voxel-based morphometry—the methods. NeuroImage **11**(6), 805–821 (2000)
5. Ashburner, J., Hutton, C., Frackowiak, R., Johnsrude, I., Price, C., Friston, K., et al.: Identifying global anatomical differences: deformation-based morphometry. Hum. Brain Mapp. **6**(5–6), 348–357 (1998)
6. Batchelor, P., Moakher, M., Atkinson, D., Calamante, F., Connelly, A.: A rigorous framework for diffusion tensor calculus. Magn. Reson. Med. **53**(1), 221–225 (2005)
7. Benjamini, Y., Hochberg, Y.: Controlling the false discovery rate: a practical and powerful approach to multiple testing. J. R. Stat. Soc. Ser. B (Methodological) **57**, 289–300 (1995)
8. Bennett, C.M., Baird, A.A., Miller, M.B., Wolford, G.L.: Neural correlates of interspecies perspective taking in the post-mortem atlantic salmon: an argument for proper multiple comparisons correction. J. Serendipitous Unexpected Results **1**, 1–5 (2010)
9. Bishop, C.M.: Pattern Recognition and Machine Learning. Springer, New York (2006)
10. Bloy, L., Ingalhalikar, M., Eavani, H., Roberts, T.P.L., Schultz, R.T., Verma, R.: HARDI based pattern classifiers for the identification of white matter pathologies. In: Fichtinger, G., Martel, A., Peters, T. (eds.) Medical Image Computing and Computer Assisted Intervention (MICCAI). Lecture Notes in Computer Science, vol. 6892, pp. 234–241. Springer, Berlin (2011)
11. Chumbley, J.R., Friston, K.J.: False discovery rate revisited: FDR and topological inference using gaussian random fields. NeuroImage **44**(1), 62–70 (2009)
12. Colby, J.B., Soderberg, L., Lebel, C., Dinov, I.D., Thompson, P.M., Sowell, E.R.: Along-tract statistics allow for enhanced tractography analysis. Neuroimage **59**(4), 3227–3242 (2012)
13. Corouge, I., Fletcher, P.T., Joshi, S., Gouttard, S., Gerig, G.: Fiber tract-oriented statistics for quantitative diffusion tensor mri analysis. Med. Image Anal. **10**(5), 786–798 (2006)
14. Cox, D.D., Savoy, R.L.: Functional magnetic resonance imaging (fMRI) "brain reading": detecting and classifying distributed patterns of fMRI activity in human visual cortex. NeuroImage **19**(2 Pt 1), 261–270 (2003)

15. Cui, Y., Wen, W., Lipnicki, D.M., Beg, M.F., Jin, J.S., Luo, S., Zhu, W., Kochan, N.A., Reppermund, S., Zhuang, L., Raamana, P.R., Liu, T., Trollor, J.N., Wang, L., Brodaty, H., Sachdev, P.S.: Automated detection of amnestic mild cognitive impairment in community-dwelling elderly adults: a combined spatial atrophy and white matter alteration approach. NeuroImage **59**, 1209–1217 (2012)
16. Cuingnet, R., Rosso, C., Chupin, M., Lehéricy, S., Dormont, D., Benali, H., Colliot, O.: Spatial regularization of SVM for the detection of diffusion alterations associated with stroke outcome. Med. Image Anal. **15**(5), 729–737 (2011)
17. Cuingnet, R., Glaunès, J.A., Chupin, M., Benali, H., Colliot, O.: Spatial and anatomical regularization of SVM: a general framework for neuroimaging data. IEEE Trans. Pattern Anal. Mach. Intell. **35**(3), 682–696 (2013)
18. Dai, Z., Yan, C., Wang, Z., Wang, J., Xia, M., Li, K., He, Y.: Discriminative analysis of early alzheimer's disease using multi-modal imaging and multi-level characterization with multi-classifier (m3). NeuroImage **59**, 2187–2195 (2012)
19. Davatzikos, C., Ruparel, K., Fan, Y., Shen, D., Acharyya, M.: Classifying spatial patterns of brain activity with machine learning methods: application to lie detection. NeuroImage **28**(3), 663–668 (2005)
20. Deshpande, G., Li, Z., Santhanam, P., Coles, C.D., Lynch, M.E., Hamann, S., Hu, X.: Recursive cluster elimination based support vector machine for disease state prediction using resting state functional and effective brain connectivity. PLOS One **5**(12), e14277 (2010)
21. Dosenbach, N.U.F., Nardos, B., Cohen, A.L., Fair, D.A., Power, J.D., Church, J.A., Nelson, S.M., Wig, G.S., Vogel, A.C., Lessov-Schlaggar, C.N., Barnes, K.A., Dubis, J.W., Feczko, E., Coalson, R.S., Pruett J.R., Jr., Barch, D.M., Petersen, S.E., Schlaggar, B.L.: Prediction of individual brain maturity using fMRI. Science **329**, 1358–1361 (2010)
22. Dryden, I.L., Koloydenko, A., Zhou, D.: Non-euclidean statistics for covariance matrices, with applications to diffusion tensor imaging. Ann. Appl. Stat. **3**(3), 1102–1123 (2009)
23. Fillard, P., Pennec, X., Arsigny, V., Ayache, N.: Clinical dt-mri estimation, smoothing, and fiber tracking with log-euclidean metrics. IEEE Trans. Med. Imaging **26**(11), 1472–1482 (2007)
24. Fletcher, P.T., Joshi, S.: Principal geodesic analysis on symmetric spaces: statistics of diffusion tensors. In: Computer Vision and Mathematical Methods in Medical and Biomedical Image Analysis, pp. 87–98. Springer, Berlin (2004)
25. Ford, J., Farid, H., Makedon, F., Flashman, L.A., McAllister, T.W., Megalooikonomou, V., Saykin, A.J.: Patient classification of fMRI activation maps. In: Medical Image Computing and Computer Assisted Intervention (MICCAI). Lecture Notes in Computer Science, vol. 2879, pp. 58–65. Springer, Berlin (2003)
26. Frackowiak, R.S., Friston, K.J., Frith, C.D., Dolan, R.J., Price, C.J., Zeki, S., Ashburner, J.T., Penny, W.D.: Human brain function. Academic, New York (2004)
27. Franke, K., Luders, E., May, A., Wilke, M., Gaser, C.: Brain maturation: predicting individual BrainAGE in children and adolescents using structural MRI. NeuroImage **63**, 1305–1312 (2012)
28. Friston, K.J., Holmes, A.P., Worsley, K.J., Poline, J.P., Frith, C.D., Frackowiak, R.S.: Statistical parametric maps in functional imaging: a general linear approach. Hum. Brain Mapp. **2**(4), 189–210 (1994)
29. Friston, K.J., Holmes, A.P., Poline, J., Grasby, P., Williams, S., Frackowiak, R.S., Turner, R.: Analysis of fmri time-series revisited. NeuroImage **2**(1), 45–53 (1995)
30. Genovese, C.R., Lazar, N.A., Nichols, T.: Thresholding of statistical maps in functional neuroimaging using the false discovery rate. NeuroImage **15**(4), 870–878 (2002)
31. Goodlett, C.B., Fletcher, P.T., Gilmore, J.H., Gerig, G.: Group analysis of dti fiber tract statistics with application to neurodevelopment. NeuroImage **45**(1), S133–S142 (2009)
32. Grosenick, L., Greer, S., Knutson, B.: Interpretable classifiers for fMRI improve prediction of purchases. IEEE Trans. Neural Syst. Rehabil. Eng. **16**(6), 539–548 (2008)
33. Haller, S., Nguyen, D., Rodriguez, C., Emch, J., Gold, G., Bartsch, A., Lovblad, K.O., Giannakopoulos, P.: Individual prediction of cognitive decline in mild cognitive impairment

using support vector machine-based analysis of diffusion tensor imaging data. J Alzheimers Dis. **22**(1), 315–327 (2010)
34. Haufe, S., Meinecke, F., Görgen, K., Dähne, S., Haynes, J.D., Blankertz, B., Bießmann, F.: On the interpretation of weight vectors of linear models in multivariate neuroimaging. NeuroImage **87**, 96–110 (2014)
35. Honorio, J., Tomasi, D., Goldstein, R.Z., Leung, H.C., Samaras, D.: Can a single brain region predict a disorder? IEEE Trans. Med. Imaging **31**(11), 2062–2072 (2012)
36. Jones, D.K., Symms, M.R., Cercignani, M., Howard, R.J.: The effect of filter size on VBM analyses of DT-MRI data. Neuroimage **26**(2), 546–554 (2005)
37. Keihaninejad, S., Zhang, H., Ryan, N.S., Malone, I.B., Modat, M., Cardoso, M.J., Cash, D.M., Fox, N.C., Ourselin, S.: An unbiased longitudinal analysis framework for tracking white matter changes using diffusion tensor imaging with application to alzheimer's disease. NeuroImage **72**, 153–163 (2013)
38. Kindlmann, G., Estepar, R.S.J., Niethammer, M., Haker, S., Westin, C.F.: Geodesic-loxodromes for diffusion tensor interpolation and difference measurement. In: Medical Image Computing and Computer-Assisted Intervention—MICCAI 2007, pp. 1–9. Springer, Heidelberg (2007)
39. Klöppel, S., Stonnington, C.M., Chu, C., Draganski, B., Scahill, R.I., Rohrer, J.D., Fox, N.C., Jack, C.R., Ashburner, J., Frackowiak, R.S.J.: Automatic classification of MR scans in alzheimer's disease. Brain **131**(3), 681–689 (2008)
40. LaConte, S., Strother, S., Cherkassky, V., Anderson, J., Hu, X.: Support vector machines for temporal classification of block design fMRI data. NeuroImage **26**(2), 317–329 (2005)
41. Lenglet, C., Rousson, M., Deriche, R., Faugeras, O.: Statistics on the manifold of multivariate normal distributions: theory and application to diffusion tensor mri processing. J. Math. Imaging Vision **25**(3), 423–444 (2006)
42. Lipton, M.L., Gellella, E., Lo, C., Gold, T., Ardekani, B.A., Shifteh, K., Bello, J.A., Branch, C.A.: Multifocal white matter ultrastructural abnormalities in mild traumatic brain injury with cognitive disability: a voxel-wise analysis of diffusion tensor imaging. J Neurotrauma **25**(11), 1335–1342 (2008)
43. Lipton, M.L., Kim, N., Park, Y.K., Hulkower, M.B., Gardin, T.M., Shifteh, K., Kim, M., Zimmerman, M.E., Lipton, R.B., Branch, C.A.: Robust detection of traumatic axonal injury in individual mild traumatic brain injury patients: intersubject variation, change over time and bidirectional changes in anisotropy. Brain Imaging Behav. **6**(2), 329–342 (2012)
44. Martínez-Ramón, M., Kltchinskii, V., Heileman, G.L., Posse, S.: fMRI pattern classification using neuroanatomically constrained boosting. NeuroImage **31**(3), 1129–1141 (2006)
45. Mitchell, T.M., Hutchinson, R., Niculescu, R.S., Pereira, F., Wang, X.: Learning to decode cognitive states from brain images. Mach. Learn. **57**, 145–175 (2004)
46. Nagy, Z., Alexander, D.C., Thomas, D.L., Weiskopf, N., Sereno, M.I.: Using high angular resolution diffusion imaging data to discriminate cortical regions. PLOS One **8**(5), e63842 (2013)
47. Nichols, T.E., Holmes, A.P.: Nonparametric permutation tests for functional neuroimaging: a primer with examples. Hum. Brain Mapp. **15**(1), 1–25 (2002)
48. O'Donnell, L., Westin, C., Golby, A.: Tract-based morphometry for white matter group analysis. NeuroImage **45**(3), 832–844 (2009)
49. O'Donnell, L.J., Golby, A.J., Westin, C.F.: Fiber clustering versus the parcellation-based connectome. NeuroImage **80**, 283–289 (2013)
50. O'Dwyer, L., Lamberton, F., Matura, S., Scheibe, M., Miller, J., Rujescu, D., Prvulovis, D., Hampel, H.: White matter differences between healthy young ApoE4 carriers and non-carriers identified with tractography and support vector machines. PLOS One **7**(4), e36024 (2012)
51. Pasternak, O., Sochen, N., Basser, P.J.: The effect of metric selection on the analysis of diffusion tensor mri data. NeuroImage **49**(3), 2190–2204 (2010)
52. Pereira, F., Mitchell, T., Botvinick, M.: Machine learning classifiers and fmri: a tutorial overview. NeuroImage **45**(1 Suppl.), S199–S209 (2009)

53. Rasmussen, P.M., Madsen, K.H., Lund, T.E., Hansen, L.K.: Visualization of nonlinear kernel models in neuroimaging by sensitivity maps. NeuroImage **55**, 1120–1131 (2011)
54. Robnik-Šikonja, M., Kononenko, I.: Theoretical and empirical analysis of relieff and rrelieff. Mach. Learn. **53**(1–2), 23–69 (2003)
55. Schlaffke, L., Lissek, S., Lenz, M., Juckel, G., Schultz, T., Tegenthoff, M., Schmidt-Wilcke, T., Brüne, M.: Shared and non-shared neural networks of cognitive and affective theory-of-mind: a neuroimaging study using cartoon picture stories. Hum. Brain Mapp. (2014). Early View. doi: 10.1002/hbm.22610
56. Schmidt-Wilcke, T., Cagnoli, P., Wang, P., Schultz, T., Lotz, A., Mccune, W.J., Sundgren, P.C.: Diminished white matter integrity in patients with systemic lupus erythematosus. NeuroImage Clin. (2014). DOI 10.1016/j.nicl.2014.07.001
57. Schölkopf, B., Smola, A.J.: Learning with Kernels. MIT Press, Massachusetts (2002)
58. Schultz, T., Fuster, A., Ghosh, A., Deriche, R., Florack, L., Lim, L.H.: Higher-order tensors in diffusion imaging. In: Westin, C.F., Vilanova, A., Burgeth, B. (eds.) Visualization and Processing of Tensors and Higher Order Descriptors for Multi-valued Data, pp. 129–161. Springer, Berlin (2014)
59. Schwartzman, A.: Random ellipsoids and false discovery rates: statistics for diffusion tensor imaging data. Ph.D. thesis, Stanford University (2006)
60. Schwartzman, A., Dougherty, R.F., Taylor, J.E.: Cross-subject comparison of principal diffusion direction maps. Magn. Reson. Med. **53**(6), 1423–1431 (2005)
61. Schwartzman, A., Dougherty, R.F., Taylor, J.E.: False discovery rate analysis of brain diffusion direction maps. Ann. Appl. Stat. **2**(1), 153–175 (2008)
62. Shenton, M.E., Kikinis, R., Jolesz, F.A., Pollak, S.D., LeMay, M., Wible, C.G., Hokama, H., Martin, J., Metcalf, D., Coleman, M., et al.: Abnormalities of the left temporal lobe and thought disorder in schizophrenia: a quantitative magnetic resonance imaging study. N. Engl. J. Med. **327**(9), 604–612 (1992)
63. Smith, S., Jenkinson, M., Johansen-Berg, H., Rueckert, D., Nichols, T., Mackay, C., Watkins, K., Ciccarelli, O., Cader, M., Matthews, P., et al.: Tract-based spatial statistics: voxelwise analysis of multi-subject diffusion data. NeuroImage **31**(4), 1487–1505 (2006)
64. Student: The probable error of a mean. Biometrika **6**(1), 1–25 (1908)
65. Van Essen, D.C., Smith, S.M., Barch, D.M., Behrens, T.E.J., Yacoub, E., Ugurbil, K.: The WU-Minn human connectome project: an overview. NeuroImage **80**, 62–79 (2013)
66. Vapnik, V.: The Nature of Statistical Learning Theory, 2nd edn. Information Science and Statistics. Springer, New York (1999)
67. Viswanath, V., Fletcher, E., Singh, B., Smith, N., Paul, D., Peng, J., Chen, J., Carmichael, O.: Impact of dti smoothing on the study of brain aging. In: 2012 Annual International Conference of the IEEE Engineering in Medicine and Biology Society (EMBC), pp. 94–97. IEEE, New York (2012). doi: 10.1109/EMBC.2012.6345879
68. Vul, E., Harris, C., Winkielman, P., Pashler, H.: Puzzlingly high correlations in fMRI studies of emotion, personality, and social cognition. Perspect. Psychol. Sci. **4**(3), 274–290 (2009)
69. Wakana, S., Jiang, H., Nagae-Poetscher, L.M., Van Zijl, P.C., Mori, S.: Fiber tract–based atlas of human white matter anatomy 1. Radiology **230**(1), 77–87 (2004)
70. Wang, X., Hutchinson, R., Mitchell, T.M.: Training fMRI classifiers to detect cognitive states across multiple human subjects. In: Thrun, S., Saul, L.K., Schölkopf, B. (eds.) Proceedings of Neural Information Processing Systems, pp. 709–716 (2003)
71. Wee, C.Y., Yap, P.T., Li, W., Denny, K., Browndyke, J.N., Potter, G.G., Welsh-Bohmer, K.A., Wang, L., Shen, D.: Enriched white matter connectivity networks for accurate identification of MCI patients. NeuroImage **54**, 1812–1822 (2011)
72. Wee, C.Y., Yap, P.T., Zhang, D., Denny, K., Browndyke, J.N., Potter, G.G., Welsh-Bohmer, K.A., Wang, L., Shen, D.: Identification of MCI individuals using structural and functional connectivity networks. NeuroImage **59**, 2045–2056 (2012)
73. Whitcher, B., Wisco, J.J., Hadjikhani, N., Tuch, D.S.: Statistical group comparison of diffusion tensors via multivariate hypothesis testing. Magn. Reson. Med. **57**(6), 1065–1074 (2007)

74. Yushkevich, P.A., Zhang, H., Simon, T.J., Gee, J.C.: Structure-specific statistical mapping of white matter tracts. NeuroImage **41**(2), 448–461 (2008)
75. Zhu, H., Styner, M., Tang, N., Liu, Z., Lin, W., Gilmore, J.H.: Frats: functional regression analysis of dti tract statistics. IEEE Trans. Med. Imaging **29**(4), 1039–1049 (2010)
76. Ziliak, S.T., McCloskey, D.N.: The Cult of Statistical Significance: How the Standard Error Costs us Jobs, Justice, and Lives. University of Michigan Press, Ann Arbor (2008)

Part V
Applications

A Clustering Method for Identifying Regions of Interest in Turbulent Combustion Tensor Fields

Adrian Maries, Timothy Luciani, P.H. Pisciuneri, Mehdi B. Nik, S. Levent Yilmaz, Peyman Givi, and G. Elisabeta Marai

Abstract Production of electricity and propulsion systems involve turbulent combustion. Computational modeling of turbulent combustion can improve the efficiency of these processes. However, large tensor datasets are the result of such simulations; these datasets are difficult to visualize and analyze. In this work we present an unsupervised statistical approach for the segmentation, visualization and potentially the tracking of regions of interest in large tensor data. The approach employs a machine learning clustering algorithm to locate and identify areas of interest based on specified parameters such as strain tensor value. Evaluation on two combustion datasets shows this approach can assist in the visual analysis of the combustion tensor field.

A. Maries
Learning Research and Development Center, University of Pittsburgh, Pittsburgh, PA 15260, USA
e-mail: adm77@pitt.edu

T. Luciani
Department of Computer Science, Brown University, Providence, RI 02912, USA
e-mail: tluciani21@gmail.com

P.H. Pisciuneri
Center for Simulation and Modeling, University of Pittsburgh, Pittsburgh, PA 15260, USA
e-mail: php8@pitt.edu

M.B. Nik • P. Givi
Department of Mechanical Engineering and Materials Science, University of Pittsburgh, Pittsburgh, PA 15260, USA
e-mail: smb51@pitt.edu; pgivi@pitt.edu

S.L. Yilmaz
Mathworks, Natick, MA 01760, USA
e-mail: levent.yilmaz@gmail.com

G.E. Marai (✉)
Department of Computer Science, University of Illinois at Chicago, Chicago, IL 60647, USA
e-mail: g.elisabeta.marai@gmail.com

I. Hotz, T. Schultz (eds.), *Visualization and Processing of Higher Order Descriptors for Multi-Valued Data*, Mathematics and Visualization,
DOI 10.1007/978-3-319-15090-1_16

1 Introduction

U.S. energy consumption is dominated by the burning of fossil fuels such as coal, natural gas and petroleum [1]. Production of electricity and propulsion systems are two primary reasons for this energy use. Both processes, at their core, involve turbulent combustion. Turbulent combustion modeling is an important area of research driven by the effort to improve the efficiency of these processes, reduce fuel consumption and reduce pollution.

In order of decreasing fidelity and costs, three computational approaches to turbulence combustion are direct numerical simulation (DNS), large eddy simulation (LES) and Reynolds-averaged Navier-Stokes (RANS). Specific tensor quantities such as stress, strain, and turbulent stress play in these computational approaches, as part of the computational modeling process; these quantities are discussed in detail in our previous work [9]. DNS requires directly capturing the wide range of length- and time-scales. This severely limits the approach to simulations of relatively simple, canonical configurations. The complex nature of turbulent reacting flows can be attributed to the non-linear convection terms and scalar transport terms appearing in the coupled set of governing equations. In DNS these terms are accounted for without modeling. However, as discussed in more detail in our previous work [9], in both LES and RANS these terms create a closure problem and require modeling. As such, an important aspect of turbulence modeling and model validation involves comparison of the subgrid scale stress tensor (for LES) or the Reynolds stress tensor (for RANS) with available DNS data.

A three-dimensional LES features millions of grid points. A DNS of the same configuration would feature several orders of magnitude more grid points. The result is that a given snapshot of the flow variables at a particular instance in time for the entire domain would range on the order of gigabytes in the LES case to terabytes for the DNS case. An entire simulation is composed of tens of thousands to hundreds of thousands of time steps. Thus data is typically only retained at specified intervals. Even in adopting this approach total datasets become cumbersome to work with. Moreover, as LES models are validated, they are in turn used for much larger flow geometries of industrial applications, resulting in snapshots that are tens or hundreds of gigabytes in size.

Producing data at this scale implies a few prerequisite conditions: that a highly scalable flow solver will be used for the simulation, and that the researcher has access to a large supercomputing environment. This introduces new complexities to the workflow. First, file I/O must be handled in parallel and is typically a costly operation relative to the time required to calculate a given time step of the simulation. Thus regularly outputting entire snapshots of simulation data will adversely impact the progress of the calculation. Second, this data is produced at a remote location. Ideally the researcher would transfer the data to a local machine for data analysis and interactive visualization. For large datasets transferring files can take a significant amount of time, adding a measurable bottleneck to the workflow.

To explore the physical phenomena from a volumetric dataset, the ability of a visualization tool to compute and track salient features is crucial. The large amount of data may severely affect the availability of the data for visualization (i.e., the simulation may not be paused to output the data for visualization), the data transfer (bandwidth limitations), and the manipulation speed. These are major challenges to the interactive visualization of tensor data. Furthermore, tensor datasets tend to be very dense, leading to clutter and occlusion problems when visualizing such datasets with existing tools.

In situ visualization aims to address some of these issues. Such approaches enable the user to connect directly to a running simulation, examine the data, do numerical queries and create graphical output while the simulation executes, bypassing the need to write data to disk. The visualization and computation can be tightly coupled (memory sharing), loosely-coupled (communication over a network), or hybrid (the data is computationally reduced and then sent out for visualization).

The feature extraction approach is an emerging in situ hybrid method. This approach extracts the meaningful and interesting regions from the datasets, showing only those parts to the researchers. Typically, only a small percentage of datasets are of interest, and the feature can be described very compactly. These abstractions lead to a sharp reduction in the amount of data processed, making an effective visualization of very large datasets possible. Another advantage of feature extraction is that it helps the users highlight and focus on regions of particular characteristics that they are interested in.

As opposed to filtering approaches, which may require expert knowledge about the structure of the flow, in this work we present an unsupervised statistical approach for the segmentation, visualization and potentially the tracking of regions of interest in large tensor data. The approach employs a machine learning clustering algorithm to locate and identify areas of interest based on specified parameters such as strain tensor value.

2 Related Work

A great deal of research has been conducted in the problems of feature extraction and tracking. While initially developed in the field of computer vision [11, 19], feature extraction and tracking have also been adopted for flow visualization—see Post et al. [15] for an extensive review. In this section we focus on the existing related feature-tracking work in flow visualization.

Most feature extraction and tracking techniques fall into one of three basic categories. The most widespread method is to extract features in each time step separately and then to track them through time. The first to employ feature extraction and tracking in flow visualization, Samtaney et al. [16] use feature attributes such as mass, centroid, volume, or moment of inertia to establish a correspondence of features across timesteps. Silver and Wang [17, 18] developed a volume tracking schema that requires that there is a certain amount of overlap between features in adjacent timesteps for them to be associated. Caban et al. [2] introduce a texture-based feature tracking technique that compares textural characteristics across timesteps to find the best match. A second approach to feature extraction and tracking is exemplified by the work of Muelder and Ma [12]. Instead of extracting features in each timestep and then establishing a correspondence, they use a prediction-correction method that makes a prediction about the location and size of the region in the next step. The region is then adjusted by growing or shrinking the border in order to extract the feature of interest. Ji et al. [5] developed a third approach to feature extraction and tracking, which uses isosurfacing in higher dimensions. Once again, as opposed to extracting isosurfaces in 3D for every timestep, their method tracks features by performing an isosurfacing process in 4D.

Other approaches use various machine learning techniques to aid in feature tracking. Tzeng and Ma [20] utilize neural networks to learn which transfer functions are most appropriate in tracking the features of interest. Noticing that tracking groups of features that exhibit similar behavior is more cost-effective than tracking the features individually, Ozer et al. [13] use a clustering algorithm to group features based on similarity measures. Our approach is similar to that of Ozer et al. in that we also utilize clustering analysis. The difference is that, rather than use it to group features, we use it to define regions of interest.

The visualization community has long been very concerned with the shock location problem. A number of techniques and algorithms for characterizing the regions of interest, detecting and visualizing shocks waves have been developed. Lovely and Haimesy [7] designed an algorithm for extracting the shock surface that uses the fact that the shock surface normal is typically aligned with the pressure gradient vector. Thus, the algorithm computes the Mach number in the direction of the pressure gradient and builds the shock surface from the points where the Mach number equals one. Another widely-used algorithm utilizes the density gradient and consists of three steps [14]. It first computes the first and second derivatives of the density in the direction of the velocity. It then builds an isosurface where the second derivative equals zero and, finally, it picks the first derivative maxima, which correspond to the shock, and discards the minima. Ma et al. [8] make the distinction between shock waves and expansion waves in the third step of the previous algorithm by using the normal Mach number rather than the first derivative of the density. Specifically, it picks regions where the Mach number is close to one. The method we present herein is novel in that it integrates machine learning with visualization for extracting and clustering regions of interest. It is thus a promising approach to apply to very large flow datasets.

3 Methods

Given the size and interaction challenges of combustion datasets, automated methods are particularly relevant to the problem of identifying regions of interest. Such an approach would allow pushing the feature extraction process in situ, to the same computational side that also processes the combustion simulation. Only the regions extracted would then be sent out to visualization, thus reducing both I/O and bandwidth usage.

Automated methods for feature identification can be provided through Machine Learning (ML), a branch of statistics and computer science which studies algorithms and architectures that learn from observed facts. Unsupervised ML algorithms are of particular interest for combustion tensor data: in an unsupervised setting, the objective is to cluster or discover structures in the data. Example algorithms and representations for unsupervised learning include K-means clustering, mixture models, hierarchical clustering, and PCA (Principal Component Analysis).

Clustering analysis is used to group data points that are similar to one another. There are various reasons for using clustering: one may wish to analyze points in the dataset that are close to one another, to reduce a high-dimensional dataset by replacing groups of dimensions with single labels, or to reduce the size of the dataset by replacing groups of data points with single labels.

From the class of clustering methods, we focus on K-means, a powerful yet computationally-effective approach. As in most Big Data applications, the ability to trade semantic meaning for performance is important in this context: more sophisticated methods like mixture models or hierarchical clustering are also significantly slower than K-means.

3.1 K-Means Clustering

K-means clustering attempts to partition a set of N observations (the number of grid points in a simulation) into K clusters; in the resulting partition each observation belongs to the cluster with the nearest mean observation. The mean is referred to as the centroid of the cluster. The cluster centroid can later be used to describe all cluster members, thus attaining data reduction.

In our case, an observation is the computed value of a tensor at a given location. A tensor is an extension of the concept of a scalar and a vector to higher orders. Scalars and vectors are 0-th and 1-order tensors, respectively. In general, a k-th order tensor can be represented by a k-dimensional array, e.g. a second order tensor is a 2D array (a matrix). For example, while a stress *vector* is the force acting on a given unit surface, a stress *tensor* is defined as the components of stress vectors acting on each coordinate surface; thus stress can be described by a symmetric 2-nd order tensor.

The velocity stress and strain tensor fields are manifested in the transport of fluid momentum, which is a vector quantity governed by the following conservation equation:

$$\frac{\partial \rho u_i}{\partial t} + \frac{\partial \rho u_i u_j}{\partial x_j} = -\frac{\partial p}{\partial x_i} + \frac{\partial \tau_{ij}}{\partial x_j}, \qquad \text{for } i = 1, 2, 3 \tag{1}$$

where the Cartesian index notation is employed in which the index $i = 1, 2, 3$ represents spatial directions along the x, y, and z Cartesian coordinates, respectively; and the repeated index j implies summation over the coordinates. t is time, ρ is the fluid density, $\mathbf{u} \equiv [u_1, u_2, u_3]$ is the Eulerian fluid velocity, p is the pressure, and $\boldsymbol{\tau}$ is the stress tensor defined as:

$$\tau_{ij} = 2\mu \left(S_{ij} - \frac{1}{3} \delta_{ij} \frac{\partial u_k}{\partial x_k} \right) \tag{2}$$

where μ is the dynamic viscosity coefficient (a fluid-dependent parameter) and S is the velocity strain tensor defined as:

$$S_{ij} = \frac{1}{2} \left(\frac{\partial u_i}{\partial x_j} + \frac{\partial u_j}{\partial x_i} \right) \tag{3}$$

We perform clustering on the six distinct values of the strain tensor in the combustion data, arranged in a six-dimensional vector $x^{(i)}$, where i ranges over the points in the dataset. Tensors used in turbulence modeling are rank 2 tensors, which for our purposes are 3×3 matrices. Additionally, strain tensors are symmetric, $s_{ij} = s_{ji}$. This means that there are a total of six distinct tensor values for each point in the grid. We compute mean values and distances using a simple, squared Euclidean distance metric. We note that in our previous work [9] the alternative approach of working in a dimensionally-reduced space—such as the space of eigenvalues and eigenvectors—revealed that in turbulent combustion modeling these reduced descriptors are small, fairly uniform and non-distinctive throughout the volume, and thus of limited value for cluster analysis. Similar prior experiments [9] have shown that reduced descriptors such as trace and determinant can act as valuable flow filters; proposing and using such descriptors requires, however, expert knowledge about the nature of a particular flow configuration.

The K-means method follows two alternative steps, one initialization step, and one assignment step. In the first step, the cluster means are initialized (for example, with K random observations $x^{(i)}$ from the set). In the second step, each of the N points is assigned to the cluster whose mean is most similar to the point. The cluster means are repeatedly recomputed based on the points assigned to each cluster, and the N points are reassigned, until convergence:

```
Tensor K-Means
```

- `Randomly initialize K cluster centroids`
  $\mu_1, \mu_2, ..\mu_K \in R^6$

- `Repeat {`

```
//cluster assignment step
For i = 1 to N
```
$c^{(i)} :=$ *index (from* 1 *to K) of cluster centroid closest to* $x^{(i)}$

```
//centroid move step
For k = 1 to K
```

$\mu_{(k)} :=$ *average (mean) of points assigned to cluster k*

```
}
```

- $J(c^{(1)}, c^{(2)}, ..c^{(N)}, \mu_1, \mu_2, ..\mu_k) = 1/N \sum ||x^{(i)} - \mu_{c(i)}||^2$

Convergence is assessed using the clustering error function J, given by the mean distance of all points to their assigned cluster centroid. The problem is NP-hard and thus computationally challenging; but the iterative approach described above can converge to local optima.

Unfortunately, ML clustering algorithms do not scale well. In our experiments, we found that datasets larger than 450,000 points cannot be clustered using K-means, and datasets larger than 250,000 points cannot be clustered using greedy agglomerative clustering in less than 24 h (Intel duo CPU at 2.26 GHz and 4 GB RAM). To reduce such run-times, it is necessary to preprocess the dataset.

A common approach for preprocessing large datasets is to use canopy clustering [10] as a pre-clustering algorithm. This pre-clustering is followed by a clustering algorithm such as K-means, hierarchical clustering or expectation maximization. The preprocessing step produces initial estimates for the dataset clusters, which are then used to speedup the clustering step. However, after a series of clustering experiments, we concluded that canopy clustering was unable to perform clustering in a reasonable amount of time. The problem was that in these tensor datasets the number of clusters of interest is typically below ten. Given this restriction and the fact that the majority of the data points in a cluster have to be in the same canopy, the size of the canopies would have had to be no smaller than 800,000 (8M/10), even for the smaller datasets. Clustering such large collections of data points is, however, unfeasible using K-means: in our further experiments we found that even 450K datasets require more than 24 h runtime to converge (Quad core Intel 5, 3.3 GHz, 16 GB RAM).

To circumvent this obstacle, we used instead a pre-clustering step in which K-means clustering was run on a sub-sampled dataset to obtain good starting cluster centers. The first dataset was sampled every $4 \times 4 \times 4$ data points and the second every $4 \times 6 \times 4$ data points regularly throughout the grid. The sampling rate was empirically selected (lower rate along larger dimensions) so that the pre-clustering step could complete in minutes. The resulting starter centroids were then used in K-means over the full datasets. Using this pre-computed cluster centroid setup, the second clustering step converges in under 50 iterations for the mixing layer dataset (8M points) and 20 iterations for the shock dataset (21M points); both datasets are described in detail in the results section. The entire approach takes on average 15–20 min to compute four clusters (8M point dataset).

3.2 Cluster Analysis

As is the standard procedure in K-means, we repeat the clustering procedure for a varying number K of clusters, from 2 to 6, and select the K value that leads to the lowest clustering error J. In our experiments, the clustering run times for different K values are fairly similar, with deviation of at most 1 h.

To more easily analyze the features of cluster centroids, we use a star-plot representation of each centroid. The star-plot is a high-dimensional visualization technique based on the parallel coordinate plot (PCP). In the PCP descriptor, dimensions are represented by parallel axes and data points are mapped to the axes; the data points are then connected by lines [4, 21]. The star-plot is a more compact representation of the PCP, in which axes are radii of a circle [3, 6].

Figure 1 shows star-plot descriptors for the centroids for $K = 5$, $K = 6$, $K = 7$, and $K = 8$; in each star-plot glyph the six centroid values are mapped to radial spokes, allowing for easy comparison of the centroid traits. Note how increasing K from 5 to 6 adds a distinctive cluster, while further increasing K to 7 and 8 introduces clusters which are fairly similar to clusters already identified. This type of analysis can be used to automate the selection of K.

To ensure consistent clustering along the time dimension for the shock dataset, the tensor centroids of the clusters can be tracked over time based on their similarity, with visual assistance where necessary to account for splitting, recombining, vanishing, and appearance phenomena.

4 Results

We evaluate this approach on two large datasets, a mixing layer dataset, and a shock dataset. Mixing layer configurations are common in combustion simulations, where two fluids flow over and against each other. Shock waves are important features in compressible flow datasets that are characterized by abrupt, nearly discontinuous changes in physical flow quantities such as density, pressure and velocity. Shock waves are of interest to researchers since they can increase drag and cause structure-failure in design problems in fluid dynamics.

4.1 Mixing-Layer Dataset

The first dataset is a temporal mixing layer and is a simple configuration where two streams of fuel and oxidizer flow over and against each other. The flow speeds are adjusted for a low Reynolds number yielding a narrow range of length scales, and this configuration can be easily tackled with DNS and then used as a benchmark. The data for the temporal mixing layer is at a snapshot in time and at the full DNS

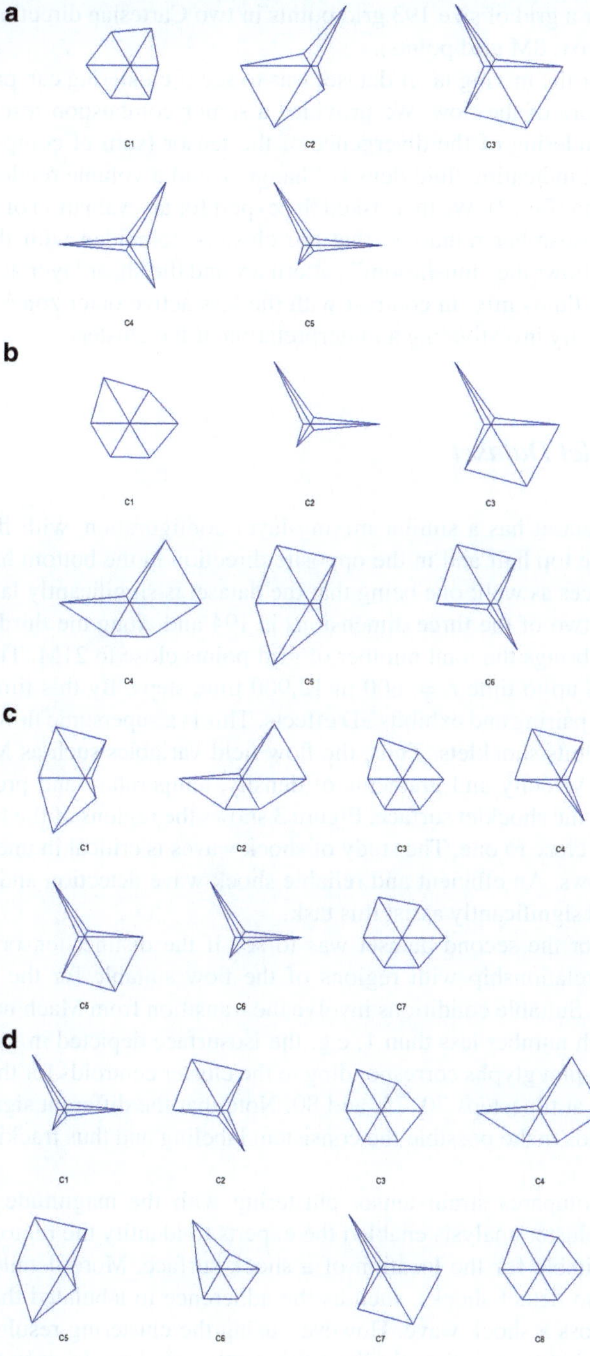

Fig. 1 Star-plot descriptors for the centroids for $K = 5$ (**a**), $K = 6$ (**b**), $K = 7$ (**c**), and $K = 8$ (**d**). Increasing K from 5 to 6 adds a distinctive cluster, while further increasing K to 7 and 8 introduces clusters which are fairly similar to clusters already identified

resolution over a grid of size 193 grid points in two Cartesian directions and 194 in the other (approx. 8M grid points).

The goal for the mixing layer dataset was to see if clustering can provide insight into the structure of the flow. We provided a senior combustion researcher with a 3D volume rendering of the divergence of the tensor (sum of components on the main diagonal, indicating fluid density changes), and a volume rendering of the 4-group clustering (Fig. 2). We then asked the expert for an evaluation of the clustering results. The researcher remarked that the clusters coincided with the interesting regions of the flow: the "mushroom" pattern around the shear layer at the mid-zone where the two fluids mix, in contrast with the less active outer zones. The domain expert is currently investigating an interpretation of the clusters.

4.2 *Shocklet Dataset*

The second dataset has a similar mixing-layer configuration, with flow from one direction in the top half and in the opposite direction in the bottom half. There are a few differences as well; one being that the dataset is significantly larger. The size of the grid in two of the three dimensions is 194 and along the third dimension it is 577, which brings the total number of grid points close to 21M. This simulation has been done up to time $t = 600$ in 12,900 time steps. By this time, the flow is going through pairing and exhibits 3D effects. This is a supersonic flow, in which the flow field exhibits shocklets. Thus, the flow field variables such as Mach number, divergence of velocity and gradients of density, temperature and pressure change sharply across the shocklet surface. Figure 3 shows the regions of the flow field with Mach number close to one. The study of shock waves is critical in understanding of high-speed flows. An efficient and reliable shock wave detection and visualization method would significantly assist this task.

The goal for the second dataset was to see if the distinct tensor field regions have a clear relationship with regions of the flow suitable for the location of a shock surface. Suitable conditions involve the transition from Mach number greater than 1 to Mach number less than 1, e.g. the isosurface depicted in Fig. 3. Figure 4 shows the star-plot glyphs corresponding to the cluster centroids for the four-cluster segmentation, at timesteps 70, 75, and 80. Note that the different signatures of the cluster centroids make possible the consistent labeling and thus tracking of clusters over time.

Figure 5 compares strain tensor clustering with the magnitude of the Mach number. The cluster analysis enabled the experts to identify the regions of the flow potentially suitable for the location of a shock surface. More detailed analysis is still required to detect shocks, such as the adherence to tabulated thermodynamic properties across a shock wave. However, using the clustering results, the domain experts were able to significantly limit this further analysis to only the regions of interest identified through the clustering.

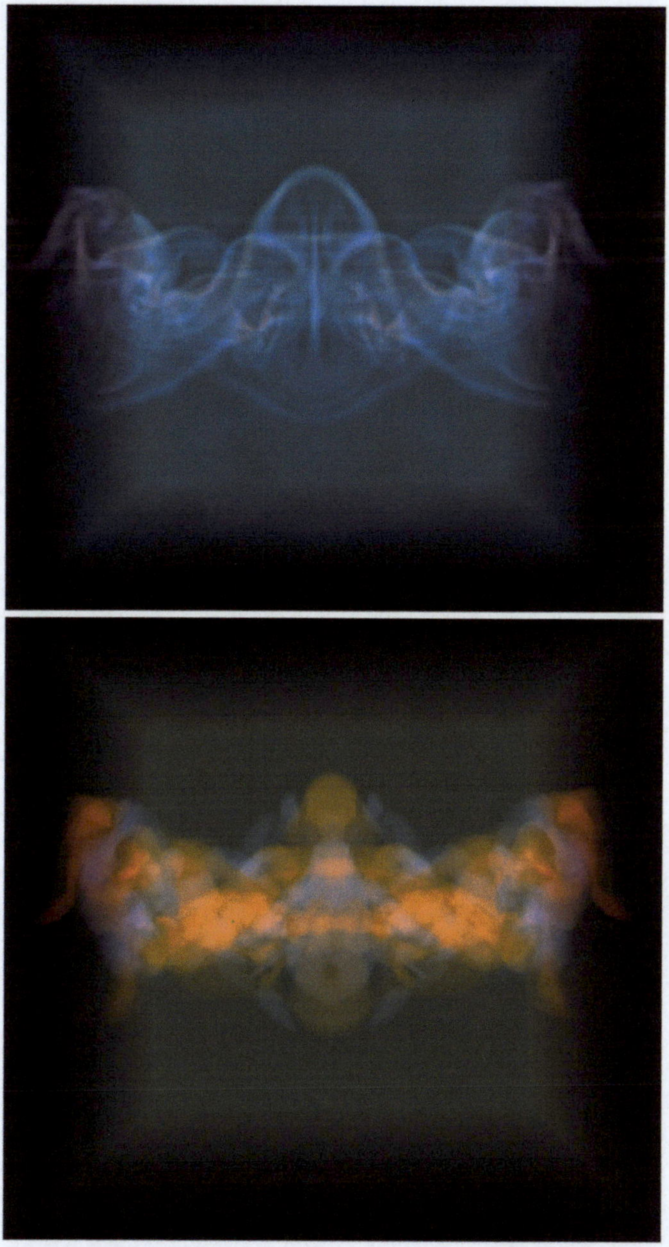

Fig. 2 Mixing-Layer Dataset: volume rendering of divergence (*top*), which can be calculated via the trace of the strain tensor, and 4-group strain tensor clustering (*bottom*), rendered by assigning each cluster an individual value and setting the transfer function monochromatically to each. The four clusters are color-encoded using a www.colorbrewer2.org qualitative scheme as shades of *orange*, *purple*, *green* and *black*. The clusters correlate well with the mixing region of interest: the "mushroom" pattern around the shear layer at the mid-zone where the two fluids mix (*orange* and *purple* clusters), in contrast with the less active outer zones (*green* and *black* clusters)

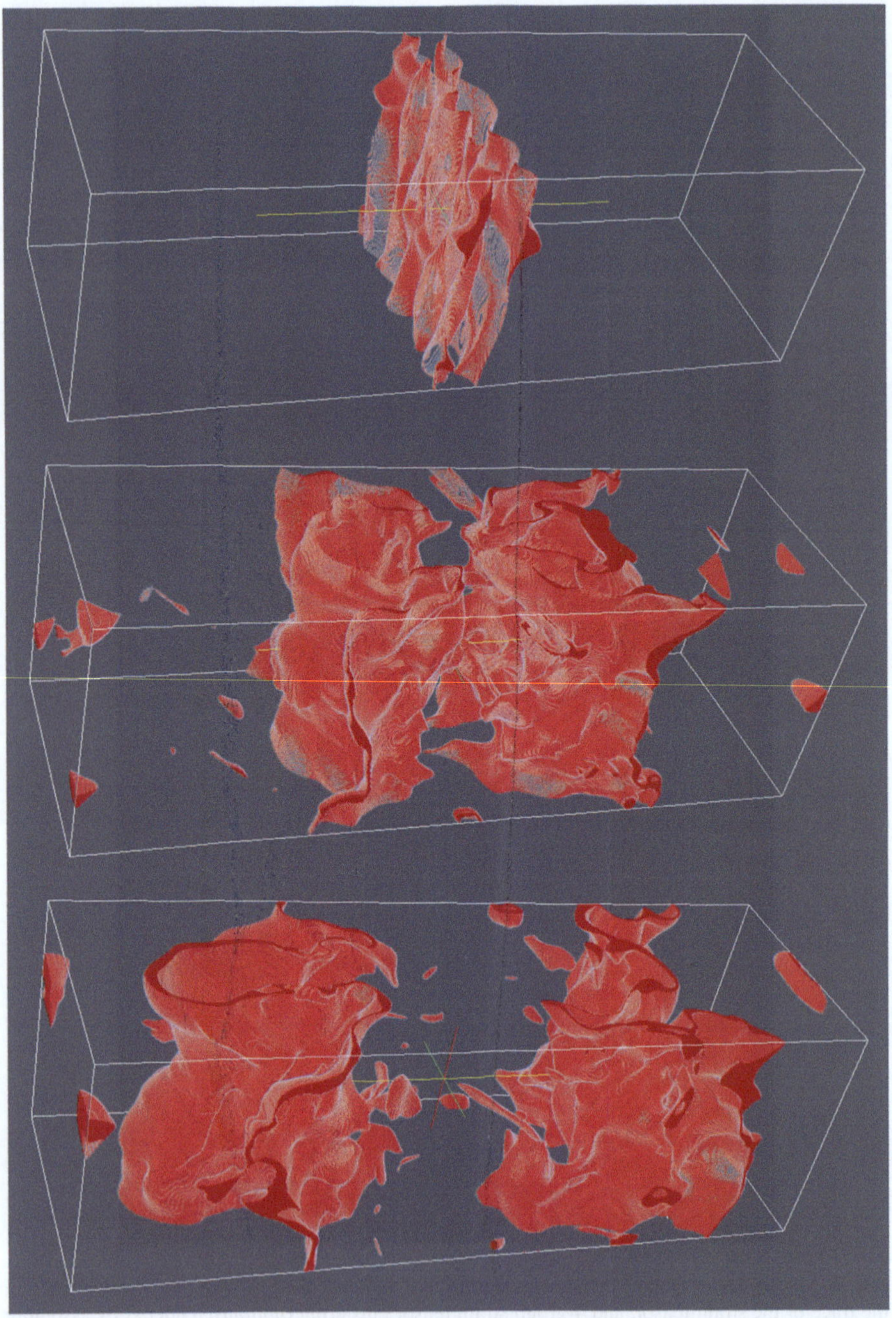

Fig. 3 Volume rendering of the region with $Ma \approx 1$. In this rendering, all points with Mach number within the 0.05 threshold of one have been replaced with 1 and the rest with 0; the resulting two-valued volume is volume-rendered in ParaView. From *top* to *bottom*; timesteps 70, 325, and 600

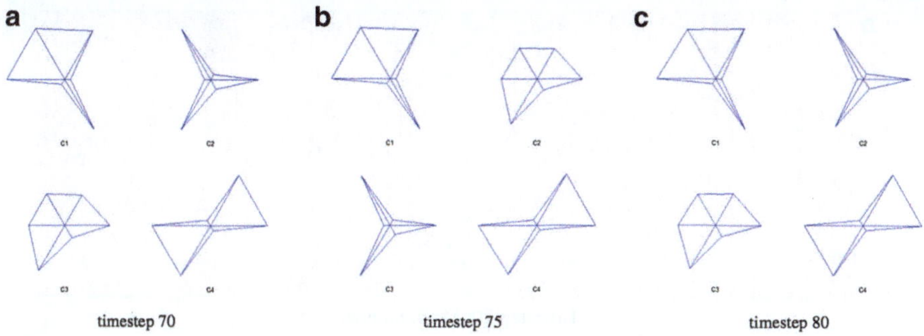

Fig. 4 Star-plot glyphs corresponding to the cluster centroids for a four-cluster segmentation of the shock datasets. From *left* to *right*: cluster centroids at timesteps (**a**) 70, (**b**) 75, and (**c**) 80. Note that the different signatures of the cluster centroids make possible the consistent labeling and thus tracking of clusters over time

5 Discussion and Conclusion

As previously discussed, the goal of this project was to examine the potential of using cluster analysis on tensor field data generated by turbulent combustion simulations. First, we were interested in finding out whether an unsupervised approach can detect structures in the data, and whether these structures correlate with the regions of interest. Second, we wanted to see whether tensor field clustering and rendering could give researchers insights into the structure of the flow through a volume. The answer to both questions is affirmative.

In summary, we found that a machine learning inspired approach, though computationally intensive, can extract and track regions of interest in large, dense tensor fields. Our K-means approach yielded interesting results: the clusters correlate well with the regions of interest. Thus, as an in situ technique, the approach has potential for compression. While the clustering itself may miss potential artifacts, the clustering approach is also a foundation for automated anomaly detection, and thus a base for further in situ benefits.

We found that the performance of machine learning algorithms is a major issue, and note that such algorithms need to be first adapted for large scale data. In our approach we adapted K-means for large data by using a pre-clustering step; this pre-clustering step was performed via sub-sampling. In an in situ setting, the pre-clustering could be coupled with the more intelligent data partitioning that is mandatory when distributing the computational simulation load across multiple processors. The run-time cost of the unsupervised machine learning approach should also decrease when distributed across multiple processors.

In our implementation and preliminary experiments, the clustering technique was run offline, not in an authentic in situ setting. We note however, that the approach was designed as an in situ technique, and could be deployed as such in a computational setting, with the added benefit of data partitioning for pre-clustering.

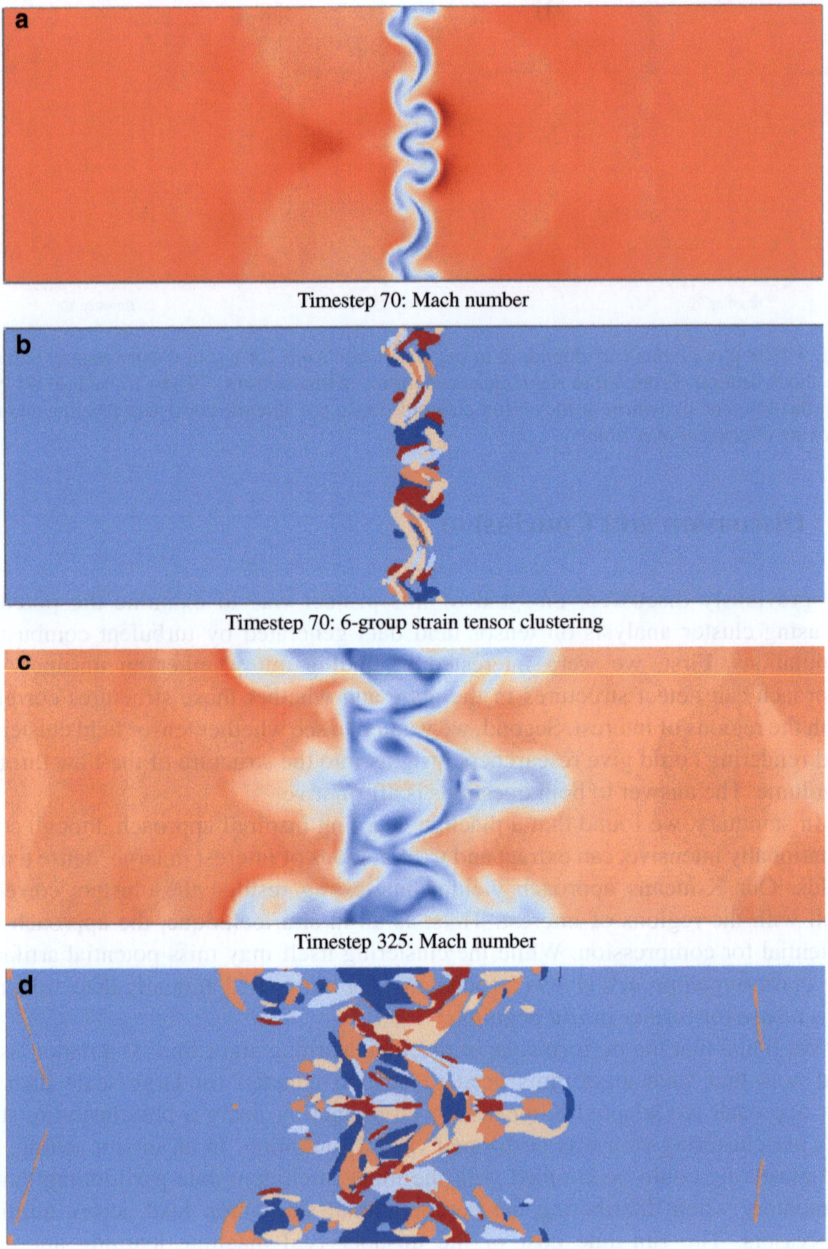

Fig. 5 Two snapshots from the shocklet dataset, same YZ slice: renderings of Mach number (**a** and **c**), and the corresponding 6-group strain tensor K-means results (**b** and **d**). The six clusters are encoded with three shades of *blue* plus three shades of *orange-red*. The clustering captures regions of interest for Mach number in both cases. The cluster analysis enables domain experts to identify the regions of the flow potentially suitable for the location of a shock surface. Cluster centroid similarity can be further used to ensure consistent cluster labeling across multiple timesteps

The only current "manual" component of the approach is the selection of K, the numbers of clusters. The K-selection step can be automated, however, as indicated in the Fig. 1 shape analysis.

In our approach, we have used a simple feature vector representation for the tensor field. Similarly, we have used a simple Euclidean distance metric to compute distances and means over the tensor field. We obtained good correlation between the cluster-based regions of interest and flow features. Nevertheless, defining more meaningful feature vectors and distance metrics for tensor similarity that have improved semantic meaning are important directions of future research.

Visualization of the tensor fields associated with turbulent combustion simulations is particularly challenging. Difficulties arise from the sheer scale and density of the data, but also from the small range of values these tensors take. Our previous attempts at visualizing these types of fields using glyphs or streamlets have had partial success—these tensors have very small, very similar eigenvalues [9]. Furthermore, many existing tensor representations do not have an intuitive equivalent in combustion turbulent flow; to combustion researchers, tensors do not have direction or shape. As the domain experts put it, "the tensor itself is very useful for computation, and pretty complete... but its individual components are not so useful to understand what is going on." This observation makes feature extraction and tracking through an unsupervised clustering approach particularly useful.

In conclusion, we have introduced an approach for the segmentation, visualization and potential tracking of regions of interest in large scale tensor field datasets generated by computational turbulent combustion simulations. The approach is novel in that it integrates machine learning with visualization—interactive volume rendering and starplots—to extract, cluster, and track regions of interest in the tensor field. Our evaluation on two rich combustion datasets shows this approach can assist in the visual analysis of the combustion tensor field.

Acknowledgements This work was supported by NSF CBET-1250171 and NSF CAREER IIS-0952720.

References

1. Anderson, J.D.J.: Modern Compressible Flow: With Historical Perspective, 3rd edn. McGraw-Hill Science/Engineering/Math (2002)
2. Caban, J.J., Joshi, A., Rheingans, P.: Texture-based feature tracking for effective time-varying data visualization. IEEE Trans. Vis. Comput. Graph. **13**(6), 1472–1479 (2007)
3. Elmqvist, N., Stasko, J., Tsigas, P.: Datameadow: a visual canvas for analysis of large-scale multivariate data. In: VAST IEEE Symposium on Visual Analytics Science and Technology, Proceedings, pp. 187–194 (2007)
4. Inselberg, A., Dimsdale, B.: Parallel coordinates: a tool for visualizing multi-dimensional geometry. In: Proceedings of the 1st Conference on Visualization '90 (VIS '90), San Francisco, pp. 361–378. IEEE Computer Society Press, Los Alamitos (1990). http://dl.acm.org/citation.cfm?id=949531.949588

5. Ji, G., Shen, H.-W., Wenger, R.: Volume tracking using higher dimensional isosurfacing. In: Proceedings of the 14th IEEE Visualization, pp. 209–216 (2003)
6. Klippel, A., Hardisty, F., Li, R., Weaver, C.: Colour-enhanced star plot glyphs: can salient shape characteristics be overcome? Cartogr.: Int. J. Geogr. Inf. Geovis. **44**(3), 217–231 (2009)
7. Lovely, D., Haimesy, R.: Shock detection from computational fluid dynamics results. In: Proceedings of the 14th AIAA Computational Fluid Dynamics Conference, 1:M2 (1999)
8. Ma, K.-L., Rosendale, J.V., Vermeer, W.: 3d shock wave visualization on unstructured grids. In: IEEE Symposium on Volume Visualization and Graphics, pp. 87–104 (1996)
9. Maries, A., Haque, M., Yilmaz, S., Nik, M., Marai, G.: Interactive exploration of stress tensors used in computational turbulent combustion. In: Laidlaw, D., Villanova, A. (eds.) New Developments in the Visualization and Processing of Tensor Fields, pp. 137–156. Springer, Heidelberg (2012)
10. McCallum, A., Nigam, K., Ungar, L.H.: Efficient clustering of high-dimensional data sets with application to reference matching. In: Proceedings of the Sixth ACM SIGKDD International Conference on Knowledge Discovery and Data Mining, KDD '00, pp. 169–178. ACM Press, New York (2000)
11. Meyer, F., Bouthemy, P.: Region-based tracking using affine motion models in long image sequences. CVGIP: Image Underst. **60**(2), 119–140 (1994)
12. Muelder, C., Ma, K.-L.: Interactive feature extraction and tracking by utilizing region coherency. In: IEEE Pacific Visualization Symposium, PacificVis '09, pp. 17–24 (2009)
13. Ozer, S., Wei, J., Silver, D., Ma, K.-L., Martin, P.: Group dynamics in scientific visualization. In: IEEE Symposium on Large Data Analysis and Visualization (LDAV), pp. 97–104 (2012)
14. Pagendarm, H.-G., Seitz, B.: An algorithm for detection and visualization of discontinuities in scientific data fields applied to flow data with shock waves. In: Scientific Visualization: Advanced Software Techniques, pp. 161–177 (1993)
15. Post, F.H., Vrolijk, B., Hauser, H., Larameeand, R.S., Doleisch, H.: The state of the art in flow visualisation: feature extraction and tracking. Comput. Graphics Forum **22**(4), 775–792 (2003)
16. Samtaney, R., Silver, D., Zabusky, N., Cao, J.: Visualizing features and tracking their evolution. Computer **27**(7), 20–27 (1994)
17. Silver, D., Wang, X.: Volume tracking. In: Proceedings of Seventh Annual IEEE Visualization '96, pp. 157–164 (1996)
18. Silver, D., Wang, X.: Tracking and visualizing turbulent 3d features. IEEE Trans. Vis. Comput. Graph. **3**(2), 129–141 (1997)
19. Smith, S.M., Brady, J.M.: Asset-2: real-time motion segmentation and shape tracking. IEEE Trans. Pattern Anal. Mach. Intell. **17**(8), 814–820 (1995)
20. Tzeng, F.-Y., Ma, K.-L.: Intelligent feature extraction and tracking for visualizing large-scale 4d flow simulations. In: Proceedings of the ACM/IEEE SC 2005 Conference Supercomputing, p. 6 (2005)
21. Wegman, E.J.: Hyperdimensional data analysis using parallel coordinates. J. Am. Stat. Assoc. **85**(411), 664–675 (1990)

Tensor Lines in Engineering: Success, Failure, and Open Questions

Marc Schöneich, Andrea Kratz, Valentin Zobel, Gerik Scheuermann, Markus Stommel, and Ingrid Hotz

Abstract Today, product development processes in mechanical engineering are almost entirely carried out via computer-aided simulations. One essential output of these simulations are stress tensors, which are the basis for the dimensioning of the technical parts. The tensors contain information about the strength of internal stresses as well as their principal directions. However, for the analysis they are mostly reduced to scalar key metrics. The motivation of this work is to put the tensorial data more into focus of the analysis and demonstrate its potential for the product development process. In this context we resume a visualization method that has been introduced many years ago, tensor lines. Since tensor lines have been rarely used in visualization applications, they are mostly considered as physically not relevant in the visualization community. In this paper we challenge this point of view by reporting two case studies where tensor lines have been applied in the process of the design of a technical part. While the first case was a real success, we could not reach similar results for the second case. It became clear that the first case cannot be fully generalized to arbitrary settings and there are many more questions to be answered before the full potential of tensor lines can be realized. In this chapter, we review our success story and our failure case and discuss some directions of further research.

M. Schöneich
University Saarbrücken, Germany

A. Kratz
Zuse Institute Berlin, Germany

G. Scheuermann
University Leipzig, Germany

M. Stommel
University Dortmund, Germany

V. Zobel
Universität Leipzig, Leipzig, Germany
e-mail: zobel@informatik.uni-leipzig.de

I. Hotz (✉)
Linköping University, Norrköping, Sweden
e-mail: ingrid.hotz@liu.se

I. Hotz, T. Schultz (eds.), *Visualization and Processing of Higher Order Descriptors for Multi-Valued Data*, Mathematics and Visualization,
DOI 10.1007/978-3-319-15090-1_17

1 Introduction

Modern product development processes in mechanical engineering are governed by computer simulations. That is, new technical systems are first optimized at the computer, which reduces the time-consuming and expensive real prototyping. Thereby, criteria for the design of new technical systems are manifold. They include part stiffness, maximum stress peaks, weight, geometrical or functional boundary conditions and also practical aspects of manufacturability. Various design options are simulated and analyzed for the validation and comparison of the performance of the technical parts under selected operating conditions. In the case of material stressing, this comparison is performed on the basis of a couple of scalar key metrics. Commonly, the visualizations used for the simulation results are limited to the methods provided by the simulation or post-processing software [11, 12]. Examples are contour plots of single tensor components or of derived scalar entities. However, the output of the simulations is much more comprehensive, e.g. containing stress and strain tensor fields. Most of this data is currently considered as an intermediate product and is not used for the evaluation and improvement of the part design. To fully exploit the power of the simulations, the data must be processed such that also non-scalar features are intuitively accessible to the engineer.

This paper investigates the usefulness of more advanced visualization methods and their value in designing *good* engineered technical parts. Our first studies are focused on the use of tensor lines to support the design process in a semi-automatic manner. A central question is the physical relevance and the interpretation of tensor lines as major load paths. Therefore, we consider some simple, well known configurations with increasing complexity. The results are very promising for some case studies, where all new structures outperformed the reference geometry. However, in other cases, we have not been able to repeat this success and it turned out that it is not sufficient to consider tensor lines as only basis for the geometry design. In this chapter, we review the success and failure cases and discuss some directions of further research.

The paper is structured as follows. First we provide the necessary basics in Sect. 2. The subsequent Sect. 3 explains the general background and setting of our case studies. The central part are the discussions of the case studies in the following sections. The paper ends with a discussion of the results in Sect. 4.

2 Basics: Stress Fields and Tensor Lines

In this section we summarize the most important definitions related to stress tensor fields and introduce the quality criteria we use to compare the performance of different component designs.

2.1 Stress Tensor Fields

External forces applied on a body cause stresses within the body. These stresses can be described in each point of the body by a *stress tensor* T_σ, given by the respective 3×3 matrix. Since the stress tensor is symmetric, there is a basis of eigenvectors of T_σ. The eigenvectors represent the *principal stress directions*. Considering planar cuts through the material orthogonal to one principal direction only normal stresses and no shear stresses can be observed. The magnitude of the normal stress, the *principal stress*, is then given by the corresponding eigenvalue, where positive values refer to tension and negative values to compression. It is an often used convention, that the principal stresses are denoted in descending order $\sigma_1 \geq \sigma_2 \geq \sigma_3$. We allude to the principal stresses and the corresponding directions as *major*, *intermediate* and *minor*, respectively. For stress tensor fields, the major and minor principal stresses are of special interest. For more details, we refer to the recent survey by Kratz et al. [5].

Strength Criteria For the evaluation of the stressing of a part mostly derived scalar quantities according to some strength theory are consulted. The objective of a strength criterion is to replace the six components of the stress tensor by one scalar value that constitutes the amount of material stressing. There are multiple material models suitable for different specific cases that are used to model the mechanical behavior of a material. For example elastoplastic material models are capable to describe elastic and plastic deformation once a characteristic scalar stress measure, e.g. the von-Mises stress, reaches a critical value: the yield strength. At this point the material starts to deform plastically. For our studies it is sufficient to use a simple linear-elastic isotropic model (Hooke's law).

Load-Deflection Curve Another important indicator for the performance of a structure is its stiffness. Stiffness is defined as the structure's resistance to deformation. The load-deflection curve is a graphical representation of the relationship between the deformation (in the direction of the applied force) of the part and the magnitude of the applied load, see Fig. 1a. The stiffness is indicated by the slope of the load deformation curve in its (linear) elastic region.

2.2 Tensor Lines

One visualization method that we have used extensively in this work are *tensor lines*. They have been introduced to the visualization community a long time ago by Delmarcelle [1] and have been utilized for visualization purposes of stress tensors in a geomechanical context [9] as well as in medical applications [2]. Tensor lines are defined as integral curves, which are tangent to one chosen eigenvector field (for stress tensors one principal direction) in each point, see Fig. 1b. As there are three principal direction fields associated with the stress tensor, there are three families

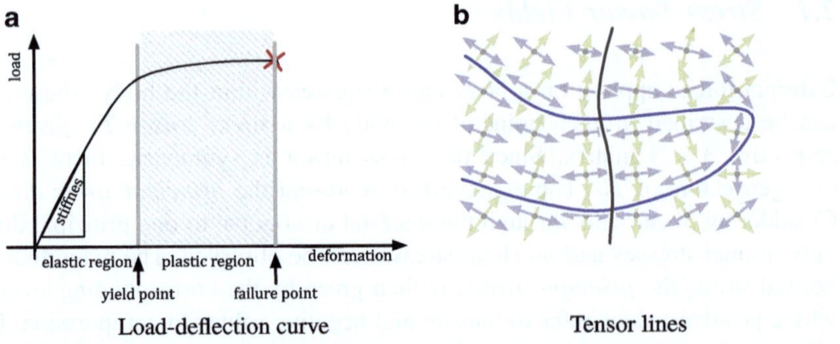

Fig. 1 (**a**) The slope of the load-deflection curve is an indicator for the stiffness of a technical part. It is used for the evaluation and comparison of the various brake levers designed in this case study. (**b**) Two-dimensional tensor field with two showcase tensor lines (major, minor). The *arrows* represent the principal directions in grid points

of tensor lines—major, intermediate, and minor. Due to the symmetry of the stress tensor, these lines are orthogonal to each other. It should be noted that tensor lines only visualize the directional information of the tensor. In this work, the tensor line computation has been embedded in a larger tensor visualization framework [4], which supports *multiple linked views*. A major question during our investigations has been whether tensor lines can be interpreted as major load paths and can be used to support the geometry design of new developed mechanical parts.

Fabric Textures To visualize 2D planar cuts through the stress tensor field, we use *fabric textures* [3]. This is a texture-based method to visualize 2D projections of the stress field applying *line integral convolution (LIC)*. The resulting texture is a dense tensor line representation showing the principal directions of the projected field in every pixel. The free texture parameters as fiber density and length can be used to encode further tensor properties as principal stresses.

3 Case Studies

3.1 Product Development Process

The product development processes in mechanical engineering are characterized by an almost completely virtually tested and optimized part design, Fig. 2. It starts with the definition of the requirements for the part. This includes the development of an initial geometry, material selection, the specification of feasibility constraints and load conditions. This also defines the design space for the subsequent structural optimization procedure based on a structure simulation. A general goal is to automate this optimization as far as possible [7, 8, 10]. While this is an interesting research direction, these methods have not yet found their way into the day-to-day

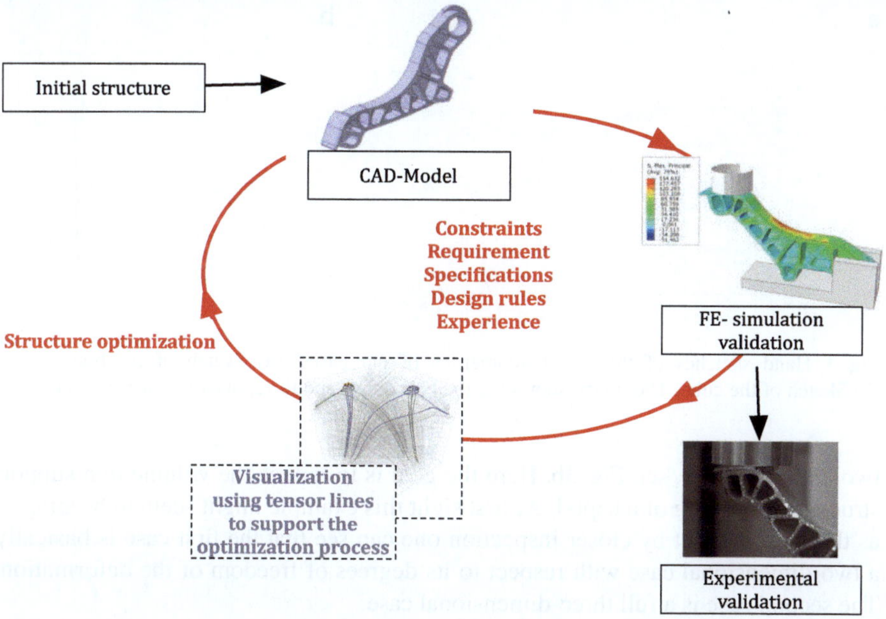

Fig. 2 Standard development cycle when designing a new part using requirements and experiences. The goal of this work is to support this process using visualization

routine workflow of an engineer. It is hard to integrate the experts knowledge to steer the optimization process. Therefore, manual solutions of experts, extending or adopting existing design solutions and design rules are still widely used. These solutions are based on general design principles [6] and the experience of the engineer from previous work. The key metrics used for the analysis and evaluation are mostly scalar measures applying an appropriate strength theory, see Sect. 2.1.

3.2 *The Test Setting*

We consider the design of two parts with increasing complexity and degrees of freedom as test cases. Thereby, we start with a filled volume from which we extract a support structure. The goal is to define a structure with reduced local stress peaks and increased stiffness compared to a reference structure (Sect. 2).

The first case is a brake lever, see Fig. 3a. This is an example, which is simple enough for a suitable validation of the method. At the same time it is a realistic component. Here, the goal is to design a rib-structure inside a given outer geometry. A rib structure is the most often used reinforcement structure for injection-molded plastic parts to efficiently increase their stiffness using as little material as possible. The design of such a rib structure comprises the definition of position, number and shape of the ribs. The second case is a solid block with three fixation points and

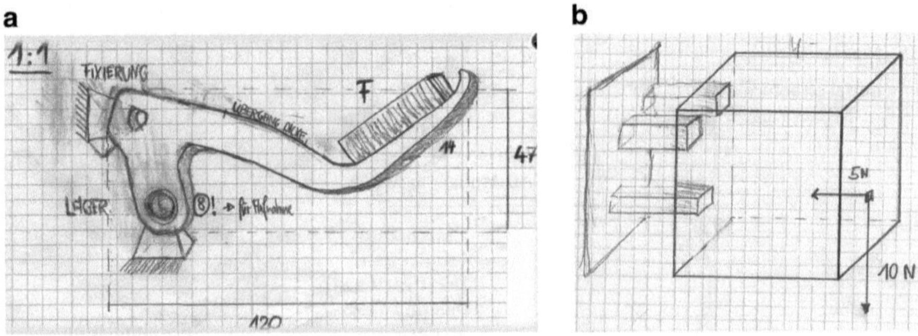

Fig. 3 Hand sketches of the initial geometries of our cases. (**a**) Sketch of the brake lever. (**b**) Sketch of the cube. The image shows the fixation points and the application of the two forces

two applied forces, see Fig. 3b. Here the goal is to reduce the volume to a support structure consisting of a tripod. At first sight this example might seem to be simpler as the first one, but by closer inspection one can see that the first case is basically a two-dimensional case with respect to its degrees of freedom of the deformation. The second case is a full three-dimensional case.

Considering the standard workflow, as shown in Fig. 2, the visualization is used as additional step after the simulation and no additional computations are performed. Thus, the results from the simulation are directly accessible to the engineer for the optimization. The hypothesis is that the major load paths follows the tensor lines, which therefore provide a good guidance for the structure design. For our test we perform one optimization cycle. To obtain comparable results we require that the total volume of the used material is the same for all considered geometries. The comparison is based on the simulation results and for the first case also on experimental tests using rapid prototyping.

The FEM-simulation analyzing the component is performed by the commercial FEM software package Abaqus [11], which is widely used in industry for finite element analysis and computer-aided engineering. The used plastic material is described by an isotropic linear-elastic material model. This simple material model is commonly used in industrial design processes of plastic parts if the loads are quasi-static and do not exceed the yield point of the material. The simulation results contain a large variety of data including the stress tensors.

Case 1: Design of a Brake Lever

Task The outer geometry of the brake lever is predefined by mechanical and ergonomic constraints. The interior part—the design space—offers flexibility for the specific rib design. The design space and the operating load are outlined in red in Fig. 4. The goal is to design an appropriate rib structure supporting the outer geometry.

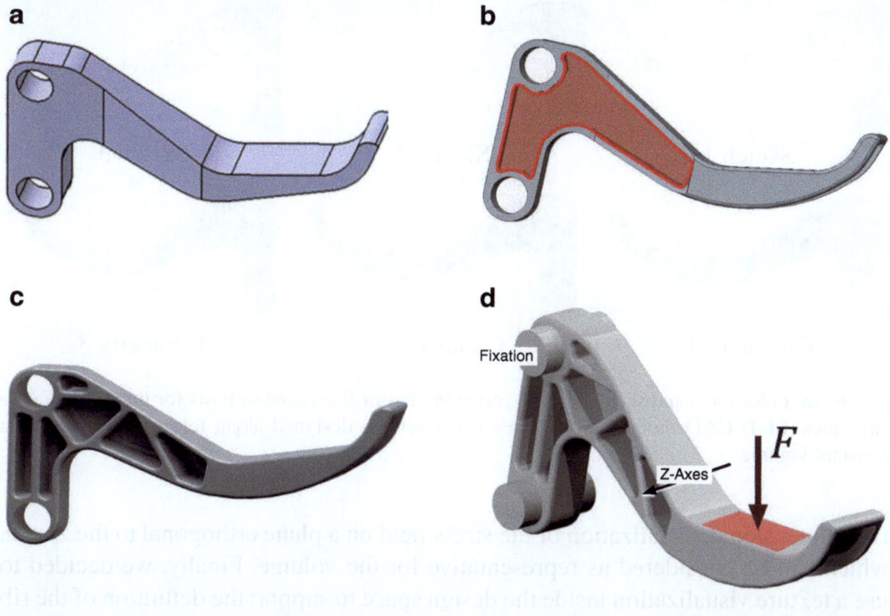

Fig. 4 (**a**) Outer geometry of the brake lever. (**b**) The design space (outlined in *red*), which is available for structural optimization. (**c**) Reference geometry of a brake lever designed according to standard design rules. (**d**) FEM-simulation model with bolts for fixation and the area of the operating force application

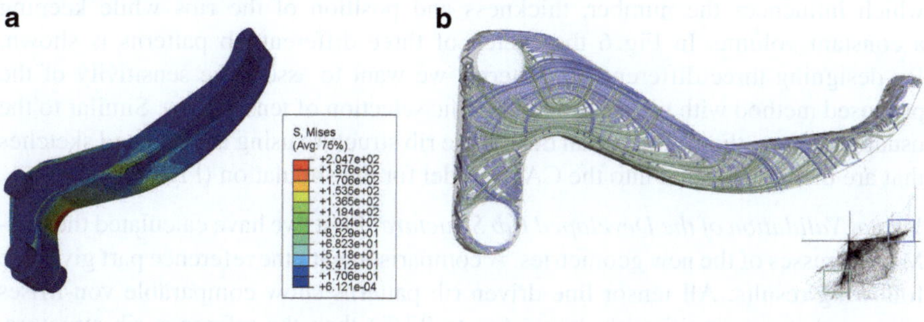

Fig. 5 Results of the first FEM-simulation of the filled geometry. (**a**) Von-Mises stress on the surface. (**b**) 3D tensor lines

Design of the Rib Structure For the first simulation the volume of the design space has been filled with a fictitious material. This allows to calculate the 'natural' flow of the load from the application point of the operating force to the fixation of the brake lever without imposing geometric restrictions. A tensor line visualization of the simulation results can be seen in Fig. 5b. This image shows that the major (purple lines) and minor (green lines) principal stresses lie in two-dimensional planes orthogonal to the z-axis. Further, it can be seen that the resulting patterns hardly change when moving a cutting plane along the z-axis. This suggests to use a

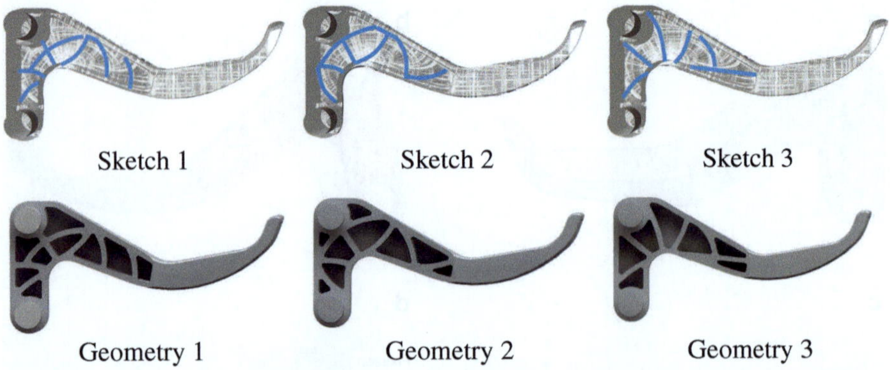

Fig. 6 **(a–c)** Simple hand sketched lines following tensor lines used as basis for three new CAD-structures. **(d–f)** CAD models of the three rib structures designed along tensor lines keeping a constant volume

two-dimensional visualization of the stress field on a plane orthogonal to the z-axis, which can be considered as representative for the volume. Finally, we decided to use a texture visualization inside the design space to support the definition of the rib structures. The texture is a dense tensor line representation of the stress field from the simulation. The density of the fabric like texture encodes the principal stresses.

In a second step, ribs are designed in accordance to the tensor line patterns. The explicit choice of the tensor lines is left to the engineer using his expert knowledge, which influences the number, thickness and position of the ribs while keeping a constant volume. In Fig. 6 the sketch of three different rib patterns is shown. By designing three different rib patterns, we want to assess the sensitivity of the proposed method with respect to the specific selection of tensor lines. Similar to the usual workflow, the engineer can design the rib structure using at first hand sketches that are then transferred into the CAD-model for the simulation (Fig. 6).

Virtual Validation of the Developed Rib Structure First, we have calculated the von-Mises stresses of the new geometries. A comparison with the reference part gives the following results: All tensor-line-driven rib patterns show comparable von-Mises stresses that are significantly lower (up to 27 %) than the reference rib structure, Fig. 7a. The finite element analysis also shows an increased stiffness of the tensor-line-driven rib pattern, with a lower deformation (on average 8 %) for the new brake-lever geometry under the same load.

Rapid Prototyping and Experimental Validation Finally, the new geometries have been experimentally tested using prototypes generated by a rapid prototyping process. Parts generated by rapid prototyping are not usable for absolute values of properties like failure but they can be used to compare the part stiffness of different part designs in a comparative way. Figure 8 shows the rapid prototype models and the test setup in a standard tensile test machine. Figure 7b shows the measured load-

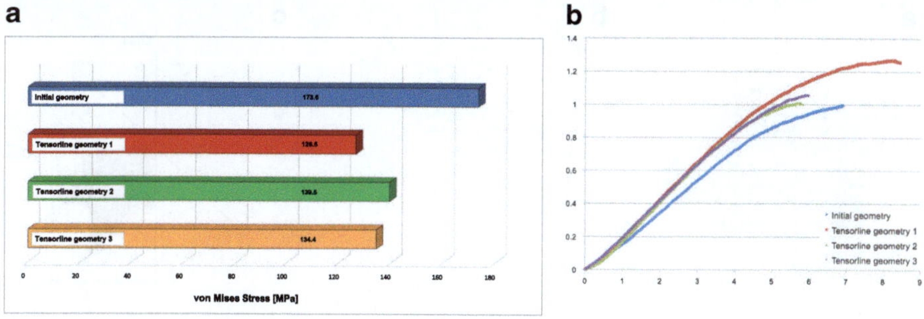

Fig. 7 (**a**) Resulting maximum von-Mises stresses based on the FEM-simulation for the three new designs (*red, green, yellow*) in comparison with the reference geometry (*blue*). (**b**) Load-deflection curves from the experimental validation: initial geometry in comparison with the three tensor line driven parts

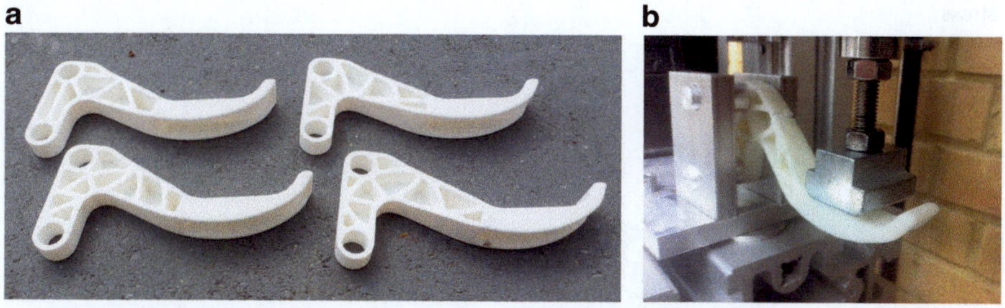

Fig. 8 The three alternative and the reference geometry have been printed using a rapid prototyping method. The right image shows the experimental setup. A load is applied and the deformation is measured until failure (**a**) Printed 3D geometries (**b**) Experimental set-up

deflection curves for the parts. It can be seen that the alternative designs show a comparable stiffness that is higher than the one of the reference part.

Case Study 2: Tripod

Task Given is a cubic design space, Fig. 3b. The goal is to define a structure inside this design space, such that the forces on the right side are transmitted in the 'best way' to the three supporting points on the left side.

Design of the Support Structure Again, the idea is to use tensor lines inside the cubic design space to define a geometrical structure. The first step is an initial simulation with a cube homogeneously filled with material. To investigate the influence of the material stiffness on the resulting tensor lines multiple runs with different material stiffness (Young's Modulus: E =500, 1,000, 2,000, 10,000 MPa) have been performed. The respective results are not distinguishable from a visual inspection of the tensor lines. Visualizations of the simulation result can be seen in

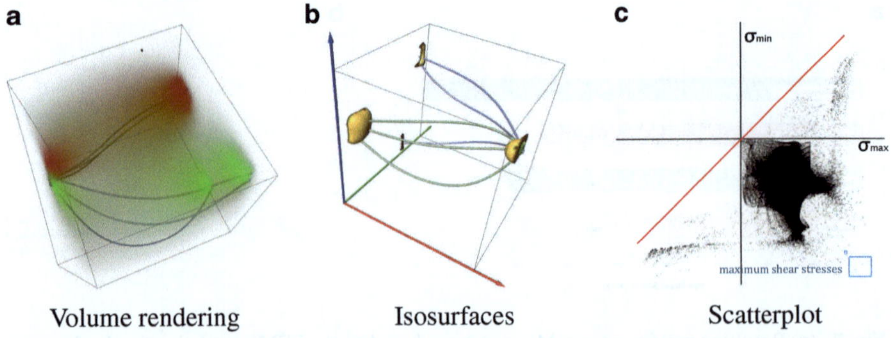

Fig. 9 (**a, b**) show tensor lines selected from the scatterplot (**c**). The *violet lines* correspond to tensile (positive) stresses and the *green lines* to compressive (negative) stresses. (**a**) shows the max respective min principal stress value as volume rendering (*red* compressive and *green* tensile forces). (**b**) shows the same tensor lines together with an isosurface of a high absolute principal stress

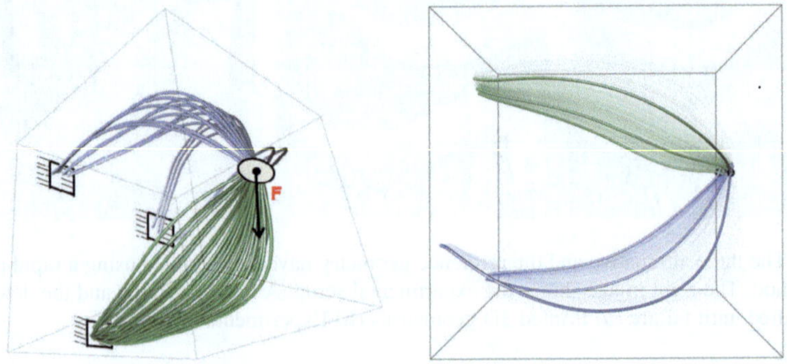

Fig. 10 Different selections of tensor lines from a dense set of tensor lines, all seeded in regions of high shear stress: There is no clear criterion to decide which lines to choose for the geometry design. Slight changes in the seeding conditions lead to different results

Fig. 9. The next step is the selection of characteristic tensor lines as basis for the generation of a CAD geometry. Looking at the visualizations one sees immediately that this task is not well-defined. There are many possibilities to connect the supporting points with the point where the forces are applied, which would result in very different geometries. Figure 10 shows some possible selections. One of the possible selection has been used two generate a new CAD geometry. Already at first sight it differs significantly from the reference geometry, which consists of straight connections. Both geometries are displayed in Fig. 11(first row). To evaluate the performance of the new geometry its behavior under load has been compared to the reference structure. The local deformations for both geometries are plotted in Fig. 11 (2nd row) and it can be observed that the straight structure is much more efficient (stiffer). To get a better understanding of this fact we run a new structure

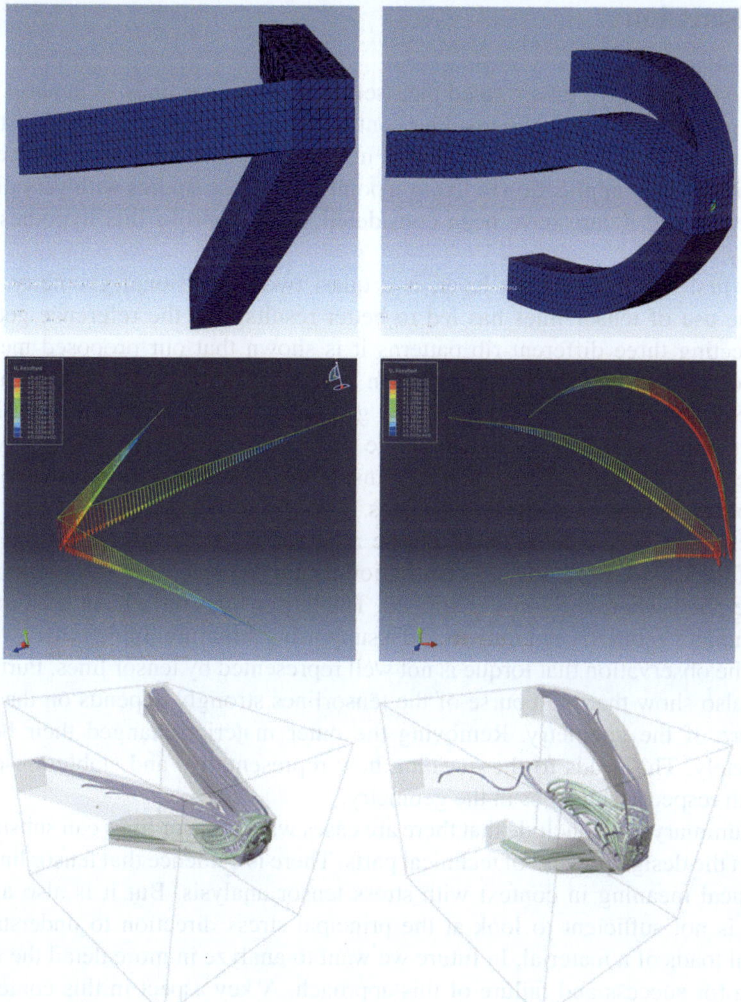

Fig. 11 Comparison of the two geometries. *Left column*: reference structure; *right column*: structure in alignment with tensor lines. *Top row*: CAD geometry; *middle row*: deformation plot; *bottom row*: tensor lines

simulation for both structures. This time the tensor lines are used to evaluate the simulation results. Surprisingly, the tensor lines inside the tripod are completely different from the tensor lines in the filled cube. For the bent structure they are far from straight and suggest a bad force transmission behavior. This shows that in this case it is not sufficient to consider tensor lines for the design of the three dimensional structure.

4 Discussion

In this work we have investigated the usefulness of tensor lines as support for the design process of technical parts. Our central hypothesis used for this investigation is that tensor lines can be interpreted as major load paths transferring forces from the points of their application to fixation points. Two case studies with very different complexity and nature have been considered. They approve this hypothesis only partially.

The first study, the break lever, is a quasi two-dimensional geometry. In this case the use of tensor lines has led to better results then the reference geometry. By selecting three different rib patterns it is shown that our proposed method is quite robust with respect to the selection of specific tensor lines in this case. For the break lever the outer outline of the geometry is predefined. Only an interior support structure had to be designed. The base configuration is already very stable with respect to torque. Thus, the loads inside the part are well represented by the principal compressive and tensile stresses.

In contrast, this success could not be repeated with the full three-dimensional case of the tripod. Here, the only restriction for the geometry are the fixation points and the point where the force is applied. The tensor lines have been used to define the complete geometry not only to add a supporting structure. In this case, we could make the observation that torque is not well represented by tensor lines. Further, we could also show that the course of the tensorlines strongly depends on the overall structure of the geometry. Removing the outer material changed their behavior completely. This leads to the question how representative and stable tensor lines are with respect to changes in the geometry.

In summary, we conclude that there are cases where tensor lines can substantially support the design process of technical parts. There is evidence that tensor lines have a physical meaning in context with stress tensor analysis. But it is also apparent that it is not sufficient to look at the principal stress direction to understand the internal loads of a material. In future we want to analyze in more detail the specific criteria for success and failure of this approach. A key aspect in this context is to investigate the stability of tensor lines with respect to changes of the geometry. Further, we consider it important to consider and represent torque effectively in the visualization.

References

1. Delmarcelle, T., Hesselink, L.: Visualizing second-order tensor fields with hyperstreamlines. IEEE Comput. Graph. Appl. **13**(4), 25–33 (1993)
2. Dick, C., Georgii, J., Burgkart, R., Westermann, R.: Stress tensor field visualization for implant planning in orthopedics. IEEE Trans. Vis. Comput. Graph. **15**(6), 1399–1406 (2009)
3. Hotz, I., Feng, L., Hagen, H., Hamann, B., Jeremic, B., Joy, K.I.: Physically based methods for tensor field visualization. In: VIS '04: Proceedings of IEEE Visualization 2004, pp. 123–130. IEEE Computer Society Press, Los Alamitos (2004)

4. Kratz, A., Meyer, B., Hotz, I.: A visual approach to analysis of stress tensor fields. In: Scientific Visualization: Interactions, Features, Metaphors, volume 2 of Dagstuhl Follow-Ups, pp. 188–211. Schloss Dagstuhl–Leibniz-Zentrum fuer Informatik, Wadern (2011)
5. Kratz, A., Auer, C., Stommel, M., Hotz, I.: Visualization and analysis of second-order tensors: moving beyond the symmetric positive-definite case. Comput. Graph. Forum—State Art Rep. **32**(1), 49–74 (2013)
6. Papalambros, P.Y., Wilde, D.J.: Principles of Optimal Design: Modeling and Computation, 2nd edn. Cambridge University Press, Cambridge (2000)
7. Paris, J., Colominas, I., Navarrina, F., Casteleiro, M.: Parallel computing in topology optimization of structures with stress constraints. Comput. Struct. **125**, 62–73 (2013)
8. Saleem, W., Yuqing, F.: Strategy for optimal configuration design of existing structures by topology and shape optimization tools. Int. J. Aerosp. Mech. Eng. **4**(4), 226–234 (2010)
9. Scheuermann, G., Frey, J., Hagen, H., Hamann, B., Jeremic, B., Joy, K.I.: Visualization of seismic soils structure interaction simulations. In: VIIP, pp. 78–83 (2001)
10. Schumacher, A.: Topology optimization of crash-loaded structures - research state of the art. In: Proceeding of the Automotive CAE Grand Challenge'13 (2013)
11. Dassault Systems: Simulia realistic simulation – ABAQUS: Software Suite for Finite Element Analysis and Computer-Aided Engineering (2014)
12. Tecplot Inc.: Software for the scientific visualizations. http://www.tecplot.com/ (1988–2014)

Contextual Diffusion Image Post-processing Aids Clinical Applications

Vesna Prčkovska, Magí Andorrà, Pablo Villoslada, Eloy Martinez-Heras, Remco Duits, David Fortin, Paulo Rodrigues, and Maxime Descoteaux

Abstract Diffusion weighted magnetic resonance imaging (dMRI) and tractography have shown great potential for the investigation of the white mater architecture in-vivo, especially with the recent advancements by using higher order techniques to model the data. Many clinical applications ranging from neurodegenerative disorders, psychiatric disorders as well as pre-surgical planning employ diffusion imaging-based analysis as an addition to conventional MRI imaging. However, despite the promising outlook, dMRI tractography confronts many challenges that complicate its use in everyday clinical practice. Some of these challenges are low test-retest accuracy, poor quantification of tracts size, poor understanding of the biological basis of the dMRI parameters, inaccuracies in the geometry of the reconstructed streamlines (especially in complex areas with curvature, bifurcations, fanning, crossings), poor alignment with the neighboring diffusion profiles, among others. Recently developed contextual processing techniques including the one presented in this work, for enhancement and well-posed geometric sharpening, have shown to result in sharper and better aligned diffusion profiles. In this paper, we present a possibility in enabling HARDI tractography on the data acquired under

V. Prčkovska (✉)
Center for Neuroimmunology, Service of Neurology, Hospital Clínic and Institut d'Investigació Biomèdica Augustí Pi I Sunyer (IDIBAPS), Barcelona, Spain
e-mail: vprchkov@clinic.ub.es

M. Andorrà • P. Villoslada • E. Martinez-Heras
IDIBAPS, Barcelona, Spain
e-mail: andorra@clinic.ub.es; pvilloslada@clinic.ub.es; emartind@clinic.ub.es

R. Duits
IST/e, Eindhoven University of Technology, Den Dolech 2, 5600 MB Eindhoven, The Netherlands
e-mail: R.Duits@tue.nl

D. Fortin • M. Descoteaux
Université de Sherbrooke, Sherbrooke, QC, J1K 2R1 Canada
e-mail: David.Fortin@USherbrooke.ca; m.descoteaux@usherbrooke.ca

P. Rodrigues
Mint Labs S.L., Barcelona, Spain
e-mail: paulo@mint-labs.com

I. Hotz, T. Schultz (eds.), *Visualization and Processing of Higher Order Descriptors for Multi-Valued Data*, Mathematics and Visualization,
DOI 10.1007/978-3-319-15090-1_18

limited diffusion tensor imaging (DTI) conditions and modeled by DTI. We enhance local features from the DTI field using operators that take 'context' information into account. Moreover, we demonstrate the potential of the contextual processing techniques in two important clinical applications: enhancing the streamlines in data acquired from patients with Multiple Sclerosis (MS) and pre-surgical planning for tumor resection. For the latter, we explore the possibilities of using this framework for more accurate neurosurgical planning and evaluate our findings with a feedback from a neurosurgeon.

1 Introduction

Diffusion weighted magnetic resonance imaging (dMRI) and fiber tractography have gained significant importance in the medical imaging community for the last decade. This novel imaging technique enables quantification of water diffusion, influenced by the structure of biological tissues from the acquisition of series of diffusion images. dMRI tractography, however, faces many challenges in clinical applications such as improvements in the geometry of the diffusion profiles on which the tractography is based including reduction of noise and alignment with the neighboring profiles. In case of Multiple Sclerosis (MS) as a white matter (WM) neurodegenerative disease, the nerve tracts are damaged due to presence of WM lesions. In the areas of the lesions the common scalar measures such as FA have low values and since they are most commonly used as a stopping criteria for the streamline tractography algorithms either do not pass through these areas if the threshold is too high (see Fig. 1), or take a random path [38]. This poses difficulties when trying to develop quantitative dMRI markers for MS.

Another challenge in the accurate reconstruction of the WM streamlines in a clinical setting is the clinical equipment. In other words, while recent advances enable to recover complex fiber geometries at a cost of diffusion measurements along various, high order, sampling schemes [36], some older MR systems work with limited gradient tables. These systems are therefore, only suitable for techniques such as diffusion tensor imaging (DTI) [6] that require low number of gradient directions (a minimum of 6). Many hospitals and research institutes in the world operate with these kind of MRI systems. Furthermore, in a clinical setting, the scanning time is often very limited, which also limits the number of applied gradients for the dMRI acquisition (often as low as 6 or 12). This is the case of neurosurgical applications, where often a fluid attenuated inversion recovery (FLAIR), functional/diffusion MRI, and gadolinium-enhanced T1 acquisitions have to fit a limited time (10–30 min depending on the hospital). Therefore, groups that want to perform state-of-the-art tractography using high angular resolution diffusion imaging (HARDI) data are unable to and can only perform DTI tractography on their old system, which is often the case in neurosurgical planning using dMRI [9]. However, the diffusion community is well aware that the Gaussian assumption of

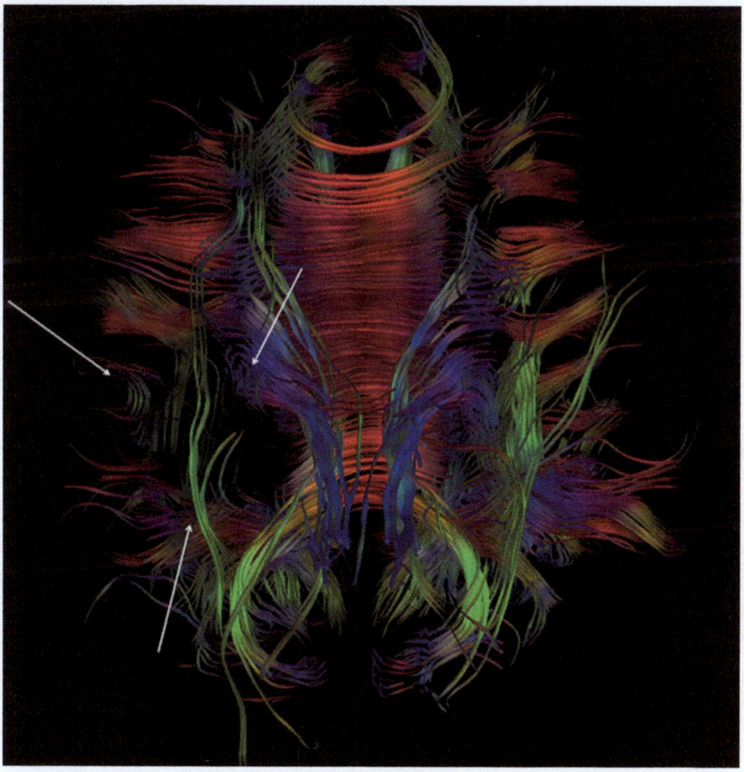

Fig. 1 DTI Fiber tractography reveals a damage of the fibers in the right hemisphere within the lesion location (*arrows*). Image borrowed from NetFormun community, Phillips cases. http://clinical.netforum.healthcare.philips.com/us_en/Explore/Case-Studies/MRI/Case-Balo-concentric-sclerosis-on-MRI-DTI-MR-Spectrocopy

the tensor model in DTI is an over simplification of the diffusion phenomenon of water molecules in the brain and is unable to resolve crossing fibers.

In this work we show that new diffusion signal modeling together with 'so called' contextual processing techniques, as explained in the next section, enable to capture complex angular structures even in the case when the data is acquired from a reduced gradient direction set arising from an older MR system. Moreover, we present novel idea to use the contextual processing in order to enhance the irregular information due to tissue damage as in the case of MS. This is important in order to set a proper processing framework for developing imaging markers for tissue damage along reconstructed streamlines in 3D.

2 Background

Contextual processing is a processing technique of the data on positions and orientations that includes a natural coupling between positions and orientations for alignment of oriented structures via rigid body motions. As dMRI data contains both spatial (3D position coordinates) and angular (diffusion properties in different directions per position) information, contextual processing techniques can use either spatial, angular, or both spatial and angular information in the data. Spatial techniques, such as (anisotropic) smoothing, use the spatial context for a given direction and, hence, are applied directly to the individual diffusion weighted images (DWIs). This has shown to be beneficial in detecting micro-structural changes [42]. Angular techniques [22] use diffusion information in all directions at each individual position, for example to sharpen diffusion profiles [18].

One can concatenate spatial and angular processing separately. However, the resulting processing does not appropriately take into account alignment of oriented structures. To this end, one should not consider the space of positions $\mathbb{R}^3$ and orientations S^2 as a (separate) Cartesian product $\mathbb{R}^3 \times S^2$ (see illustration of Fig. 2c) where the importance of coupling of position and orientation for the purpose of contextual processing of the data is illustrated).

Indeed, alignment is done via rigid body motions on the set $\mathbb{R}^3 \times S^2$ via rigid body motions $(\mathbf{x}, \mathbf{R})$ where $\mathbf{x}$ denotes translation and $\mathbf{R}$ rotation:

$$(\mathbf{x}, \mathbf{R})(\mathbf{y}, \mathbf{n}) = (\mathbf{R}\mathbf{y} + \mathbf{x}, \mathbf{R}\mathbf{n}) \tag{1}$$

As the product of two rigid body motions is again a rigid body motion, we get the following non-commutative group product:

$$(\mathbf{x}, \mathbf{R})(\mathbf{x}', \mathbf{n}') = (\mathbf{R}\mathbf{x}' + \mathbf{x}, \mathbf{R}\mathbf{n}') \tag{2}$$

which reflects the interaction between positions and rotations. The group of rigid body motions commonly denoted as $SE(3)$ motions is therefore denoted by the semi-direct product structure $SE(3) = \mathbb{R}^3 \rtimes SO(3)$, where $SO(3)$ is the group of 3D rotations.[1] Similarly the coupled space between position and orientations (embedded in $SE(3)$) is denoted as $\mathbb{R}^3 \rtimes S^2$. Likewise we denote its elements by $(\mathbf{y}, \mathbf{n}) \in \mathbb{R}^3 \rtimes S^2$ with $\mathbf{y} \in \mathbb{R}^3$ and $\mathbf{n} \in S^2$

Several techniques have been developed for contextual processing on $\mathbb{R}^n \rtimes S^{n-1}$ (with n denoting the dimension), including [4, 7, 8, 13, 21, 30] for n = 2 and [5, 11, 15, 16, 19, 20, 23, 31–33] for n = 3. These techniques come from probability theory on $\mathbb{R}^n \rtimes S^{n-1}$ where stochastic processes for alignment of oriented structures provide stochastic partial differential equation (PDE) for contour completion [19, 31] and contour enhancement [19]. Here we will not consider the details on how to derive the

[1]The coupled space of positions and orientations $\mathbb{R}^3 \rtimes S^2$ is formally defined as $\mathbb{R}^3 \rtimes S^2 := \mathbb{R}^3 \rtimes SO(3)/(\{0\} \times SO(2))$.

PDEs from these stochastic processes. The explicit formulation of these stochastic processes can be found in [20]. Intuitively, the solutions of the PDEs are the average of all stochastic trajectories starting from the initial condition as depicted in Fig. 2a (for more details see [12, 20]).

In this work we use convolution implementations via analytic kernels of the stochastic PDEs for contour enhancement as contextual operators that include alignment of the oriented structures in the data. As such, we both reduce high frequency noise and incoherent/non-aligned structures in the data. The PDEs are approximately solved with shift-twist convolutions with the analytic kernels depicted in the illustration of Fig. 2b. The kernel represents the probabilistic fiber propagation model.

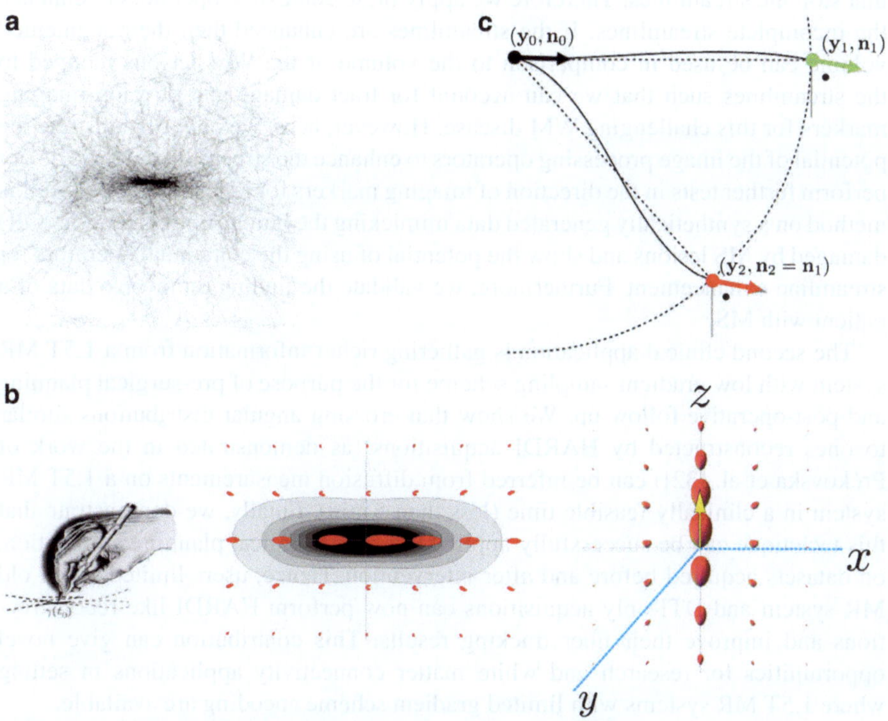

Fig. 2 (**a**) Illustration of the stochastic heat kernel where 300 sample paths show the enhancement kernel on 2D. (**b**) Illustration of the kernels used in this work in 2D with the enhancement property, as well as the visualization of the 3D kernel. (**c**) Coupling of position and orientation. Particle 0 has position and orientation $(\mathbf{y}_0, \mathbf{n}_0)$, particle 1 and 2 are denoted by $(\mathbf{y}_1, \mathbf{n}_1)$ and $(\mathbf{y}_2, \mathbf{n}_1)$ respectively, thus having a the same spatial distance from particle 0 (denoted by the *dashed circle*) and the same orientation (indicated by the *dots*). The *solid lines* represent the sub-Riemannian distance (here the unique global minimizing geodesic in $\mathbb{R}^3 \rtimes S^2$ connecting two oriented particles) from particle 0 to particle 1 and 2 respectively, where the former is shorter. In fact using this Sub-Riemannian distance (see [20, eq.43] on $\mathbb{R}^3 \rtimes S^2$ we quantify the fact that the *red and black arrows* are better aligned than the *green and black* which would not be quantified when using Euclidean norm on the Cartesian product $\mathbb{R}^3 \times S^2$

We apply these contextual operators in two clinical applications. The first application is a proposition of a novel technique for streamline reconstruction in patients with Multiple Sclerosis (MS) whose white matter (WM) tracts are damaged by MS lesions thus preventing proper segmentation of the WM streamlines. Segmenting the WM streamlines is important in order to follow and report percentage of tract damage in MS. MS is a dynamic disease and the lesions change over the course of time, especially in the case of relapsing remitting MS (RRMS) where demyelination and remyelination processes happen dynamically along time. The lesions can be segmented using other MRI imaging modalities such as T1 and FLAIR as described in the following sections, however, once mapped in diffusion space it is impossible to follow the tract damage if we do not segment the WM streamlines properly. In the case of MS this is particularly challenging due to the presence of the lesions that stop the streamlines. Therefore we apply these contextual operators to enhance the incomplete streamlines. If the streamlines are enhanced then their segmented volume can be used in comparison to the volume of the WM lesions mapped to the streamlines such that we can account for tract damage and develop imaging markers for this challenging WM disease. However, here, we only demonstrate the potential of the image processing operators to enhance the streamlines and we do not perform further tests in the direction of imaging markers for MS. Finally, we test this method on a synthetically generated data mimicking the human optic radiation (OR) damaged by MS lesions and show the potential of using the contextual operators for streamline enhancement. Furthermore, we validate the finding on in-vivo data of a patient with MS.

The second clinical application is gathering richer information from a 1.5T MR system with low gradient sampling scheme for the purpose of pre-surgical planning and post-operative follow up. We show that crossing angular distributions similar to ones reconstructed by HARDI acquisitions (as demonstrated in the work of Prčkovska et al. [32]) can be inferred from diffusion measurements on a 1.5T MR system in a clinically feasible time (less than 5 min). Finally, we demonstrate that this technique can be successfully applied to a neurosurgical planning application, on datasets acquired before and after intervention. Hence, users limited by an old MR system and DTI-only acquisitions can now perform HARDI-like reconstructions and improve their fiber tracking results. This contribution can give novel opportunities for research and white matter connectivity applications in setting where 1.5T MR systems with limited gradient scheme encoding are available.

3 Methods

In order to infer crossing angular distributions from diffusion tensors (DT) we implemented a method presented by Prčkovska et al. [32] which employs an accelerated contextual techniques studied in the work of Rodrigues et al. [34] and Duits et al. [19, 20] in HARDI imaging rather than in DTI imaging. For completeness we describe it in this section.

3.1 *Inferring Crossing Angular Distributions from Diffusion Tensors*

In DTI, the signal decay is assumed to be mono-exponential [6], and yields the equation for each position in $\mathbf{y} \in \mathbb{R}^3$:

$$S_g(\mathbf{y}) = S_0 \exp(-b\mathbf{g}^T\mathbf{D}\mathbf{g}), \tag{3}$$

where S_g is the signal in the presence of diffusion sensitizing gradient, and S_0 is the zero-weighted baseline signal, b-value is the parameter of the scanner closely related to the effective diffusion time, and the strength of the gradient field, $\mathbf{g}$ are the unit-norm diffusion gradient vectors, and $\mathbf{D}$ is the second order symmetric, positive definite diffusion tensor (DT). Once the DT is calculated per voxel $\mathbf{y} \in \mathbb{R}^3$, the spherical orientation distribution function (ODF) can be reconstructed, and sampled on the sphere

$$ODF(\mathbf{y}, \mathbf{n}) = \mathbf{n}^T\mathbf{D}(\mathbf{y})\mathbf{n}, \tag{4}$$

where $\mathbf{n}$ is the direction vector defined by the tessellation. Note that there are other ways to get the orientation density function [1, 20]. Figure 3 shows a typical linear DT and the correspondent diffusivity profile sampled on a sphere (in our case icosahedron of order 4,642 points on a sphere).

Following Rodrigues et al. [34], from a tensor field, we create an ODF field, i.e., a HARDI-like dataset $\mathscr{U}$, a coupled space of positions and orientations:

$$\mathscr{U} : \mathbb{R}^3 \rtimes S^2 \to \mathbb{R}^+ : \mathscr{U}(\mathbf{y}, \mathbf{n}(\beta, \gamma)) \tag{5}$$

This means that on every position $\mathbf{y} \in \mathbb{R}^3$, the probability of a water particle diffusing in a certain direction

$$\mathbf{n}(\beta, \gamma) = (\sin\beta, -\cos\beta \sin\gamma, \cos\beta\cos\gamma)^T \in S^2, \tag{6}$$

is given as a positive scalar. Here $\mathbf{n}(\beta, \gamma)$ is a point on the sphere parameterized by $\beta \in [-\pi, \pi)$ and $\gamma \in [-\frac{\pi}{2}, \frac{\pi}{2})$. This parameterization of S^2 ensures that the singularities in the coordinate frame are at the x-axis and far away from the z-axis where the unity element is placed (the z-axis serves as reference axis along which our convolution kernels are aligned). Throughout this article we consider DTI-data as the initial condition, which means that we set $\mathscr{U}(\mathbf{y}, \mathbf{n}) = \mathbf{n}^T\mathbf{D}(\mathbf{y})\mathbf{n}$.

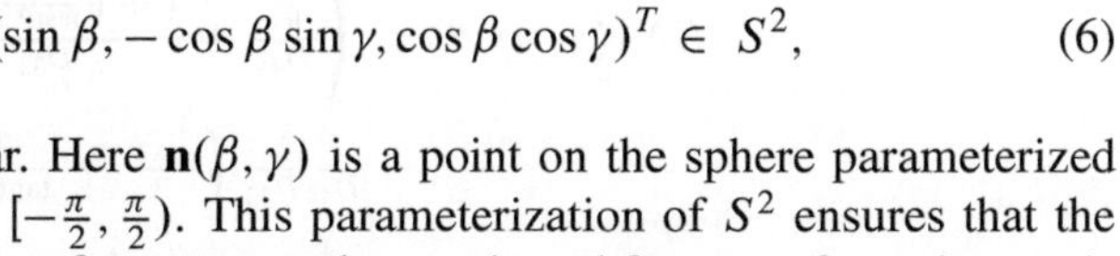

Fig. 3 A linear diffusion tensor (*left*) and the corresponding tessellated ODF (*right*)

3.2 Contextual Enhancement of $\mathscr{U}$ via Shift-Twist Convolution

Duits et al. [19] proposed a kernel implementation that solves the diffusion equation on $\mathbb{R}^3 \rtimes S^2$, more precisely the solution of:

$$\begin{aligned} &\tfrac{\partial W}{\partial t}(\mathbf{y},\mathbf{n},t) = D_{33}(\mathbf{n}\cdot\nabla)^2 W(\mathbf{y},\mathbf{n},t) + D_{44}\Delta_{S^2} W(\mathbf{y},\mathbf{n},t) \\ &W(\mathbf{y},\mathbf{n},0) = \mathscr{U}(\mathbf{y},\mathbf{n}), \end{aligned} \tag{7}$$

where Δ_{S^2} denotes the Laplace Beltrami-operator on the sphere $S^2 = \{\mathbf{x} \in \mathbb{R}^3 \mid \|\mathbf{x}\| = 1\}$, is given by shift-twist convolution with the (line-spread) kernel

$$\begin{aligned} W(\mathbf{y},\mathbf{n},t) &= (p_{3D,t}^{D_{33},D_{44}} *_{\mathbb{R}^3\rtimes S^2} U)(\mathbf{y},\mathbf{n}) \\ &= \int_{\mathbb{R}^3}\int_{S^2} p_{3D,t}^{D_{33},D_{44}}(R_{\mathbf{n}'}^T(\mathbf{y}-\mathbf{y}'), R_{\mathbf{n}'}^T\mathbf{n})\mathscr{U}(\mathbf{y}',\mathbf{n}')\,\mathrm{d}\mathbf{y}'\mathrm{d}\sigma(\mathbf{n}') \end{aligned} \tag{8}$$

This kernel represents the Brownian motion kernel, on the coupled space $\mathbb{R}^3 \rtimes S^2$ of positions and orientations. With some heuristics it can be approximated by:

$$\begin{aligned} &p_{3D,t}^{D_{33},D_{44}}((x,y,z)^T, \mathbf{n}(\beta,\gamma)) \approx \\ &N(D_{33},D_{44},t)\cdot p_{2D,t}^{D_{33},D_{44}}((z/2,x),\beta)\cdot p_{2D,t}^{D_{33},D_{44}}((z/2,-y),\gamma), \end{aligned} \tag{9}$$

where $\mathbf{y} = (x,y,z)^T$, and $N(D_{33},D_{44},t) \approx \frac{8}{\sqrt{2}}\sqrt{\pi t}\sqrt{tD_{33}}\sqrt{D_{33}D_{44}}$ takes care that the total integral over positions and orientations is 1. The $2D$ kernel is given by:

$$p_{2D}^{D_{33},D_{44},t}(x,y,\theta) \equiv \frac{1}{32\pi t^2 c^4 D_{44} D_{33}} e^{-\frac{\sqrt{\mathrm{EN}((x,y),\theta)}}{4c^2 t}} \tag{10}$$

where we use short notation

$$\begin{aligned} EN((x,y),\theta) = &\left(\frac{\theta^2}{D_{44}} + \frac{\left(\frac{\theta y}{2} + \frac{\theta/2}{\tan(\theta/2)}x\right)^2}{D_{33}}\right)^2 \\ &+ \frac{1}{D_{44}D_{33}}\left(\frac{-x\theta}{2} + \frac{\theta/2}{\tan(\theta/2)}y\right)^2 \end{aligned}$$

where one can use the estimate $\frac{\theta/2}{\tan(\theta/2)} \approx \frac{\cos(\theta/2)}{1-(\theta^2/24)}$ for $|\theta| < \frac{\pi}{10}$ to avoid numerical errors. $c \approx 1$ is a positive constant for rescaling the time t that we typically set to 1. Equation (9) is one analytic approximation. Other (more accurate) approximations can be found in [19, 20]. For more details see the work of Duits et al. [19] and Rodrigues et al. [34].

The diffusion parameters $D_{33} > 0$ and $D_{44} > 0$ and stopping time t allow the adaptation of the kernels to different purposes: t determines the overall neighborhood size in $\mathbb{R}^3 \rtimes S^2$; $D_{33} > 0$ determines spatial extent of the kernel; and the quotient D_{44}/D_{33} models the bending of the fiber propagation model underlying the diffusion.

We can now convolve this kernel with the ODF image $\mathcal{U}$, using the HARDI convolution, as expressed in Eq. (11).

$$\Phi(\mathcal{U})[\mathbf{y}, \mathbf{n}_k] = \sum_{\mathbf{y}' \in P} \sum_{\mathbf{n}' \in T} p_{\mathbf{y},\mathbf{n}_k}(\mathbf{y}', \mathbf{n}')\, \mathcal{U}(\mathbf{y}', \mathbf{n}')\; \Delta\mathbf{y}'\, \Delta\mathbf{n}' \tag{11}$$

where $p_{\mathbf{y},\mathbf{n}_k}$ is the kernel centered at position $\mathbf{y}$ and orientation $\mathbf{n}_k$, $\Delta\mathbf{y}'$ is the discrete volume measure and $\Delta\mathbf{n}'$ the discrete surface measure, which in case of (nearly) uniform sampling of the sphere, such as tessellations of icosahedrons, can reasonably be approximated by $\frac{4\pi}{|T|}$. P is the set of lattice positions neighbouring to $\mathbf{y}$ and T is the set of tessellation vectors. Kernel $p_{\mathbf{y},\mathbf{n}_k}$ is the rotated and translated correlation kernel (such that it is aligned with $(\mathbf{y}, \mathbf{n}_k)$) given by

$$p_{\mathbf{y},\mathbf{n}_k}(\mathbf{y}', \mathbf{n}') = p_{3D,t}^{D_{33},D_{44}}(R_{\mathbf{n}'}^{-1}(\mathbf{y}' - \mathbf{y}), R_{\mathbf{n}'}^{-1}\mathbf{n}) \tag{12}$$

where $R_{\mathbf{n}'}$ is any rotation mapping $\mathbf{e}_z$ onto $\mathbf{n}'$.

The convolution with such a kernel will result on the extrapolation of crossing profiles where the neighbourhood information so indicates, i.e., the E-ODFs.

3.3 Data

Synthetic Data

We created a simplified artificial optic radiation (OR) dataset following the known anatomy of the human OR including curvature in the track (however, excluding the Meyer's loop for simplicity purposes). The OR can be damaged in diseases such as multiple sclerosis (MS) by the presence of focal lesions, diffuse direct damage or trans-synaptic axonal degeneration [24, 29]. Moreover, it is one of the most interesting tracks to be investigated in research in MS due to the possibility to correlate the lesion load in the track with the visual field deficits that is easy to measure.

We use our framework [35] to create artificial synthetic data with the following parameters:

- a healthy OR tract, based on in-vivo data, with eigenvalues λ_i = [367; 496; 1442] x 10^6 mm^2/s, b-value of 1,000 s/mm^2 as commonly found tensor values in the OR at this b-value and added Rician noise with SNR of 10;
- added a lesion to the artificial tract, with respective eigenvalues of λ_i= [287; 517; 1224] x 10^6 mm^2/s, sampled from a real OR lesion [38]. (Note the lower anisotropy tensors in Fig. 5 yellow circle.)

After creating the artificial phantom data we performed deterministic tracking on the diffusion tensors (DT) with stopping criteria FA=0.6 (see Fig. 5a). We furthermore used the kernels to enhance the lesion in diffusion space and afterwards apply the same deterministic tracking with the same stopping criteria. Note that the FA threshold is slightly higher than the frequently used thresholds in streamline tracking in real data since here we are working on a synthetic dataset. In this data the SNR is higher than in real data and the FA values are therefore significantly increased (see Fig. 5e).

We performed a parameter search to determine the effect of the convolution on the artificial OR dataset. To lower the parameter space we fixed $D_{33} = 1$, since the artificial OR tube has radius of 2 voxels, and iterated $D_{44} = \{0.01, 0.13, 0.23\}$ and $t = \{1.5, 2.5, 3.5\}$.

MS Patient Data Acquired at 3T Siemens MR System

The scan of the MS patient was performed on a 3T Siemens Trio MRI scanner (Erlangen, Germany), using a 32 channel head coil for radio-frequency transmission and signal reception. The MRI protocol included the following sequences: (a) 3D structural T1-weighted Magnetization Prepared Rapid Acquisition Gradient Echo (MPRAGE) sequence: Repetition Time (TR): 2,050 ms, Echo Time (TE): 2.4 ms, Inversion recovery time (TI): 1,050 ms, Flip angle: 9°, FOV: 220 mm and voxel size of $0.9 \times 0.9 \times 0.9\,\text{mm}^3$ (b) DTI sequence with the following parameters: TR/TE, 6900/89 ms; acquisition matrix, 96×96; 55 contiguous axial slices; $2.5\,\text{mm}^3$ isotropic voxel size; diffusion weighted images in 30 diffusion directions; b value, $1{,}000\,\text{s/mm}^2$. We identified an artifact in the diffusion-weighted images due to vibration of the MRI platform (for more information see the work of Gallichan et al. [25]), which was found in parietal regions and mainly in six directions. In order to remove this artifact we discarded the affected directions following the work of Ling et al. [28].

Since the DTI images were obtained by Echo Planar Imaging (EPI) method they suffered from field inhomogeneities caused by susceptibility changes at the interfaces between tissues and therefore it is useful to acquire gradient field maps. By means of field maps, the B_0 correction was applied for unwarping images in order to make the registration step more accurate. Then, the reconstruction of the diffusion tensors was performed using software toolkit MrTrix [40] with eddy current correction to carry out the registration to the reference volume. The T1-MPRAGE sequence was used by two experienced neurologists to manually generate a lesion map for the patient using ITK-SNAP software [45]. The T1-weighted images were used because they more accurately identify axonal damage than T2-weighted images [14]. The manually detected lesions were furthermore mapped from the T1 space to the diffusion space. This way we know the exact position of the lesions of the T1 space and the reasons for the fibers to stop due to the track's impairment.

We calculated FA maps that help us register DTI images with high resolution anatomical images to carry out the fitting of lesion mask into T1 space. The presence of white matter (WM) lesions distort and bias the output of the registration. For this reason we removed this bias by filling such lesions to make the brain look like a 'healthy brain' before the registration step. MS lesions were filled by replacing the lesion voxel intensity values with values that were randomly sampled from an intensity distribution that measured from surrounding WM voxels [10].

From the original tensor field, we created a seeding mask with $FA > 0.3$ as threshold. From the tensor data estimated by the above mentioned pipeline we created the ODFs as described in Sect. 3.1. Spherical harmonics of order 2 were estimated and given as an input for the deterministic fibertracking using MRTrix with the default stopping criteria by the software (0.1 of the spherical harmonics amplitude), however with thresholded seeding mask.

On the field of ODFs, the kernels were applied in order to enhance the data, with the parameters $D_{33} = 1$, $D_{44} = 0.1$ and $t = 2$ (these were determined by a refined parameter search based on the results from the artificial OR parameter iteration as best fit for the real data).

A new tensor field was estimated with a least square fit method from the E-ODFs field. A new mask is obtained with the new FA map, applying the same threshold of $FA > 0.3$. The same fiber tracking algorithm with identical tracking parameters was applied, however with the new thresholded seeding mask.

To summarize, 100,000 streamlines were estimated using deterministic tracking applied over:

- original ODF field (order 2 equivalent to the DTI model), masked with original FA map;
- Constrained Spherical Deconvolution (CSD) [39] (or order 4 as maximum fitting order given our sampling scheme of 30 gradient direction) from original DWI, masked with original FA map;
- E-ODFs field, masked with new FA map.

Healthy and Tumor Data Acquired at 1.5T Siemens MR System

Healthy Subject Datasets

The data was acquired from an old 1.5T Siemens MR system with hard coded gradient tables fixed at 6 or 12 directions on the sphere. Systems like this one, without a research key, typically have fixed configurations for the gradient scheme that cannot be altered. On this system, good quality DTI dataset is achievable by acquiring DW images at b=1,000 s/mm^2, 2 mm^3 isotropic, TE/TR=95/12500 ms, and 12 gradient directions. For improving the signal to noise ration (SNR) they can be additionally averaged at most three times (NEX=3). The DTI with 12 directions and a single measurement takes 2 min and 46 s, whereas the one with NEX=3 takes 8 min and 18 s, which is acceptable in our clinical setting. Note that this is slower than actual systems as we do not have parallel imaging possibilities on the system. A healthy subject was thus imaged with DTI NEX=1 and NEX=3. The same subject went into a recent 1.5T SIEMENS Magnetom (Vision) system and was furthermore imaged with a 64 directions HARDI protocol, b=1,000 s/mm^2, 2 mm^3 isotropic, TE/TR=40/3864 ms, which takes 12 min due to a parallel imaging acceleration factor of 2. The E-ODFs method was applied on the old DTI reconstruction from log-Euclidean reconstruction [2], whereas state-of-the-art local reconstruction was applied on the HARDI data. We have performed analytical q-ball [17] with spherical harmonics (SH) of order 8, constant solid angle (CSA) q-ball [1, 41] with SH of order 8, and fiber orientation distributions (FOD)s from spherical deconvolution [39] of order 8 as well.

Tumor Subject Datasets

The tumor data is acquired with the same DTI NEX= 1 described above. We have imaged a glioblastoma patient before the operation providing pre-operative data and after the intervention providing post-operative data. From this approximately 3 min acquisition, we have reconstructed the diffusion tensors using the log-Euclidean framework [2] and have applied the E-ODFs technique on the pre-operative and post-operative data. Then, our in-house deterministic streamline tractography was applied to the 'original' tensor and the maxima extracted from E-ODFs, using a step size of 0.5 voxel, maximum curvature angle of 60° in an FA (from the original tensor field) thresholded mask above 0.10 (similar to the technique of [26] and as advised by [9] for neurosurgical planning).

We chose the parameters for the kernel in order to give a high relevance to the diffusion along the principal axis $D33 = 0.9$, $D44 = 0.001$ and $t = 3.4$, a third order tessellation on the sphere and a $3 \times 3 \times 3$ neighboring lattice. Intuitively, with these parameters, crossings will be inferred in a given voxel when there are nearby highly anisotropic profiles oriented towards that voxel.

For all the real data we applied a squaring of the E-ODFs in order to sharpen the data. This yields better results given the typical blurring of data in convolutions.

4 Results

4.1 Artificial Optic Radiation Phantom Data

In Fig. 4 we illustrate the effect of convolving the artificial OR data with kernels of different parameters. We fix $D33 = 1.0$ and vary $D44 = \{0.03, 0.13, 0.23\}$ and $t = \{1.5, 2.5, 3.5\}$. We show the glyphs corresponding to the E-ODFs versus the original tensor main direction. We can observe the influence of a bigger t portraying a bigger contextual neighborhood and thus a better fit to the original

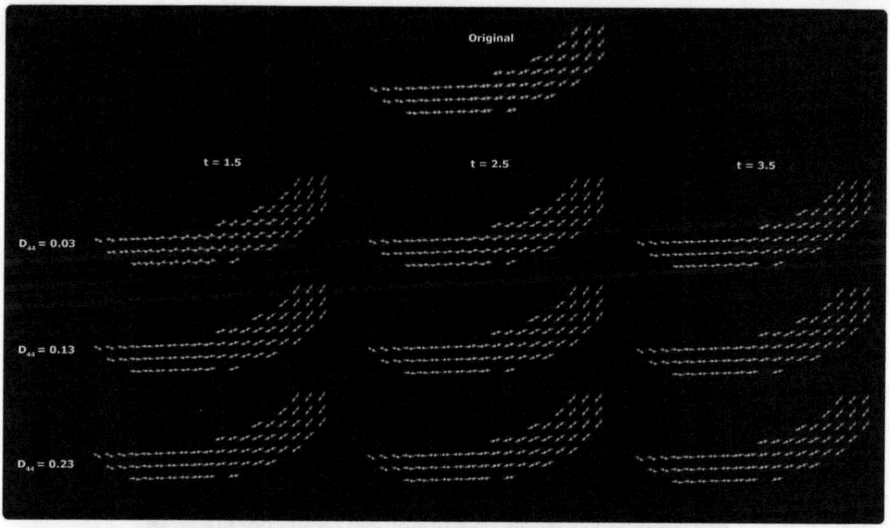

Fig. 4 Original tensor field of the artificial optic radiation tract. E-ODFs field result of convolutions with varying parameters

tensor directions. Increasing $D44$ conveys a closer recovery of the curve structure of the OR, as observed with a better fit to the original tensor directions.

Figure 5a shows the streamlines clearly being stopped due to the lesion in the tract (with lower FA, see change of FA along the streamlines in Fig. 5e), whereas in Fig. 5b the full streamlines are completely recovered. The tracking in both cases has been seeded from the same region and done with the same stopping criteria. Here, in this artificial example, we show the power of the contextual kernels to be used for streamline enhancement in diffusion space (see Fig. 5e). This shows the potential of the kernels to be used as markers for tract damage by simply comparing the original and enhanced data.

4.2 MS Patient Data: DTI Versus E-ODFs

Figure 6 shows the results of an axial plane with deterministic fiber tracking (whole brain seeding with white matter mask) including the OR tract in the three different modeling techniques (DTI, CSD (order 4), E-ODF). The MS lesions that have been previously segmented by the neurologists are highlighted in white. We observe that in case of DTI Fig. 6a all of the streamlines are stopped by the presence of the lesions. In Fig. 6b we observe that CSD of order 4 performs better managing to recover some of the streamlines that contain MS lesions (violet square) however, still having problems in different areas (red squares). In Fig. 6c we observe the enhancement effect on the data after being post-processed with the kernels. All the streamlines that include lesions are enhanced and pass through the area with lesions.

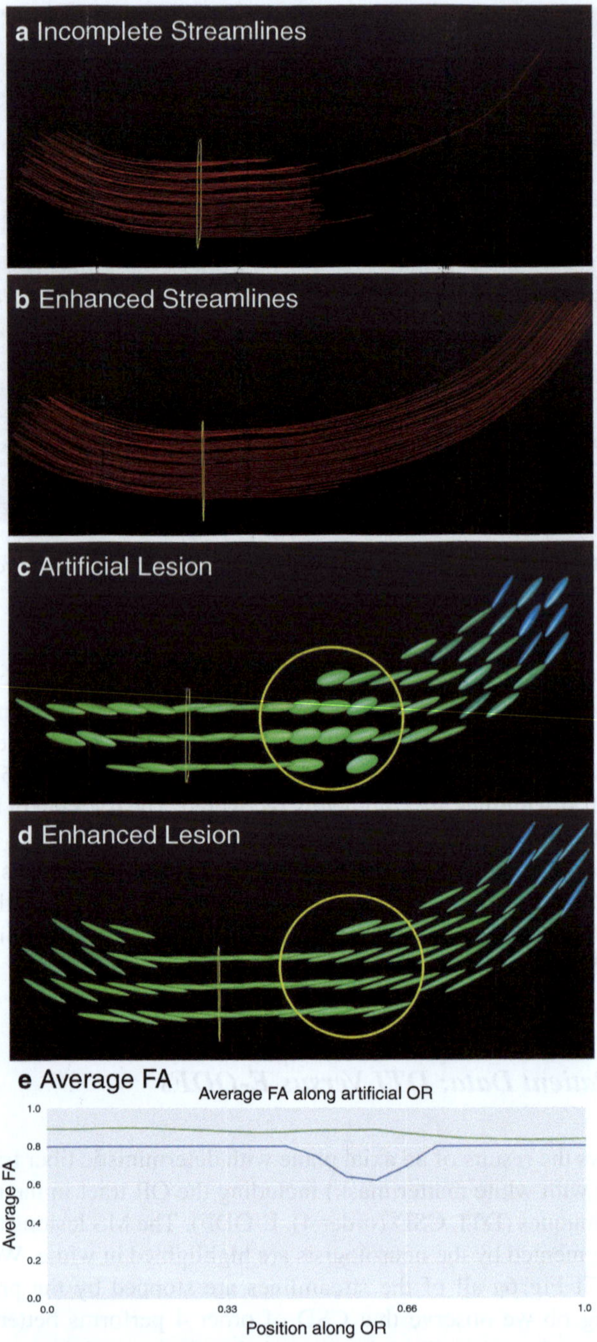

Fig. 5 Artificial optic radiation tract. (**a**) Incomplete streamline due to lesion. (**b**) Full streamline in the enhanced data. In both tracking FA=0.6 as a tracking threshold. (**c**) Tensors with highlighted lower diffusivity in the lesion. (**d**) Enhanced lesions. (**e**) Average FA along the damaged tract (*blue*) and enhanced tract (*green*)

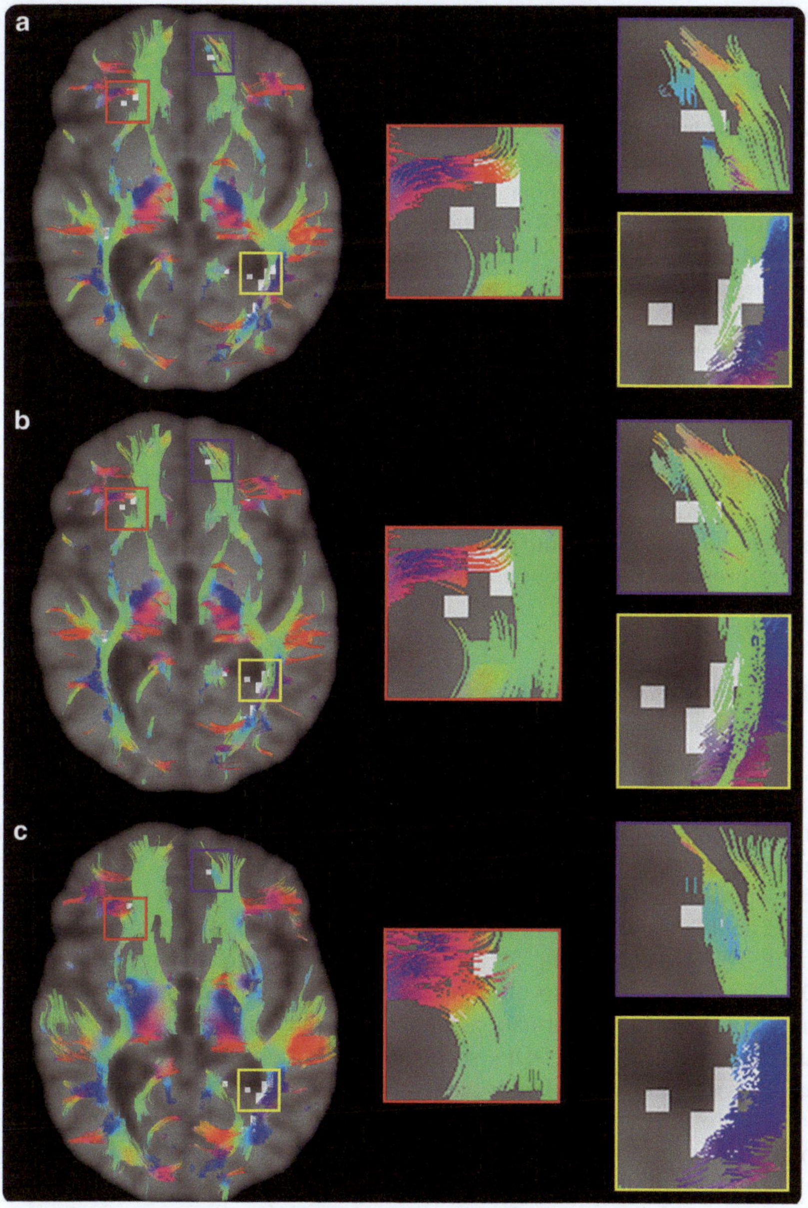

Fig. 6 Data from the MS patient with (**a**) DTI streamline fiber tracking. (**b**) CSD streamline fiber tracking of order 4. (**c**) Tracking on the estimated E-ODF field. The zoomed-in areas illustrate how different techniques behave in the areas of the lesions (*white squares*)

4.3 Healthy Subject: DTI Versus HARDI Versus E-ODFs

Figure 7 shows local reconstructions from the healthy subject from a 64 direction HARDI acquisition, a single average and three averages DTI acquisition with 12 directions. First, we note that E-ODFs for the 12 direction acquisition are able to recover crossing fiber configurations similar to q-ball and CSA q-ball reconstructed from the HARDI dataset. The E-ODFs are much smoother than the q-ball reconstructions, due to the particular choice of the convolutions kernels. However, their maxima coincide with the maxima calculated from the HARDI techniques applied on the high angular acquisition datasets, as can be observed in the second row of Fig. 7.

As expected, compared to state-of-the-art spherical deconvolution, we see that FODs have a much better angular resolution than respective E-ODFs and q-ball/CSA reconstructions. To obtain as sharp ODFs, one could always apply a spherical wavelet transform [27] or a sharpening deconvolution transform [18]. Overall, one concludes that a single average DTI acquisition, taking approximately 3 min, can be transformed into a dataset similar to a 64 direction HARDI-like dataset with the proposed post-processing routine.

4.4 Tumor Patient

Figure 8 shows local fiber crossings recovered with the E-ODFs technique. The field of E-ODFs is relatively smooth and is aligned with the anatomy. Nonetheless, as for the healthy subject, E-ODFs are able to capture crossing fibers in the centrum semiovale and elsewhere. In particular, one can see the famous crossing between the corticospinal tract (CST), projections from the corpus callosum (CC) and superior longitudinal fasciculs (SLF) coming out of the plane as in the case of the healthy patient. This is encouraging for enhanced fiber tractography from these local E-ODFs. Finally, we can also observe more structure around and within the tumor using the extrapolated information. Glioblastomas are known to infiltrate the white matter and the field of E-ODFs seem to suggest some remaining coherent local structure within the tumor. This is purely qualitative analysis, furthermore confirmed by the neurosurgeon at our institute who states that this structure can exist and is coherent as seen in the images. In fact, when given the choice, our neurosurgeon prefers looking at the field of E-ODFs more than the tensor field since they better depict the underlying expected anatomy.

Figure 9 shows the advantage of tracking on a field of E-ODFs reconstructed fiber crossings. Lateral projections from the CC and CST are neatly recovered from the ODF tracking algorithm whereas most of these structures are missing on the tracking applied on the DTI dataset. A similar observation is seen in the post-op data (see additional material).

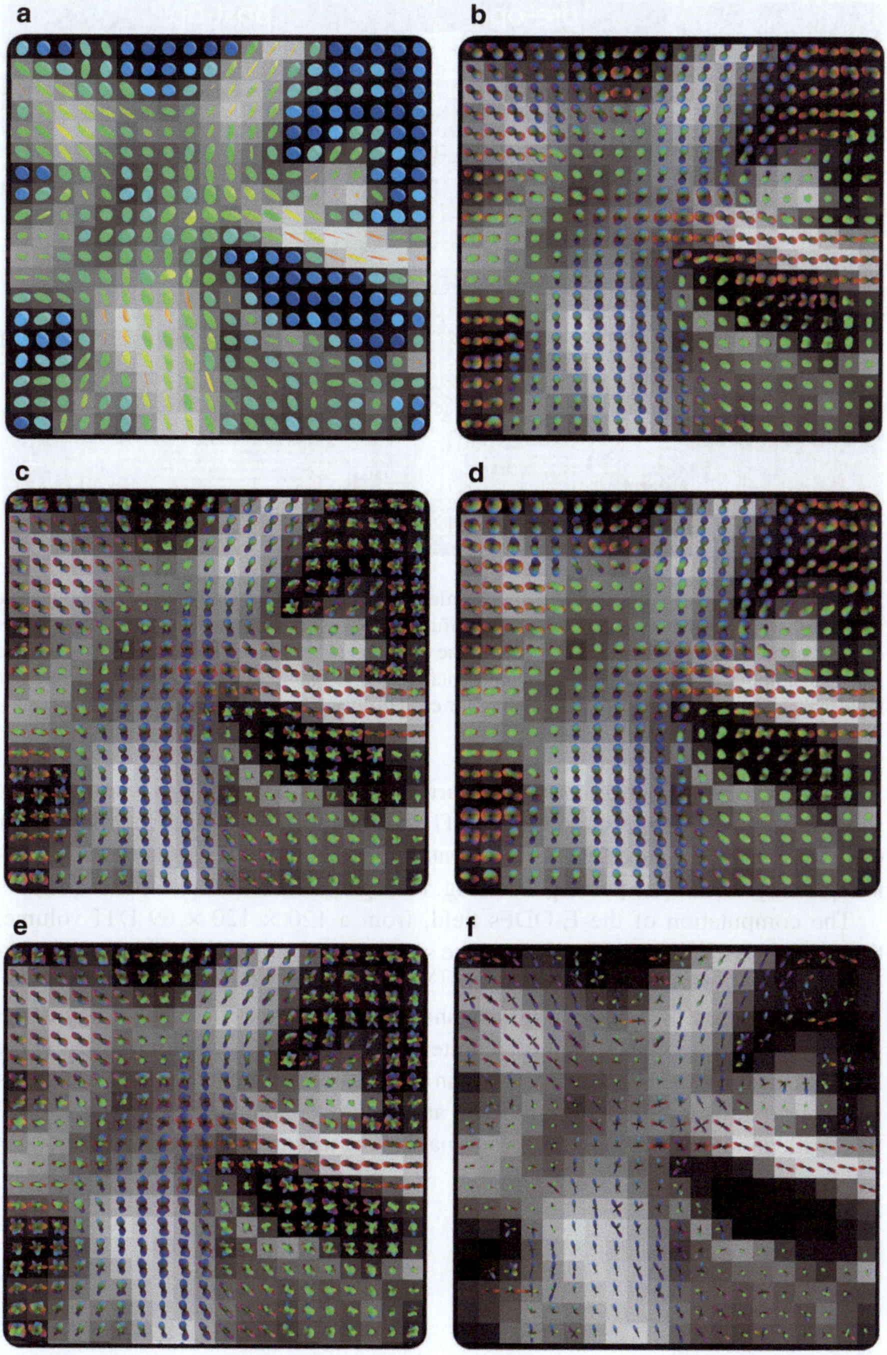

Fig. 7 Healthy dataset. DTI, analytical q-ball, constant solid-angle q-ball, spherical deconvolution versus E-ODFs (**a**) DTI. (**b**) E-ODFs from DTI. (**c**) Analytical q-ball. (**d**) E-ODFs from DTI three averages. (**e**) CSA-qball. (**f**) Spherical deconvolution

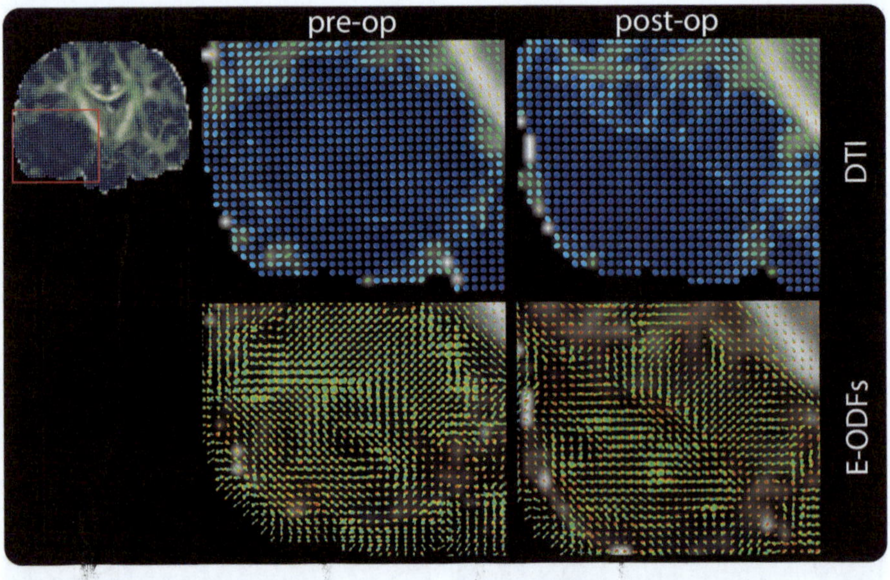

Fig. 8 Diffusion tensors versus E-ODFs of order 8 in a zoomed region (coronal slice anterior view) of the pre-op and post-op dataset. The diffusion tensors are color coded by FA values (*blue* encodes low FA values, *red* high FA values). The color in each point on the E-ODF glyph surface maps the amplitude of the E-ODF for that orientation where *blue* encodes low amplitude and *red* high. This is the standard glyph color coding for each diffusion technique

When selecting fiber tracts on the surrounding boundary of the tumor, Fig. 10 shows differences also seen between DTI and ODF tracking. The fiber density of ODF tracking is enhanced and high curvature fibers seem to be tracked easier. This is especially seen in the post-op data (Fig. 10 (right)).

The computation of the E-ODFs field, from a $120 \times 120 \times 69$ DTI volume, third order tessellation (162 points on the sphere), took approximately 10 min, on a standard desktop computer Intel ®Core ^{TM}i3-3220 with 8 GB of RAM.

Even though, it is a difficult task to show the differences in a single figure, the pre-op and post-op fiber structure is quite different, as seen when navigated using FiberNavigator,[2] and seen interactively in 3D. As for the local tensor and E-ODFs field, our neurosurgeon prefers looking at fiber tracts computed from the E-ODF field with an interactive tool that allows navigating around and inside the tumor.

[2]https://github.com/scilus/fibernavigator.

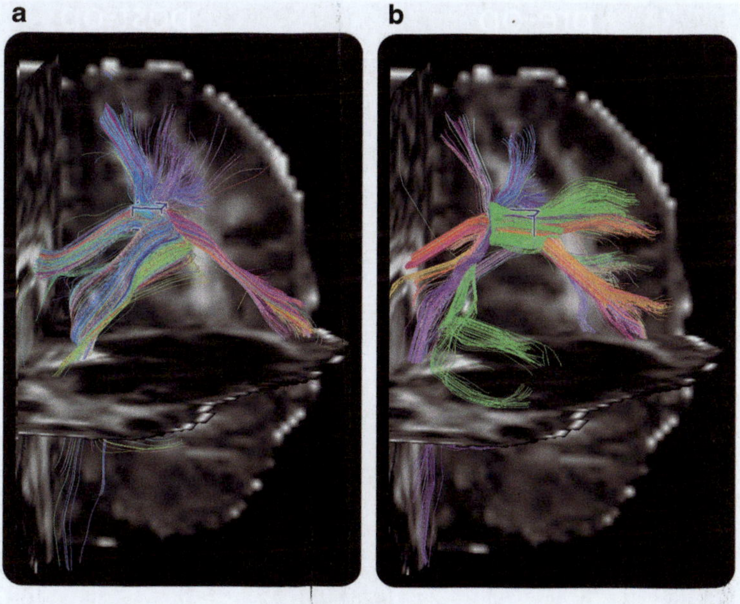

Fig. 9 Diffusion tensor tracking versus ODF tracking on E-ODFs of order 8 on the pre-op dataset. Only fibers that pass through the blue selection cube are shown (**a**) DTI tracking pre-op. (**b**) ODF tracking from E-ODFs pre-op

5 Discussion

This work is among the first attempts to use analytic kernel implementations for enhancement of dMRI data in clinical applications. As such these results should be taken with care and used to motivate more work in this direction given the limitations of the dMRI acquisitions (low SNR, antipodal symmetrically acquired signal, among others).

Moreover, this work for the first time motivates the use of extrapolated contextual information for development of novel dMRI markers for applications such as MS. Hopefully it can be applied to indicate the tract damage of specific fiber bundles, but as well to improve the connectomic-based analysis for these types of applications. In MS connectivity matrices are corrupted by the presence of the lesions and incorrectly reconstructed diffusion streamlines. However, further work is needed in this direction since here we only present preliminary results.

Regarding the neurosurgical application, the more data that can be presented to support evidence of the current state of the imaged patient, the better. Neurosurgeons would like to see not only the original DTI data (that has to be taken with care given the limitations) but as well maps of residuals where the DTI model performs the worst (i.e., in non-Gaussian areas) or, as in our case, fields of E-ODFs that show the actual missing crossings structures inferred from the surrounding data.

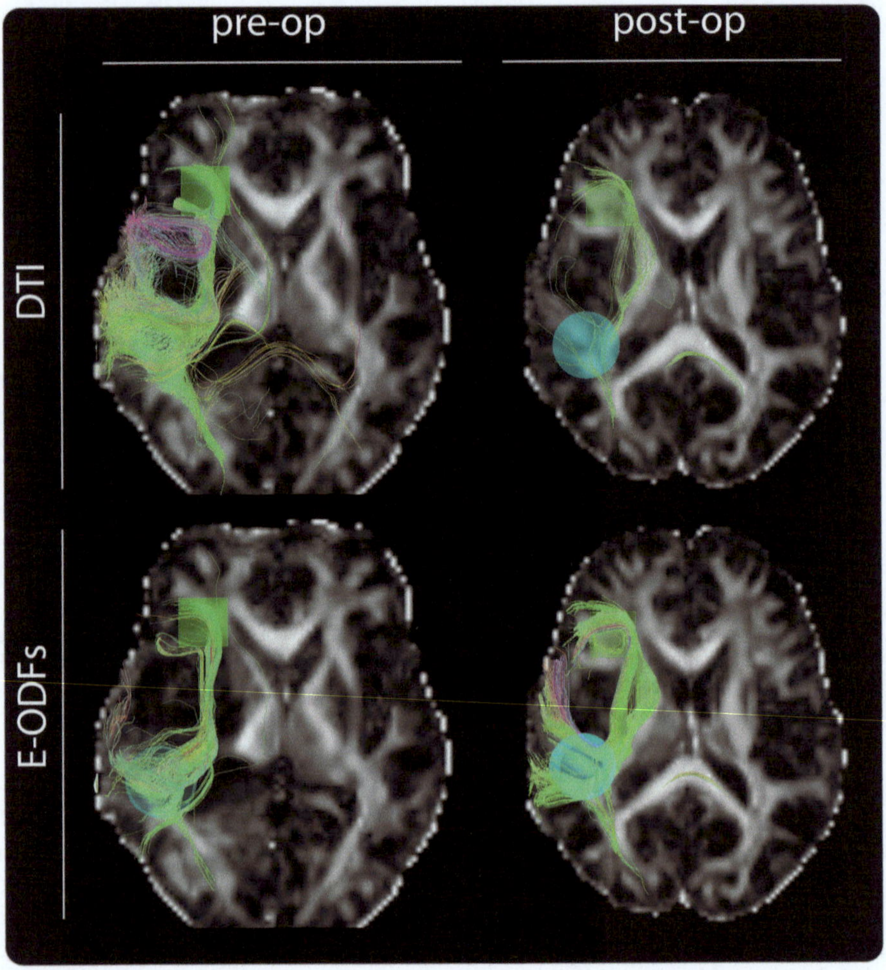

Fig. 10 Diffusion tensor tracking versus E-ODF tracking on pre-op versus post-op dataset. Only fibers that pass simultaneously through the blue selection sphere and green cube are shown

This rich multi-modal data gives additional information, re-assures and encourages the strategies for pre-surgical planning and performing the resection in the most non-invasive and efficient way. To the best of our knowledge the contextual enhancement does not destroy information but rather enhances the directional information that is well-aligned ('in agreement') with neighboring directional information (via Brownian motion on the coupled space of positions and orientations). Since our neighborhood is relatively small, subtle local information is being preserved. One must also keep in mind that E-ODFs complement the DTI tensors and metrics derived from the original data such as FA and MD. E-ODFs are greatly used to enhance the tracking. Of course more thorough validation in this direction is necessary. In our experience, neurosurgeons would like to gather as much as possible data in order to make crucial decisions for the planning of the surgery.

Future work should be done on a simplified visualization of the E-ODFs, as the neurosurgeon finds it important to see the extracted maxima rather than the glyph profiles. Finally, the neurosurgeon finds that E-ODFs are more informative than diffusion tensors in the regions of the edema.

6 Conclusion

In this work we demonstrated the importance of using contextual image processing operators in a clinical setting, for (a) MS application (b) improving the quality of the DWI data acquired on an old MR system for the purpose of pre-surgical planning for tumor resection as well as follow-up. The results are done on a small sample of subjects (healthy control and two patients) and only show first proof of concept as and as such should stimulate more research in this direction in a larger patient cohort and different clinical applications.

More work is needed in the direction of novel imaging markers for MS. This pilot study illustrates the possibilities of improving the fiber tracking in MS such that the patient can have its own control template in order to be able to follow the changes (with respect to lesion load on the streamline) along time. Also the possibility of having its personal template can minimize the problems with co-registrations and deformations compared to what is currently done in literature in cases of voxel-based morphometry (VBM) [3, 43, 44] and track-based spatial statistics (TBSS) [37].

Moreover, in a clinical setting, the dMRI acquisitions are often limited by two factors: the gradient strength of the MRI scanner and the scanning time. Both of these factors are very difficult to compromise given that the first requires replacement of the whole system and the latter less saturated scanning schedule. These are often limiting conditions to the research in dMRI, often driven towards improving the accuracy of the reconstructed fibertracks.

In this paper, we demonstrated and evaluated a unique possibility of not only performing advanced research on an old MRI machine, but also improving the tractography results compared to the ones normally obtained by simple DTI tractography. We started by studying a healthy volunteer scanned with both modest and more advanced acquisitions (w.r.t. number of gradients) and we performed several typical reconstruction techniques. We demonstrated that DTI acquisition can be transformed into a HARDI-like acquisition.

Furthermore, we presented a study of a pre and post-op tumor patient where the advances from this novel post-processing technique not only will improve the pre-surgical planning but also give good indications for the post-surgical recovery of the patient. The feedback given from our neurosurgeon was very positive, given that the reconstructed profiles with this technique gave more accurate anatomical information compared to the pure DTI tracking.

The fiber propagation is now based on a global probabilistic Brownian motion model on $\mathbb{R}^3 \rtimes S^2$ which sometimes lacks data-adaptability in practice. Here local adaptation of the diffusion in Eq. (7) is needed. More precisely, local differential geometric and/or statistical adaptation of the entire diffusivity matrix is needed, beyond the relatively limited Perona and Malik extension (with diffusion scalar adaptation only) presented in [15].

Further work is both necessary and recommended for the validation of the benefits of this technique in the clinical setting in both neurology and neurosurgery with a larger cohort of patients.

Acknowledgements This work was supported by the FP7 Marie Curie Intra-European Fellowship, project acronym: ConnectMS, project number: 328060. Moreover, the research leading to the results of this article has received funding from the European Research Council under the European Community's 7th Framework Program (FP7/2014) ERC grant agreement no. 335555.

References

1. Aganj, I., Lenglet, C., Sapiro, G., Yacoub, E., Ugurbil, K., Harel, N.: Reconstruction of the orientation distribution function in single- and multiple-shell q-ball imaging within constant solid angle. Magn. Reson. Med. **64**(2), 554–566 (2010). doi: 10.1002/mrm.22365
2. Arsigny, V., Fillard, P., Pennec, X., Ayache, N.: Log-Euclidean metrics for fast and simple calculus on diffusion tensors. Magn. Reson. Med. **56**(2), 411–421 (2006)
3. Ashburner, J., Friston, K.J.: Voxel-based morphometrythe methods. NeuroImage **11**(6), 805–821 (2000). doi:http://dx.doi.org/10.1006/nimg.2000.0582. http://www.sciencedirect.com/science/article/pii/S1053811900905822
4. Barbieri, D., Citti, G., Sanguinetti, G., Sarti, A.: An uncertainty principle underlying the functional architecture of V1. J. Physiol. Paris (2012). doi:10.1016/j.jphysparis.2012.03.001. http://dx.doi.org/10.1016/j.jphysparis.2012.03.001
5. Barmpoutis, A., Vemuri, B.C., Howland, D., Forder, J.R.: Extracting tractosemas from a displacement probability field for tractography in dw-mri. In: Metaxas, D.N., Axel, L., Fichtinger, G., Szkely, G. (eds.) Medical Image Computing and Computer- Assisted Intervention (1). Lecture Notes in Computer Science, vol. 5241, pp. 9–16. Springer, Heidelberg (2008). http://dblp.uni-trier.de/db/conf/miccai/miccai2008-1.html#BarmpoutisVHF08
6. Basser, P.J., Mattiello, J., LeBihan, D.: {MR} Diffusion tensor spectroscopy and imaging. Biophys. J. **66**(1), 259–267 (1994)
7. Ben-Yosef, G., Ben-Shahar, O.: Tangent bundle curve completion with locally connected parallel networks. Neural Comput. **24**(12), 3277–3316 (2012). http://dblp.uni-trier.de/db/journals/neco/neco24.html#Ben-YosefB12
8. Boscain, U., Charlot, G., Rossi, F.: Minimization of length and curvature on planar curves. In: Proceedings of the 48th IEEE Conference on Decision and Control, 2009 held jointly with the 2009 28th Chinese Control Conference. CDC/CCC 200, pp. 1062–1067 (2009). doi:10.1109/CDC.2009.5399749
9. Castellano, A., Bello, L., Michelozzi, C., Gallucci, M., Fava, E., Iadanza, A., Riva, M., Casaceli, G., Falini, A.: Role of diffusion tensor magnetic resonance tractography in predicting the extent of resection in glioma surgery. Neuro Oncol. **14**(2), 192–202 (2011). doi:10.1093/neuonc/nor188. http://neuro-oncology.oxfordjournals.org/content/14/2/192.full

10. Chard, D.T., Jackson, J.S., Miller, D.H., Wheeler-Kingshott, C.A.: Reducing the impact of white matter lesions on automated measures of brain gray and white matter volumes. J. Magn. Reson. Imaging **32**(1), 223–228 (2010). doi:10.1002/jmri.22214. http://dx.doi.org/10.1002/jmri.22214
11. Chirikjian, G.S.: Engineering Applications of Noncommutative Harmonic Analysis: With Emphasis on Rotation and Motion Groups. CRC Press, Abingdon (2000)
12. Chirikjian, G.S.: Stochastic Models, Information Theory, and Lie groups. Classical Results and Geometric Methods. Applied and Numerical Harmonic Analysis, vol. 1. Birkhuser, Berlin (2009). http://opac.inria.fr/record=b1132711. Document avec exercices
13. Citti, G., Sarti, A.: A cortical based model of perceptual completion in the roto-translation space. J. Math. Imaging Vision **24**(3), 307–326 (2006). http://dblp.uni-trier.de/db/journals/jmiv/jmiv24.html#CittiS06
14. Comi, G., Rovaris, M., Leocani, L., Martinelli, V., Filippi, M.: Assessment of the damage of the cerebral hemispheres in MS using neuroimaging techniques. J. Neurol. Sci. **172**(Suppl. 1), S63–S66 (2000)
15. Creusen, E.J., Duits, R., Haije, T.C.J.D.: Numerical schemes for linear and non-linear enhancement of dw-mri. In: Bruckstein, A.M., ter Haar Romeny, B.M., Bronstein, A.M., Bronstein, M.M. (eds.) Scale Space and Variational Methods in Computer Vision. Lecture Notes in Computer Science, vol. 6667, pp. 14–25. Springer, Berlin (2011). http://dblp.uni-trier.de/db/conf/scalespace/ssvm2011.html#CreusenDH11
16. Dela Haije, T., Duits, R., Tax, C.: Sharpening fibers in diffusion weighted mri via erosion. In: Visualization and Processing of Tensors and Higher Order Descriptors for Multi-Valued Data, Dagstuhl Seminar, vol. 11501. Springer, Heidelberg (2012)
17. Descoteaux, M., Angelino, E., Fitzgibbons, S., Deriche, R.: Regularized, fast, and robust analytical Q-ball imaging. Magn. Reson. Med. **58**(3), 497–510 (2007)
18. Descoteaux, M., Deriche, R., Knosche, T.R., Anwander, A.: Deterministic and probabilistic tractography based on complex fibre orientation distributions. IEEE Trans. Med. Imaging **28**(2), 269–286 (2009)
19. Duits, R., Franken, E.: Left-invariant diffusions on the space of positions and orientations and their application to crossing-preserving smoothing of HARDI images. Int. J. Comput. Vis. **92**(3), 231–264 (2010). doi:10.1007/s11263-010-0332-z. http://www.springerlink.com/content/511j713042064t35/
20. Duits, R., Dela Haije, T.C.J., Creusen, E., Ghosh, A.: Morphological and linear scale spaces for fiber enhancement in DW-MRI. J. Math. Imaging Vision **46**(3), 326–368 (2013). doi:10.1007/s10851-012-0387-2. http://bmia.bmt.tue.nl/people/RDuits/JMIVDuits2011final.pdf
21. Felsberg, M.: Adaptive filtering using channel representations. In: Florack, L., Duits, R., Jongbloed, G., van Lieshout, M.C., Davies, L. (eds.) Mathematical Methods for Signal and Image Analysis and Representation. Computational Imaging and Vision, vol. 41, pp. 31–48. Springer, London (2012). doi:10.1007/978-1-4471-2353-8_2. http://dx.doi.org/10.1007/978-1-4471-2353-8_2
22. Florack, L.: Codomain scale space and regularization for high angular resolution diffusion imaging. In: IEEE Computer Society Conference on Computer Vision and Pattern Recognition Workshops, 2008. CVPRW '08, pp. 1–6 (2008). doi:10.1109/CVPRW.2008.4562967
23. Franken, E.M.: Enhancement of crossing elongated structures in images. Ph.D. thesis, Eindhoven University of Technology (2008)
24. Gabilondo, I., Martinez-Lapiscina, E.H., Martinez-Heras, E., Fraga-Pumar, E., Llufriu, S., Ortiz, S., Bullich, S., Sepulveda, M., Falcon, C., Berenguer, J., Saiz, A., Sanchez-Dalmau, B., Villoslada, P.: Trans-synaptic axonal degeneration in the visual pathway in multiple sclerosis. Ann. Neurol. (2013). doi: 10.1002/ana.24030. http://www.biomedsearch.com/nih/Trans-synaptic-axonal-degeneration-in/24114885.html
25. Gallichan, D., Andersson, J.L., Jenkinson, M., Robson, M.D., Miller, K.L.: Reducing distortions in diffusion-weighted echo planar imaging with a dual-echo blip-reversed sequence. Magn. Reson. Med. **64**(2), 382–390 (2010)

26. Hagmann, P., Cammoun, L., Gigandet, X., Meuli, R., Honey, C.J., Wedeen, V.J., Sporns, O.: Mapping the structural core of human cerebral cortex. PLoS Biol. **6**(7), e159 (2008). doi:10.1371/journal.pbio.0060159. http://dx.plos.org/10.1371/journal.pbio.0060159
27. Kezele, I., Descoteaux, M., Poupon, C., Poupon, F., Mangin, J.F.: Spherical wavelet transform for ODF sharpening. Med. Image Anal. **14**(3), 332–342 (2010). doi:10.1016/j.media.2010.01.002. http://www.sciencedirect.com/science/article/B6W6Y-4Y95V7F-1/2/96f995a6e324481afd598ed0f4fd9fcd
28. Ling, J., Merideth, F., Caprihan, A., Pena, A., Teshiba, T., Mayer, A.R.: Head injury or head motion? Assessment and quantification of motion artifacts in diffusion tensor imaging studies. Hum. Brain Mapp. **33**(1), 50–62 (2012)
29. Llufriu, S., Martinez-Heras, E., Fortea, J., Blanco, Y., Berenguer, J., Gabilondo, I., Ibarretxe-Bilbao, N., Falcon, C., Sepulveda, M., Sola-Valls, N., Bargallo, N., Graus, F., Villoslada, P., Saiz, A.: Cognitive functions in multiple sclerosis: impact of gray matter integrity. Mult. Scler. **20**(4), 424–432 (2014)
30. Medioni, G., Lee, M.S., Tang, C.K.: A Computational Framework for Segmentation and Grouping. Elsevier, Burlington, MA (2000)
31. MomayyezSiahkal, P., Siddiqi, K.: 3D stochastic completion fields for mapping connectivity in diffusion mri. IEEE Trans. Pattern Anal. Mach. Intell. **35**(4), 983–995 (2013). http://dblp.uni-trier.de/db/journals/pami/pami35.html#MomayyezSiahkalS13
32. Prčkovska, V., Duits, R.R., Vilanova, A., ter Haar Romenij, B.: Extrapolating fiber crossings from DTI data. Can we gain similar information as HARDI? In: Workshop Computational Diffusion MRI, MICCAI, Beijing (2010)
33. Reisert, M., Skibbe, H.: Left-invariant diffusion on the motion group in terms of the irreducible representations of so(3). CoRR abs/1202.5414 (2012). http://dblp.uni-trier.de/db/journals/corr/corr1202.html#abs-1202-5414
34. Rodrigues, P., Duits, R., ter Haar Romenij, B., Bartroli, A.V.: Accelerated diffusion operators for enhancing DW-MRI. In: Visual Computing for Biology and Medicine (VCBM), Leipzig (2010)
35. Rodrigues, P., Prčkovska, V., Pullens, W., Strijkers, G., Vilanova, A., ter Haar Romeny, B.: Validating validators: an analysis of dwmri hardware and software phantoms. In: Proceedings 18th Scientific Meeting, International Society for Magnetic Resonance in Medicine, Stockholm, p. 3964 (2010)
36. Seunarine, K.K., Alexander, D.C.: Multiple fibres: beyond the diffusion tensor. In: Behrens, T.E.B., Johansen-Berg, H. (eds). Diffusion MRI. Elsevier, London (2009)
37. Smith, S.M., Jenkinson, M., Johansen-Berg, H., Rueckert, D., Nichols, T.E., Mackay, C.E., Watkins, K.E., Ciccarelli, O., Cader, M.Z., Matthews, P.M., Behrens, T.E.: Tract-based spatial statistics: voxelwise analysis of multi-subject diffusion data. NeuroImage **31**(4), 1487–1505 (2006)
38. Tehrani, M.A.: Diffusion tensor imaging study for multiple sclerosis in the optic radiation. Master's thesis, Universitat Pumpeu Fabra, Health and Life Science Department (2012)
39. Tournier, J.D., Calamante, F., Connelly, A.: Robust determination of the fibre orientation distribution in diffusion MRI: non-negativity constrained super-resolved spherical deconvolution. NeuroImage **35**(4), 1459–1472 (2007)
40. Tournier, J.D., Calamante, F., Connelly, A.: Mrtrix: Diffusion tractography in crossing fiber regions. Int. J. Imaging Syst. Technol. **22**(1), 53–66 (2012). doi:10.1002/ima.22005. http://dx.doi.org/10.1002/ima.22005
41. Tristán-Vega, A., Aja-Fernández, S.: DWI filtering using joint information for DTI and HARDI. Med. Image Anal. **14**(2), 205–218 (2010). doi:10.1016/j.media.2009.11.001. http://www.sciencedirect.com/science/article/B6W6Y-4XPB73V-1/2/a3676e62ab8bf29fabf348def2a7f539
42. Van Hecke, W., Leemans, A., De Backer, S., Jeurissen, B., Parizel, P.M., Sijbers, J.: Comparing isotropic and anisotropic smoothing for voxel-based DTI analyses: a simulation study. Hum. Brain Mapp. **31**(1), 98–114 (2010)

43. Whitwell, J.L.: Voxel-based morphometry: an automated technique for assessing structural changes in the brain. J. Neurosci. **29**(31), 9661–9664 (2009)
44. Wright, I.C., McGuire, P.K., Poline, J.B., Travere, J.M., Murray, R.M., Frith, C.D., Frackowiak, R.S., Friston, K.J.: A voxel-based method for the statistical analysis of gray and white matter density applied to schizophrenia. NeuroImage **2**(4), 244–252 (1995)
45. Yushkevich, P.A., Piven, J., Cody Hazlett, H., Gimpel Smith, R., Ho, S., Gee, J.C., Gerig, G.: User-guided 3D active contour segmentation of anatomical structures: significantly improved efficiency and reliability. NeuroImage **31**(3), 1116–1128 (2006)

Index

ActiveAx, 193
Adaptive Boosting, 308
Aggregation scheme, 176
Algebraic lattices, 111
Algebraic opening, 111, 116
Anti-extensive, 111
Aperture angle, 182
Atlas, 289
Averaging, 275
AxCaliber, 193
Axon diameter, 192, 209

Ball-and-Sticks, 295
Ball tensor, 167
Ball tensor voting, 171
Beucher gradient, 140, 141
Bézout theorem, 85
Bifurcation, 89, 90
Big Data, 327
Black top-hat, 140, 141
Bloch-Torrey equation, 3, 7, 8, 25, 149
Bloch-Torrey propagator, 8
Bonferroni correction, 303
Brain connectivity, 309
Brain Microstructure, 271
Brownian motion, 21
BSQ-format, 143
B-value, 192, 195

Capped cylinders, 5
Carr-Purcell-Meiboom-Gill experiment, 5
Cell size, 4
Christoffel symbols, 24
Classical tensor voting, 166
Classification, 307
Closing, 140, 141
Clustering, 325, 327–329, 335, 337
Combustion, 324, 326, 330, 335, 337
Complete Mixture Model, 275
Complexity, 4
Compressed sensing, 191
Concentration parameter, 46
Condensation graph, 118, 124
Connectomis, 194
Context-based decomposition, 174
Contextual processing, 355
    techniques, 353
Contextual voting, 173
Convexity, 23, 31
    measure, 31
Convolution, 172, 357
Corpus callosum, 204, 206
Corticospinal tract, 204, 206
Cross validation, 311
Crossing, 204, 206, 208

DAG. *See* graph, directed acyclic
Decomposition
    of a tensor (*see* tensor decomposition)
Degenerate curve, 73, 74, 77, 78, 81, 85, 86, 89, 90
Demyalination, 192, 358
Diffusion anisotropy, 4
Diffusion Compartment Imaging (DCI), 272
    atlas, 289

I. Hotz, T. Schultz (eds.), *Visualization and Processing of Higher Order Descriptors for Multi-Valued Data*, Mathematics and Visualization,
DOI 10.1007/978-3-319-15090-1

estimation, 295
filterting, 290
microstructure-preserving simplification, 282
population studies, 291
registration, 287
signal-preserving simplification, 276
smoothing, 290
tractography, 291
Diffusion MR, 3
Diffusion MRI, 109, 121, 300
Diffusion propagator, 192, 194, 195, 201–204, 206
Diffusion Spectrum Imaging (DSI), 150, 194, 197, 199, 203, 314
Diffusion tensor, 24
Diffusion Tensor Imaging (DTI), 25, 149, 192, 195, 197, 201, 203, 207, 271, 300
limitations, 272
Diffusion-weighted magnetic resonance imaging (DwMRI), 235–252, 354
Dilation, 140, 141
Directed acyclic graph. *See* graph, directed acyclic
Distance metric, 328, 337
Distribution function, 7
Distribution of phases, 6
dMRI, 353
Double diffusion encoding (DDE), 4
DSI. *See* Diffusion Spectrum Imaging (DSI)
DTI. *See* Diffusion Tensor Imaging (DTI)
DwMRI. *See* diffusion-weighted magnetic resonance imaging (DwMRI)

Echo, 6, 8
Eckart-Young Theorem, 96
Efficient tensor voting, 172
Eigendecomposition. *See* tensor decomposition, 166, 176
Einstein velocity
addition, 139
subtraction, 139
Erosion, 140, 141
Estimation, 295
Euler's theorem for homogeneous functions, 152
Exchange, 274
Exponentials of matrices, 10
External gradient, 140, 141

Fabric tensor
first kind, 39
second kind, 40
third kind, 40
Fabric texture, 342
Feature extraction, 309
Feature selection, 310, 313
Feature vector, 337
Fiber orientation, 38
Fiber tracts, 235–252
brain connectivity, 244
bundles, 246–248
lines, 236, 238
probabilistic tracking, 247–249
tract bundles, 246–248
tubes, 236, 237, 241–242, 244
Fibre crossings, 121
Fibre tractography. *See* tractography
Figuratrix, 152
Filtering, 275, 290
Finsler
framework, 25
function, 22
geodesic, 23
equation, 24
geometry, 22
Laplace–Beltrami, 28
manifold, 22
metric tensor, 23
Finsler-DTI, 151
Fourier transform, 194
Frame rate, 206
Free diffusion, 13
Fundamental tensor, 23

Gabor filter, 182
Gaussian diffusion, 13
General gradient waveforms, 3
General Linear Model, 303
Generalized diffusion tensor imaging (GDTI), 25
Glyph, 192, 204, 206
GPU computing, 172
Gradient echo condition, 14
Gradient waveforms, 3
Graph
condensation (*see* condensation graph)
directed acylic, 112, 115
of tangent vectors, 113
Greyscale image analysis, 181

HARDI. *See* High angular resolution diffusion imaging (HARDI)
Heat equation, 259

Heat kernel, 260
  signature, 257
Hessian normal form, 136
High angular resolution diffusion imaging (HARDI), 150, 192, 195, 314
Higher-order data, 183
Higher-order SVD, 99
Higher-order tensor, 175
HOOI, 101
HOSVD, 103
Hybrid diffusion imaging (HYDI), 195
Hypercube, 131
Hypersimplex, 131
Hyperspectral image, 130

Idempotent, 111
Identity tensor, 112, 119
Illustrative visualization, 235–252
  abstraction, 249–251
  ambient occlusion, 239–241, 243–245
  applications
    brain connectivity, 235
    diffusion-weighted magnetic resonance imaging, 235–252
    muscle tissue, 237
    neurosciences, 235
  contours (*see* silhouettes)
  depth-dependent halos, 237–238, 240, 241, 243, 246, 251
  depth perception, 237–240, 245
  focus+context, 242–249
  halftoning, 244
  halos, 237–239, 243, 246, 247
  hatching, 243
  lineAO, 239–241, 246
  line style, 238, 239
  multi-modal visualizations, 243–245, 249
  patterns, 241
  screen-space ambient occlusion, 239, 241, 243
  shadows, 237, 239
  silhouettes, 237, 244–247, 249
  stippling, 243, 248–249
  unsharp masking, 241
Increasing, 111
Indicatrix, 28, 152
Infimum, 138
Infinitesimal propagators, 9
Infinitesimally short pulses, 4
Inhibitory voting field, 182
Inner product, 112
in situ, 325, 326, 335
Inter-scale information, 180
Internal gradient, 140, 141
Interpolation, 149–160, 199, 275
  dyadic-tensor, 150
  geodesic, 150
  geodesic-loxodrome, 150
  inpainting, 150
  linear invariant, 150
  log-Euclidean, 150
  pde-based, 150
  quaternion-based, 150
  tensor spline, 150
Isocontour, 195
Isosurface, 195, 200
Isotropic transport process, 26
Iterative algorithm, 179

Laplace Beltrami-operator, 360
Lattices. *See* algebraic lattices
Layered structures, 5
Linear interpolation, 150
Linear tensor field, 73, 78, 80–83, 85, 86, 88, 89
Load-deflection curve, 341
Load path, 340
Locally Minkowski manifold, 24
Loewner cone, 130
Loewner order, 130
Longest path problem, 115
Low gradient, 358

Machine learning, 307
Magnetic field gradients, 4
Magnetic resonance, MR, 3
Magnetization density, 5
Mathematical morphology, 129
Matricization, 96
Matrix representation, 3
MCF. *See* multiple correlation function (MCF)
Medical image analysis, 163
Metric partial derivative, 184
Microstructure, 3, 193
Mixture Model Simplification, 276
Modal vector, 46
Morphological Laplacian, 140, 141
MRI dataset, 103
Multi-band, 191
Multilinear rank, 101
Multilinear SVD, 99
Multiple comparisons, 303
Multiple correlation function (MCF), 3, 5
Multiple correlator, 9
Multiple propagator, 3, 4, 10

Multiple Sclerosis, 354
Multiple-PFG, 4
Multiplicative calculus, 151
Multiplicative update rules, 119
Multiscale image analysis, 177
Multi-scale visualization, 259
Multispectral image, 130
Multi-Tensor Model, 274
Multivariate analysis, 314

Neural networks, 308
Neurodegenerative, 354
Neuroimaging, 299
Noise, 198, 208
Non-negative least squares, 119
Non-photorealistic rendering (NPR). *See* illustrative visualization
Null hypothesis, 302

ODF, 38, 195, 199, 203, 206, 209
- antipodally symmetric, 38
- axially symmetric, 44
- cubic, 59, 60
- de la Vallée Poussin, 54
- hexagonal, 62
- isotropic, 42
- orthotropic, 56
- scores, 109, 117, 119
- singular-kernel, 51
- tensor, 38
- transversely hemitropic, 44
- transversely isotropic, 45
- von Mises-Fisher, 46
- Watson, 49

Opening. *See* algebraic opening, 140, 141
Opposite (tangent vector), 115, 118
Orientation distribution function (ODF). *See* ODF
Oscillating gradient, 4
Osculating figuratrix, 152
Osculating indicatrix, 152

Parallel imaging, 191
Path integral, 3, 11
Path opening, 116, 124
- on acyclic graphs, 115
- properties, 116

PCA. *See* Principal component analysis (PCA)
Perceptual organisation, 164
Permutation testing, 304, 311
Phase shift, 6
Plate tensor, 167
Plate tensor voting, 170
Poincaré-Hopf theorem, 87
Polarity vector, 179
Population Studies, 291
Porous materials, 3
Power method, symmetric higher order. *See* tensor decomposition
Principal Component Analysis, 310
Principal component analysis (PCA), 201, 202, 207
Principal stress, 341
Principle of good continuation, 169
Principle of proximity, 169
Product development process, 340
Propagator, 8
Pulsed field gradient (PFG), 4, 13

Q-space, 193
Quadratic scaling assumption, 31
Quantum mechanics, 5
Quasi-Newton, 103

Random forests, 308
Randomized SVD, 97
Rank-one decomposition. *See* tensor decomposition
Region Of Interest, 304, 312
Registration, 287
Regression, 308
Reinforcement structure, 343
Restricted diffusion, 12
Rib structure, 344
Riemann-Finsler geometry, 151
Riemannian
- framework, 24
- Laplace–Beltrami operator, 25
- metric, 258, 306

Riemannian metric, 177
Rotation invariant operators, 112

Sampling, 200
Self-dual top-hat, 140, 141
Sensitivity analysis, 314
Simplified Mixture Model, 275
Size distribution, 4
Smallest enclosing base
- pyramid, 137
- triangle, 136

Smoothing, 290
Spherical deconvolution, 121, 195

Spherical harmonics, 66, 199
Statistical hypothesis testing, 301
Statistical Parametric Mapping, 303
Statistical significance, 302
Stejskal-Tanner formula, 151
Stick tensor, 167
Stick tensor voting, 169
Stick voting field, 170
Stochastic processes, 356
Strain tensor, 262
Stress tensor, 324, 327, 341
  field, 261
Strongly connected components. *See* condensation graph
Superior longitudinal fasciculus, 204, 206
Superquadrics, 195, 197, 202
Support Vector Machine, 308
Supremum, 138
Susceptibility gradients, 5
Symmetric higher order power method. *See* tensor decomposition
Symmetric tensors, 111

Tensor
  higher-order, 41
  orientation, 38
  totally symmetric, 38
Tensor decomposition
  rank-one decomposition, 112, 113, 117, 119
  symmetric higher order power method, 119
Tensor field, 183
Tensor fields, 76–78, 111, 235
  3D, 235
  topology, 74, 75, 90
Tensor interpolation, 306
Tensorial orientation scores. *See* orientation scores
Tensor line, 340, 341
Test statistic, 302
Time-slicing, 9
Topological ordering, 112, 118
Tract-Based Spatial Statistics, 305, 309
Tractography, 110, 113, 122, 155, 159, 160, 192, 208, 291, 305, 309, 353
Transfer function, 200
Transition point, 77, 81, 89
Transverse magnetization, 6
Triangular pores, 5
Trisector, 76, 77, 81–83, 87, 89, 90
Tucker mode product, 95
Tumor resection, 354
Turbulent flow, 324, 337

Uncertainty, 159
  visualization, 241, 247–249

Vector field, 183
Visualization
  illustrative (*see* illustrative visualization)
  uncertainty (*see* uncertainty visualization)
Volume rendering, 192, 199, 206, 209
von-Mises stress, 341
Voting field, 168
Voxel-Based Morphometry, 303

Watson distribution, 306
Wedge, 76, 77, 81–83, 87, 89, 90
Weight vector, 313
Weighted Combinations, 275
White top-hat, 140, 141
Wiener functional, 6
Wishart matrix, 98

MIX
Papier aus verantwortungsvollen Quellen
Paper from responsible sources
FSC® C105338

If you have any concerns about our products,
you can contact us on
ProductSafety@springernature.com

In case Publisher is established outside the EU,
the EU authorized representative is:
Springer Nature Customer Service Center GmbH
Europaplatz 3, 69115 Heidelberg, Germany

Printed by Libri Plureos GmbH
in Hamburg, Germany